Localized Waves

The Wiley Bicentennial—Knowledge for Generations

Each generation has its unique needs and aspirations. When Charles Wiley first opened his small printing shop in lower Manhattan in 1807, it was a generation of boundless potential searching for an identity. And we were there, helping to define a new American literary tradition. Over half a century later, in the midst of the Second Industrial Revolution, it was a generation focused on building the future. Once again, we were there, supplying the critical scientific, technical, and engineering knowledge that helped frame the world. Throughout the 20th Century, and into the new millennium, nations began to reach out beyond their own borders and a new international community was born. Wiley was there, expanding its operations around the world to enable a global exchange of ideas, opinions, and know-how.

For 200 years, Wiley has been an integral part of each generation's journey, enabling the flow of information and understanding necessary to meet their needs and fulfill their aspirations. Today, bold new technologies are changing the way we live and learn. Wiley will be there, providing you the must-have knowledge you need to imagine new worlds, new possibilities, and new opportunities.

Generations come and go, but you can always count on Wiley to provide you the knowledge you need, when and where you need it!

William J. Pesce
President and Chief Executive Officer

Peter Booth Wiley
Chairman of the Board

Localized Waves

Edited by

HUGO E. HERNÁNDEZ-FIGUEROA
MICHEL ZAMBONI-RACHED
ERASMO RECAMI

IEEE PRESS

BICENTENNIAL 1807 WILEY 2007 BICENTENNIAL

WILEY-INTERSCIENCE

A JOHN WILEY & SONS, INC., PUBLICATION

Published by John Wiley & Sons, Inc., Hoboken, New Jersey.
Published simultaneously in Canada.

Wiley Bicentennial Logo: Richard J. Pacifico

Library of Congress Cataloging-in-Publication Data:

Localized waves / edited by Hugo E. Hernández-Figueroa, Michel Zamboni-Rached, Erasmo Recami.
p. cm.
ISBN 978-0-470-10885-7 (cloth)
1. Localized waves–Research. I. Hernández-Figueroa, Hugo E.
II. Zamboni-Rached, Michel. III. Recami, Erasmo.
QC157.L63 2007
532′.0593—dc22

2007002548

Printed in the United States of America

10 9 8 7 6 5 4 3 2 1

Contents

Contributors

Ioannis M. Besieris, The Bradley Department of Electrical and Computer Engineering, Virginia Polytechnic Institute and State University, Blacksburg, Virginia

Martin Bock, Max-Born-Institute for Nonlinear Optics and Short-Pulse Spectroscopy, Berlin, Germany

Zdeněk Bouchal, Department of Optics, Palacký University, Olomouc, Czech Republic

Radek Čelechovský, Department of Optics, Palacký University, Olomouc, Czech Republic

Claudio Conti, Research Center Enrico Fermi, Rome, Italy, and Research Center SOFT INFM-CNR, University La Sapienza, Rome, Italy

Paolo Di Trapani, Dipartimento di Fisica e Matematica, Università degli Studi dell'Insubria sede di Como, Como, Italy

Ari T. Friberg, Department of Microelectronics and Applied Physics, Royal Institute of Technology, Kista, Sweden

Uwe Griebner, Max-Born-Institute for Nonlinear Optics and Short-Pulse Spectroscopy, Berlin, Germany

Ruediger Grunwald, Max-Born-Institute for Nonlinear Optics and Short-Pulse Spectroscopy, Berlin, Germany

Hugo E. Hernández-Figueroa, Faculdade de Engenharia Elétrica e de Computação, Departamento de Microonda e Óptica, Universidade Estadual de Campinas, Campinas, SP, Brazil

Wei Hu, Laboratory of Photonic Information Technology, School for Information and Optoelectronic Science and Technology, South China Normal University, Guangzhou, P. R. China

Davide Janner, Dipartimento di Fisica, Politecnico di Milano, Milan, Italy

Volker Kebbel, Automation and Assembly Technologies GmbH, Bremen, Germany

Sergei V. Kukhlevsky, Department of Experimental Physics, Institute of Physics, University of Pécs, Pécs, Hungary

Stefano Longhi, Dipartimento di Fisica, Politecnico di Milano, Milan, Italy

Jian-yu Lu, Ultrasound Laboratory, Department of Bioengineering, The University of Toledo, Toledo, Ohio

Iacopo Mochi, Nello Carrara Institute of Applied Physics–CNR, Florence Research Area, Sesto Fiorentino, Italy

Daniela Mugnai, Nello Carrara Institute of Applied Physics–CNR, Florence Research Area, Sesto Fiorentino, Italy

Uwe Neumann, Max-Born-Institute for Nonlinear Optics and Short-Pulse Spectroscopy, Berlin, Germany

Miguel A. Porras, Departamento de Física Aplicada, Escuela Técnica Superior de Ingenieros de Minas, Universidad Politécnica de Madrid, Madrid, Spain

Erasmo Recami, Facoltà di Ingegneria, Università degli Studi di Bergamo, Bergamo, Italy and INFN–Sezione di Milano, Milan, Italy

Kaido Reivelt, Institute of Physics, University of Tartu, Tartu, Estonia

Peeter Saari, Institute of Physics, University of Tartu, Tartu, Estonia

Janne Salo, Laboratory of Physics, Helsinki University of Technology, Espoo, Finland

Amr M. Shaarawi, Department of Physics, The American University of Cairo, Cairo, Egypt

Günter Steinmeyer, Max-Born-Institute for Nonlinear Optics and Short-Pulse Spectroscopy, Berlin, Germany

Gero Stibenz, Max-Born-Institute for Nonlinear Optics and Short-Pulse Spectroscopy, Berlin, Germany

Grover A. Swartzlander, Jr., College of Optical Sciences, University of Arizona, Tucson, Arizona

Stefano Trillo, Department of Engineering, University of Ferrara, Ferrara, Italy, and Research Center SOFT INFM-CNR, University La Sapienza, Rome, Italy

Michel Zamboni-Rached, Centro de Ciências Naturais e Humanas, Universidade Federal do ABC, Santo André, SP, Brazil

Preface

Diffraction and dispersion effects have been well known for centuries and are recognized to be limiting factors in many industrial and technology applications based, for example, on electromagnetic beams and pulses. Diffraction is an always-present phenomenon, affecting two- or three-dimensional waves traveling in nonguiding media. Arbitrary pulses and beams contain plane-wave components that propagate in different directions, causing a progressive increase in their spatial width along propagation. Dispersion is due to the dependence of the material media (refractive index) with frequency: therefore, each pulse's spectral component propagates with a different phase velocity, so that an electromagnetic pulse will suffer a progressive increase in its temporal width along propagation. It is clear that these two effects may be a serious restriction for applications where it is highly desirable that the beam keeps its transverse localization or the pulse keeps its transverse localization and/or temporal width along propagation, which might be desirable, for example, in free-space microwave, millimetric wave, terahertz and optical communications, microwave and optical images, optical lithography, and optical tweezers. As a consequence, the development of techniques capable of alleviating signal degradation effects caused by these two effects is of crucial importance.

Localized waves, also known as nondiffractive waves, arose initially as an attempt to obtain beams and pulses capable of resisting diffraction in free space for long distances. Such waves were obtained initially theoretically as solutions to the wave equation in the early 1940s (J. A. Stratton, *Electromagnetic Theory*, McGraw-Hill, New York, 1941), and were demonstrated experimentally in 1987 (J. Durnin, J. J. Miceli, and J. H. Eberly, Diffraction-free beams, *Phys. Rev. Lett.*, vol. 58, pp. 1499–1501, 1987). Nowadays localized waves constitute a growing and dynamic research topic, not only in relation to nondispersive free space (or vacuum), but also for dispersive, nonlinear, and lossy nonguiding media. Taking into account the significant amount of exciting and impressive results published especially in the last five years or so, we decided to edit a book on this topic, the first of its kind in the literature. The book is composed of 13 chapters authored by the most productive researchers in the field, with a well-balanced presentation of theory and experiment.

In Chapter 1, Recami et al. present a thorough review of localized waves, emphasizing the theoretical foundations along with historical aspects and the interconnections of this subject with other technology and scientific areas.

In Chapter 2, Zamboni-Rached et al. discuss in detail the theoretical structure of localized waves, and some applications are presented, among which frozen waves are of particular interest.

In Chapter 3, Besieris and Shaarawi present a hybrid spectral representation method which permits a smooth transition between two seemingly disparate classes of finite-energy spatiotemporally localized wave solutions to the three-dimensional scalar wave equation in free space: superluminal (X-shaped) and luminal (FWM-type) pulsed waves. An additional advantage of the hybrid form is that it obviates the presence of backward wave components, propagating at the luminal speed c, that have to be minimized in practical applications. A modified hybrid spectral representation method has also been presented which permits a seamless transition from superluminal localized waves to exact luminal splash modes. Within the framework of a certain parametrization, the latter are rendered indistinguishable from the paraxial luminal finite-energy-focused pulsed beam solutions.

In Chapter 4, Jian-yu Lu describes X-waves in depth, providing generalized methods for obtaining such waves through proper transformations, related primarily to the Lorentz transformation. X-wave solutions to Schrödinger and Klein–Gordon equations are also provided. In addition, the potential application of X-waves in medical ultrasound imaging is demonstrated experimentally.

In Chapter 5, Salo and Friberg show theoretically that diffraction-free wave propagation can also be achieved in anisotropic crystalline media. They explicitly analyze the effect of arbitrary anisotropies on both continuous-wave and pulsed nondiffracting fields. Due to beam steering and other effects, generation of nondiffracting waves in anisotropic media poses new challenges, and the authors propose an efficient scheme for the generation and detection of a continuous-wave beam in a crystal wafer.

In Chapter 6, Mugnai and Mochi explore Bessel X-waves' ability to provide localized energy and to exhibit superluminal propagation in both phase and group velocities (as verified experimentally). The authors also describe the ability of such waves to travel through a classically forbidden region (tunneling region) with no shift in the direction of propagation, which makes them different and unique with respect to ordinary waves.

In Chapter 7, Reivelt and Saari focus on the physical nature and experimental implementation or generation of localized waves. The authors demonstrate that the angular spectrum representation and the tilted pulse representation of localized waves are suitable tools for achieving these purposes. They explain the concepts and results of their experiments, where the realizability of Bessel X-waves and focus wave modes was verified for the first time.

In Chapter 8, Porras et al. present an interesting discussion of linear bullets, three-dimensional localized waves or particlelike waves propagating across a host medium, defeating diffraction spreading and dispersion broadening. Special attention is given to the generation of these bullets in practical settings by optical devices or by nonlinear means, showing the intimate relation between linear and nonlinear approaches to wave bullets, as in light filaments. The advantage of linear bullets with respect to standard wave packets (Gaussian-like) is also demonstrated for a variety of applications, such as laser writing in thick media, ultraprecise microhole drilling for photonic-crystal fabrication, and laser micromachining.

In Chapter 9, the theory of X-waves in nonlinear materials is discussed thoroughly by Conti and Trillo. Potential applications in light-matter interactions at high laser intensities in quantum optics and on the theoretical prediction of X-waves in Bose–Einstein condensates are pointed out.

In Chapter 10, by Kukhlevsky, the problem of spatial localization of light in free space on a nanometer scale is presented in detail. The author shows that a sub-wavelength nanometer-sized beam propagating without diffractive broadening can be produced by the interference of multiple beams of a Fresnel light source of the respective material waveguide. The results demonstrate theoretically the feasibility of diffraction-free subwavelength-beam optics on a nanometer scale for both continuous waves and ultrashort (near-single-cycle) pulses. The approach extends the operational principle of near-field subwavelength-beam optics, such as near-field scanning optical microscopy, to the "not-too-distant" field regime (up to about 0.5 wavelength). The chapter includes theoretical illustrations to facilitate an understanding of the natural spatiotemporal broadening of light waves and the physical mechanisms that contribute to the diffraction-free propagation of subwavelength beams in free space.

In Chapter 11, Grunwald et al. show experimentally that ultraflat thin-film axicons enable the real physical approximation of nondiffracting beams and X-pulses of extremely narrow angular spectra. By self-apodized truncation of Bessel–Gauss pulses (coincidence of first field minimum with the rim of an aperture), needle-shaped propagation zones of large axial extension can be obtained without additional diffraction effects. The signature of undistorted progressive waves was indicated for such needle beams by the fringe-free propagation characteristics and ultrabroadband spatio-spectral transfer functions.

In Chapter 12, Longhi and Janner provide a general overview of wave localization (in a weak sense) for an important and novel class of inhomogeneous periodic dielectric media (i.e., in photonic crystals), which have received increasing attention in recent years. Compared to wave localization in homogeneous media, such as in a vacuum, the presence of a periodic dielectric permittivity strongly modifies the space–time dispersion surfaces and hence the types of localized waves that may be observed in photonic crystals.

In Chapter 13, Bouchal et al. focus on theoretical and experimental problems of nondiffracting and singular optics. Particular attention is devoted to physical properties, methods of experimental realization, and potential applications of single and composed vortex fields carried by a pseudo-nondiffracting background beam. The unique propagation effects of vortex fields are pointed out, and consequences of their spiral phase singularities manifested by a transfer of the orbital angular momentum are also discussed. The complex vortex structures whose parameters and properties are controlled dynamically by a spatial light modulation provide advanced methods of encoding and recording of information and can be utilized effectively in optical manipulations. Spatially localized vortex structures can be extended into nonstationary optical fields where novel spatiotemporal effects and applications can be expected.

Acknowledgments

The editors are grateful to all the contributors to this volume for their efforts in producing stimulating high-quality chapters in an area that is not yet well known outside the community of experts, always with the aim of making the area more easily accessible to interested physicists and engineers. For useful discussions they are grateful to, among others, R. Bonifacio, M. Brambilla, R. Chiao, C. Cocca, C. Conti, A. Friberg, G. Degli Antoni, F. Fontana, G. Kurizki, M. Mattiuzzi, P. Milonni, P. Saari, A. Shaarawi, R. Ziolkowski, M. Tygel, and L. Ambrosio.

Preparation of the manuscript was facilitated greatly by George J. Telecki, Rachel Witmer, and Angioline Loredo from Wiley; we thank them for their fine professional work. We are indebted to Kai Chang, Ioannis M. Besieris, and Richard W. Ziolkowski for their crucial and inspirational encouragement. We would also like to thank our wives: Marli de Freitas Gomes Hernández, Jane Marchi Madureira, and Marisa T. Vasconselos, for their continuous loving support.

The Editors

CHAPTER ONE

Localized Waves: A Historical and Scientific Introduction

ERASMO RECAMI
Università degli Studi di Bergamo, Bergamo, Italy, and INFN–Sezione di Milano, Milan, Italy

MICHEL ZAMBONI-RACHED
Centro de Ciências Naturais e Humanas, Universidade Federal do ABC, Santo André, SP, Brazil

HUGO E. HERNÁNDEZ-FIGUEROA
Universidade Estadual de Campinas, Campinas, SP, Brazil

In the first part of this introductory chapter, we present general and formal (simple) introductions to the ordinary Gaussian waves and to the Bessel waves, by explicitly separating the cases of the beams from the cases of the pulses; and, finally, an analogous introduction is presented for the localized waves (LW), pulses or beams. Always we stress the very different characteristics of the Gaussian with respect to the Bessel waves and to the LWs, showing the numerous and important properties of the latter w.r.t. the former ones: Properties that may find application in all fields in which an essential role is played by a wave-equation (like electromagnetism, optics, acoustics, seismology, geophysics, gravitation, elementary particle physics, etc.). In the second part of this chapter (namely, in its Appendix), we recall at first how, in the seventies and eighties, the geometrical methods of special relativity (SR) predicted—in the sense below specified—the existence of the most interesting LWs, i.e., of the X-shaped pulses. At last, in connection with the circumstance that the X-shaped waves

Localized Waves, Edited by Hugo E. Hernández-Figueroa, Michel Zamboni-Rached, and Erasmo Recami

are endowed with superluminal group-velocities (as carefully discussed in the first part of this chapter), we mention briefly the various experimental sectors of physics in which superluminal motions seem to appear: In particular, a bird's-eye view is presented of the experiments till now performed with evanescent waves (and/or tunneling photons), and with the "localized superluminal solutions" to the wave equations.

1.1 GENERAL INTRODUCTION

Diffraction and dispersion have long been known as phenomena that limit the application of (e.g., optical) beams or pulses. *Diffraction* is always present, affecting any waves that propagate in two- or three-dimensional media, even when homogeneous. Pulses and beams comprise waves traveling along different directions which produce gradual *spatial* broadening [6]. This effect is a limiting factor whenever a pulse is needed that maintains its transverse localization (e.g., in free-space communications [7], image forming [8], optical lithography [9,10], electromagnetic tweezers [11,12]).

Dispersion acts on pulses propagating in material media, causing mainly temporal broadening: an effect known to be due to the variation in refraction index with frequency, so that each spectral component of the pulse possesses a different phase velocity. This entails gradual temporal widening, which constitutes a limiting factor when a pulse is needed that maintains its *time* width (e.g., in communication systems [13]).

It is important, therefore, to develop any techniques able to reduce those phenomena. *Localized waves*, also known as *nondiffracting waves*, are indeed able to resist diffraction for a long distance in free space. Such solutions to the wave equations (and, in particular, to Maxwell's equations, under weak hypotheses) were predicted theoretically long ago [14–17] (cf. also [18] and the Appendix to this chapter), constructed mathematically in more recent times [19,20] and soon after produced experimentally [21–23]. Today, localized waves are well established both theoretically and experimentally and are being used in innovative applications not only in vacuum but also in material (linear or nonlinear) media, showing to be able to resist also dispersion. As we mentioned, their potential applications are being explored intensively, always with surprising results, in such fields as acoustics, microwaves, and optics, and are also promising in mechanics, geophysics, and even in gravitational waves and elementary particle physics. Also worth noting are the possible applications of the "frozen waves," discussed in Chapter 2, and the ones already obtained, for instance, in high-resolution ultrasound scanning of moving organs in the human body [24,25].

Restricting ourselves to electromagnetism, we cite present-day studies on electromagnetic tweezers [26–29], optical (or acoustic) scalpels, optical guiding of atoms or (charged or neutral) corpuscles [30–32], optical litography [26,33], optical (or acoustic) images [34], communications in free space [19,35–37], remote optical alignment [38], and optical acceleration of charged corpuscles, among others.

Next we describe briefly the theory and applications of localized beams and pulses.

Localized (Nondiffracting) Beams The word *beam* refers to a monochromatic solution to a wave equation, with transverse localization of its field. To fix our ideas, we refer explicitly to the optical case, but our considerations hold for any wave equation (vectorial, spinorial, scalar—in particular, for the acoustic case).

The most common type of optical beam is the *Gaussian beam*, whose transverse behavior is described by a Gaussian function. But all the common beams suffer a diffraction, which spoils the transverse shape of their field, widening it gradually during propagation. As an example, the transverse width of a Gaussian beam doubles when it travels a distance $z_{\rm dif} = \sqrt{3}\,\pi\,\Delta\rho_0^2/\lambda_0$, where $\Delta\rho_0$ is the beam initial width and λ_0 is its wavelength. One can verify that a Gaussian beam with an initial transverse aperture of the order of its wavelength will double its width after having traveled only a few wavelengths.

It was generally believed that the only wave devoid of diffraction was the plane wave, which does not undergo any transverse change, but some authors had shown that it is not the only one. For instance, in 1941, Stratton [15] obtained a monochromatic solution to the wave equation whose transverse shape was concentrated in the vicinity of its propagation axis and represented by a Bessel function. Such a solution, now called a *Bessel beam*, was not subject to diffraction, since no change in its transverse shape took place with time. Later, Courant and Hilbert [16] demonstrated how a large class of equations (including the wave equations) admit "nondistorted progressing waves" as solutions; and as early as 1915, Bateman [17] and others [39] showed the existence of soliton-like, wavelet-type solutions to Maxwell's equations. But all such literature did not attract the attention it deserved. In Stratton's work [15] this can be partially justified since the Bessel beam was endowed with infinite energy (as much as the plane waves). An interesting problem, therefore, was that of investigating what would happen to the ideal Bessel beam solution when truncated by a finite transverse aperture.

Not until 1987 did a heuristical answer came from an actual experiment, when Durnin et al. [40] showed that a realistic Bessel beam endowed with wavelength $\lambda_0 = 0.6328$ μm and central spot† $\Delta\rho_0 = 59$ μm, passing through an aperture with radius $R = 3.5$ mm, is able to travel about 85 cm keeping its transverse intensity shape approximately unchanged (in the region $\rho << R$ surrounding its central peak). In other words, it was shown experimentally that the transverse intensity peak as well as the field surrounding it do not undergo any appreciable change in shape all along a large depth of field. By comparison, let us recall once more that a Gaussian beam with the same wavelength and with the central "spot"‡ $\Delta\rho_0 = 59$ μm, when passing through an aperture with the radius $R = 3.5$ mm, doubles its transverse width after 3 cm, and after 6 cm its intensity has already diminished by a factor of 10. Therefore, in the case considered, a Bessel beam can travel, approximately without deformation, a distance 28 times longer than that traveled by a Gaussian beam.

Such a remarkable property is due to the fact that, when diffracting, the transverse intensity fields (whose value decreases with increasing ρ), associated with the rings that constitute the (transverse) structure of the Bessel beam, end up reconstructing the beam itself all along a large depth of field. All this depends on the Bessel beam

† Let us define the *central spot* of a Bessel beam as the distance, along the propagation axis $\rho = 0$, at which occurs the first zero of the Bessel function, characterizing its transverse shape.

‡ In the case of a Gaussian beam, let us define its *central spot* as the distance along $\rho = 0$ at which its transverse intensity has decayed by the factor $1/e$.

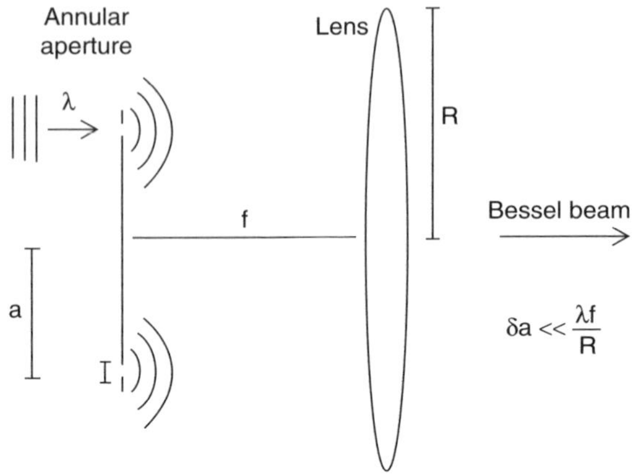

FIGURE 1.1 Simple experimental setup used by Durnin et al. to generate a Bessel beam.

spectrum (wave number and frequency) [34,38,41], as explained in detail in [42]. Let us stress that given a Bessel and a Gaussian beam, both with the same spot $\Delta\rho_0$ and passing through apertures with the same radius R in the plane $z = 0$ and with the same energy E, the percentage of the total energy E contained inside the central peak region ($0 \leq \rho \leq \Delta\rho_0$) is smaller for a Bessel than for a Gaussian beam. This different energy distribution on the transverse plane is responsible for reconstruction of the Bessel beam central peak even at large distances from the source (and even for an obstacle smaller than the aperture [71,78,117]: a nice property also possessed by the localized pulses [117] we examine below).

It may be worth mentioning that most experiments carried on in this area have been performed rapidly and, often, using rather simple apparatus. For example, Durnin et al.'s setup, used to generate a Bessel beam, had a laser source, an annular slit, and a lens, as depicted in Fig. 1.1. In a sense, such apparatus produces what can be regarded as the cylindrically symmetric generalization of a couple of plane waves emitted at angles θ and $-\theta$ with respect to the z-direction (in which case the plane-wave intersection moves along z with the speed $c/\sin\theta$). Of course, these nondiffracting beams can also be generated by a conic lens (an *axicon*) (see, e.g., [34]), or by other means, such as holographic elements (cf., e.g., [38,44]).

As mentioned earlier, a lot of interesting applications of nondiffracting beams are being investigated in addition to Lu et al.'s work in acoustics. In the optical sector, one example is the use of Bessel beams as optical tweezers able to confine small particles or move then around. In such theoretical and application areas, a noticeable contribution is the one presented in [45,46,77], wherein, by suitable superposition of Bessel beams endowed with the same frequency but different longitudinal wave numbers, *stationary* fields have been constructed mathematically in closed form, which possess a high transverse localization and, more important, a longitudinal intensity shape that can be chosen freely inside a predetermined space interval $0 \leq z \leq L$. For instance, a high-intensity field, with a static envelope, can be created within a tiny region, with negligible intensity elsewhere. Chapter 2 deals with such "frozen waves."

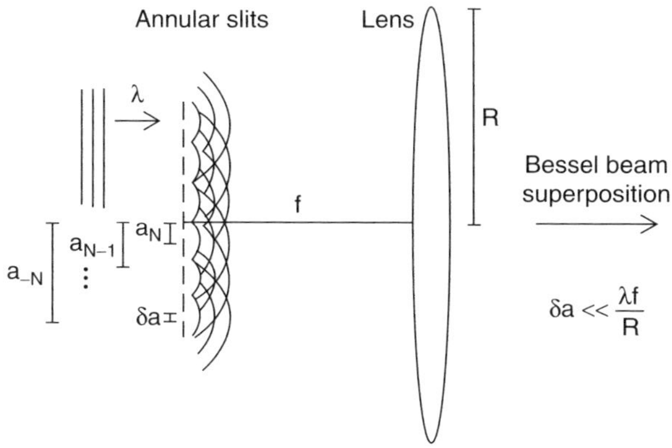

FIGURE 1.2 One of the simplest experimental setups for generating various kinds of Bessel beam superpositions.

Localized (Nondiffracting) Pulses As we have seen, the existence of nondiffractive (or localized) pulses was long predicted; see [16,17], and also [14,18], as well as more recent articles (e.g., [47,48]). Modern studies on nondiffractive pulses (confining ourselves to those that attracted more attention) developed somewhat independently of those on nondiffracting beams, even if both phenomena are part of the same sector of physics: localized waves.

In 1983, Brittingham [49] set forth a luminal ($V = c$) solution to the wave equation (more particularly, to Maxwell's equations), which travels rigidly (i.e., without diffraction). The solution proposed [49] possessed infinite energy, however, and once more the problem arose of overcoming such a problem. A way out was first obtained by Sezginer [50], who showed how to construct finite-energy luminal pulses, which, however, do not propagate without distortion for an infinite distance, but, as expected, travel with constant speed and approximately without deformation for a certain (long) depth of field: much longer, in this case, too, than that of ordinary pulses such as the Gaussian pulses. In a series of subsequent papers [35,36,51–54], a simple theoretical method was developed, called *bidirectional decomposition* by those authors, for constructing a new series of nondiffracting luminal pulses.

Finally, at the beginning of the 1990s, Lu et al. [19,21] constructed, both mathematically and experimentally, new solutions to the wave equation in free space: namely, an X-shaped localized pulse, with the form predicted by the so-called extended special relativity [1,14]; for the connection between what Lu et al. called *X-waves* and extended relativity, see, for example, [18]; brief excerpts of that theory can also be found, for example, in [20,55–58]. Lu et al.'s solutions were continuous superpositions of Bessel beams with the same phase velocity (i.e., with the same axicon angle [1,19,20,59]; cf. Fig. 1.2) so they could keep their shape for long distances. Such X-shaped waves resulted in interesting and flexible localized solutions, even if their velocity V is supersonic or superluminal ($V > c$), and have been studied in a number of papers. Actually, when the phase velocity does not depend on the

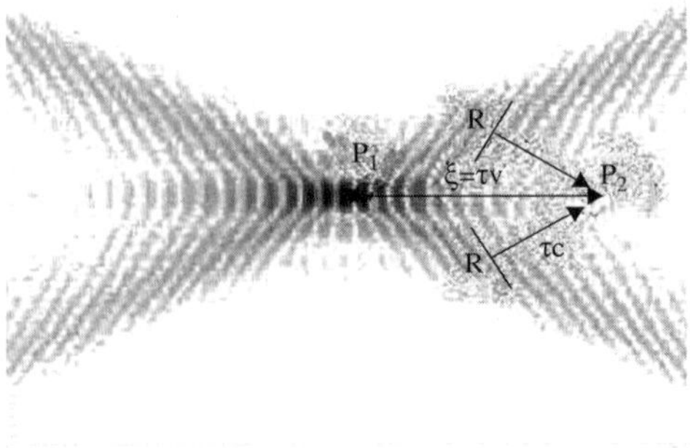

FIGURE 1.3 An X-shaped wave: that is, a localized superluminal pulse. It is an X-wave, possessing the velocity $V > c$, and the figure illustrates the fact that if its vertex or central spot is located at P_1 at time t_0, it will reach position P_2 at time $t + \tau$, where $\tau = |P_2 - P_1|/V < |P_2 - P_1|/c$. This is different from the illusory "scissor effect," *even if the feeding energy, coming from the regions R, has traveled with the ordinary speed c* (which is the speed of light in the electromagnetic case, the sound speed in acoustics, etc.).

frequency, it is known that such a phase velocity becomes the group velocity! Remembering how a superposition of Bessel beams is generated (e.g., by a discrete or continuous set of annular slits or transducers), it is clear that the energy forming the X-waves coming from those rings travels at the ordinary speed c of plane waves in the medium considered [20,60–62]. (Here, c, representing the velocity of the plane waves in the medium, is the speed of sound in the acoustic case, the speed of light in the electromagnetic case, and so on.) Nevertheless, the peak of the X-shaped waves is faster than c.

It is possible to generate (in addition to the "classic" X-wave produced by Lu et al. in 1992) infinite sets of new X-shaped waves, with their energy concentrated more and more in a spot corresponding to the vertex region [42]. It may therefore appear rather intriguing that such a spot (even if no violation of special relativity is obviously implied: all the results come from Maxwell's equations or from the wave equations [73,74]) travels superluminally when the waves are electromagnetic. We shall call all the X-shaped waves *superluminal* even when, for example, the waves are acoustic. In Fig. 1.3, we illustrate the fact that if its vertex or central spot is located at P_1 at time t_1, it will reach position P_2 at time $t + \tau$, where $\tau = |P_2 - P_1|/V < |P_2 - P_1|/c$. We discuss all these points below.

Soon after having constructed their "classic" acoustic X-wave mathematically and experimentally, Lu et al. started applying them to ultrasonic scanning, obtaining, as we already said, very high quality images. Subsequently, in a 1996 e-print and report, Recami et al. (see, e.g., [20] and references therein) published the analogous X-shaped solutions to Maxwell's equations: by constructing scalar superluminal localized solutions for each component of the Hertz potential. That showed, by the way, that localized solutions to scalar equation can also be used, under very weak conditions, for obtaining localized solutions to Maxwell's equations (actually,

Ziolkowski et al. [43] had found something similar, which they called *slingshot pulses*, for the simple scalar case, but their solution had gone practically unnoticed). In 1997, Saari and Reivelt [22] announced, in an important paper, the production in the lab of an X-shaped wave in the optical realm, thus proving experimentally the existence of superluminal electromagnetic pulses. Three years later, in 2000, Mugnai et al. [23] produced, experimentally, superluminal X-shaped waves in the microwave region (their paper aroused various criticisms, to which the authors responded).

1.2 MORE DETAILED INFORMATION

Let us consider [5] the differential equation known as *homogeneous wave equation:* simple, but so important in acoustics, electromagnetism (microwaves, optics, etc.), geophysics, and even, as we said, gravitational waves and elementary particle physics:

$$\left(\frac{\partial^2}{\partial x^2}+\frac{\partial^2}{\partial y^2}+\frac{\partial^2}{\partial z^2}-\frac{1}{c^2}\frac{\partial^2}{\partial t^2}\right)\psi(x,y,z;t)=0. \tag{1.1}$$

Let us write it in the cylindrical coordinates (ρ, ϕ, z) and, for simplicity's sake, confine ourselves to axially symmetric solutions $\psi(\rho, z; t)$. Then, Eq. (1.1) becomes

$$\left(\frac{\partial^2}{\partial \rho^2}+\frac{1}{\rho}\frac{\partial}{\partial \rho}+\frac{\partial^2}{\partial z^2}-\frac{1}{c^2}\frac{\partial^2}{\partial t^2}\right)\psi(\rho,z;t)=0. \tag{1.2}$$

In free space, the solution $\psi(\rho, z; t)$ can be written in terms of a Bessel–Fourier transform with respect to the variable ρ, and two Fourier transforms with respect to variables z and t:

$$\psi(\rho,z,t)=\int_0^\infty\int_{-\infty}^\infty\int_{-\infty}^\infty k_\rho\, J_0(k_\rho\,\rho)\, e^{ik_z z}\, e^{-i\omega t}\,\bar{\psi}(k_\rho,k_z,\omega)\,dk_\rho\,dk_z\,d\omega, \tag{1.3}$$

where $J_0(\cdot)$ is an ordinary zero-order Bessel function and $\bar{\psi}(k_\rho, k_z, \omega)$ is the transform of $\psi(\rho, z, t)$.

Substituting Eq. (1.3) into Eq. (1.2), one finds that the relation

$$\frac{\omega^2}{c^2}=k_\rho^2+k_z^2 \tag{1.4}$$

among ω, k_ρ, and k_z has to be satisfied. In this way, by using condition (1.4) in Eq. (1.3), any solutions to the wave equation (1.2) can be written

$$\psi(\rho,z,t)=\int_0^{\omega/c}\int_{-\infty}^\infty k_\rho\, J_0(k_\rho\,\rho)\, e^{iz\sqrt{\omega^2/c^2-k_\rho^2}}\, e^{-i\omega t}\, S(k_\rho,\omega)\,dk_\rho\,d\omega, \tag{1.5}$$

where $S(k_\rho, \omega)$ is the spectral function chosen.

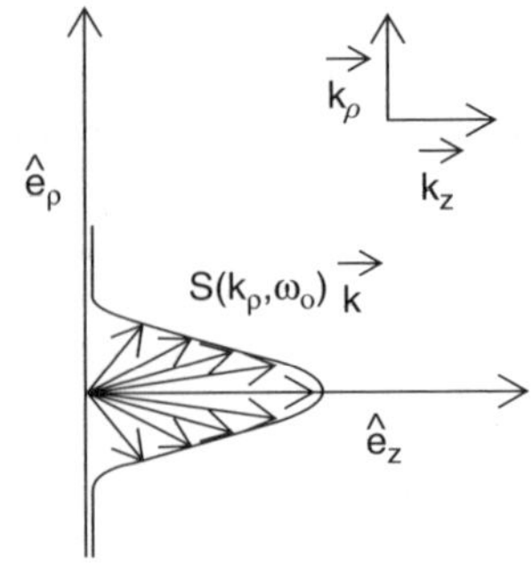

FIGURE 1.4 Visual interpretation of the integral solution (1.5) with the spectrum (1.6) in terms of a superposition of plane waves.

The general integral solution (1.5) yields, for instance, (nonlocalized) Gaussian beams and pulses, to which we shall refer to illustrate the differences between localized waves and them.

Gaussian Beam A very common (nonlocalized) beam is the Gaussian beam [76], corresponding to the spectrum

$$S(k_\rho, \omega) = 2a^2 e^{-a^2 k_\rho^2}\, \delta(\omega - \omega_0). \tag{1.6}$$

In Eq. (1.6), a is a positive constant that will be shown to depend on the transverse aperture of the initial pulse.

Figure 1.4 illustrates the interpretation of integral solution (1.5) with spectral function (1.6) as a superposition of plane waves. From Fig. 1.4 one can easily realize that this case corresponds to plane waves propagating in all directions (always with $k_z \geq 0$), the most intense being those directed along (positive) z. Notice that in the plane-wave case, $\vec{k}_z$ is the longitudinal component of the wave vector, $\vec{k} = \vec{k}_\rho + \vec{k}_z$, where $\vec{k}_\rho = \vec{k}_x + \vec{k}_y$.

On substituting Eq. (1.6) into Eq. (1.5) and adopting the paraxial approximation, one obtains the Gaussian beam:

$$\psi_{\text{Gauss}}(\rho, z, t) = \frac{2a^2 \exp[-\rho^2/4(a^2 + iz/2k_0)]}{2(a^2 + iz/2k_0)}\, e^{ik_0(z-ct)}, \tag{1.7}$$

where $k_0 = \omega_0/c$. We can verify that such a beam, which suffers transverse diffraction, doubles the initial width $\Delta\rho_0 = 2a$ after having traveled the distance $z_{\text{dif}} = \sqrt{3}\, k_0\, \Delta\rho_0^2/2$, called the *diffraction length*. The more concentrated a Gaussian beam happens to be, the more rapidly it gets spoiled.

Gaussian Pulse The most common (nonlocalized) pulse is the Gaussian pulse, which is obtained from Eq. (1.5) using the spectrum [75]

$$S(k_\rho, \omega) = \frac{2ba^2}{\sqrt{\pi}} e^{-a^2 k_\rho^2} e^{-b^2(\omega-\omega_0)^2}, \tag{1.8}$$

where a and b are positive constants. Indeed, such a pulse is a superposition of Gaussian beams of different frequencies.

Now, on substituting Eq. (1.8) into Eq. (1.5), and again adopting the paraxial approximation, one gets the Gaussian pulse

$$\psi(\rho, z, t) = \frac{a^2 \exp[-\rho^2/4(a^2 + iz/2k_0)]\exp[-(z-ct)^2/4c^2b^2]}{a^2 + iz/2k_0}, \tag{1.9}$$

endowed with speed c and temporal width $\Delta t = 2b$, and suffering a progressive enlargement of its transverse width so that its initial value is already doubled at position $z_{\text{dif}} = \sqrt{3}\, k_0\ \Delta\rho_0^2/2$, with $\Delta\rho_0 = 2a$.

1.2.1 Localized Solutions

Finally, let's look at the construction of the two most renowned localized waves: the Bessel beam and the ordinary X-shaped pulse [5].

It is interesting, first, to observe that, when superposing (axially symmetrical) solutions of the wave equation in vacuum, three spectral parameters (ω, k_ρ, k_z) come into play which have however to satisfy constraint (1.4), deriving from the wave equation itself. Consequently, only two of them are independent, and here we choose† ω and k_ρ. Such freedom in choosing ω and k_ρ was apparent in the spectral functions generating Gaussian beams and pulses, which are the product of two functions, one depending only on ω and the other on k_ρ.

We are going to see, moreover, that particular relations can be imposed between ω and k_ρ (or analogously, between ω and k_z) in order to get interesting and unexpected results, such as localized waves.

Bessel Beam Let us start by imposing a linear coupling between ω and k_ρ (it could actually be shown [41] that it is the unique coupling leading to localized solutions). Let us consider the spectral function

$$S(k_\rho, \omega) = \frac{\delta(k_\rho - (\omega/c)\sin\theta)}{k_\rho}\ \delta(\omega - \omega_0), \tag{1.10}$$

which implies that $k_\rho = (\omega \sin\theta)/c$, with $0 \le \theta \le \pi/2$, a relation that can be regarded as a space–time coupling. Let us add that this linear constraint between ω and k_ρ,

† Elsewhere we chose ω and k_z.

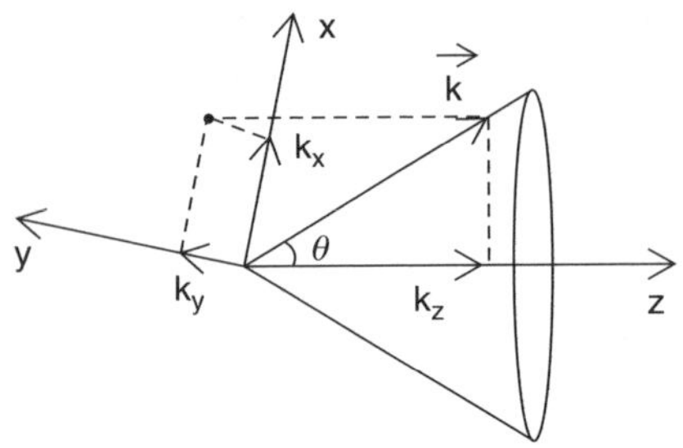

FIGURE 1.5 An axially symmetric Bessel beam is created by the superposition of plane waves whose wave vectors lay on the surface of a cone having the propagation axis as its symmetry axis and angle equal to θ (axicon angle).

together with relation (1.4), yields $k_z = (\omega \cos\theta)/c$. This is an important fact, since it has been shown elsewhere [42] that an *ideal* localized wave *must* contain a coupling of the type $\omega = Vk_z + b$, where V and b are arbitrary constants.

The interpretation of the integral function (1.5), this time with the spectrum (1.10), as a superposition of plane waves is visualized in Fig. 1.5, which shows that an axially symmetric Bessel beam is nothing but the result of the superposition of plane waves whose wave vectors lay on the surface of a cone having the propagation axis as its symmetry axis and an angle equal to θ, the *axicon angle*.

By inserting Eq. (1.10) into Eq. (1.5), one gets the mathematical expression of a Bessel beam:

$$\psi(\rho, z, t) = J_0\left(\frac{\omega_0}{c}\sin\theta\,\rho\right)\exp\left[i\,\frac{\omega_0}{c}\cos\theta\left(z - \frac{c}{\cos\theta}t\right)\right]. \tag{1.11}$$

This beam possesses phase velocity $v_{\text{ph}} = c/\cos\theta$ and field transverse shape represented by a Bessel function $J_0(\cdot)$, so that its field is concentrated in the area surrounding the propagation axis z. Moreover, Eq. (1.11) tells us that the Bessel beam keeps its transverse shape (which is therefore invariant) while propagating, with central spot $\Delta\rho = 2.405c/(\omega\sin\theta)$.

The ideal Bessel beam is, however, not a square-integrable function and thus possesses infinite energy (i.e., it cannot be produced experimentally).

But we can have recourse to truncated Bessel beams, generated by finite apertures. In this case the (truncated) Bessel beams are still able to travel a long distance while maintaining their transfer shape, as well as their speed, approximately unchanged [40,69,70]; that is, they still possess a large field depth. For instance, the depth of field of a Bessel beam generated by a circular finite aperture with radius R is given by

$$Z_{\max} = \frac{R}{\tan\theta}, \tag{1.12}$$

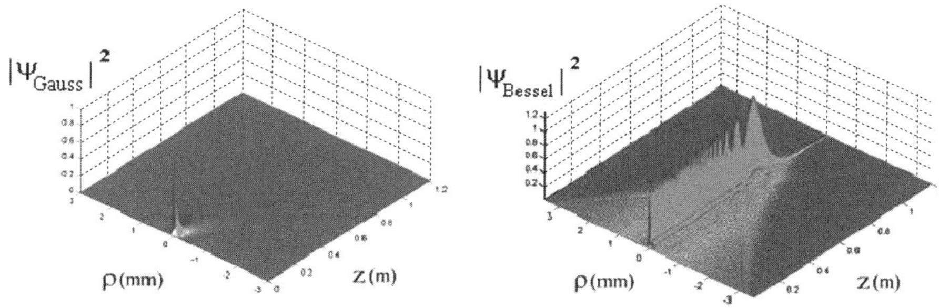

(a) (b)

FIGURE 1.6 Comparison of a Gaussian (a) and a truncated Bessel beam (b). The Gaussian beam doubles its initial transverse width after 3 cm, and after 6 cm its intensity decays by a factor of 10. By contrast, the Bessel beam keeps its transverse shape approximately up to a distance of 85 cm.

where θ is the beam axicon angle. In the finite-aperture case, the Bessel beam can no longer be represented by Eq. (1.11). One has to calculate it by the scalar diffraction theory: using, for example, Kirchhoff's or Rayleigh–Sommerfeld's diffraction integrals. But until reaching Z_{max}, one may still use Eq. (1.11) for describing it approximately, at least in the surrounding of the axis $\rho = 0$: namely, for $\rho << R$. To realize how much a truncated Bessel beam succeeds in resisting diffraction, let us also consider a Gaussian beam, with the same frequency and central spot, and compare their field depths. In particular, let us assume for both beams $\lambda = 0.63$ μm, and an initial central spot size $\Delta\rho_0 = 60$ μm. The Bessel beam will possess axicon angle $\theta = \arcsin(2.405c/(\omega\,\Delta\rho_0)) = 0.004$ rad. Figure 1.6 depicts the behavior of the two beams for a Bessel beam circular aperture with radius 3.5 mm. We can see how the Gaussian beam doubles its initial transverse width after 3 cm, and after 6 cm its intensity becomes an order of magnitude smaller. By contrast, the truncated Bessel beam keeps its transverse shape until distance $Z_{\text{max}} = R/\tan\theta = 85$ cm. Afterward, the Bessel beam decays rapidly as a consequence of the sharp cut performed on its aperture (which is also responsible for the intensity oscillations suffered by the beam along its propagation axis and for the fact that eventually the feeding waves coming from the aperture get faint at a certain point).

The zeroth-order (axially symmetric) Bessel beam is simply one example of a localized beam. Further examples are higher-order (not cylindrically symmetric) Bessel beams

$$\psi(\rho,\phi,z;t) = J_\nu\left(\frac{\omega_0}{c}\sin\theta\;\rho\right)\exp(i\nu\phi)\exp\left[i\;\frac{\omega_0}{c}\cos\theta\left(z - \frac{c}{\cos\theta}t\right)\right], \tag{1.13}$$

and Mathieu beams [68].

Ordinary X-Shaped Pulse Following the procedure adopted earlier, let us construct pulses by using spectral functions of the type

$$S(k_\rho, \omega) = \frac{\delta(k_\rho - (\omega/c)\sin\theta)}{k_\rho} F(\omega), \tag{1.14}$$

where this time the Dirac delta function furnishes the spectral space–time coupling $k_\rho = (\omega \sin\theta)/c$. The function $F(\omega)$ is, of course, the frequency spectrum; for the moment it is left undetermined.

On plugging Eq. (1.14) into Eq. (1.5) we obtain

$$\psi(\rho, z, t) = \int_{-\infty}^{\infty} F(\omega) J_0\left(\frac{\omega}{c}\sin\theta\ \rho\right) \exp\left[\frac{\omega}{c}\cos\theta\left(z - \frac{c}{\cos\theta}t\right)\right] d\omega. \tag{1.15}$$

It is easy to see that ψ will be a pulse of the type

$$\psi = \psi(\rho, z - Vt) \tag{1.16}$$

with a speed $V = c/\cos\theta$ independent of the frequency spectrum $F(\omega)$. Such solutions are known as X-shaped pulses and are *localized* (nondiffractive) waves in the sense that they maintain their spatial shape during propagation (see., e.g., [19,20,42] and references therein).

At this point, some remarkable observations are in order:

1. When a pulse consists of a superposition of waves (in this case, Bessel beams) all endowed with the same phase velocity V_{ph} (in this case, with the same axicon angle) independent of their frequency, it is known that the phase velocity (in this case $V_{\text{ph}} = c/\cos\theta$) becomes the group velocity [64,72] V; that is, $V = c/\cos\theta > c$. In this sense, the X-shaped waves are called *superluminal localized pulses* (cf., e.g., [20] and references therein).

2. Even if their group velocity is superluminal, such pulses do not contradict standard physics, having been found in what precedes on the basis of the wave equations—in particular, Maxwell's equations [20,35]—only. Indeed, as we shall see more clearly in the Appendix, their existence can be understood within special relativity itself [14,18,20,55–57], on the basis of its ordinary postulates [1]. Actually, let us repeat that such pulses are fed by waves originating at the aperture and carrying energy at the standard speed c (the velocity of light in the electromagnetic case and the velocity of sound in the particular medium considered in the acoustic [21] case). We can become convinced of the possibility of realizing superluminal X-shaped pulses by imagining the simple ideal case of a negligibly sized superluminal source S endowed with speed $V > c$ in vacuum and emitting electromagnetic waves W (each traveling with the invariant speed c). The electromagnetic waves will be internally tangent to an enveloping cone C having S as its vertex and the propagation line z of the source as its axis [1,18]: *This is completely analogous to what happens for an airplane that*

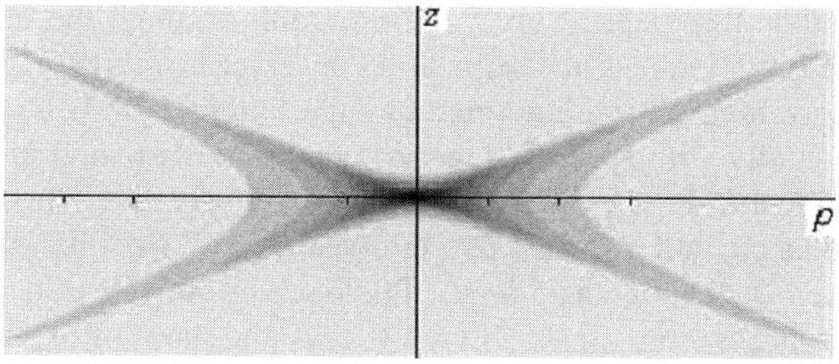

FIGURE 1.7 The truncated X-waves considered in this chapter, as predicted by SR (all wave equations have an intrinsic relativistic structure), must have a leading cone in addition to the rear cone, such a leading cone having a role for the peak stability [19]: For example, when producing a finite conic wave truncated in both space and time, SR theory suggested, in the simplest case, use of a dynamic antenna emitting radiation cylindrically symmetrical in space and symmetric in time, for a better approximation to what Courant and Hilbert [16] called an undistorted progressing wave (see below).

moves in air with constant supersonic speed. The waves W interfere mainly negatively inside the cone C and constructively on its surface. We can place a plane detector orthogonal to z and record magnitude and direction of the W waves that hit it, as (cylindrically symmetric) functions of position and time. It will be enough, then, to replace the plane detector with a plane antenna that emits (instead of recording) exactly the same (axially symmetric) space–time pattern of waves W, for constructing a cone-shaped electromagnetic wave C that will propagate with superluminal speed V (of course, no longer with a source at its vertex), even if each wave W travels with the invariant speed c. Once more, *this is exactly what would happen in the case of a supersonic airplane* (in which case, c is the sound speed in air; for simplicity, assume the observer to be at rest with respect to the air). For further details, see the references cited. Actually, by suitable superpositions and interference of speed-c waves, one can obtain pulses that are more and more localized in the vertex region [42], that is, very localized field "blobs" traveling with superluminal group velocity. This has nothing to do with the illusory "scissors effect," since such blobs, along their depth of field, are a priori able, for example, to get two successive (weak) detectors, located at distance L, clicking after a time smaller than L/c. Incidentally, an analysis of the above-mentioned case (that of a supersonic plane or a superluminal charge) led, as expected [1], to the simplest type of X-shaped pulse [18]. It might be useful, finally, to recall that special relativity (SR) (even the wave equations have an internal relativistic structure!) implies also considering the forward cone (see Fig. 1.7). The truncated X-waves considered in this chapter, for instance, must have a leading cone in addition to the rear cone, such a leading cone having a role for the peak stability [19]. For example, in the approximate case in which we produce a finite conic wave truncated in both space and time, SR theory suggested the biconic shape (symmetrical in space with respect to the vertex S) to be a better approximation to a rigidly traveling wave (so that SR suggests using a dynamic antenna emitting a radiation cylindrically symmetrical in space and symmetrical in time, for a better approximation to an undistorted progressing wave).

3. Any solutions that depend on z and on t only through the quantity $z - Vt$, such as Eq. (1.15), will appear the same to an observer traveling along z with speed V, whatever the value of V (subluminal, or superluminal). That is, such a solution will propagate rigidly with speed V (and in fact, there exist superluminal, luminal, and subluminal localized waves). This explains further why the X-shaped pulses, after having been produced, will travel almost rigidly at speed V (in this case, a faster-than-light group velocity) all along their depth of field. To be even clearer, let us consider a generic function, depending on $z - Vt$ with $V > c$, and show, by explicit calculations involving Maxwell's equations only, that it obeys the scalar wave equation. Following Selleri [103], let us consider the wave function

$$\Phi(x, y, z, t) = \frac{a}{\sqrt{[b - ic(z - Vt)]^2 + (V^2 - c^2)(x^2 + y^2)}}, \tag{1.17}$$

with a and b nonzero constants, c the ordinary speed of light, and $V > c$ (incidentally, this wave function is simply the classic X-shaped wave in Cartesian coordinates).

Let us verify that it is a solution to the wave equation

$$\nabla^2\Phi(x, y, z, t) - \frac{1}{c^2}\frac{\partial^2\Phi(x, y, z, t)}{\partial^2 t} = 0. \tag{1.18}$$

On setting

$$R \equiv \sqrt{[b - ic(z - Vt)]^2 + (V^2 - c^2)(x^2 + y^2)}, \tag{1.19}$$

one can write $\Phi = a/R$ and evaluate the second derivatives

$$\begin{aligned}
\frac{1}{a}\frac{\partial^2\Phi}{\partial^2 z} &= \frac{c^2}{R^3} - \frac{3c^2}{R^5}[b - ic(z - Vt)]^2 \\
\frac{1}{a}\frac{\partial^2\Phi}{\partial^2 x} &= -\frac{V^2 - c^2}{R^3} + 3\left(V^2 - c^2\right)^2\frac{x^2}{R^5} \\
\frac{1}{a}\frac{\partial^2\Phi}{\partial^2 y} &= -\frac{V^2 - c^2}{R^3} + 3\left(V^2 - c^2\right)^2\frac{y^2}{R^5} \\
\frac{1}{a}\frac{\partial^2\Phi}{\partial^2 t} &= \frac{c^2V^2}{R^3} - \frac{3c^2V^2}{R^5}[b - ic(z - Vt)]^2,
\end{aligned}$$

from which

$$\begin{aligned}
\frac{1}{a}\left(\frac{\partial^2\Phi}{\partial^2 z} - \frac{1}{c^2}\frac{\partial^2\Phi}{\partial^2 t}\right) &= -\frac{V^2 - c^2}{R^3} + 3\left(V^2 - c^2\right)^2\frac{[b - ic(z - Vt)]^2}{R^5} \\
\frac{1}{a}\left(\frac{\partial^2\Phi}{\partial^2 x} + \frac{\partial^2\Phi}{\partial^2 y}\right) &= -2\frac{V^2 - c^2}{R^3} + 3\left(V^2 - c^2\right)^2\frac{x^2 + y^2}{R^5}.
\end{aligned}$$

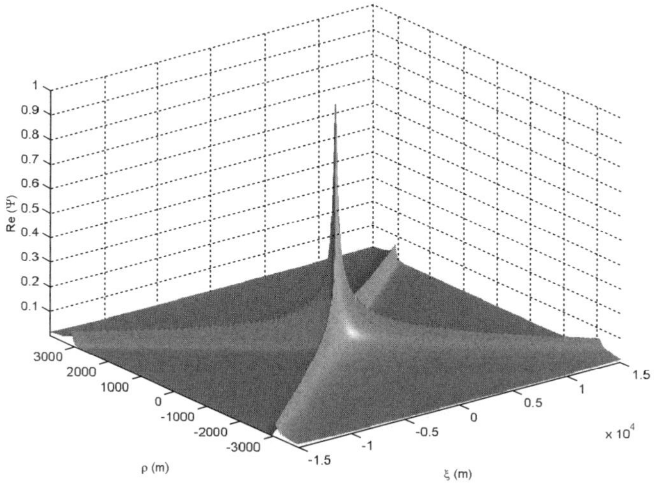

FIGURE 1.8 Plot of the real part of an ordinary X-wave evaluated for $V = 1.1c$ with $a = 3$ m.

From these two equations, based on the previous definition, one finally gets

$$\frac{1}{a}\left(\frac{\partial^2\Phi}{\partial^2 z} + \frac{\partial^2\Phi}{\partial^2 x} + \frac{\partial^2\Phi}{\partial^2 y} - \frac{1}{c^2}\frac{\partial^2\Phi}{\partial^2 t}\right) = 0,$$

which is simply the (d'Alembert) wave equation (1.18). In conclusion, the function Φ is a solution of the wave equation even if it represents a pulse (Selleri says a "signal") propagating with superluminal speed.

After the three important observations that we made above, let us return to our evaluations with regard to X-type solutions to the wave equations. Let us now consider, for example, the particular frequency spectrum $F(\omega)$ in Eq. (1.15), given by

$$F(\omega) = H(\omega)\,\frac{a}{V}\,\exp\left(-\frac{a}{V}\,\omega\right), \tag{1.20}$$

where $H(\omega)$ is the Heaviside step function and a is a positive constant. Then, Eq. (1.15) yields

$$\psi(\rho, z - Vt) \equiv X = \frac{a}{\sqrt{(a - i\zeta)^2 + \left[(V^2/c^2) - 1\right]\rho^2}}, \tag{1.21}$$

with $\zeta \equiv z - Vt$. This solution is the well-known ordinary, or "classic," X-wave, which is a simple example of an X-shaped pulse [19,20]. Notice that function (1.20) contains mainly low frequencies, so that the classic X-wave is suitable for low frequencies only. Figure 1.8 depicts the real part of an ordinary X-wave with $V = 1.1c$ and $a = 3$ m.

Solutions (1.15), and in particular the pulse (1.21), have infinite depth of field as well as infinite energy. Therefore, as done in the Bessel beam case, we should go on to truncated pulses originating from a finite aperture. Afterward, our truncated pulses will keep their spatial shape (and their speed) all along the depth of field

$$Z = \frac{R}{\tan\theta}, \tag{1.22}$$

where, as before, R is the aperture radius and θ is the axicon angle.

Further Observations It is not strictly correct to call localized waves nondiffractive, since diffraction more or less affects all waves obeying Eq. (1.1). *However*, all localized waves (both beams and pulses) possess the remarkable self-reconstruction property: That is, when diffracting during propagation, localized waves rebuild their shape immediately [71,78,117] (even after obstacles with size much larger than the implied wavelengths, provided, of course, that it is smaller than the aperture size), due to their particular spectral structure, as shown in more detail in other chapters. In particular, ideal localized waves (those with infinite energy and depth of field) are able to rebuild for an infinite time, whereas, as we have seen, finite-energy (truncated) waves can rebuild, and thus resist diffraction effects, only along a certain depth of field.

Let us stress again that interest in localized waves (especially from the point of view of applications) lies in the fact that they are almost nondiffractive, and not in their group velocity. From this point of view, superluminal, luminal, and subluminal localized solutions are equally interesting and suited to important applications.

Actually, localized waves are not all restricted to the X-shaped, superluminal waves corresponding to the integral solution (1.15) of the wave equation; and, as we said earlier, three classes of localized pulses exist: superluminal (with speed $V > c$), luminal ($V = c$), and subluminal ($V < c$), all of them with or without axial symmetry, and all corresponding to a single unified mathematical background. This issue is touched on again in the book. Incidentally, elsewhere we have addressed such topics as (1) the construction of infinite families of generalizations of the classic X-shaped wave, with energy concentrated more and more around the vertex, as shown in Fig. 1.9; (2) the behavior of some finite total-energy superluminal localized solutions (SLS); (3) a way to build up a new series of SLSs to Maxwell's equations suitable for arbitrary frequencies and bandwidths; and (4) questions related to dispersive media: In Chapter 2 we return to some of these points. We add that X-shaped waves have also been produced easily in nonlinear media [4], as described in Chapter 9. A more technical introduction to the subject of localized waves (particularly with respect to superluminal X-shaped waves) may be found in [55].

APPENDIX: THEORETICAL AND EXPERIMENTAL HISTORY

In this mainly historical section, we first describe from a theoretical point of view the most intriguing localized solutions to the wave equation: the superluminal solutions

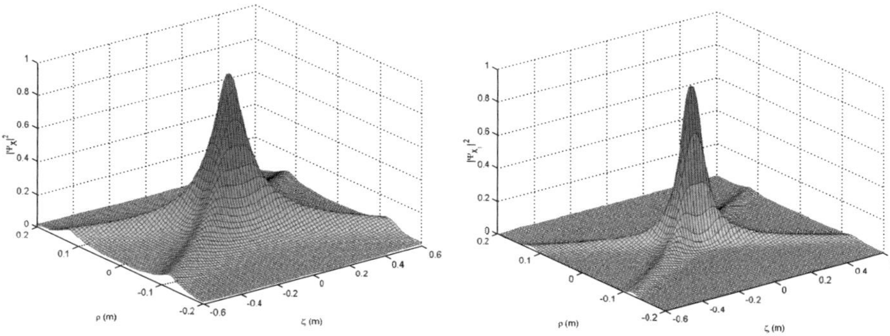

(a) (b)

FIGURE 1.9 (a) Square magnitude (arbitrary units) of the classic, X-shaped superluminal localized solution (SLS) to the wave equation, with $V = 5c$ and $a = 0.1$ m. Families of infinite SLSs exist, which generalize the classic X-shaped solution; for instance, a family of SLSs obtained [42] by suitably differentiating the classic X-wave. (b) The first SLS (corresponding to the first differentiation) with the same parameters. Successsive solutions in such a family are more localized around their vertex. The quantity ρ is the distance in meters from the propagation axis z, while the quantity ζ is the V-cone variable [42] (also in meters) $\zeta \equiv z - Vt$, with $V \geq c$. Since all these solutions depend on z only via the variable ζ, they propagate "rigidly" (i.e., without distortion and thus are called localized, or nondispersive). Here we assume propagation in the vacuum (or in a homogeneous medium).

(SLSs), in particular the X-shaped pulses. To start with, we recall their geometrical interpretation within special relativity (SR). Afterward, to help resolve possible doubts, we present a bird's-eye view of the various experimental sectors of physics in which superluminal motions seem to appear: in particular, of the experiments with evanescent waves (and/or tunneling photons) or with the SLSs we are more interested in here. In some parts of this appendix the motion line is called x rather than z, but that should not present any problems of interpretation.

The subject of superluminal ($V^2 > c^2$) objects or waves has a long history, beginning prior to special relativity in papers by J. J. Thomson and A. Sommerfeld, among others. However, with the development of special relativity, the conviction spread that the speed c of light in a vacuum was the upper limit of speed possible. For example, R. C. Tolman (in 1917) believed that he had shown by his "paradox" that the existence of particles endowed with speeds higher than c would allow sending information into the past. The problem was not tackled again until the 1950s and 1960s, in particular after the work by Bilaniuk et al. [89] and later by one of the present authors with Mignani et al. [80,81], as well as (confining ourselves at present to theoretical research) by H. C. Corben and others. The first experimental attempts were performed by T. Alväger et al.

We wish to face the still unusual issue of the possible existence of superluminal wavelets and objects within standard physics and SR, since at least four different experimental sectors of physics seem to support such a possibility (apparently confirming some long-standing theoretical predictions [1,14,81,104]). The experimental

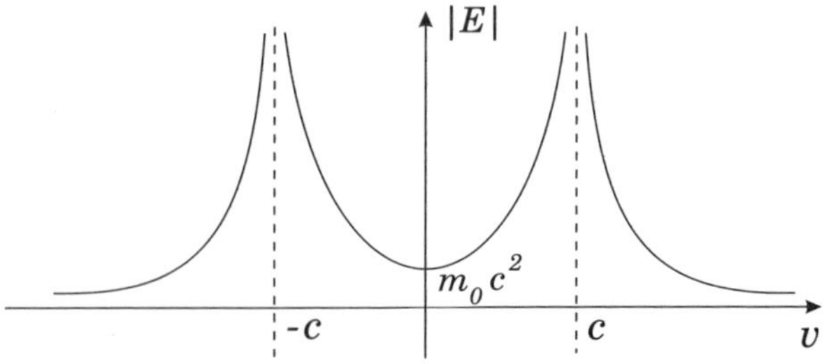

FIGURE 1A.1 Energy of a free object as a function of its speed [1,80,104].

review will be necessarily short, but we provide the reader with further bibliographical information (limited for the sake of brevity to the twentieth century).

Historical Recollections: Theory

Long ago a simple theoretical framework based on the space–time geometrical methods of SR was proposed [1,80] which appears to incorporate superluminal waves and objects and to predict [14] superluminal X-shaped waves without violating the principles of relativity. A suitable choice of the postulates of SR (equivalent, of course, to other, more common choices) consists in the standard principle of relativity and space–time homogeneity and space isotropy. It follows that one and only one invariant speed exists; and experience shows invariant speed to be the speed of light, c, in vacuum: The essential role of c in SR is due simply to its invariance, not to the fact that it is a maximal or minimal speed. No sub- or superluminal objects or pulses can be endowed with an invariant speed, so in SR their speed cannot play the same essential role played by the light-speed c. Indeed, c turns out to be a limiting speed, but any limit possesses two sides and can be approached a priori from both below and above (see Fig. 1A.1). As Sudarshan put it, from the fact that no one can climb over the Himalayas, people in India should not conclude that there are no people north of the Himalayas; actually, speed-c photons exist, which are born, live, and die "at the top of the mountain," without any need to performing the impossible task of accelerating from rest to the speed of light. (Actually, the ordinary formulation of SR is too restricted: Even leaving superluminal speeds aside, it can easily be broadened to include antimatter [1,57,58].)

An immediate consequence is that the quadratic form $c^2\,dt^2 - d\mathbf{x}^2 \equiv dx_\mu\,dx^\mu$, called ds^2, with $d\mathbf{x}^2 \equiv dx^2 + dy^2 + dz^2$, is invariant except for its sign. The quantity ds^2 is a four-dimensional length-element square along the space–time path of any object. Corresponding to a positive (negative) sign, we have subluminal (superluminal) *Lorentz transformations* [LTs]. Ordinary subluminal LTs are known to leave exactly the quadratic forms $dx_\mu\,dx^\mu$, $dp_\mu\,dp^\mu$, and $dx_\mu\,dp^\mu$ invariant (where the p_μ are components of the energy-impulse four-vector), whereas superluminal LTs have to change only the sign of such quadratic forms. This is enough to deduce some important consequences, such as the fact that a superluminal charge has to behave as a magnetic monopole (in the sense specified in [1] and references therein).

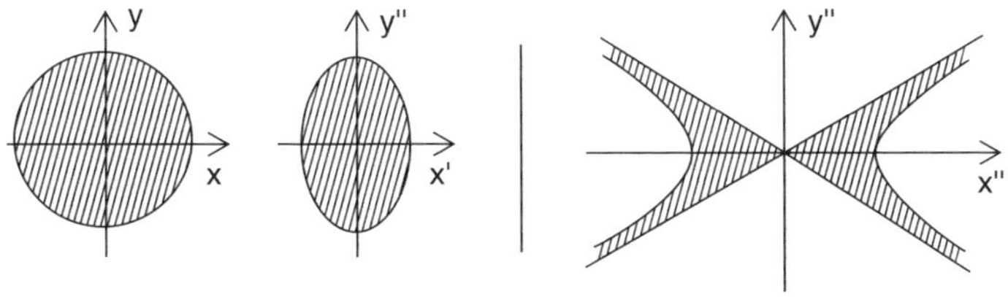

FIGURE 1A.2 An intrinsically spherical (or pointlike, at the limit) object appears in the vacuum as an ellipsoid contracted along the direction of motion when endowed with a speed $v < c$. By contrast, if endowed with a speed $V > c$ (even if the c-speed barrier cannot be crossed from neither the left nor the right), it would no longer appear [1,14] as a particle but, rather, as an X-shaped wave traveling rigidly: namely, as occupying a region delimited by a double cone and a two-sheeted hyperboloid, or at the limit as a double cone, moving without distortion in vacuum or in a homogeneous medium with superluminal speed V [the cotangent square of the cone semiangle being $(V > c)^2 - 1$]. For simplicity, a space axis is omitted. (From [1,14].)

A more important consequence for us is that the simplest subluminal object (Fig. 1A.2), a spherical particle at rest (which appears ellipsoidal due to Lorentz contraction, at subluminal speeds v) will appear [1,14,20] to occupy the cylindrically symmetrical region bounded by a two-sheeted rotation hyperboloid and an indefinite double cone (Fig. 1A.2d) for superluminal speeds V. In the figure the motion is along the x-axis. In the limiting case of a pointlike particle, we obtain only a double cone. Such a result is reached simply by writing down the equation of the *world-tube* of a subluminal particle and transforming it merely by changing the sign of the quadratic forms entering that equation. Thus, in 1980–1982, it was predicted [14] that the simplest superluminal object appears not as a particle, but as a field or, rather, as a wave: namely, as an X-shaped pulse, the cone semiangle α being given (with $c = 1$) by $\cot\alpha = \sqrt{V^2 - 1}$. Such X-shaped pulses will move rigidly with speed V along their motion direction. In fact, any X-pulse can be regarded at each instant of time as the superluminal Lorentz transform of a spherical object, which of course moves without any deformation in vacuum, or in a homogeneous medium, as time elapses. A three-dimensional picture of Fig. 1A.2d appears in Fig. 1A.3, where its annular intersections with a transverse plane are shown (see [14]). The X-shaped waves considered here are merely the simplest ones: If one started not from an intrinsically spherical or pointlike object, but from a nonspherically symmetric particle, a pulsating (contracting and dilating) sphere, or a particle oscillating back and forth along the direction of motion, their superluminal Lorentz transforms would be more and more complicated. The X-waves above are typical for a superluminal object, however, as the spherical or pointlike shape are typical for a subluminal object.

Incidentally, it has been believed for a long time that superluminal objects would have allowed sending information into the past, but such problems with causality seem to be solvable within SR. Apparently, once SR is generalized to include superluminal

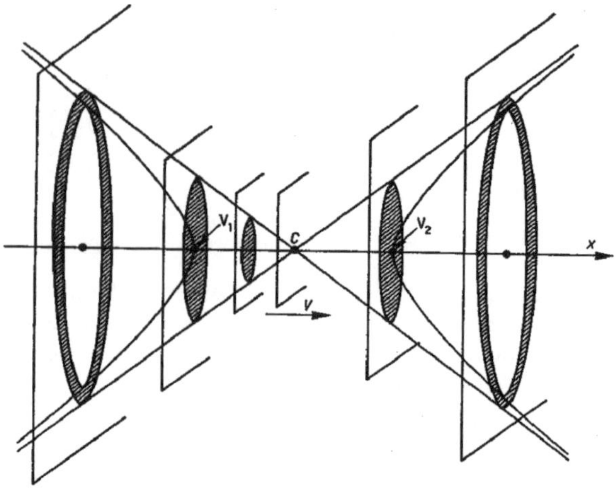

FIGURE 1A.3 Intersections of the superluminal object T in Fig. 1A.2d with planes P orthogonal to its line of motion (the x-axis). For simplicity, we again assumed the object to be spherical in its rest frame and the cone vertex C to coincide with the origin O for $t = 0$. Such intersections evolve in time so that the same pattern appears on a second plane, shifted by Δx, after the time $\Delta t = \Delta x / V$. On each plane, as time elapses, the intersection is therefore predicted by extended SR to be a circular ring which, for negative times, goes on shrinking until it is reduced to a circle and then to a point (for $t = 0$); afterward, such a point again becomes a circle and then a circular ring that goes on broadening [1,14,20]. [Notice that if the object is not spherical when at rest (but, e.g., is ellipsoidal in its own rest frame), the axis of T will no longer coincide with x, but its direction will depend on the speed V of the tachyon itself.] For the case in which the space extension of the superluminal object T is finite, see [14]. (From [1,14].)

objects or pulses, no signal traveling backward in time is left. For a solution of those causal paradoxes, see [58,104] and references therein.

To address the problem, even within this elementary context, of the production of an X-shaped pulse such as the one depicted in Fig. 1A.3 (perhaps truncated in space and time by use of a finite antenna radiating for a finite time), all the considerations described under observation 2 of the section "Ordinary X-Shaped Pulse" (Section 1.2.1) come into order: And, here, we simply refer to them. Those considerations, together with the present ones (related, e.g., to Fig. 1A.3), suggest that the simplest antenna consists of a series of concentric annular slits or transducers (as in Fig. 1.2), which suitably radiate following specific time patterns (see, e.g., [102] and references therein). Incidentally, the process above can lead to a very simple type of X-shaped wave.

From the present point of view, it is rather interesting to note that, during the last 15 years, X-shaped waves have *actually* been found as solutions to Maxwell and wave equations (recall once more that the form of any wave equations is intrinsically relativistic). To see more deeply the connection existing between what predicted by SR (see, e.g., Figs. 1A.2 and 1A.3) and the localized X-waves (mathematically and experimentally constructed in recent times), below we look in detail at the problem of the X-shaped field created by a superluminal electric charge, by following a recent paper [18].

X-Shaped Field Associated with a Superluminal Charge

It is well known by now that Maxwell's equations admit of wavelet-type solutions endowed with arbitrary group velocities ($0 < v_g < \infty$). We again confine ourselves, as above, to localized solutions moving rigidly, in particular to superluminal solutions (SLSs), the most interesting of which turned out to be X-shaped. As we already know, SLSs have been produced in a number of experiments, always by suitable interference of ordinary-speed waves. Here we show, by contrast, that even a superluminal charge creates an electromagnetic X-shaped wave, in agreement with what has been predicted within SR [1,14]: namely, that, on the basis of Maxwell equations, one is able to evaluate the field associated with a superluminal charge (at least under the rough approximation of pointlikeness). As noted earlier, it will result to a very simple example of *true* X-wave.

Indeed, when based on the ordinary postulates but not restricted to subluminal waves and objects (i.e., in its extended version), SR theory predicted the simplest X-shaped wave to be the wave corresponding to the electromagnetic field created by a superluminal charge [18,79]. It seems really important to evaluate such a field, at least approximately, by following [18].

Toy Model of a Pointlike Superluminal Charge We begin by considering, formally, a pointlike superluminal charge, even if the hypothesis of pointlikeness (already unacceptable in the subluminal case) is totally inadequate in the superluminal case [1]. Then we consider the ordinary vector potential A^μ and a current density $j^\mu \equiv (0, 0, j_z; j^o)$ flowing in the z-direction (notice that the motion line here is the z-axis). On assuming the fields to be generated by the sources only, we have that $A^\mu \equiv (0, 0, A_z; \phi)$, which, when adopting the Lorentz gauge, obeys the equation $A^\mu = j^\mu$. We can write such a nonhomogeneous wave equation in cylindrical coordinates $(\rho, \theta, z; t)$; for axial symmetry [which requires a priori that $A^\mu = A^\mu(\rho, z; t)$], when choosing the V-cone variables $\zeta \equiv z - Vt,\ \eta \equiv z + Vt$, with $V^2 > c^2$, we arrive [18] at the equation

$$\left[-\rho\frac{\partial}{\partial\rho}\left(\rho\frac{\partial}{\partial\rho}\right) + \frac{1}{\gamma^2}\frac{\partial^2}{\partial\zeta^2} + \frac{1}{\gamma^2}\frac{\partial^2}{\partial\eta^2} - 4\frac{\partial^2}{\partial\zeta\,\partial\eta}\right]A^\mu(\rho, \zeta, \eta) = j^\mu(\rho, \zeta, \eta), \tag{1A.1}$$

where μ assumes the two values $\mu = 3, 0$ only, so that $A^\mu \equiv (0, 0, A_z; \phi)$ and $\gamma^2 \equiv (V^2 - 1)^{-1}$. (Notice that, whenever convenient, we set $c = 1$.) Let us now suppose A^μ to be actually independent of η: namely, $A^\mu = A^\mu(\rho, \zeta)$. Due to Eq. (1A.1), we also have $j^\mu = j^\mu(\rho, \zeta)$, and therefore, $j_z = Vj^0$ (from the continuity equation) and $A_z = V\phi/c$ (from the Lorentz gauge). Then, by setting $\psi \equiv A_z$, we end with two

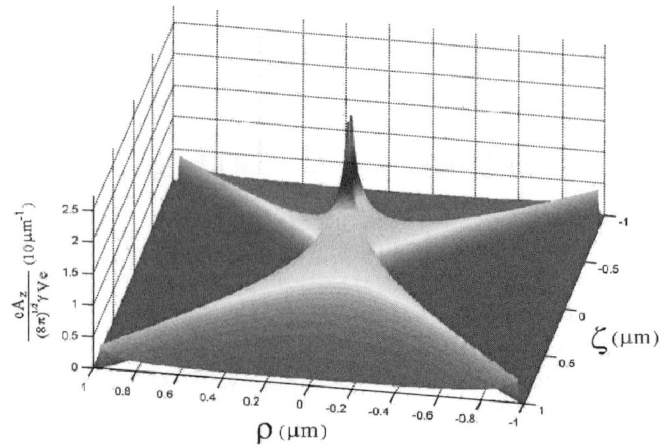

FIGURE 1A.4 Behavior of the field $\psi \equiv A_z$ generated by a charge supposed to be superluminal, as a function of ρ and $\zeta \equiv z - Vt$, evaluated for $\gamma = 1$ (i.e., for $V = c\sqrt{2}$), according to [18]. (Of course, we skipped the points in which ψ must diverge: namely, the vertex and the cone surface.)

equations which allow us to analyze the possibility and consequences of having a superluminal pointlike charge, e, traveling with constant speed V along the z-axis ($\rho = 0$) in the positive direction, in which case $j_z = eV\delta(\rho)/\rho\delta(\zeta)$. Indeed, one of those two equations becomes the hyperbolic equation

$$\left[-\frac{1}{\rho}\frac{\partial}{\partial\rho}\left(\rho\frac{\partial}{\partial\rho}\right) + \frac{1}{\gamma^2}\frac{\partial^2}{\partial\zeta^2}\right]\psi = eV\,\frac{\delta(\rho)}{\rho}\,\delta(\zeta), \tag{1A.2}$$

which can be solved [18] in a few steps: first, by applying (with respect to the variable ρ) the Fourier–Bessel (FB) transformation $f(x) = \int_0^\infty \Omega f(\Omega) J_0(\Omega x)\,d\Omega$, the quantity $J_0(\Omega x)$ being the ordinary zero-order Bessel function; second, by applying the ordinary Fourier transformation with respect to the variable ζ (passing from ζ to the variable ω); and third, by performing the corresponding inverse Fourier and FB transformations. Afterward it is enough to have recourse to formulas (3.723.9) and (6.671.7) of [82], still with $\zeta \equiv z - Vt$, to enable writing the solution of Eq. (1A.2) in the form

$$\psi(\rho,\zeta) = \begin{cases} 0 & \text{for } 0 < \gamma\,|\zeta| < \rho \\ e\dfrac{V}{\sqrt{\zeta^2 - \rho^2(V^2-1)}} & \text{for } 0 \le \rho < \gamma\,|\zeta|. \end{cases} \tag{1A.3}$$

In Fig. 1A.4 we show our solution, $A_z \equiv \psi$, as a function of ρ and ζ, evaluated for $\gamma = 1$ (i.e., for $V = c\sqrt{2}$). Of course, we skipped the points in which A_z *must* diverge: namely, the vertex and the cone surface.

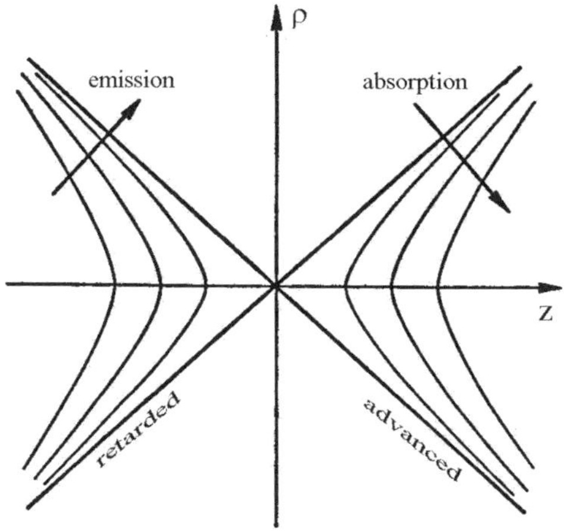

FIGURE 1A.5 The spherical equipotential surfaces of an electrostatic field created by a charge at rest get transformed into two-sheeted rotation hyperboloids contained inside an unlimited double cone when the charge travels at superluminal speed (see [1,18]). This figures shows that a superluminal charge traveling at constant speed in a homogeneous medium such as a vacuum does not lose energy [79]. Note that this double cone has nothing to do with the Cherenkov cone. (From [1].)

For comparison, one may recall that the classic X-shaped solution [19] of the homogeneous wave equation (shown, e.g., in Figs. 1.8, 1.9, and 1A.3) has the form (with $a > 0$)

$$X = \frac{V}{\sqrt{(a - i\zeta)^2 + \rho^2(V^2 - 1)}}. \qquad (1A.4)$$

The second of Eqs. (1A.3) includes the expression (1A.4), given by the spectral parameter [42,63] $a = 0$, which indeed corresponds to the nonhomogeneous case (the fact that for $a = 0$ these equations differ for an imaginary unit will be discussed elsewhere).

It is rather important at this point to note that such a solution, Eq. (1A.3), does represent a wave existing only inside the (unlimited) double cone $\mathcal{C}$ generated by the rotation around the z-axis of the straight lines $\rho = \pm\gamma\zeta$: This, too, is in full agreement with the predictions of extended SR theory. For an explicit evaluation of the electromagnetic fields generated by the superluminal charge (and of their boundary values and conditions), we confine ourselves here to quoting [18]. Incidentally, the same results found by following the procedure described above can be obtained by starting from the four-potential associated with a subluminal charge (e.g., an electric charge at rest) and applying to it the suitable superluminal Lorentz “transformation”. One should also notice that this double cone does not have much to do with the Cherenkov cone [1,79]; and a superluminal charge traveling at constant speed, in the vacuum, does not lose energy (see, e.g., Fig. 1A.5).

Outside cone $\mathcal{C}$ (i.e., for $0 < \gamma \mid \zeta \mid < \rho$) we get, as expected, no field, so that one meets a field discontinuity when crossing the double-cone surface. Nevertheless, the boundary conditions imposed by Maxwell's equations are satisfied by our solution (1A.3), since at each point of the cone surface the electric and magnetic fields are both tangent to the cone (for a discussion of this point, see [18]).

Here, let us stress that, when $V \to \infty$, $\gamma \to 0$, the electric field tends to vanish while the magnetic field tends to the value $H_\phi = -\pi e/\rho^2$. This does agree with what we expect from extended SR, which predicts that superluminal charges behave (in a sense) as magnetic monopoles. We refer interested readers to [1,2,80,81], and references therein.

A Glance at the Experimental State of the Art

Extended relativity can also provide a better understanding of many aspects of ordinary physics [1], even if superluminal objects (tachyons) did not exist in our cosmos as asymptotically free objects. In any case, at least three or four different experimental sectors of physics seem to suggest the possible existence of faster-than-light motion, or at least of superluminal group velocities. Next, we provide some experimental results obtained in two sectors and mention two others.

Neutrinos A long series of experiments begun in 1971 seem to show that the square m_0^2 of the mass m_0 of muon-neutrinos, and more recently of electron-neutrinos, is negative; which, if confirmed, would mean that such neutrinos possess an "imaginary mass" and are therefore tachyonic or mainly tachyonic [1,83,84]. (Actually, in extended SR, the dispersion relation for a free superluminal object becomes $\omega^2 - \boldsymbol{k}^2 = -\Omega^2$, or $E^2 - \boldsymbol{p}^2 = -m_0^2$, and there is no need at all, therefore, for imaginary masses.)

Galactic Microquasars As for the apparent superluminal expansions observed in the cores of quasars [85] and, recently, in galactic microquasars [86], they are outside the range of this chapter. Moreover, we note that there exist orthodox interpretations for such astronomical observations, based on [87], that are accepted by the majority of astrophysicists (for a theoretical discussion, see [88]). Here, we mention only that simple geometrical considerations in Minkowski space show that a single superluminal source of light would appear [1,88]: (1) initially, in the "optical boom" phase (analogous to the acoustic boom produced by an airplane traveling at a constant supersonic speed), as an intense source that suddenly comes into view; and (2) which afterward seems to split into two objects receding from one another with speed $V > 2c$ (all of this being similar to what is actually observed according to [86]).

Evanescent Waves and Tunneling Photons Within quantum mechanics (specifically, in the tunneling processes), it has been shown that tunneling time (first evaluated as a simple Wigner's "phase time" and later calculated through the analysis of wave

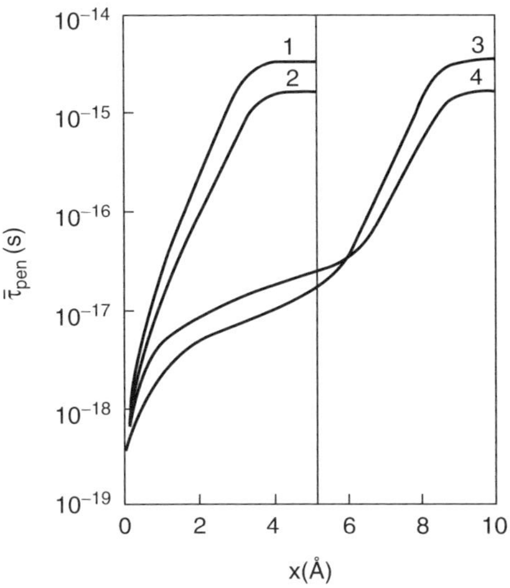

FIGURE 1A.6 Behavior of the average penetration time (in seconds) spent by a tunneling wave packet as a function of the penetration depth (in angstroms) down a potential barrier. According to the predictions of quantum mechanics, the wave spacket speed inside the barrier increases in an unlimited way for opaque barriers, and the total tunneling time does not depend on the barrier width [90,92]. (From Olkhovsky et al. [92].)

packet behavior) does not depend [90] on barrier width in the case of opaque barriers (the *Hartman effect*). This implies superluminal and arbitrarily large group velocities V inside sufficiently long barriers (see Fig. 1A.6). Experiments that might verify this prediction using, say, electrons or neutrons are difficult and rare [65,91]. Luckily enough, however, the Schrödinger equation in the presence of a potential barrier is mathematically identical to the Helmholtz equation for an electromagnetic wave propagating, for instance, down a metallic waveguide (along the x-axis): as shown, for example, in [118]; and a barrier height U greater than the electron energy E corresponds (for a given wave frequency) to a waveguide of transverse size lower than the cutoff value. In the context of this example, a segment of "undersized" guide therefore behaves as a barrier for the wave (photonic barrier) as well as any other photonic bandgap filters. Thus, like a particle inside a quantum barrier, the wave therein assumes an imaginary momentum or wave number and so is damped exponentially along x (see, e.g., Fig. 1A.7). It becomes an *evanescent wave* (returning to normal propagation, even if with reduced amplitude, when the narrowing ends and the guide returns to its initial transverse size). Thus, a tunneling experiment can be simulated by having recourse to evanescent waves (for which the concept of group velocity can be properly extended: see the first of [57]).

The fact that evanescent waves can travel with superluminal speeds (see, e.g., Fig. 1A.8) has actually been verified in a series of famous experiments. Work performed from 1992 onward by Nimtz et al. in Cologne [106], Steinberg et al. at Berkeley

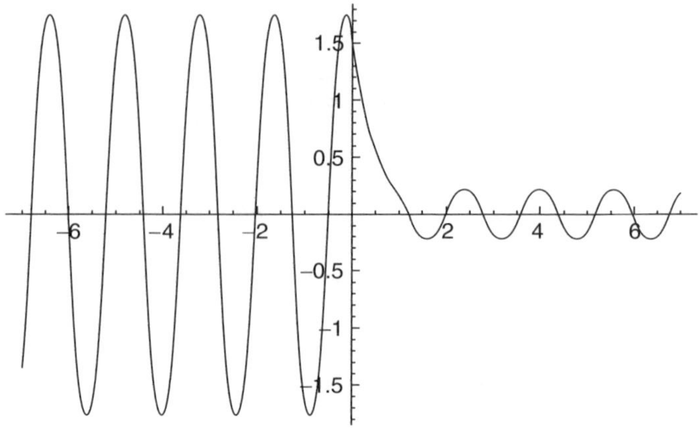

FIGURE 1A.7 The damping that takes place inside a barrier reduces the amplitude of a tunneling wave packet, imposing a practical limit on the barrier length. (From [57].)

[105], Mugnai et al. in Florence [23], and by others in Vienna, Orsay, Rennes, and other locations [95,107] verified that tunneling photons travel with superluminal group velocities. Such experiments even raised a great deal of interest [93,96,108] within the nonspecialized press and were reported in *Scientific American, Nature, New Scientist*, and other journals. We should add that, since also extended SR had predicted [109]

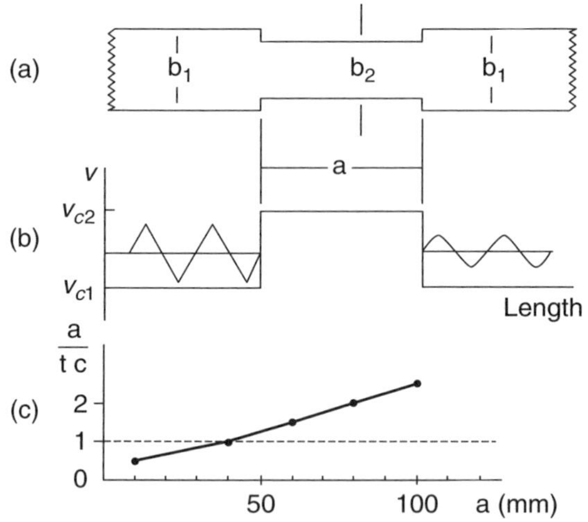

FIGURE 1A.8 Simulation of tunneling by experiments with evanescent classical waves (see the text), which were also predicted to be superluminal on the basis of extended SR [109]. The figure shows one of the measurement results by Nimtz et al. [94]: the average beam speed while crossing the evanescent region (i.e., segment of undersized waveguide, or "barrier") as a function of its length. As predicted theoretically [90,109], such an average speed exceeds c for sufficently long barriers. Further results appeared in [99] and are reported below (see Figs. 1A.11 and 1A.12).

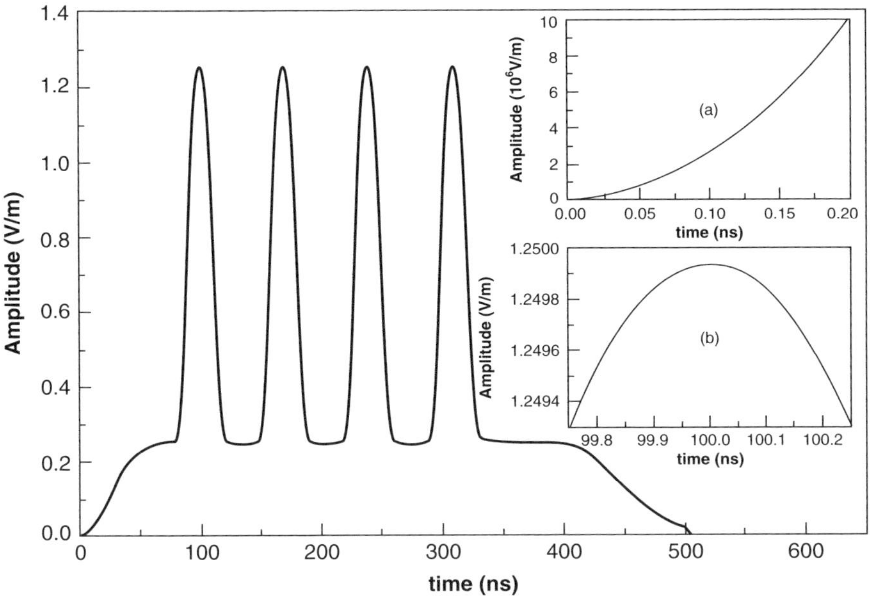

FIGURE 1A.9 The delay of a wave packet crossing a barrier (cf., e.g., Fig. 1A.8) is due to the initial discontinuity. Suitable numerical simulations [73] considerered an (indefinite) undersized waveguide (therefore eliminating any geometric discontinuity in its cross section). The figure shows the envelope of the *initial* signal. Inset (a) depicts the initial part of this signal as a function of time, and inset (b) depicts the Gaussian pulse peak centered at $t = 100$ ns.

evanescent waves to be endowed with faster-than-c speeds, the entire matter therefore appears to be theoretically self-consistent. The debate in the current literature does not refer to the experimental results (which can be reproduced correctly even by numerical simulations [73,74] based on Maxwell's equations only; see Figs.1A.9 and 1A.10), but rather, to the question of whether they do or do not allow signals or information to be sent with superluminal speed (see, e.g., [66]).

In the experiments mentioned above, there is substantial attenuation of the pulses (see Fig. 1A.7) during tunneling (or during propagation in an absorbing medium). However, by employing "gain doublets," undistorted pulses have been observed to propogate at superluminal group velocity with only a small change in amplitude (see, e.g., [97]).

Let us emphasize that some of the most interesting experiments in this series seem to be those with *two* or more barriers (e.g., with two gratings in an optical fiber or with two segments of undersized waveguide separated by a piece of normal-sized waveguide; Fig. 1A.11). For suitable frequency bands (i.e., for tunneling far from resonances) it was found that the total crossing time does not depend on the length of the intermediate (normal) guide, that is, that the beam speed along it is infinite [91,100,111]. This does agree with what we predicted within quantum mechanics for nonresonant tunneling through two successive opaque barriers [100] (Fig. 1A.12).

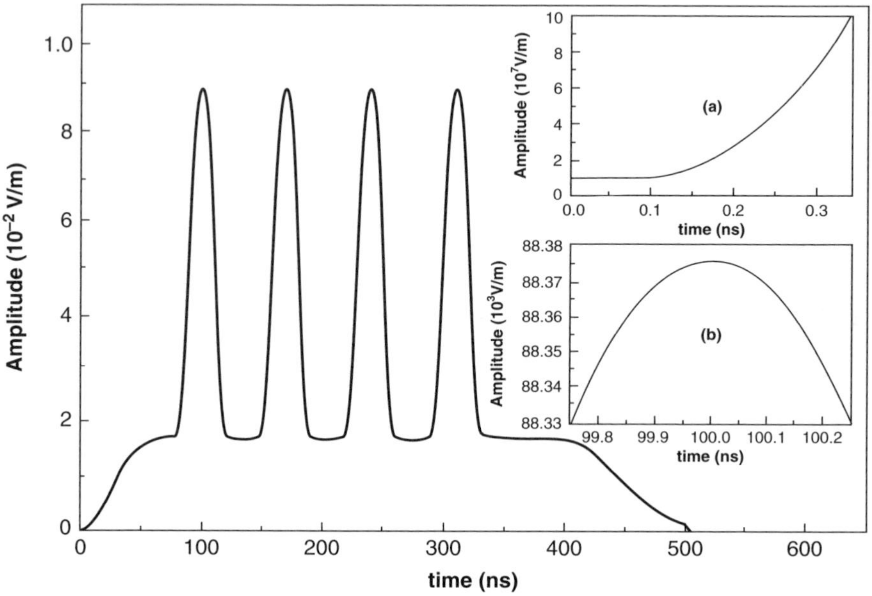

FIGURE 1A.10 Envelope of the signal in (Fig. 1A.9) after traveling a distance $L = 32.96$ mm through an undersized waveguide. Inset (a) shows the initial part (in time) of such arriving signal, and inset (b) shows the peak of the Gaussian pulse that had initially been modulated by centering it at $t = 100$ ns. One can see that its propagation took zero time, so that the signal traveled with infinite speed. The numerical simulation is based on Maxwell's equations only. Going from Fig. 1A.9 to Fig. 1A.10, one verifies that the signal amplitude is reduced greatly. However, the width of each peak did not change (and this might have some relevance when dealing with a Morse alphabet transmission; see the text).

Such a prediction has been verified first, theoretically, by Aharanov et al. [100], and then experimentally by taking advantage of the circumstance [111] that evanescence regions can consist of a variety of photonic bandgap materials or gratings (from multilayer dielectric mirrors, or semiconductors, to photonic crystals). Indeed, the best experimental confirmation has come by having recourse to two gratings in an optical fiber [99]; see Figs. 1A.13 and 1A.14, in particular the rather peculiar (and quite interesting) results represented by the latter.

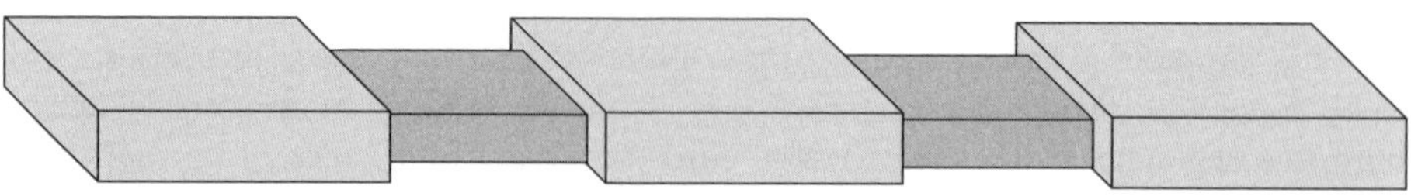

FIGURE 1A.11 Very interesting experiments have been performed with two successive barriers (i.e., with two evanescence regions): for example, with two gratings in an optical fiber. This figure refers to the interesting experiment [111] performed with microwaves traveling along a metallic waveguide, the waveguide being endowed with two classical barriers (undersized guide segments). See the text. (From [57].)

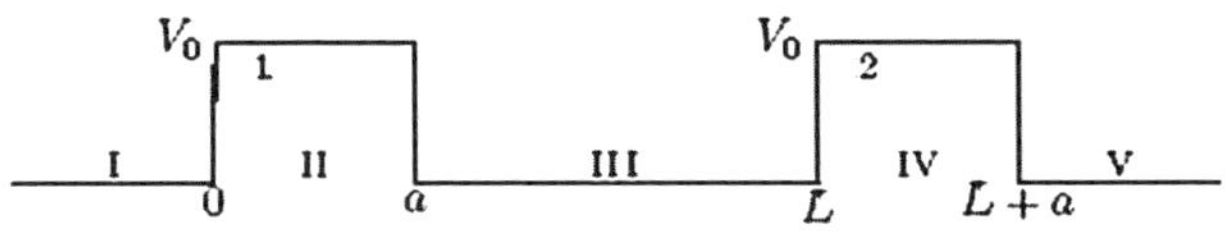

FIGURE 1A.12 Scheme of a nonresonant tunneling process through two successive (opaque) quantum barriers. Far from resonances, the (total) phase time for tunneling through the two potential barriers depends on neither the barrier widths nor the *distance* between the barriers (this is "the generalized Hartman effect") [91,98,100]. See the text.

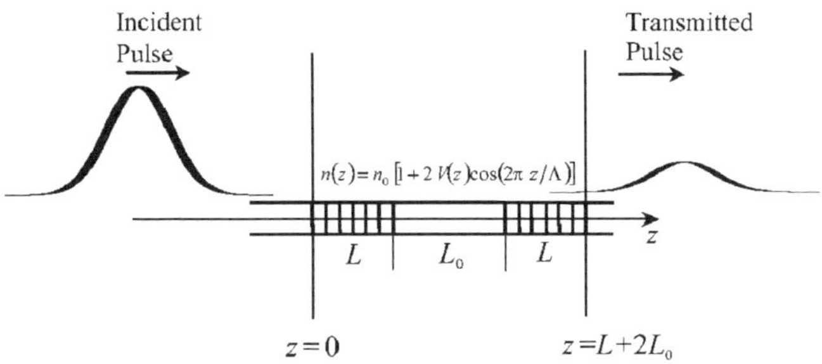

FIGURE 1A.13 Realization of the quantum-theoretical setup represented in Fig. 1A.12 using as classical (photonic) barriers two gratings in an optical fiber [98]. The corresponding experiment has been performed by Longhi et al. [99].

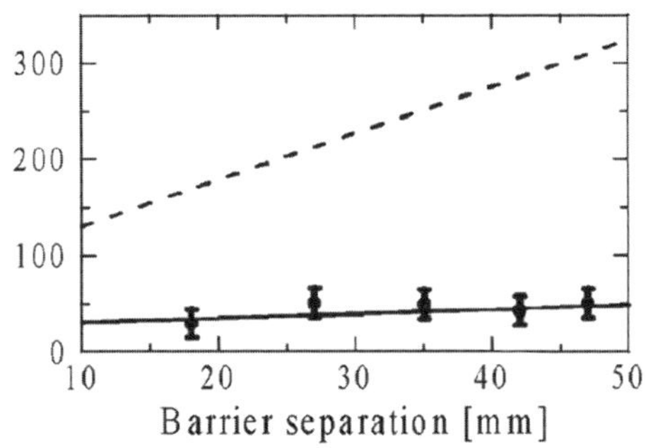

FIGURE 1A.14 Off-resonance tunneling time versus barrier separation for the rectangular symmetric double-barrier frequency bandgap (FBG) structure considered in [99] (see Fig. 1A.13). The solid line is the theoretical prediction based on group delay calculations; the dots are the experimental points obtained by time delay measurements (the dashed curve is the transit time expected from the input to the output planes for a pulse tuned far away from the stopband of the FBGs). The experimental results [99], as well as the early results in [111], confirm the theoretical prediction of a generalized Hartman effect: in particular, the independence of the total tunneling time from the distance between the two barriers.

We may also note another topic that is arousing more and more interest [97]. Even if all the ordinary causal paradoxes seem to be solvable [1,57,58], one must also bear in mind that, whenever an object $\mathcal{O}$ traveling with superluminal speed is encountered, we may have to deal with negative contributions to the tunneling times [1,91,112], and this should not be regarded as nonphysical. In fact, whenever an object $\mathcal{O}$ (e.g., particle, electromagnetic pulse) *overcomes* [1,58] the infinite speed with respect to a certain observer, it will appear later to the same observer as the anti-object $\overline{\mathcal{O}}$ traveling in the opposite direction in space [1,58,104]. For instance, when going from the lab frame to a frame $\mathcal{F}$ moving in the same direction as the particles or waves entering the barrier region, the object $\mathcal{O}$ penetrating the final part of the barrier (with almost infinite speed [73,90–92], as in Figs. 1A.6) will appear in frame $\mathcal{F}$ as an anti-object $\overline{\mathcal{O}}$ crossing that portion of the barrier in the opposite space direction [1,58,104]. In the new frame $\mathcal{F}$, therefore, such an anti-object $\overline{\mathcal{O}}$ would make a negative contribution to the tunneling time, which could even result in a negative total tunneling time. For clarification, see the references that we have cited. Let us stress here that even the appearance of negative tunneling times has been predicted within extended SR, on the basis of its ordinary postulates, and has been confirmed recently by quantum-theoretical evaluations [3,91]. (In the case of a nonpolarized beam, the wave anti-packet coincides with the initial wave packet; if, however, a photon is endowed with helicity $\lambda = +1$, the anti-photon will bear the opposite helicity, $\lambda = -1$.) For the theoretical point of view, see the papers cited above (in particular, [90,91], and, more specifically, [113]). On the very interesting *experimental* side, see the intriguing papers cited in [101,114].

Let us add here that it is possible to obtain, via quantum interference effects, dielectrics with refraction indices varying very rapidly as a function of frequency, and also three-level atomic systems, with almost complete absence of light absorption (i.e., with quantum-induced transparency) [115]. The group velocity of a light pulse propagating in such a medium can decrease to very low values, either positive or negative, with no pulse distortion. It is known that experiments have been performed both in atomic samples at room temperature and in Bose–Einstein condensates, which showed the possibility of reducing the speed of light to a few meters per second. Similar, but negative group velocities, interpreted as implying propagation with superluminal speeds thousands of times higher than those that had been considered previously, have also been predicted in the presence of such an electromagnetically induced transparency for light moving in a rubidium condensate [116]. Finally, let us recall that faster-than-c propagation of light pulses can also be (and was, in same cases) observed by taking advantage of the anomalous dispersion near an absorbing line, or nonlinear and linear gain lines (as already seen), or nondispersive dielectric media, or inverted two-level media, as well as of some parametric processes in nonlinear optics (see, e.g., the work of Kurizki et al.).

Superluminal Localized Solutions to Wave Equations: X-Shaped Waves This fourth sector is no less important. It came into fashion again when in a series of remarkable works it was rediscovered that any wave equation (e.g., in the electromagnetic

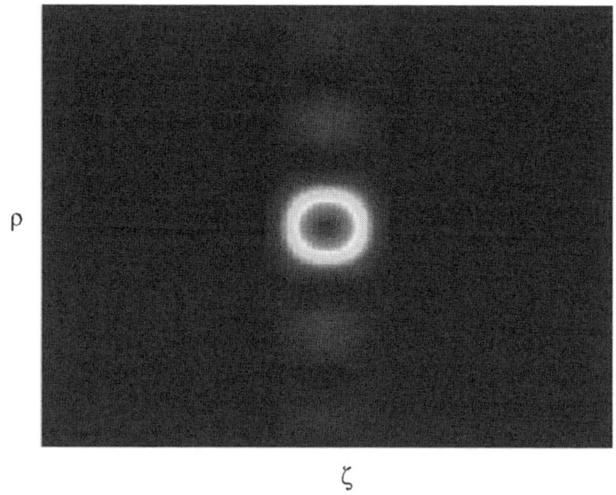

FIGURE 1A.15 The wave equations possess pulse-type solutions that in the subluminal case are ball-like, in agreement with Fig. 1A.2. For additional comments, see the text.

case) provides solutions as much subluminal as superluminal (in addition to the luminal solutions, which have speed c/n). Starting with pioneering work such as that of Bateman, it slowly became known that all wave equations admit soliton-like (or rather, wavelet-type) solutions with subluminal group velocities. Subsequently, superluminal solutions began to be written down (in one case [39] by the mere application of a superluminal Lorentz "transformation" [1]).

As we know, a remarkable feature of some of these new solutions (which attracted much attention for their possible applications) is that they propagate as localized, nondispersive pulses even because of their self-reconstruction property. It is easy to realize the practical importance, for instance, of a radio transmission carried out by localized beams (independently of their speed); but nondispersive wave packets can be of use even in theoretical physics for a reasonable representation of elementary particles; and so on. Incidentally, from the point of view of elementary particles, the fact that wave equations possess pulse-type solutions that are ball-like in the subluminal case (see Fig 1A.15), can be a source of meditation, as this can have a bearing on the corpuscle–wave duality problem met in quantum physics (besides agreeing, for example, with Fig. 1A.2).

At the cost of repeating ourselves, let us reemphasize that within extended SR it had been found that, whereas the simplest subluminal object conceivable is a small sphere (or a point in the limiting case), the simplest superluminal object (see [14] and Figs. 1A.2 and 1A.3) is an X-shaped wave (or a double cone as its limit), which travels in a homogeneous medium without deforming (i.e., rigidly). It is not without meaning that the most interesting localized solutions to the wave equations happen to be the superluminal solutions, and with the predicted shape. Even more, since from Maxwell's equations under simple hypotheses one goes on to the usual scalar wave equation for each electric or magnetic field component, one expects the same solutions to exist also in the sectors of acoustic waves, seismic waves, and gravitational

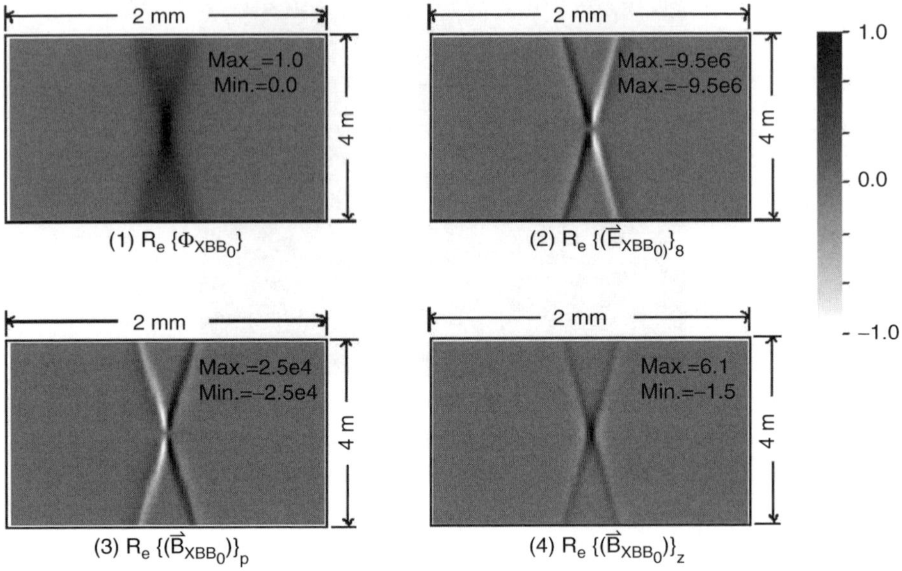

FIGURE 1A.16 Real part of the Hertz potential and of the field components of the localized electromagnetic ("classic," axially symmetric) X-shaped wave predicted and first constructed mathematically for the electromagnetic case in [20]. For the meaning of the various panels, see the references cited. The dimensions of each panel are 4 m (in the radial direction) × 2 mm (in the propagation direction). [The values shown in the top-right corner of each panel represent the maxima and minima of the images before normalization for display (MKSA units).]

waves: and this has already been demonstrated in the literature for the acoustic case. Actually, such pulses (as suitable superpositions of Bessel beams) were constructed mathematically for the first time by Lu et al. in the field of acoustics and were then called *X-waves* or *X-shaped waves*.

It is important for us that X-shaped waves have been indeed produced in experiments with both acoustic and electromagnetic waves; that is, X-pulses were produced that traveled undistorted in their medium with a speed greater than the speed of sound, in the first case, and than the speed of light, in the second case. The first experiment in acoustics was performed by Lu et al. in 1992 at the Mayo Clinic (and their papers received the first IEEE award). In the electromagnetic case, which is certainly more intriguing, superluminal localized X-shaped solutions were first constructed mathematically (cf., e.g., Fig. 1A.16) in [20], and later produced experimentally by Saari and Reivelt [22] using visible light (Fig. 1A.17), and more recently by Mugnai et al. using microwaves [23]. In the theoretical sector the activity has not been less intense: with the goal to build up, for example, analogous new solutions with finite total energy or more suitable for high frequencies, on the one hand, and localized solutions superluminally propagating even along a normal waveguide (see Fig. 1A.18), on the other hand.

Let us recall the problem of producing an X-shaped superluminal wave like the one in Fig. 1A.13, but truncated, of course, in space and time (using a finite antenna

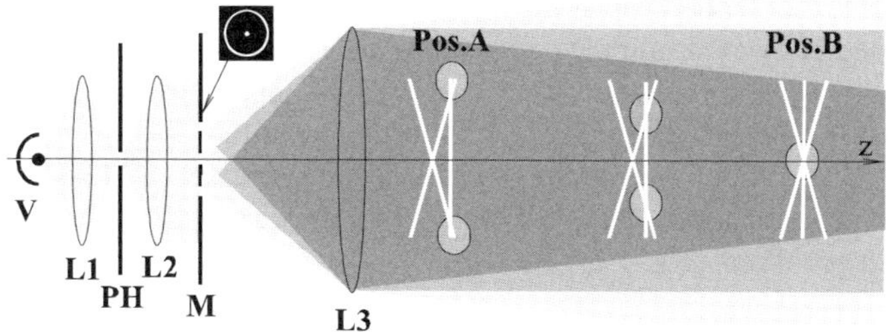

FIGURE 1A.17 Scheme of the experiment by Saari and Reivelt [22], who announced the production in optics of the beams depicted in Fig. 1A.16. In the present figure one can see what was shown by the experimental results: namely, that the X-shaped waves are superluminal. Indeed, running after the plane waves (the latter regularly traveling with speed c), they do catch up with the plane waves. Later, an analogous experiment was performed with microwaves by Mugnai et al. [23].

radiating for a finite time). In such a situation, the pulse is known to keep its localization and superluminality only up to a certain depth of field [i.e., as long as it is fed by the waves arriving (with speed c) from the antenna], decaying abruptly afterward [20,40,42]. Various authors, taking account of the time needed to foster such superluminal waves, have concluded that these localized superluminal pulses are unable to transmit information faster than c. Many such questions have been discussed in what precedes; for further details, see the second of [20].

In any case, the existence of X-shaped superluminal (or supersonic) pulses seems to constitute, together with, for example, the superluminality of evanescent waves,

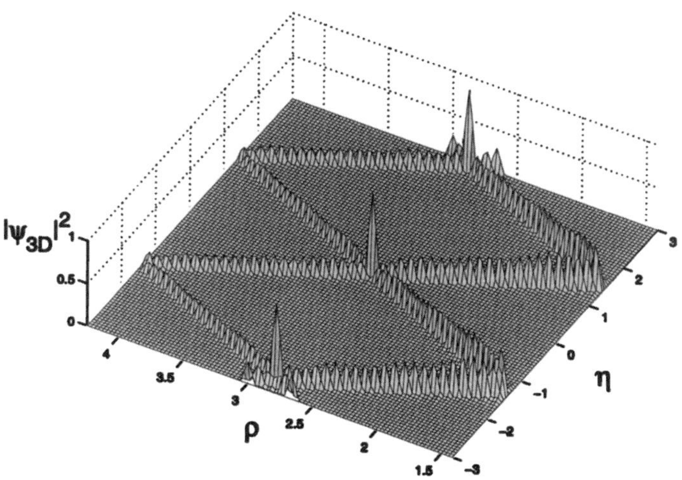

FIGURE 1A.18 Elements of one of the trains of X-shaped pulses constructed mathematically in [67], which propagate down a coaxial guide (in the transverse magnetic case): Analogous X-pulses exist (with infinite or finite total energy) for propagation along a normal-sized cylindrical metallic waveguide. [From (67).]

a confirmation of extended SR, a theory [1] based on the ordinary postulates of SR that consequently does not appear to violate any of the fundamental principles of physics. It is curious, moreover, that one of the first applications of X-waves (which takes advantage of their propagation without deformation) has been accomplished in the field of medicine, specifically in ultrasound scanners [24,25], whereas the most important applications of (subluminal!) frozen waves will probably again affect human health problems (e.g., the cancer).

Acknowledgments

This work was partially supported by FAPESP (Brazil) and by MIUR and INFN (Italy).

REFERENCES

1. E. Recami, Classical tachyons and possible applications, *Rivista Nuovo Cimento* **9**(6), 1–178 (1986), issue 6, and refs. therein.
2. E. Recami and R. Mignani, Magnetic monopoles and tachyons in special relativity, *Phys. Lett.* B **62**, 41–43 (1976).
3. V. Petrillo and L. Refaldi, *Phys. Rev. A* **67**, 012110 (2003); V. Petrillo and L. Refaldi, *Opt. Commun.* **186**, 35 (2000); see also L. Refaldi, M.Sc. thesis, Physics Department, Milan University, Milan, Italy, 2000.
4. C. Conti, S. Trillo, G. Valiulis, A. Piskarkas, O. van Jedrkiewicz, J. Trull, and P. Di Trapani, Non-linear electromagnetic X-waves, *Phys. Rev. Lett.* **90**, 170406 (2003).
5. See, e.g., M. Zamboni-Rached, Localized waves in diffractive/dispersive media, Ph.D. disertation, DMO/FEEC, Universidade Estadual de Campinas, Campinos, Brazil, Aug. 2004.
6. M. Born and E. Wolf, *Principles of Optics: Electromagnetic Theory of Propagation, Interference and Diffraction of Light*, 6th ed., Cambridge University Press; Cambridge, 1998.
7. H. A. Willebrand and B. S. Ghuman, Fiber optics without fiber, *IEEE Spectrum* **38**(8) (2001).
8. J. W. Goodman, *Introduction to Fourier Optics*, 2nd ed., McGraw-Hill, New York, 1996.
9. S. Okazaki, Resolution limits of optical lithography, *J. Vac. Sci. Technol. B* **9**, 2829–2833 (1991).
10. T. Ito and S. Okazaki, Pushing the limits of lithography, *Nature* **406**, 1027–1031 (2000).
11. A. Ashkin, J. M. Dziedzic, J. E. Bjorkholm, and S. Chu, Observation of a single-beam gradient force optical trap for dielectric particles, *Opt. Lett.* **11**, 288–290 (1986).
12. J. E. Curtis, B. A. Koss, and D. G. Grier, Dynamic holographic optical tweezers, *Opt. Commun.* **207**, 169–175 (2002).
13. G. P. Agrawal, *Nonlinear Fiber Optics*, 2nd ed., Academic Press, San Diego, CA, 1995.
14. A. O. Barut, G. D. Maccarrone, and E. Recami, On the shape of tachyons, *Nuovo Cimento A* **71**, 509–533 (1982), and refs. therein; see also E. Recami and G. D. Maccarrone, *Lett. Nuovo Cimento* **28**, 151–157 (1980) and **37**, 345–352 (1983); P. Caldirola,

G. D. Maccarrone, and E. Recami, *Lett. Nuovo Cimento* **29**, 241–250 (1980); G. D. Maccarrone, M. Pavsic, and E. Recami, *Nuovo Cimento B* **73**, 91–111 (1983).

15. J. A. Stratton, *Electromagnetic Theory*, McGraw-Hill, New York, 1941, p. 356.
16. R. Courant and D. Hilbert, *Methods of Mathematical Physics*, Vol. 2, Wiley, New York, 1966, p. 760.
17. H. Bateman, *Electrical and Optical Wave Motion*, Cambridge University Press, Cambridge, 1915.
18. E. Recami, M. Zamboni-Rached, and C. A. Dartora, The X-shaped, localized field generated by a superluminal electric charge, LANL archives e-print physics/0210047, *Phys. Rev. E*, **69**, 027602 (2004).
19. J.-y. Lu and J. F. Greenleaf, Nondiffracting X-waves: exact solutions to free-space scalar wave equation and their finite aperture realizations, *IEEE Trans. Ultrason. Ferroelectr. Freq. Control* **39**, 19–31 (1992), and refs. therein.
20. J.-y. Lu, J. F. Greenleaf, and E. Recami, Limited diffraction solutions to Maxwell (and Schroedinger) equations, LANL archives e-print physics/9610012, Report INFN/FM–96/01, Instituto Nazionale de Fisica Nucleare, Frascati, Italy Oct. 1996; E. Recami: On localized X-shaped superluminal solutions to Maxwell equations, *Physica A* **252**, 586–610 (1998), and refs. therein; see also R. W. Ziolkowski, I. M. Besieris, and A. M. Shaarawi, *J. Opt. Soc. Am. A* **10**, 75 (1993); *J. Phys. A Math. Gen.* **33**, 7227–7254 (2000).
21. J.-y. Lu and J. F. Greenleaf, Experimental verification of nondiffracting X-waves, *IEEE Trans. Ultrason. Ferroelectr. Freq. Control* **39**, 441–446 (1992).
22. P. Saari and K. Reivelt, Evidence of X-shaped propagation-invariant localized light waves, *Phys. Rev. Lett.* **79**, 4135–4138 (1997).
23. D. Mugnai, A. Ranfagni, and R. Ruggeri, Observation of superluminal behaviors in wave propagation, *Phys. Rev. Lett.* **84**, 4830–4833 (2000).
24. J.-y. Lu, H.-H. Zou, and J. F. Greenleaf, Biomedical ultrasound beam forming, *Ultrasound Med. Bio.* **20**, 403–428 (1994).
25. J.-y. Lu, H.-H. Zou, and J. F. Greenleaf, Producing deep depth of field and depth independent resolution in NDE with limited diffraction beams, *Ultrason. Imag.* **15**, 134–149 (1993).
26. V. Garcés-Chavez, D. McGloin, H. Melville, W. Sibbett and K. Dholakia, Simultaneous micromanipulation in multiple planes using a self-reconstructing light beam, *Nature* **419**, 145–147 (2002).
27. D. McGloin, V. Garcés-Chavez, and K. Dholakia, Interfering Bessel beams for optical micromanipulation, *Opt. Lett.* **28**, 657–659 (2003).
28. M. P. MacDonald et al., Creation and manipulation of three-dimensional optically trapped structures, *Science* **296**, 1101–1103 (2002).
29. J. Arlt, V. Garcés-Chavez, W. Sibbett, and K. Dholakia, Optical micromanipulation using a Bessel light beam, *Opt. Commun.* **197**, 239–245 (2001).
30. D. P. Rhodes, G. P. T. Lancaster, J. Livesey, D. McGloin, J. Arlt, and K. Dholakia, Guiding a cold atomic beam along a co-propagating and oblique hollow light guide, *Opt. Commun.*, **214**, 247–254 (2002).
31. J. Fan, E. Parra, and H. M. Milchberg, Resonant self-trapping and absorption of intense Bessel beams, *Phy. Rev. Lett.* **84**, 3085–3088 (2000).
32. J. Arlt, T. Hitomi and K. Dholakia, Atom guiding along Laguerre–Gaussian and Bessel light beams, *Appl. Phys. B* **71**, 549–554 (2000).

33. M. Erdélyi, Z. L. Horvath, G. Szabo, Z. Bor, F. K. Tittell, J. R. Cavallaro, and M. C. Smayling, Generation of diffraction-free beams for applications in optical microlithography, *J. Vac. Sci. Technol. B* **15**, 287–292 (1997), and refs. therein.
34. R. M. Herman and T. A. Wiggins, Production and uses of diffractionless beams, *J. Opt. Soc. Am. A* **8** 932–942 (1991).
35. R. W. Ziolkowski, Localized transmission of electromagnetic energy, *Phy. Rev. A* **39**, 2005–2033 (1989).
36. R. W. Ziolkowski, Localized wave physics and engineering, *Phys. Rev. A* **44**, 3960–3984 (1991).
37. J.-y. Lu and H. Shiping, Optical X wave communications, *Opt. Commun.* **161**, 187–192 (1999).
38. A. Vasara, J. Turunen, and A. T. Friberg, Realization of general nondiffracting beams with computer-generated holograms, *J. Opt. Soc. Am. A* **6**, 1748–1754 (1989); J. Salo, J. Fagerholm, A. T. Friberg, and M. M. Salomaa, Nondiffracting bulk-acoustic X-waves in crystals, *Phys. Rev. Lett.* **83**, 1171–74 (1990); J. Salo, A. T. Friberg, and M. M. Salomaa, Orthogonal X-waves, *J. Phys. A Math. Gen.* **34**, 9319–9327 (2001).
39. See, e.g., A. O. Barut and H. C. Chandola, Localized tachyonic wavelet solutions of the wave equation, *Phy. Lett. A* **180**, 5–8 (1993), and refs. therein.
40. J. Durnin, J. J. Miceli and J. H. Eberly, Diffraction-free beams, *Phy. Rev. Lett.* **58**, 1499–1501 (1987).
41. J. Durnin, Exact solutions for nondiffracting beams: I. The scalar theory, *J. Opt. Soc. Am. A* **4**, 651–654 (1987).
42. M. Zamboni-Rached, E. Recami, and H. E. Hernández-Figueroa, New localized superluminal solutions to the wave equations with finite total energies and arbitrary frequencies, *Eur. Phy. J. D* **21**, 217–228 (2002).
43. W. Ziolkowski, I. M. Besieris, and A. M. Shaarawi, Aperture realizations of exact solutions to homogeneous wave-equations, *J. Opt. Soc. Am. A* **10**, 75 (1993), Sec. 5 and 6.
44. R. P. MacDonald, J. Chrostowski, S. A. Boothroyd, and B. A. Syrett, Holographic formation of a diode laser non-diffracting beam, *Appl. Opt.* **32**, 6470–6474 (1983).
45. M. Zamboni-Rached, Stationary optical wave fields with arbitrary longitudinal shape by superposing equal frequency Bessel beams: frozen waves, *Opt. Express* **12**, 4001–4006 (2004).
46. M. Zamboni-Rached, E. Recami and H. E. Hernández-Figueroa, Theory of "frozen waves", LANL archives e-print physics/0502105, *J. Opt. Soc. Am. A*, **11**, 2465–2475 (2005).
47. A. O. Barut and A. Grant, Quantum particle-like configurations of the electromagnetic field, *Found. Phys. Lett.* **3**, 303–310 (1990).
48. A. O. Barut and A. J. Bracken, Particle-like configurations of the electromagnetic field: an extension of de Broglie's ideas, *Found. Phys.* **22**, 1267–1285 (1992).
49. J. N. Brittingham, Focus wave modes in homogeneous Maxwell's equations: transverse electric mode, *J. Appl. Phys.* **54**, 1179–1189 (1983).
50. A. Sezginer, A general formulation of focus wave modes, *J. Appl. Phys.* **57**, 678–683 (1985).
51. I. M. Besieris, A. M. Shaarawi, and R. W. Ziolkowski, A bidirectional traveling plane wave representation of exact solutions of the scalar wave equation, *J. Math. Phys.* **30**, 1254–1269 (1989).

52. A. M. Shaarawi, I. M. Besieris, and R. W. Ziolkowski, A novel approach to the synthesis of nondispersive wave packet solutions to the Klein–Gordon and Dirac equations, *J. Math. Phys.* **31**, 2511–2519 (1990).
53. R. W. Ziolkowski, I. M. Besieris, and A. M. Shaarawi, Aperture realizations of exact solutions to homogeneous wave equations, *J. Opt. Soc. Am. A* **10**, 75–87 (1993).
54. R. Donnelly and R. W. Ziolkowski, Designing localized waves, *Proc. R. Soc. London A* **440**, 541–565 (1993).
55. E. Recami, M. Zamboni-Rached, K. Z. Nobrega, C. A. Dartora, and H. E. Hernández-Figueroa, On the localized superluminal solutions to the Maxwell equations, *IEEE J. Sel.Top. Quantum Electron.* **9**, 59–73 (2003).
56. E. Recami, Superluminal motions? A bird's-eye view of the experimental status-of-the-art, *Found. Phys.* **31**, 1119–1135 (2001).
57. E. Recami, F. Fontana, and R. Garavaglia, Superluminal motions and special relativity: a discussion of recent experiments, *Int. J. Mod. Phys. A* **15**, 2793–2812 (2000); see also R. Garavaglia, thesis work, Dipartimento di Scientifics Informazióne, Università statale di Milano, Milan, Italy, 1998; M. Pavsic and E. Recami, Charge conjugation and internal space-time symmetries, *Lett. Nuovo Cimento* **34**, 357–362 (1982); E. Recami and W. A. Rodriguez, Antiparticles from special relativity with ortho-chronous and anti-chronous Lorentz transformations, *Found. Phys.* **12**, 709–718 (1982); **13**, 533 (1983).
58. E. Recami, Tachyon mechanics and causality: A systematic thorough analysis of the tachyon causal paradoxes, *Found. Phys.* **17**, 239–296 (1987); see also E. Recami, The Tolman–Regge antitelephone paradox: Its solution by tachyon mechanics, *Lett. Nuovo Cimento* **44** 587–593 (1985); G. D. Maccarrone and E. Recami, Two-body interactions through tachyon exchange, *Nuovo Cimento A* **57**, 85–101 (1980).
59. A. T. Friberg, J. Fagerholm, and M. M. Salomaa, Space-frequency analysis of nondiffracting pulses, *Opt. Commun.* **136**, 207–212 (1997).
60. R. W. Ziolkowski, I. M. Besieris, and A. M. Shaarawi, Aperture realizations of exact solutions to homogeneous wave equations, *J. Opt. Soc. Am. A* **10**, 75–87 (1993).
61. A. M. Shaarawi and I. M. Besieris, On the superluminal propagation of X-shaped localized waves, *J. Phys. A* **33**, 7227–7254 (2000).
62. A. M. Shaarawi and I. M. Besieris, Relativistic causality and superluminal signalling using X-shaped localized waves, *J. Phys. A* **33**, 7255–7263 (2000), and refs. therein.
63. M. Zamboni-Rached, K. Z. Nobrega, H. E. Hernández-Figueroa, and E.Recami, Localized superluminal solutions to the wave equation in (vacuum or) dispersive media, for arbitrary frequencies and with adjustable bandwidth, *Opt. Commun.* **226**, 15–23 (2003).
64. M. Zamboni-Rached, E. Recami, and F. Fontana, Localized superluminal solutions to Maxwell equations propagating along a normal-sized waveguide, *Phys. Rev. E* **64**, 066603 (2001).
65. V. S. Olkhovsky, E. Recami, and A. K. Zaichenko, Resonant and non-resonant tunneling through a double barrier, *Europhys. Lett.* **70**, 712–718 (2005).
66. See, e. g., P. W. Milonni, *J. Phys. B* **35**, R31–R56 (2002); G. Nimtz and A. Haibel, *Ann. Phys.* **11**, 163–171 (2002); R. W. Ziolkowski, *Phys. Rev. E* **63**, 046604 (2001); A. M. Shaarawi and I. M. Besieris, *J. Phys. A* **33**, 7227–7254, 7255–7263 (2000); E. Recami, F. Fontana, and R. Garavaglia, ref. [57], and refs. therein.
67. M. Zamboni-Rached, K. Z. Nóbrega, E. Recami, and H. E. Hernández-Figueroa, Superluminal X-shaped beams propagating without distortion along a coaxial guide, *Phys. Rev. E* **66**, 036620 (2002), and refs. therein.

68. C. A. Dartora, M. Zamboni-Rached, K. Z. Nóbrega, E. Recami, and H. E. Hernández-Figueroa, General formulation for the analysis of scalar diffraction-free beams using angular modulation: Mathieu and Bessel beams, *Opt. Commun.* **222**, 75–80 (2003).
69. J. Durnin, J. J. Miceli, and J. H. Eberly, Comparison of Bessel and Gaussian beams, *Opt. Lett.* **13**, 79–80 (1988).
70. P. L. Overfelt and C. S. Kenney, Comparison of the propagation characteristics of Bessel, Bessel–Gauss, and Gaussian beams diffracted by a circular aperture, *J. Opt. Soc. Am. A* **8**, 732–745 (1991).
71. Z. Bouchal, J. Wagner, and M. Chlup, Self-reconstruction of a distorted nondiffracting beam, *Opt. Communi.* **151**, 207–211 (1998).
72. S. Esposito, E. Majorana, Jr., A. van der Merwe, and E. Recami, Eds., *Ettore Majorana: Notes on Theoretical Physics*, Kluwer, New York, Nov. 2003.
73. A. P. L. Barbero, H. E. Hernández, Figuerao, and E. Recami, On the propagation speed of evanescent modes, *Phys. Rev. E* **62**, 8628 (2000), and refs. therein; see also A. M. Shaarawi and I. M. Besieris, *Phys. Rev. E* **62**, 7415 (2000).
74. H. M. Brodowsky, W. Heitmann, and G. Nimtz, *Phys. Lett. A* **222**, 125 (1996).
75. M. Zamboni-Rached, H. E. Hernández-Figueroa, and E. Recami, Chirped optical X-shaped pulses in material media, *J. Opt. Soc. Am. A* **21**, 2455–2463 (2004).
76. A. C. Newell and J. V. Molone, *Nonlinear Optics*, Addison-Wesley, Reading, MA., 1992.
77. M. Zamboni-Rached, Diffraction-attenuation resistant beams in absorbing media, *Opt. Express* **14**, 1804–1809 (2006); also chosen for mention in the *Virtual Journal for Biomedical Optics*, and in *Laser Focus World*, section "New Bracks".
78. R. Grunwald et al., Generation and characterization of spatially and temporally localized few-cycle optical wavepackets, *Phys. Rev. A* **67**, 063820 (2003); R. Grunwald, U. Griebner, U. Neumann, and V. Kebbel, Self-reconstruction of ultrashort-pulse Bessel-like X-waves, presented at the CLEO/QELS Conference, San Francisco, CA, 2004, paper CMQ7; R. Grunwald et al., Ultrabroadband spectral transfer in extended focal zones: truncated few-cycle Bessel–Gauss beams, presented at the CLEO Conference, Baltimore, MD, 2005, paper CTuAA6.
79. See E. Recami, ref. [1], pp. 80–81; figure 27, and refs. therein; R. Folman and E. Recami, *Found. Phys. Lett.* **8**, 127–134 (1995).
80. E. Recami and R. Mignani, *Riv. Nuovo Cimento.* **4**, 209–290 (1974), E398.
81. See, e.g., E. Recami, Ed., *Tachyons, Monopoles, and Related Topics*, North-Holland, Amsterdam, The Netherland, 1978.
82. I. S. Gradshteyn and I. M. Ryzhik, *Integrals, Series and Products*, 4th ed., Academic Press, New York, 1965.
83. E. Giannetto, G. D. Maccarrone, R. Mignani, and E. Recami, *Phys. Lett. B* **178**, 115–120 (1986), and refs. therein.
84. See. M. Baldo Ceolin, Review of neutrino physics, invited talk at the 13th International Sympasium on Multiparticle Dynamics, Aspen, CO, Sept. 1993; E. W. Otten, *Nucl. Phys. News* **5**, 11 (1995). From the theoretical point of view, see, e.g., E. Giannetto et al., ref. [83] and refs. therein; S. Giani, Experimental evidence of superluminal velocities in astrophysics and proposed experiments, CP458, in M. S. El-Genk, Ed., *Space*

Technology and Applications: International Forum 1999, A.I. Press, Melville, NY 1999, pp. 881–888.

85. See, e.g., J. A. Zensus and T. J. Pearson, Eds., *Superluminal Radio Sources*, Cambridge University Press, Cambridge, 1987.
86. I. F. Mirabel and L. F. Rodriguez, A superluminal source in the galaxy, *Nature* **371**, 46 (1994), with an editorial comment, A galactic speed record, by G. Gisler, on page 18 of the same issue; S. J. Tingay et al., Relativistic motion in a nearby bright x-ray source, *Nature* **374**, 141 (1995).
87. M. J. Rees, *Nature* **211**, 46 (1966); A. Cavaliere, P. Morrison, and L. Sartori, *Science* **173**, 525 (1971).
88. E. Recami, A. Castellino, G. D. Maccarrone, and M. Rodonò, Considerations about the apparent superluminal expansions observed in astrophysics, *Nuovo Cimento B* **93**, 119 (1986); see also R. Mignani and E. Recami, *Gen. Relat. Grav.* **5**, 615 (1974).
89. See, e.g., O. M. Bilaniuk, V. K. Deshpande, and E. C. G. Sudarshan, *Am. J. Phys.* **30**, 718 (1962).
90. V. S. Olkhovsky and E. Recami, Recent developments in the time analysis of tunnelling processes, *Phys. Rep.* **214**, 339–357 (1992), issue 6, and refs. therein: in particular T. E. Hartman, *J. Appl. Phys.* **33**, 3427 (1962); L. A. MacColl, *Phys. Rev.* **40**, 621 (1932). See also V. S. Olkhovsky, E. Recami, F. Raciti, and A. K. Zaichenko, *J. Phys. I* **5**, 1351–1365 (1995); V. S. Olkhovsky, E. Recami, and J. Jakiel, Unified time analysis of photon and nonrelativistic particle tunneling, Report INFN/FM-01/01, Instituto Nationale de Fisica Nucleare, Frascati, Italy, Apr. 2001, LANL archives e-print quant-ph/0102007; P. W. Milonni, *J. Phys. B* **35**, R31–R56 (2002).
91. V. S. Olkhovsky, E. Recami, and J. Jakiel, Unified time analysis of photon and nonrelativistic particle tunneling, *Phys. Rep.* **398**, 133–178 (2004).
92. V. S. Olkhovsky, E. Recami, F. Raciti, and A. K. Zaichenko, *J. Phys. I* **5**, 1351–1365 (1995).
93. A. M. Steinberg, P. G. Kwiat, and R. Y. Chiao, *Phys. Rev. Lett.* **71**, 708 (1993), and refs. therein; *Sci. Am.* **269**(2), 38 (1993), issue 2. For a review, see, e.g., R. Y. Chiao and A. M. Steinberg, in E. Wolf, Ed., *Progress in Optics*, Elsevier; Amsterdam, The Netherland, 1997, p. 345; see also Y. Japha and G. Kurizki, *Phys. Rev. A* **53**, 586 (1996).
94. G. Nimtz and A. Enders, *J. Phys. I* **2**, 1693 (1992), **3**, 1089 (1993), **4**, 1379 (1994), and *Phys. Rev. E* **48**, 632 (1993); H. M. Brodowsky, W. Heitmann, and G. Nimtz, *J. Phys. I* **4**, 565 (1994), and *Phys. Lett. A* **222**, 125 (1996), **196**, 154 (1994). For a review, see, e.g., G. Nimtz and W. Heitmann, *Prog. Quantum Electron* **21**, 81 (1997).
95. A. Ranfagni, P. Fabeni, G. P. Pazzi, and D. Mugnai, *Phys. Rev. E* **48**, 1453 (1993); C. Spielmann, R. Szipocs, A. Stingl, and F. Krausz, *Phys. Rev. Lett.* **73**, 2308 (1994); P. Balcou and L. Dutriaux, *Phys. Rev. Lett.* **78**, 851 (1997); V. Laude and P. Tournois, *J. Opt. Soc. Am. B* **16**, 194 (1999).
96. *Scientific American*, Aug. 1993; *Nature*, Oct. 21, 1993; *New Scientist*, Apr. 1995; *Newsweek*, June 19, 1995.
97. L. J. Wang, A. Kuzmich, and A. Dogariu, *Nature* **406**, 277 (2000).
98. E. Recami, Superluminal motions? A birds-eye view of the experimental situation, *Found. Phys.* **31**, 1119–1135 (2001); V. S. Olkhovsky, E. Recami, and G. Salesi, Tunneling through two successive barriers and the Hartman (superluminal) effect, LANL archives e-print quant-ph/0002022, *Europhys. Lett.* **57**, 879–884 (2002); see also E. Recami, M. Z.

Rached, K. Z. Nobrega, C. A. Dartora, and H. E. Hernández Figueroa, On the localized superluminal solutions to the Maxwell equations, *IEEE J. Sel. Top. Quantum Electron.* **9**(1), 59–73 (2003); E. Recami, Superluminal tunnelling through successive barriers: Does quantum mechanics predict infinite group velocities? *J. Mod. Opt.* **51**, 913–923 (2004).

99. S. Longhi, P. Laporta, M. Belmonte, and E. Recami, Measurement of superluminal optical tunneling times in double-barrier photonic bandgaps, *Phys. Rev. E* **65**, 046610 (2002).

100. V. S. Olkhovsky, E. Recami, and G. Salesi, *Europhys. Lett.* **57**, 879–884 (2002); Tunneling through two successive barriers and the Hartman (superluminal) effect, LANL archives e-print quant-ph/0002022; Y. Aharonov, N. Erez and B. Reznik, *Phys. Rev. A* **65**, 052124 (2002); S. Esposito, *Phys. Rev. E* **67**, 016609 (2003); see also E. Recami, Superluminal tunnelling through successive barriers: Does quantum mechanics predict infinite group velocities? *J. Mod. Opt* **51**, 913–923 (2004).

101. L. J. Wang, A. Kuzmich, and A. Dogariu, *Nature* **406**, 277 (2000); M. W. Mitchell and R. Y. Chiao, *Phys. Lett. A* **230**, 133–138 (1997); see also S. Chu and W. Wong, *Phys. Rev. Lett.* **48**, 738 (1982); B. Segard and B. Macke, *Phys. Lett. A* **109**, 213–216 (1985); B. Macke et al., *J. Phys.*, **48**, 797–808 (1987); G. Nimtz, *Eur. Phys. J. B* (to appear as a Rapid Note).

102. M. Zamboni-Rached, Analytical expressions for the longitudinal evolution of non-diffracting pulses truncated by finite apertures, *J. Opt. Soc. Am. A* **23**, 2166–2176 (2006).

103. F. Selleri, talk presented at the 2nd Cracow–Clausthal workshop on Extensions of Quantum Theory, Krakòw, Poland, 2000 (unpublished); and in F. Selleri, *Lezioni di relatività da Einstein all'etere di Lorentz*, Progedit, Bari, Italy, 2003.

104. See, e.g., O. M. Bilaniuk, V. K. Deshpande, and E. C. G. Sudarshan, *Am. J. Phys.* **30**, 718 (1962).

105. A. M. Steinberg, P. G. Kwiat, and R. Y. Chiao, *Phys. Rev. Lett.* **71**, 708 (1993), and refs. therein; *Sci. Am.* **269**(2), 38 (1993), issue 2. For a review, see, e.g., R. Y. Chiao and A. M. Steinberg, in, E. Wolf, Ed., *Progress in Optics*, Elsevier, Amsterdam, The Netherlands 1997, 345; see also Y. Japha and G. Kurizki *Phys. Rev. A* **53**, 586 (1996).

106. G. Nimtz and A. Enders, *J. Phys. I* **2**, 1693 (1992), **3**, 1089 (1993), **4**, 1379 (1994), and *Phys. Rev. E* **48**, 632 (1993); H. M. Brodowsky, W. Heitmann, and G. Nimtz, *J. Phys. I* **4**, 565 (1994); *Phys. Lett. A* **222**, 125 (1996), **196**, 154 (1994). For a review, see, e.g., G. Nimtz and W. Heitmann, *Prog. Quantum. Electron* **21**, 81 (1997).

107. A. Ranfagni, P. Fabeni, G. P. Pazzi, and D. Mugnai, *Phys. Rev. E* **48**, 1453 (1993); C. Spielmann, R. Szipocs, A. Stingl, and F. Krausz, *Phys. Rev. Lett.* **73**, 2308 (1994); P. Balcou and L. Dutriaux, *Phys. Rev. Lett.* **78**, 851 (1997); V. Laude and P. Tournois, *J. Opt. Soc. Am. B* **16**, 194 (1999).

108. *Scientific American*, Aug. 1993; *Nature*, Oct. 21, 1993; *New Scientist*, Apr. 1995; *Newsweek*, June 19, 1995.

109. Ref. [1], pp. 158, 116–117; see also D. Mugnai, A. Ranfagni, R. Ruggeri, A. Agresti, and E. Recami, *Phys. Lett. A* **209**, 227 (1995).

110. H. M. Brodowsky, W. Heitmann, and G. Nimtz, *Phys. Lett. A* **222**, 125 (1996).

111. G. Nimtz, A. Enders, and H. Spieker, in A. van der Merwe and A. Garuccio, Eds., *Wave and Particle in Light and Matter*, Plenum Press, New York, 1993; *J. Phys. I* **4**, 565 (1994); see also A. Enders and G. Nimtz, *Phys. Rev. B* **47**, 9605 (1993).

112. V. S. Olkhovsky, E. Recami, F. Raciti, and A. K. Zaichenko, ref. [92], p. 1361 and refs. therein; E. Recami, F. Fontana, and R. Garavaglia, ref. [57], p. 2807; and ref. [91], p. 152.

113. R. Y. Chiao, A. E. Kozhekin, and G. Kurizki, *Phys. Rev. Lett.* **77**, 1254 (1996); E. L. Bolda et al., *Phys. Rev. A* **48** 3890 (1993); C. G. B. Garret and D. E. McCumber *Phys. Rev. A* **1**, 305 (1970).

114. S. Chu and W. Wong, *Phys. Rev. Lett.* **48**, 738 (1982); B. Segard and B. Macke, *Phys. Lett. A* **109**, 213–216 (1985); B. Macke et al., *J. Phys.* **48**, 797–808 (1987); M. W. Mitchell and R. Y. Chiao, *Phys. Lett. A* **230**, 133–138 (1997); L. J. Wang, A. Kuzmich, and A. Dogariu, *Nature* **406**, 277 (2000); G. Nimtz, *Eur. Phys. J. B* **7**, 523–525 (1999).

115. G. Alzetta, A. Gozzini, L. Moi, and G. Orriols, *Nuovo Cimento B* **36**, 5 (1976).

116. M. Artoni, G. C. La Rocca, F. S. Cataliotti, and F. Bassani, *Phys. Rev. A* **63**, 023805 (2001).

117. M. Zamboni-Rached: private communications (2003).

118. See, e.g., R. Y. Chiao, P. G. Kwiat and A. M. Steinberg, Analogies between electron and photon tunneling: A proposed experiment to measure photon tunneling times, *Physica B* **175**, 257–262 (1991); A. Ranfagni, D. Mugnai, P. Fabeni and G. P. Pazzi, Delay-time measurements in narrowed wave-guides as a test of tunneling, *Appl. Phys. Lett.* **58**, 774–776 (1991); Th. Martin and R. Landauer, Time delay of evanescent electromagnetic waves and the analogy to particle tunneling, *Phys. Rev. A* **45**, 2611–2617 (1992); Y. Japha and G. Kurizki, Superluminal delays of coherent pulses in nondissipative media: A universal mechanism, *Phys. Rev. A* **53**, 586–590 (1996).

CHAPTER TWO

Structure of Nondiffracting Waves and Some Interesting Applications

MICHEL ZAMBONI-RACHED
Centro de Ciências Naturais e Humanas, Universidade Federal do ABC, Santo Andre, SP, Brazil

ERASMO RECAMI
Università degli Studi di Bergamo, Bergamo, Italy, and INFN–Sezione di Milano, Milan, Italy

HUGO E. HERNÁNDEZ-FIGUEROA
Universidade Estadual de Campinas, Campinas, SP, Brazil

2.1 INTRODUCTION

Since early work [1–4] on *nondiffracting waves* (also called *localized waves*), a great deal has been published on this important subject from either the theoretical or experimental point of view. Initially, the theory was developed taking into account only free space; however, in recent years it has been extended to deal with more complex media, exhibiting effects such as dispersion [5–7], nonlinearty [8], anisotropy [9], and losses [10]. Such extensions have been carried out along with the development of efficient methods for obtaining nondiffracting beams and pulses in the subluminal, luminal, and superluminal regimes [11–18].

In this chapter we address some theoretical methods related to nondiffracting solutions of linear wave equations in unbounded homogeneous media and to interesting

Localized Waves, Edited by Hugo E. Hernández-Figueroa, Michel Zamboni-Rached, and Erasmo Recami

applications of these waves. The usual cylindrical coordinates (ρ, ϕ, z) are used here. In these coordinates the linear wave equation is written as

$$\frac{1}{\rho}\frac{\partial}{\partial\rho}\left(\rho\frac{\partial\Psi}{\partial\rho}\right)+\frac{1}{\rho^2}\frac{\partial^2\Psi}{\partial\phi^2}+\frac{\partial^2\Psi}{\partial z^2}-\frac{1}{c^2}\frac{\partial^2\Psi}{\partial t^2}=0. \tag{2.1}$$

In Section 2.2 we analyze the general structure of localized waves, develop generalized bidirectional decomposition, and use it to obtain several luminal and superluminal nondiffracting wave solutions of Eq. (2.1). In Section 2.3 we develop a space–time focusing method through a continuous superposition of X-shaped pulses of different velocities. In Section 2.4 we address the properties of chirped optical X-shaped pulses propagating in material media without boundaries. Finally, in Section 2.5 we show how a suitable superposition of Bessel beams can be used to obtain stationary localized wave fields with high transverse localization whose longitudinal intensity pattern can assume any shape desired within a chosen interval $0 \leq z \leq L$ of the propagation axis.

Important Observation We call the reader's attention to the fact that the sections in this chapter are essentially independent, and due to this fact, the notation is sometimes not standardized. For example, in Section 2.2 the longitudinal wave number is represented by k_z and one of the spectra parameters of the generalized bidirectional decomposition is denoted by β, whereas in Sections 2.4 and 2.5, β represents the longitudinal wave number.

2.2 SPECTRAL STRUCTURE OF LOCALIZED WAVES

An effective way to understand the concept of (ideal) nondiffracting waves is through a precise mathematical definition of these solutions so that we can extract from them the necessary spectral structure. Intuitively, an *ideal nondiffracting wave* (beam or pulse) can be defined as a wave capable of maintaining its spatial form (except for local variations) indefinitely while propagating. We can express this intuitive property by saying that a localized wave has to possess the property [12,13]

$$\Psi(\rho,\phi,z,t)=\Psi\left(\rho,\phi,z+\Delta z_0,t+\frac{\Delta z_0}{V}\right), \tag{2.2}$$

where Δz_0 is a certain length and V is the pulse propagation speed, which can assume any value here: $0 \leq V \leq \infty$. Using a Fourier–Bessel expansion, we can express a function $\Psi(\rho, \phi, z, t)$ as

$$\Psi(\rho,\phi,z,t) = \sum_{n=-\infty}^{\infty}\left[\int_0^{\infty} dk_\rho \int_{-\infty}^{\infty} dk_z \int_{-\infty}^{\infty} d\omega\, k_\rho A'_n(k_\rho,k_z,\omega)J_n(k_\rho\rho)e^{ik_z z}e^{-i\omega t}e^{in\phi}\right]. \tag{2.3}$$

Using the translation property of the Fourier transforms $T[f(x+a)] = \exp(ika)T[f(x)]$, we have that $A'_n(k_\rho, k_z, \omega)$ and $\exp[i(k_z \Delta z_0 - \omega \Delta z_0/V)] \times A'_n(k_\rho, k_z, \omega)$ are the Fourier–Bessel transforms of the left- and right-hand-side functions in Eq. (2.2). From this same equation we can get [12,13] the fundamental constraint linking the angular frequency ω and the longitudinal wave number k_z:

$$\omega = V k_z + 2m\pi \frac{V}{\Delta z_0} \tag{2.4}$$

with m an integer. Obviously, this constraint can be satisfied through the spectral functions $A'_n(k_\rho, k_z, \omega)$.

Now let us mention explicitly that constraint (2.4) does not imply any breakdown of the wave equation validity. In fact, when inserting expression (2.3) in wave equation (2.2), we find that

$$\frac{\omega^2}{c^2} = k_z^2 + k_\rho^2. \tag{2.5}$$

So, to obtain a solution of the wave equation from (2.3), the spectrum $A'_n(k_\rho, k_z, \omega)$ must have the form

$$A'_n(k_\rho, k_z, \omega) = A_n(k_z, \omega)\, \delta\left(k_\rho^2 - \left(\frac{\omega^2}{c^2} - k_z^2\right)\right), \tag{2.6}$$

where $\delta(\cdot)$ is the Dirac delta function. With this we can write a solution of the wave equation as

$$\Psi(\rho, \phi, z, t) = \sum_{n=-\infty}^{\infty} \left[\int_0^\infty d\omega \int_{-\omega/c}^{\omega/c} dk_z A_n(k_z, \omega) J_n\left(\rho \sqrt{\frac{\omega^2}{c^2} - k_z^2}\right) e^{ik_z z} e^{-i\omega t} e^{in\phi} \right], \tag{2.7}$$

where we have considered positive angular frequencies only. Equation (2.7) is a superposition of Bessel beams and it is understood that the integrations in the $\omega - k_z$ plane are confined to the region $0 \le \omega \le \infty$; $-\omega/c \le k_z \le \omega/c$.

To obtain an ideal nondiffracting wave, the spectra $A_n(k_z, \omega)$ must obey the fundamental constraint (2.4), so we write

$$A_n(k_z, \omega) = \sum_{m=-\infty}^{\infty} S_{nm}(\omega) \delta\left(\omega - (V k_z + b_m)\right), \tag{2.8}$$

where the b_m are constants representing the terms $2m\pi V/\Delta z_0$ in Eq. (2.4) and the $S_{nm}(\omega)$ are arbitrary frequency spectra. With (2.8) in (2.7) we get a general integral form of an ideal nondiffracting wave defined by Eq. (2.2):

$$\Psi(\rho, \phi, z, t) = \sum_{n=-\infty}^{\infty} \sum_{m=-\infty}^{\infty} \psi_{nm}(\rho, \phi, z, t), \tag{2.9}$$

with

$$\psi_{nm}(\rho, \phi, z, t) = e^{-ib_m z/V} \int_{(\omega_{\min})_m}^{(\omega_{\max})_m} d\omega \, S_{nm}(\omega) \times J_n\left(\rho\sqrt{\left(\frac{1}{c^2} - \frac{1}{V^2}\right)\omega^2 + \frac{2b}{V^2}\omega - \frac{b^2}{V^2}}\right) e^{i(\omega/V)(z-Vt)} e^{in\phi}, \tag{2.10}$$

where $\omega_{\min}$ and $\omega_{\max}$ depend on the values of V:

- For subluminal ($V < c$) localized waves: $b_m > 0$, $(\omega_{\min})_m = cb_m/(c+V)$ and $(\omega_{\max})_m = cb_m/(c-V)$.
- For luminal ($V = c$) localized waves: $b_m > 0$, $(\omega_{\min})_m = b_m/2$, and $(\omega_{\max})_m = \infty$.
- For superluminal ($V > c$) localized waves: $b_m \geq 0$, $(\omega_{\min})_m = cb_m/(c+V)$, and $(\omega_{\max})_m = \infty$; or $b_m < 0$, $(\omega_{\min})_m = cb_m/(c-V)$, and $(\omega_{\max})_m = \infty$.

It is important to notice that each $\psi_{nm}(\rho, \phi, z, t)$ in the superposition (2.9) is a truly nondiffracting wave (beam or pulse) and their superposition (2.9) is just the most general form available to represent a nondiffracting wave defined by Eq. (2.2). Due to this fact, the search for methods capable of providing analytical solutions for $\psi_{nm}(\rho, \phi, z, t)$, Eq. (2.10), becomes an important task.

Keep in mind that Eq. (2.10) is also a Bessel beam superposition but with the constraint (2.4) between their angular frequencies and longitudinal wave numbers. Despite the fact that expression (2.10) represents ideal nondiffracting waves, it is difficult to use it to obtain closed analytical solutions. Because of this, we are going to develop a method able to overcome this limitation, providing several interesting localized wave solutions (luminal and superluminal) in arbitrary frequencies, including some exhibiting finite energy.

2.2.1 Generalized Bidirectional Decomposition

For reasons that will soon become clear, instead of dealing with the integral expression (2.9), our starting point is the general expression (2.7). Here, for simplicity, we restrict ourselves to axially symmetric solutions assuming the spectral functions as

$$A_n(k_z, \omega) = \delta_{n0} A(k_z, \omega), \tag{2.11}$$

where δ_{n0} is the Kronecker delta. In this way, we get the following general solution (considering positive angular frequencies only) describing axially symmetric waves:

$$\Psi(\rho, \phi, z, t) = \int_0^\infty d\omega \int_{-\omega/c}^{\omega/c} dk_z\, A(k_z, \omega) J_0 \left(\rho \sqrt{\frac{\omega^2}{c^2} - k_z^2} \right) e^{ik_z z} e^{-i\omega t}. \quad (2.12)$$

As we have seen, ideal nondiffracting waves can be obtained since the spectrum $A(k_z, \omega)$ satisfies the linear relationship (2.4). In this way, it is natural to adopt new spectral parameters in place of (ω, k_z) that make it easier to implement that constraint [12,13].

With this in mind, we choose the new spectral parameters (α, β) through

$$\alpha = \frac{1}{2V}(\omega + Vk_z), \qquad \beta = \frac{1}{2V}(\omega - Vk_z). \quad (2.13)$$

Let us consider here only luminal $(V = c)$ and superluminal $(V > c)$ nondiffracting pulses.

With the change of variables (2.13) in the integral solution (2.12), and considering $V \geq c$, the integration limits on α and β have to satisfy the three inequalities:

$$\begin{bmatrix} 0 < \alpha + \beta < \infty \\ \alpha \geq \dfrac{c - V}{c + V}\beta \\ \alpha \geq \dfrac{c + V}{c - V}\beta. \end{bmatrix} \quad (2.14)$$

Let us suppose both α and β to be positive $[\alpha, \beta \geq 0]$. The first inequality in (2.14) is then satisfied, whereas the coefficients $(c - V)/(c + V)$ and $(c + V)/(c - V)$ entering relations (2.14) are both negative (since $V > c$). As a consequence, the other two inequalities in (2.14) are satisfied automatically. In other words, the integration limits in $0 \leq \alpha \leq \infty$ and $0 \leq \beta \leq \infty$ are *contained* in the limits (2.14) and are therefore acceptable. Indeed, they constitute a rather suitable choice for facilitating all subsequent integrations.

Therefore, instead of Eq. (2.12) we consider the (more easily integrable) Bessel beam superposition in the new variables (with $V \geq c$)

$$\Psi(\rho, \zeta, \eta) = \int_0^\infty d\alpha \int_0^\infty d\beta\, A(\alpha, \beta) \times J_0 \left(\rho \sqrt{\left(\frac{V^2}{c^2} - 1 \right)(\alpha^2 + \beta^2) + 2\left(\frac{V^2}{c^2} + 1 \right)\alpha\beta} \right) e^{i\alpha\zeta}\, e^{-i\beta\eta}, \quad (2.15)$$

where we have defined

$$\zeta \equiv z - Vt, \qquad \eta \equiv z + Vt. \tag{2.16}$$

The present procedure is a generalization of the bidirectional decomposition technique, which was described earlier [11] for $V = c$.

From the new spectral parameters defined in transformation (2.13), it is easy to see that constraint (2.4) (i.e., $\omega = Vk_z + b$) is implemented just by making

$$A(k_z, \omega) \rightarrow A(\alpha, \beta) = S(\alpha)\delta(\beta - \beta_0), \tag{2.17}$$

with $\beta_0 = b/2V$. The delta function $\delta(\beta - \beta_0)$ in the spectrum (2.17) means that we are integrating Bessel beams along the continuous line $\omega = Vk_z + 2V\beta_0$, and in this way the function $S(\alpha)$ will give the frequency dependence of the spectrum: $S(\alpha) \rightarrow S(\omega/V - \beta_0)$. This method is a natural way of obtaining pulses with field concentration on $\rho = 0$ and $\zeta = 0 \rightarrow z = Vt$.

It is important to stress [13] that when $\beta_0 > 0$ in (2.17), the superposition (2.15) has contributions of both backward and forward Bessel beams in the frequency intervals $V\beta_0 \leq \omega < 2V\beta_0$ (where $k_z < 0$) and $2V\beta_0 \leq \omega \leq \infty$ (where $k_z \geq 0$), respectively. Nevertheless, we can obtain physical solutions when making the contribution of the backward components negligible by choosing suitable weight functions $S(\alpha)$. It is also important to note that we use the new spectral parameters α and β just to obtain (closed) analytical localized wave solutions, as the spectral characteristics of these new solutions can be brought into evidence just by using transformations (2.13) and writing the correspondent spectrum in terms of the usual ω and k_z spectral parameters.

In the following, we consider some cases with $\beta_0 = 0$ and $\beta_0 > 0$.

Closed Analytical Expressions Describing Some Ideal Nondiffracting Pulses Let us first consider, in Eq. (2.15), the following spectra of the type (2.17) with $\beta_0 = 0$:

$$A(\alpha, \beta) = \begin{cases} aV\delta(\beta)e^{-aV\alpha} & (2.18) \\ aV\delta(\beta)J_0(2d\sqrt{\alpha})e^{-aV\alpha} & (2.19) \\ \delta(\beta)\dfrac{\sin d\alpha}{\alpha}e^{-aV\alpha}, & (2.20) \end{cases}$$

with $a > 0$ and d being constants. One can obtain from these spectra the following superluminal localized wave solutions. From spectrum (2.18), we can use the identity (6.611.1) in [19] to obtain the well-known ordinary X-wave solution (also called an X-shaped pulse)

$$\Psi(\rho, \zeta) \equiv X = \frac{aV}{\sqrt{(aV - i\zeta)^2 + [(V^2/c^2) - 1]\rho^2}}. \tag{2.21}$$

Using spectrum (2.19) and identity (6.644.4) of [19], we get

$$\Psi(\rho,\zeta) = X \cdot J_0\left(\sqrt{\frac{V^2}{c^2}-1}\ (aV)^{-2}d^2X^2\rho\right)\exp[-(aV-i\zeta)(aV)^{-2}d^2\,X^2]. \tag{2.22}$$

The superluminal nondiffracting pulse

$$\Psi(\rho,\zeta) = \sin^{-1}\left[2\frac{d}{aV}\left(\sqrt{X^{-2}+(d/aV)^2+2\rho d(aV)^{-2}\sqrt{V^2/c^2-1}}\right.\right.$$
$$\left.\left.+\sqrt{X^{-2}+(d/aV)^2-2\rho d(aV)^{-2}\sqrt{V^2/c^2-1}}\right)^{-1}\right] \tag{2.23}$$

is obtained from spectrum (2.21) using identity (6.752.1) of [19] for $a > 0$ and $d > 0$.

From the previous discussion we know that any solution obtained from spectra of the type (2.17) with $\beta_0 = 0$ is free of noncausal (backward) components. In addition, when $\beta_0 = 0$, we can see that the pulsed solutions depend on z and t through $\zeta = z - Vt$ only, and so propagate rigidly (i.e., without distortion). Such pulses can be localized transversally only if $V > c$, because if $V = c$, the function Ψ has to obey the Laplace equation on transverse planes [12,13].

Many others superluminal localized waves can easily be constructed [13] from the solutions above just by taking the derivatives (of any order) with respect to ζ. It is also possible to show [13] that new solutions obtained in this way have their spectra shifted toward higher frequencies.

Now, let us consider, in Eq. (2.15), a spectrum of the type (2.17) with $\beta_0 > 0$:

$$A(\alpha,\beta) = aV\delta(\beta-\beta_0)e^{-aV\alpha}, \tag{2.24}$$

with a a positive constant. As we have seen, the presence of the delta function with the constant $\beta_0 > 0$ implies that we are integrating (summing) Bessel beams along the continuous line $\omega = Vk_z + 2V\beta_0$. The function $S(\alpha) = aV\exp(-aV\omega)$ means that we are considering a frequency spectrum of the type $S(\omega) \propto \exp(-a\omega)$ and therefore with a bandwidth given by $\Delta\omega = 1/a$.

Since $\beta_0 > 0$, the interval $V\beta_0 \le \omega < 2V\beta_0$ (or, equivalently in this case, $0 \le \alpha < \beta_0$) corresponds to backward Bessel beams (i.e., negative values of k_z). However, we can get physical solutions when making the contribution of this frequency interval negligible. In this case, it can be done by making $a\beta_0 V << 1$, so that the exponential decay of the spectrum S with respect to ω is very slow and the contribution of the interval $\omega \ge 2V\beta_0$ (where $k_z \ge 0$) overruns the $V\beta_0 \le \omega < 2V\beta_0$ (where $k_z < 0$) contribution.

We note that once we ensure the causal behavior of the pulse by making $aV\beta_0 << 1$ in (2.24), we have that the $\Delta\alpha = 1/aV >> \beta_0$, so we can simplify the argument of the Bessel function in the integrand of superposition (2.15) by neglecting the term $(V^2/c^2-1)\beta_0^2$. With this, the superposition (2.15), with the spectrum (2.24), can be written as

$$\Psi(\rho, \zeta, \eta) \approx aVe^{-i\beta_0\eta} \times \int_0^\infty d\alpha\, J_0\left(\rho\sqrt{\left(\frac{V^2}{c^2}-1\right)\alpha^2 + 2\left(\frac{V^2}{c^2}+1\right)\alpha\beta_0}\right) e^{i\alpha\zeta} e^{-aV\alpha}. \tag{2.25}$$

We can now use identity (6.616.1) of [19] and obtain the new localized superluminal solution called the *superluminal focus wave mode* (SFWM) [13]:

$$\Psi_{\text{SFWM}}(\rho, \zeta, \eta) = \mathrm{e}^{-i\beta_0\eta} X \exp\left[\frac{\beta_0(V^2+c^2)}{V^2-c^2}\left((aV - i\zeta) - aVX^{-1}\right)\right], \tag{2.26}$$

where, as before, X is the ordinary X pulse (2.21). The center of the SFWM is localized on $\rho = 0$ and $\zeta = 0$ (i.e., in $z = Vt$). The intensity, $|\Psi|^2$, of this pulse propagates rigidly, as a function of ρ and ζ only. However, the complex function Ψ_{SFWM} (i.e., its real and imaginary parts) propagates just with local variations, recovering their entire three-dimensional form after each space and time interval given by $\Delta z_0 = \pi/\beta_0$ and $\Delta t_0 = \pi/\beta_0 V$.

The SFWM solution given above for $V \longrightarrow c^+$ reduces to the well-known *focus wave mode* (FWM) solution [11] traveling with speed c:

$$\Psi_{\text{FWM}}(\rho, \zeta, \eta) = ac\frac{\mathrm{e}^{-i\beta_0\eta}}{ac - i\zeta} \exp\left(-\frac{\beta_0\rho^2}{ac - i\zeta}\right). \tag{2.27}$$

Let us also emphasize that since $\beta_0 > 0$, the resulting spectrum (2.24) is comprised of angular frequencies $\omega \geq V\beta_0$. Thus, our new solution can be used to construct high-frequency pulses.

Finite-Energy Nondiffracting Pulses Next, we show how to get finite-energy localized wave pulses. These new waves can propagate long distances while maintaining their spatial resolution (i.e., they possess a large depth of field).

As we have seen, ideal nondiffracting waves can be constructed by superposing Bessel beams [Eq. (2.12) for cylindrical symmetry] with a spectrum $A(\omega, k_z)$ that satisfies a linear relationship between ω and k_z. In the general bidirectional decomposition method, this can be made by using spectra of the type (2.17) in superposition (2.15).

Solutions of this type possess an infinite depth of field; however, they also exhibit infinite energy [11,13]. To overcome this problem, we can truncate an ideal nondiffracting wave by a finite aperture, and the resulting pulse will have finite energy with finite field depths. Even so, these depths may be very large compared with those of ordinary waves. The problem in this case is that the resulting field has to be calculated from the diffraction integrals (such as the well-known Rayleigh–Sommerfeld formula), and, in general, a closed analytical formula for the resulting pulse cannot be obtained.

However, there is another way to construct localized pulses with finite energy [13]: by using spectra $A(\omega, k_z)$ in (2.12) whose domains are not restricted to being defined exactly over the straight line $\omega = Vk_z + b$, but around that line, where the spectra should concentrate their main values. In other words, the spectrum has to be well localized in the vicinity of the line.

Similarly, in terms of the generalized bidirectional decomposition given in (2.15), finite-energy nondiffracting wave pulses can be constructed based on well-localized spectral functions $A(\alpha, \beta)$ in the vicinity of the line $\beta = \beta_0$, β_0 being a constant.

To exemplify this method, let us consider the spectrum

$$A(\alpha, \beta) = \begin{cases} aqVe^{-aV\alpha}e^{-q(\beta-\beta_0)} & \text{for } \beta \geq \beta_0 \\ 0 & \text{for } 0 \leq \beta < \beta_0 \end{cases} \tag{2.28}$$

in the superposition (2.15), a and q being free positive constants and V the peak's pulse velocity (here, $V \geq c$).

It is easy to see that the spectrum above is zero in the region above the $\beta = \beta_0$ line, while it decays in the region below (as well as along) such a line. We can concentrate this spectrum on $\beta = \beta_0$ by choosing values of q such that $q\beta_0 >> 1$. The faster the spectrum decay takes place in the region below the $\beta = \beta_0$ line, the larger the field depth of the corresponding pulse.

Once we choose $q\beta_0 >> 1$ to obtain pulses with a large depth of field, we can also minimize the contribution of the noncausal (backward) components by choosing $aV\beta_0 << 1$, in analogy with the results we obtained for the SFWM case.

Still in analogy with the SFWM case, when we choose $q\beta_0 >> 1$ (i.e., long depth of field) and $aV\beta_0 << 1$ (minimal contribution of backward components), we can simplify the argument of the Bessel function in the integrand of superposition (2.15) by neglecting the term $(V^2/c^2 - 1)\beta_0^2$.

With the comments above, we can write the superposition (2.15) with the spectrum (2.28) as

$$\Psi(\rho, \zeta, \eta) \approx aqV \int_{\beta_0}^{\infty} d\beta \int_0^{\infty} d\alpha\, J_0\left[\rho\sqrt{\left(\frac{V^2}{c^2} - 1\right)\alpha^2 + 2\left(\frac{V^2}{c^2} + 1\right)\alpha\beta}\right]$$
$$\times e^{-i\beta\eta}e^{i\alpha\zeta}e^{-q(\beta-\beta_0)}e^{-aV\alpha}, \tag{2.29}$$

and using identity (6.616.1) in [19], we get

$$\Psi(\rho, \zeta, \eta) \approx qX \int_{\beta_0}^{\infty} d\beta\, e^{-q(\beta-\beta_0)}e^{-i\beta\eta} \exp\left[\beta\frac{V^2 + c^2}{V^2 - c^2}\left(aV - i\zeta - aVX^{-1}\right)\right], \tag{2.30}$$

which can be viewed as a superposition of the SFWM pulses [see Eq. (2.26)].

The integration above can be made easily and results [13] in the *superluminal modified power spectrum* (SMPS) pulse:

$$\Psi_{\mathrm{SMPS}}(\rho, \zeta, \eta) = qX \frac{\exp[(Y - i\eta)\beta_0]}{q - (Y - i\eta)}, \tag{2.31}$$

where X is the ordinary X pulse (2.21) and Y is defined as

$$Y \equiv \frac{V^2 + c^2}{V^2 - c^2} \left[(aV - i\zeta) - aVX^{-1}\right]. \tag{2.32}$$

The SMPS pulse is a superluminal localized wave with field concentration around $\rho = 0$ and $\zeta = 0$ (i.e., in $z = Vt$) and with finite total energy. We will show that the depth of field, Z, of this pulse is given by $Z_{\mathrm{SMPS}} = q/2$.

An interesting property of the SMPS pulse is related to its transverse width (the transverse spot size at the pulse center). It can be shown from (2.31) that for cases where $aV << 1/\beta_0$ and $q\beta_0 >> 1$ (i.e., for the cases considered by us), the transverse spot size, $\Delta\rho$, of the pulse center ($\zeta = 0$) is dictated by the exponential function in (2.31) and is given by

$$\Delta\rho = c\sqrt{\frac{aV}{\beta_0(V^2 + c^2)} + \frac{V^2 - c^2}{4\beta_0^2(V^2 + c^2)^2}}, \tag{2.33}$$

which clearly does not depend on z, and so remains constant during the propagation. In other words, despite the fact that the SMPS pulse suffers an intensity decrease during the propagation, it preserves its transverse spot size. This interesting characteristic is not verified in ordinary pulses such as Gaussian pulses, where the amplitude of the pulse decreases and the width increases by the same factor.

Figure 2.1 shows a SMPS pulse intensity with $\beta_0 = 33\,\mathrm{m}^{-1}$, $V = 1.01c$, $a = 10^{-12}$ s, and $q = 10^5$ m at two different moments, for $t = 0$ and after 50 km of propagation, where as we can see, the pulse becomes less intense (half of its initial peak intensity). It can be noted that despite the intensity decrease, the pulse maintains its transverse width, as we can see from the two-dimensional plots in Fig. 2.1, which show the field intensities in transverse sections at $z = 0$ and $z = q/2 = 50$ km.

Three other important well-known finite-energy nondiffracting solutions can be obtained directly from the SMPS pulse. The first one, obtained from (2.31) by making $\beta_0 = 0$, is the *superluminal splash pulse* (SSP) [13],

$$\Psi_{\mathrm{SSP}}(\rho, \zeta, \eta) = \frac{qX}{q + i\eta - Y}. \tag{2.34}$$

The other two are luminal pulses. By taking the limit $V \to c^+$ in the SMPS pulse (2.31) we get the well-known *luminal modified power spectrum* (MPS) pulse [11],

$$\Psi_{\mathrm{MPS}}(\rho, \zeta, \eta) = \frac{aqce^{-i\beta_0\eta}}{(q + i\eta)(ac - i\zeta) + \rho^2} \exp\left(\frac{-\beta_0\rho^2}{ac - i\zeta}\right). \tag{2.35}$$

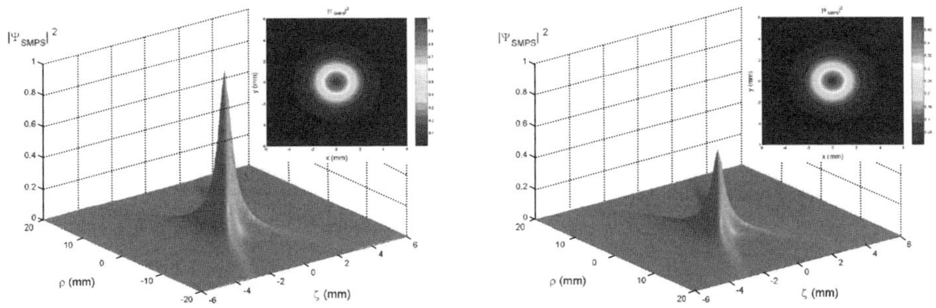

(a) (b)

FIGURE 2.1 Representation of a superluminal modified power spectrum pulse, Eq. (2.31). Its total energy is finite (even without truncation), so it gets deformed while propagating, since its amplitude decreases with time. In (a) we represent for $t = 0$, the pulse corresponding to $\beta_0 = 33\,\mathrm{m}^{-1}$, $V = 1.01c$, $a = 10^{-12}$ s, and $q = 10^5$ m. The same pulse is depicted in (b) after having traveled 50 km.

Finally, by taking the limit $V \to c^+$ and making $\beta_0 = 0$ in the SMPS pulse [or equivalently, by making $\beta_0 = 0$ in the MPS pulse (2.35), or by taking the limit $V \to c^+$ in the SSP (2.34)], we obtain the well-known *luminal splash pulse* (SP) solution [11],

$$\Psi_{\mathrm{SP}}(\rho, \zeta, \eta) = \frac{aqc}{(q + i\eta)(ac - i\zeta) + \rho^2}. \tag{2.36}$$

It is also interesting to note that the X and SFWM pulses can be obtained from the SSP and SMPS pulses (respectively) by making $q \to \infty$ in Eqs. (2.34) and (2.31). As a matter of fact, the solutions SSP and SMPS can be viewed as finite-energy versions of the X and SFWM, pulses, respectively.

Some Characteristics of the SMPS Pulse Let us examine the on-axis ($\rho = 0$) behavior of the SMPS pulse. On $\rho = 0$ we have

$$\Psi_{\mathrm{SMPS}}(\rho = 0, \zeta, \eta) = aqVe^{-i\beta_0 z}[(aV - i\zeta)(q + i\eta)]^{-1}. \tag{2.37}$$

From this expression we can show that the longitudinal localization Δz, for $t = 0$, of the SMPS pulse square magnitude is

$$\Delta z = 2aV. \tag{2.38}$$

If we now define the field depth Z as the distance over which the pulse's peak intensity is 50% at least of its initial value,[†] we can obtain, from (2.37), the depth of field

$$Z_{\mathrm{SMPS}} = \frac{q}{2}, \tag{2.39}$$

[†] We can expect that while the pulse peak intensity is maintained, so is its spatial form.

which depends solely on q, as we expected, since q regulates the concentration of the spectrum around the line $\omega = Vk_z + 2V\beta_0$.

Now, let us examine the maximum amplitude M of the real part of (2.37), which for $z = Vt$ is writen ($\zeta = 0$ and $\eta = 2z$)

$$M_{\text{SMPS}} \equiv \text{Re}[\Psi_{\text{SMPS}}(\rho = 0, z = Vt)] = \frac{\cos 2\beta_0 z - 2(z/q)\sin 2\beta_0 z}{1 + 4(z/q)^2}. \tag{2.40}$$

Initially, for $z = 0$, $t = 0$, one has $M = 1$ and can also infer that:

1. When $z/q << 1$ (i.e., when $z << Z$), Eq. (2.40) becomes

$$M_{\text{SMPS}} \approx \cos 2\beta_0 z \qquad \text{for } z << Z \tag{2.41}$$

   and the pulse's peak oscillates harmonically with "wavelength" $\Delta z_0 = \pi/\beta_0$ and "period" $\Delta t_0 = \pi/V\beta_0$, all along its field depth.
2. When $z/q >> 1$ (i.e., $z >> Z$), Eq. (2.40) becomes

$$M_{\text{SMPS}} \approx -\frac{\sin 2\beta_0 z}{2z/q} \qquad \text{for } z >> Z. \tag{2.42}$$

   Therefore, beyond its depth of field, the pulse goes on oscillating with the same Δz_0, but its maximum amplitude decays proportionally to z.

In the next two sections we look at an interesting application of localized wave pulses.

2.3 SPACE–TIME FOCUSING OF X-SHAPED PULSES

In this section we show how one can in general use any known superluminal solutions to obtain a large number of analytic expressions for space–time focused waves, endowed with a very strong intensity peak at the location desired. The method presented here is a natural extension of that developed by A. Shaarawi et al. [20], where space–time focusing was achieved by superimposing a discrete number of ordinary X-waves, characterized by different values of axicon angles θ.

In this section, we go on to more efficient superpositions [21] for varying velocities V, related to θ through the known [3,4] relation $V = c/\cos\theta$. This enhanced focusing scheme has the advantage of yielding analytic (closed-form) expressions for the spatiotemporally focused pulses.

Let us start by considering an axially symmetric ideal nondiffracting superluminal wave pulse $\psi(\rho, z - Vt)$ in a dispersionless medium, where $V = c/\cos\theta > c$ is the pulse velocity, with θ being the axicon angle. As we saw in Section 2.2, pulses like these can be obtained by a suitable frequency superposition of Bessel beams.

Suppose that we now have N waves of the type $\psi_n(\rho, z - V_n(t - t_n))$, with different velocities, $c < V_1 < V_2 < \cdots < V_N$, emitted at (different) times t_n, quantities t_n being constants, while $n = 1, 2, \ldots, N$. The center of each pulse is localized at $z = V_n(t - t_n)$. To obtain a highly focused wave, we need all wave components $\psi_n(\rho, z - V_n(t - t_n))$ to reach the point $z = z_f$ at the same time $t = t_f$. On choosing $t_1 = 0$ for the slowest pulse ψ_1, it is easily seen that the peak of this pulse reaches the point $z = z_f$ at the time $t_f = z_f / V_1$. So for each ψ_n, the instant of emission t_n must be

$$t_n = \left(\frac{1}{V_1} - \frac{1}{V_n} \right) z_f. \tag{2.43}$$

With this we can construct other exact solutions to the wave equation given by [21]

$$\Psi(\rho, z, t) = \int_{V_{\min}}^{V_{\max}} \mathrm{d}V \, A(V) \, \psi \left[\rho, z - V \left(t - \left(\frac{1}{V_{\min}} - \frac{1}{V} \right) z_f \right) \right], \tag{2.44}$$

where V is the velocity of the wave $\psi(\rho, z - Vt)$ in the integrand of (2.44). In the integration, V is considered as a continuous variable in the interval $[V_{\min}, V_{\max}]$. In Eq. (2.44), $A(V)$ is the velocity-distribution function that specifies the contribution to the integration of each wave component (with velocity V). The resulting wave $\Psi(\rho, z, t)$ can have a more or less strong amplitude peak at $z = z_f$ at time $t_f = z_f / V_{\min}$, depending on $A(V)$ and on the difference $V_{\max} - V_{\min}$. Let us notice that the resulting wave field will also propagate with a superluminal peak velocity, also depending on $A(V)$. When the velocity-distribution function is well concentrated around a certain velocity value, one can expect the wave (2.44) to increase its magnitude and spatial localization while propagating. Finally, as we know, the pulse peak acquires its maximum amplitude and localization at the chosen point $z = z_f$ and at time $t = z_f / V_{\min}$. Afterward the wave suffers progressive spreading and decrease in amplitude.

2.3.1 Focusing Effects Using Ordinary X-Waves

Here we present a specific example by integrating (2.44) over ordinary standard X-waves [40], $X = aV[(aV - i(z - Vt))^2 + (V^2/c^2 - 1)\rho^2]^{-1/2}$. When using this ordinary X-wave, the largest spectral amplitudes are obtained for low frequencies. For this reason, one may expect that the solutions considered below will be suitable mainly for low-frequency applications. Let us choose, then, the function ψ in the integrand of Eq. (2.44) to be $\psi(\rho, z, t) \equiv X(\rho, z - V(t - (1/V_{\min} - 1/V)z_f))$, that is,

$$\psi(\rho, z, t) \equiv X = \frac{aV}{\sqrt{\left[aV - i\left(z - V\left(t - (1/V_{\min} - 1/V)\, z_f\right)\right)\right]^2 + \left(V^2/c^2 - 1\right)\rho^2}}. \tag{2.45}$$

After some manipulations, we obtain the analytic integral solution

$$\Psi(\rho, z, t) = \int_{V_{\min}}^{V_{\max}} \frac{aVA(V)}{\sqrt{PV^2 + QV + R}} dV, \tag{2.46}$$

with

$$\begin{aligned} P &= \left[a + i\left(t - z_f/V_{\min}\right)\right]^2 + \rho^2/c^2 \\ Q &= 2\left(t - z_f/V_{\min} - ai\right)(z - z_f) \\ R &= -(z - z_f)^2 - \rho^2. \end{aligned} \tag{2.47}$$

In what follows we illustrate the behavior of some new spatiotemporally focused pulses by taking into consideration some different velocity distributions $A(V)$. These new pulses are closed analytical *exact* solutions of the wave equation.

Example 1 Let us consider our integral solution (2.46) with $A(V) = 1$ s/m. In this case the contribution of the X-waves is the same for all velocities in the range allowed, $[V_{\min}, V_{\max}]$. Using identity (2.264.2) of [19], we get the particular solution

$$\begin{aligned} \Psi(\rho, z, t) = {} & \frac{a}{P}\left(\sqrt{PV_{\max}^2 + QV_{\max} + R} - \sqrt{PV_{\min}^2 + QV_{\min} + R}\right) \\ & + \frac{aQ}{2P^{3/2}} \ln \frac{2\sqrt{P(PV_{\min}^2 + QV_{\min} + R)} + 2PV_{\min} + Q}{2\sqrt{P(PV_{\max}^2 + QV_{\max} + R)} + 2PV_{\max} + Q}, \end{aligned} \tag{2.48}$$

where P, Q, and R are as given in Eq. (2.47). A three-dimensional plot of this function is provided in Fig. 2.2, where we have chosen $a = 10^{-12}$ s, $V_{\min} = 1.001c$, $V_{\max} = 1.005c$, and $z_f = 200$ cm. It can be seen that this solution exhibits a rather evident space–time focusing. An initially spread-out pulse (shown for $t = 0$) becomes highly localized at $t = t_f = z_f/V_{\min} = 6.66$ ns, the pulse peak amplitude at z_f being 40.82 times greater than the initial one. In addition, at the focusing time t_f, the field is much more localized than at any other times. The velocity of this pulse is approximately $V = 1.003c$.

Example 2 In this case we choose $A(V) = 1/V$ (s/m), and using the identity (2.261) of [19], Eq. (2.46) gives

$$\Psi(\rho, z, t) = \frac{a}{\sqrt{P}} \ln \frac{2\sqrt{P(PV_{\max}^2 + QV_{\max} + R)} + 2PV_{\max} + Q}{2\sqrt{P(PV_{\min}^2 + QV_{\min} + R)} + 2PV_{\min} + Q}. \tag{2.49}$$

Other exact closed solutions can be obtained [21] considering, for instance, velocity distributions as $A(V) = 1/V^2$ and $A(V) = 1/V^3$.

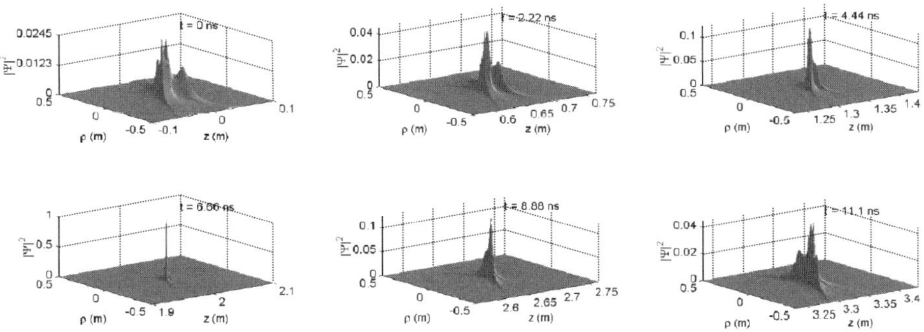

FIGURE 2.2 Space–time evolution of the superluminal pulse represented by Eq. (2.48); the parameter chosen values are $a = 10^{-12}$ s, $V_{\min} = 1.001c$, and $V_{\max} = 1.005c$, and the focusing point is at $z_f = 200$ cm. One can see that this solution is associated with rather good spatiotemporal focusing. The field amplitude at $z = z_f$ is 40.82 times larger than the initial amplitude. The field amplitude is normalized at the space–time point $\rho = 0, z = z_f, t = t_f$.

Actually, we can construct many others spatiotemporally focused pulses from the foregoing solutions just by taking time derivatives (of any order). It is also possible to show [21] that the new solutions obtained in this way have their spectra shifted toward higher frequencies.

2.4 CHIRPED OPTICAL X-TYPE PULSES IN MATERIAL MEDIA

The theory of the localized waves was developed initially for free space (vacuum). In 1996, Sõnajalg and Saari [5] showed that the localized wave theory can be extended to include (unbounded) dispersive media. This was obtained by making the axicon angle of Bessel beams (BBs) vary with the frequency [5–7] in such a way that a suitable frequency superposition of these beams compensates for the material dispersion. Soon after this idea was reported, many interesting nondiffracting/nondispersive pulses were obtained theoretically [5–7] and experimentally [5].

Despite this extended method being of remarkable importance, working well in theory, its experimental implementation is not so simple.† In 2004, Zamboni-Rached et al. [22] developed a simpler way to obtain pulses capable of recovering their spatial shape, both transversally and longitudinally, after some propagation. It consists of using chirped optical X-type pulses while keeping the axicon angle fixed. Let us recall that, by contrast, chirped Gaussian pulses in unbounded material media may recover only their longitudinal shape, since they undergo a progressive transverse spreading while propagating. The present section is devoted to this approach.

Let us start with an axially-symmetric Bessel beam in a material medium with refractive index $n(\omega)$:

$$\psi(\rho, z, t) = J_0(k_\rho \rho) \exp(i\beta z) \exp(-i\omega t), \qquad (2.50)$$

†We refer interested readers to [5–7] to obtain a description, theoretical and experimental, of this extended method.

which must obey the condition $k_\rho^2 = n^2(\omega)\omega^2/c^2 - \beta^2$, which connects the transverse and longitudinal wave numbers k_ρ and β and the angular frequency ω. In addition, we impose the conditions that $k_\rho^2 \geq 0$ and $\omega/\beta \geq 0$, to avoid the nonphysical behavior of the Bessel function $J_0(\cdot)$ and to confine ourselves to forward propagation.

Once the conditions above are satisfied, we have the liberty of writing the longitudinal wave number as $\beta = [n(\omega)\omega\cos\theta]/c$, and therefore, $k_\rho = [n(\omega)\omega \sin\theta]/c$, where (as in the free-space case) θ is the axicon angle of the Bessel beam.

Now we can obtain a X-shaped pulse by performing a frequency superposition of these BBs, with β and k_ρ given by the previous relations:

$$\Psi(\rho, z, t) = \int_{-\infty}^{\infty} S(\omega)\, J_0\left(\frac{n(\omega)\omega}{c}\sin\theta\,\rho\right) \exp[i\beta(\omega)z]\, \exp(-i\omega t)\, d\omega, \qquad (2.51)$$

where $S(\omega)$ is the frequency spectrum and the axicon angle is kept constant.

One can see that the phase velocity of each BB in our superposition (2.51) is different and is given by $V_{\text{phase}} = c/[n(\omega)\cos\theta]$, so the pulse given by (2.51) will suffer dispersion during its propagation.

As we said, the method developed by Sõnajalg and Saari [5] and explored by others [6,7] to overcome this problem consisted of regarding the axicon angle θ as a function of the frequency, in order to obtain a linear relationship between β and ω. Here, however, we wish to work with a *fixed* axicon angle, and we have to find another way to avoid dispersion and diffraction along a certain propagation distance. To do that, we might choose a chirped Gaussian spectrum $S(\omega)$ in Eq. (2.51):

$$S(\omega) = \frac{T_0}{\sqrt{2\pi(1+iC)}} \exp[-q^2(\omega-\omega_0)^2] \qquad \text{with } q^2 = \frac{T_0^2}{2(1+iC)}, \qquad (2.52)$$

where ω_0 is the central frequency of the spectrum, T_0 is a constant related to the initial temporal width, and C is the chirp parameter (we chose as a temporal width the half-width of the relevant Gaussian curve when its height equals $1/e$ times its full height). Unfortunately, there is no analytical solution to Eq. (2.51) with $S(\omega)$ given by Eq. (2.52), so some approximations must be made.

Then, let us assume that the spectrum $S(\omega)$ surrounding the carrier frequency ω_0 is narrow enough that $\Delta\omega/\omega_0 << 1$, to ensure that $\beta(\omega)$ can be approximated by the first three terms of its Taylor expansion in the vicinity of ω_0; that is, $\beta(\omega) \approx \beta(\omega_0) + \beta'(\omega)|_{\omega_0}(\omega-\omega_0) + (1/2)\beta''(\omega)|_{\omega_0}(\omega-\omega_0)^2$, where after using $\beta = [n(\omega)\omega\cos\theta]/c$,

$$\frac{\partial\beta}{\partial\omega} = \frac{\cos\theta}{c}\left[n(\omega) + \omega\frac{\partial n}{\partial\omega}\right], \qquad \frac{\partial^2\beta}{\partial\omega^2} = \frac{\cos\theta}{c}\left(2\frac{\partial n}{\partial\omega} + \omega\frac{\partial^2 n}{\partial\omega^2}\right). \qquad (2.53)$$

As we know, $\beta'(\omega)$ is related to the pulse group velocity by the relation $Vg = 1/\beta'(\omega)$. Here we can see the difference between the group velocity of an X-type pulse (with a fixed axicon angle) and that of a standard Gaussian pulse. Such a difference is due to the factor $\cos\theta$ in Eq. (2.53). Because of it, the group velocity of our X-type pulse is always greater than that of a Gaussian pulse. In other words, $(V_g)_X = (1/\cos\theta)(V_g)_{\text{gauss}}$.

We also know that the second derivative of $\beta(\omega)$ is related to the group-velocity dispersion (GVD) β_2 by $\beta_2 = \beta''(\omega)$. The GVD is responsible for the temporal (longitudinal) spreading of the pulse. Here one can see that the GVD of the X-type pulse is always smaller than that of standard Gaussian pulses, due to the factor $\cos\theta$ in Eq. (2.53): namely, $(\beta_2)_X = \cos\theta(\beta_2)_{\text{Gauss}}$.

Using the results above, we can write

$$\Psi(\rho, z, t) = \frac{T_0 \exp[i\beta(\omega_0)z] \exp(-i\omega_0 t)}{\sqrt{2\pi(1+iC)}} \int_{-\infty}^{\infty} d\omega\, J_0\left[\frac{n(\omega)\omega}{c} \sin\theta\, \rho\right]$$
$$\times \exp\left[i\frac{\omega-\omega_0}{V_g}(z - V_g t)\right] \exp\left[(\omega-\omega_0)^2\left(\frac{i\beta_2}{2}z - q^2\right)\right]. \quad (2.54)$$

The integral in Eq. (2.54) cannot be solved analytically, but it is enough for us to obtain the pulse behavior. Let us analyze the pulse for $\rho = 0$. In this case we obtain

$$\Psi(\rho = 0, z, t) = \frac{T_0 \exp[i\beta(\omega_0)z] \exp(-i\omega_0 t)}{\sqrt{T_0^2 - i\beta_2(1+iC)z}} \exp\left[\frac{-(z - V_g t)^2(1+iC)}{2V_g^2[T_0^2 - i\beta_2(1+iC)z]}\right]. \quad (2.55)$$

From Eq. (2.55) we can see immediately that the initial temporal width of the pulse intensity is T_0 and that after some propagation distance z, the time width T_1 becomes

$$\frac{T_1}{T_0} = \left[\left(1 + \frac{C\beta_2 z}{T_0^2}\right)^2 + \left(\frac{\beta_2 z}{T_0^2}\right)^2\right]^{1/2}. \quad (2.56)$$

Relation (2.56) describes the pulse spreading behavior. One can easily show that such behavior depends on the sign (positive or negative) of product $\beta_2 C$, as is well known for standard Gaussian pulses [23].

When $\beta_2 C > 0$, the pulse will become broader and broader monotonically with distance z. On the other hand, if $\beta_2 C < 0$, in the first stage, the pulse will suffer a narrowing and then will spread during the remainder of its propagation. So there will be a certain propagation distance in which the pulse will recover its initial temporal width ($T_1 = T_0$). From relation (2.56) we can find this distance $Z_{T_1=T_0}$ (given that $\beta_2 C < 0$) to be

$$Z_{T_1=T_0} = \frac{-2CT_0^2}{\beta_2(C^2+1)}. \quad (2.57)$$

One may notice that the maximum distance at which our chirped pulse, with given T_0 and β_2, may recover its initial temporal width can be evaluated easily from Eq. (2.57) and results in $L_{\text{disp}} = T_0^2/\beta_2$. We call such a maximum value L_{disp} the dispersion length. It is an maximum distance an X-type pulse may travel while recovering its initial longitudinal shape. Obviously, if we want the pulse to reassume its longitudinal shape at some desired distance $z < L_{\text{disp}}$, we have simply to choose a suitable value of the chirp parameter.

Let us emphasize that the property of recovering its own initial temporal (or longitudinal) width may also be verified to exist in the case of chirped standard Gaussian pulses. However, the latter will suffer a progressive transverse spreading which will not be reversible. The distance at which a Gaussian pulse doubles its initial transverse width w_0 is $z_{\text{diff}} = \sqrt{3}\,\pi w_0^2/\lambda_0$, where λ_0 is the carrier wavelength. Thus, we can see that optical Gaussian pulses with great transverse localization will get spoiled in a few centimeters or even less.

Next we show that it is also possible to recover the transverse shape of the chirped X-type pulse intensity; actually, it is possible to recover its entire spatial shape after a distance $Z_{T_1=T_0}$. To see this, let us go back to our integral solution (2.54), and perform the change of coordinates $(z,t) \to (\Delta z, t_c = z_c/V_g)$, with

$$z = z_c + \Delta z, \qquad t = t_c \equiv \frac{z_c}{V_g}, \tag{2.58}$$

where z_c is the center of the pulse (Δz is the distance from such a point) and t_c is the time at which the pulse center is located at z_c. In this way, $\Psi(\rho, z, t)$ becomes $\Psi(\rho, t_c + \Delta z, z_c/V_g)$; that is, $\Psi(\rho, z_c, \Delta z)$. We are going to compare our integral solution (2.54) when $z_c = 0$ (initial pulse) with that when $z_c = Z_{T_1=T_0} = -2CT_0^2/[\beta_2(C_2+1)]$. In this way, the solution (2.54) can be written, when $z_c = 0$, as

$$\begin{aligned}\Psi(\rho, z_c = 0, \Delta z) = {} & \frac{T_0 \exp(i\beta_0\,\Delta z)}{\sqrt{2\pi(1+iC)}} \int_{-\infty}^{\infty} d\omega\, J_0(k_\rho(\omega)\rho) \exp\left[\frac{-T_0^2\,(\omega-\omega_0)^2}{2(1+C^2)}\right] \\ & \times \exp\left\{ i \left[\frac{(\omega-\omega_0)\,\Delta z}{V_g} + \frac{(\omega-\omega_0)^2 \beta_2\,\Delta z}{2} + \frac{(\omega-\omega_0)^2 T_0^2 C}{2(1+C^2)} \right] \right\},\end{aligned} \tag{2.59}$$

where we have taken the value q given by (2.52).

To verify that the pulse intensity recovers its entire original form at $z_c = Z_{T_1=T_0} = -2\,CT_0^2/[\beta_2(C^2+1)]$, we can analyze our integral solution at that point, obtaining

$$\begin{aligned}\Psi(\rho, z_c = Z_{T_1=T_0}, \Delta z) & \\ = {} & \frac{T_0 \exp\left\{ i\beta_0 \left[z_c - \Delta z' - \dfrac{c z_c}{\cos\theta\, n(\omega_0) V_g} \right] \right\}}{\sqrt{2\pi(1+iC)}} \\ & \times \int_{-\infty}^{\infty} d\omega\, J_0(k_\rho(\omega)\rho) \exp\left[\frac{-T_0^2\,(\omega-\omega_0)^2}{2(1+C^2)}\right] \\ & \times \exp\left\{ -i \left[\frac{(\omega-\omega_0)\,\Delta z'}{V_g} + \frac{(\omega-\omega_0)^2 \beta_2\,\Delta z'}{2} + \frac{(\omega-\omega_0)^2 T_0^2 C}{2(1+C^2)} \right] \right\},\end{aligned} \tag{2.60}$$

where we put $\Delta z = -\Delta z'$. In this way, we see immediately that

$$|\Psi(\rho, z_c = 0, \Delta z)|^2 = |\Psi(\rho, z_c = Z_{T_1=T_0}, -\Delta z)|^2. \tag{2.61}$$

Therefore, from Eq. (2.61) it is clear that the chirped optical X-type pulse intensity will recover its original three-dimensional form, with just a longitudinal inversion at the pulse center: the present method being, in this way, a simple and effective procedure for compensating the effects of diffraction and dispersion in an unbounded material medium and a method simpler than that of varying the axicon angle with the frequency. Let us stress that we can choose the distance $z = Z_{T_1=T_0} \leq L_{\text{disp}}$ at which the pulse will again take on its spatial shape by choosing a suitable value of the chirp parameter.

Up to now we have shown that the chirped X-type pulse recovers its three-dimensional shape after some distance, and we have also obtained an analytical description of the pulse longitudinal behavior (for $\rho = 0$) during propagation, by means of Eq. (2.56). However, we do not get the same information about pulse transverse behavior: We just know that it is recovered at $z = Z_{T_1=T_0}$.

So, to complete the picture, it would be interesting if we could also find the transverse behavior in the plane of the pulse center $z = V_g t$. In that way we would obtain quantitative information about the evolution of the pulse shape during its entire propagation. We do not examine the mathematical details here; we just affirm that transverse behavior of the pulse (in the plane $z = z_c = V_g t$) during its entire propagation can be described approximately by

$$\begin{aligned}\Psi&\left(\rho, z = z_c, t = \frac{z_c}{V_g}\right)\\ &\approx \frac{T_0 \exp[i\beta(\omega_0)z]\exp(-i\omega_0 t)}{\sqrt{2\pi(1+iC)}} \frac{\exp\left[(-\tan^2\theta\rho^2)/8V_g^2(-i\beta_2 z_c/2 + q^2)\right]}{\sqrt{-i\beta_2 z_c/2 + q^2}}\\ &\times \left[\Gamma(1/2)J_0\left(\frac{n(\omega_0)\omega_0\sin\theta\rho}{c}\right) I_0\left(\frac{\tan^2\theta\rho^2}{8\,V_g^2(-i\beta_2 z_c/2 + q^2)}\right)\right.\\ &+ 2\sum_{p=1}^{\infty} \frac{2^p\Gamma(p+1/2)\Gamma(p+1)}{\Gamma(2p+1)} J_{2p}\left(\frac{n(\omega_0)\,\omega_0\,\sin\theta\,\rho}{c}\right)\\ &\left.\times\, I_{2p}\left(\frac{\tan^2\theta\,\rho^2}{8\,V_g^2(-i\beta_2 z_c/2 + q^2)}\right)\right],\end{aligned} \tag{2.62}$$

where $I_p(\cdot)$ is the modified Bessel function of the first kind of order p, the quantity $\Gamma(\cdot)$ being the gamma function and q given by (2.52). The interested reader may consult [22] for details on how to obtain Eq. (2.62) from Eq. (2.54). At first glance this solution could appear to be very complicated, but *the series on its right-hand side provides a negligible contribution*. This fact renders our solution (2.62) of important practical interest and we use it in the following.

For additional information about the transverse pulse evolution [extracted from Eq. (2.62)] reader's are referred to [22], where the effects of finite aperture generation on chirped X-type pulses are analyzed.

2.4.1 Example: Chirped Optical X-Type Pulse in Bulk Fused Silica

For bulk fused silica, the refractive index $n(\omega)$ can be approximated by the Sellmeier equation [23],

$$n^2(\omega) = 1 + \sum_{j=1}^{N} \frac{B_j\,\omega_j^2}{\omega_j^2 - \omega^2}, \tag{2.63}$$

where ω_j are the resonance frequencies, B_j the strength of the j th resonance, and N the total number of material resonances that appear in the frequency range of interest. For our purposes it is appropriate to choose $N = 3$, which yields for bulk fused silica [23] the values $B_1 = 0.6961663$, $B_2 = 0.4079426$, $B_3 = 0.8974794$, $\lambda_1 = 0.0684043\ \mu\text{m}$, $\lambda_2 = 0.1162414\ \mu\text{m}$, and $\lambda_3 = 9.896161\ \mu\text{m}$.

Now, let us consider in this medium a chirped X-type pulse with $\lambda_0 = 0.2\ \mu\text{m}$, $T_0 = 0.4$ ps, $C = -1$, and an axicon angle $\theta = 0.00084$ rad, which correspond to an initial central spot with $\Delta\rho_0 = 0.117$ mm. From Eqs. (2.56) and (2.62) we get the longitudinal and transverse pulse evolution, which are represented in Fig. 2.3.

From Fig. 2.3a we note that initially, the pulse suffers a longitudinal narrowing with increased intensity up to the position $z = T_0^2/2\beta_2 = 0.186$ m. After this point the pulse starts to broaden, decreasing its intensity and recovering its entire longitudinal shape (width and intensity) at the point $z = T_0^2/\beta_2 = 0.373$ m, as was predicted. At the same time, from Fig. 2.3b we note that the pulse maintains its transverse width $\Delta\rho = 2.4\,c/[n(\omega_0)\omega_0 \sin\theta] = 0.117$ mm (because $T_0\omega_0 >> 1$) during its entire propagation; however, the same does not occur with pulse intensity. Initially, the

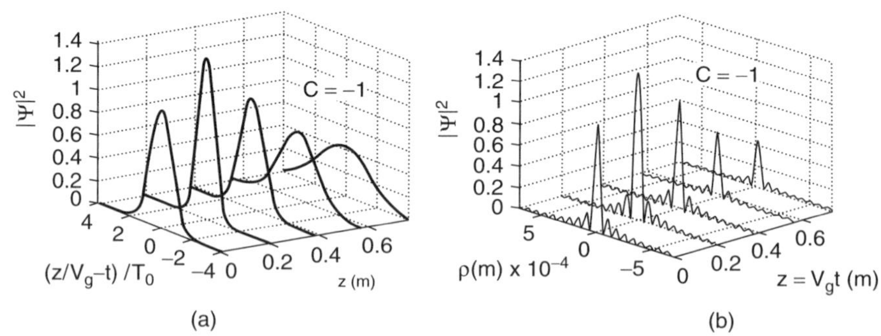

FIGURE 2.3 (a) Longitudinal-shape evolution of a chirped X-type pulse propagating in fused silica with $\lambda_0 = 0.2\ \mu\text{m}$, $T_0 = 0.4$ ps, $C = -1$, and axicon angle $\theta = 0.00084$ rad, which correspond to an initial transverse width of $\Delta\rho_0 = 0.117$ mm. (b) Transverse-shape evolution for the same pulse.

pulse suffers increased intensity up to the position $z_c = T_0^2/2\beta_2 = 0.186$ m; after this point, the intensity begins to decrease, and the pulse recovers its entire transverse shape at the point $z_c = T_0^2/\beta_2 = 0.373$ m, as we expected. Here we have skipped the series on the right-hand side of Eq. (2.62), because as we said earlier, it yields a negligible contribution.

Summarizing, from Fig. 2.3 we can see that the chirped X-type pulse recovers its longitudinal and transverse shape totally at the position $z = L_{\text{disp}} = T_0^2/\beta_2 = 0.373$ m, as we expected. Let us remember that a chirped Gaussian pulse may just recover its longitudinal width, but with an intensity decrease, at the position given by $z = Z_{T_1=T_0} = L_{\text{disp}} = T_0^2/\beta_2$. Its transverse width, on the other hand, suffers progressive and irreversible spreading.

2.5 MODELING THE SHAPE OF STATIONARY WAVE FIELDS: FROZEN WAVES

In this section we develop a very simple method [10,16,17] using superposition of forward propagating and equal-frequency Bessel beams that allows us to control the beam-intensity longitudinal shape within a chosen interval $0 \leq z \leq L$, where z is the propagation axis and L can be much greater than the wavelength λ of the monochromatic light (or sound) that is being used. Inside such a space interval, indeed, we succeed in constructing a stationary envelope whose longitudinal intensity pattern can assume any approximate shape desired, including, for instance, one or more high-intensity peaks, which is also naturally endowed with a good transverse localization. Since the intensity envelopes remains static (i.e., with velocity $V = 0$), we call such new solutions [10,16,17] to the wave equations *frozen waves*.

Although we are dealing here with exact solutions of the scalar wave equation, vectorial solutions of the same type can be obtained for the electromagnetic field, since solutions to Maxwell's equations follow naturally from scalar wave equation solutions [24,25]. We first present a method involving lossless media [16,17] and then extend the method to absorbing media [10].

2.5.1 Stationary Wave Fields with Arbitrary Longitudinal Shape in Lossless Media Obtained by Superposing Equal-Frequency Bessel Beams

We begin with the well-known axially-symmetric zeroth-order Bessel beam solution to the wave equation,

$$\psi(\rho, z, t) = J_0(k_\rho \rho) e^{i\beta z} e^{-i\omega t}, \tag{2.64}$$

with

$$k_\rho^2 = \frac{\omega^2}{c^2} - \beta^2, \tag{2.65}$$

where ω, k_ρ, and β are the angular frequency and the transverse and longitudinal wave numbers, respectively. We also impose the conditions

$$\frac{\omega}{\beta} > 0 \quad \text{and} \quad k_\rho^2 \geq 0 \tag{2.66}$$

(which imply that $\omega/\beta \geq c$) to ensure forward propagation only (with no evanescent waves), as well as the physical behavior of the Bessel function J_0.

Now let us make a superposition of $2N + 1$ Bessel beams with the same frequency ω_0 but with different (and still unknown) longitudinal wave numbers β_m:

$$\Psi(\rho, z, t) = e^{-i\,\omega_0\,t} \sum_{m=-N}^{N} A_m\, J_0(k_{\rho\,m}\rho)\, e^{i\beta_m z}, \tag{2.67}$$

where m represents integer numbers and the A_m are constant coefficients. For each m, the parameters ω_0, $k_{\rho\,m}$, and β_m must satisfy (2.65), and, because of conditions (2.66), when considering $\omega_0 > 0$, we must have

$$0 \leq \beta_m \leq \frac{\omega_0}{c}\,. \tag{2.68}$$

Let us now suppose that we wish $|\Psi(\rho, z, t)|^2$, given by Eq. (2.67), to assume on the axis $\rho = 0$ the pattern represented by a function $|F(z)|^2$ inside the chosen interval $0 \leq z \leq L$. In this case, the function $F(z)$ can be expanded, as usual, in a Fourier series:

$$F(z) = \sum_{m=-\infty}^{\infty} B_m\, e^{i(2\pi/L)mz}\,,$$

where

$$B_m = \frac{1}{L}\int_0^L F(z)\, e^{-i(2\pi/L)mz}\, dz\,.$$

More precisely, our goal is to find the values of the longitudinal wave numbers β_m and the coefficients A_m of (2.67) in order to reproduce approximately, within the interval $0 \leq z \leq L$ (for $\rho = 0$), the predetermined longitudinal intensity pattern $|F(z)|^2$. That is, we wish to have

$$\left|\sum_{m=-N}^{N} A_m e^{i\beta_m z}\right|^2 \approx |F(z)|^2 \qquad \text{with } 0 \leq z \leq L. \tag{2.69}$$

Looking at Eq. (2.69), one might be tempted to take $\beta_m = 2\pi m/L$, thus obtaining a truncated Fourier series, expected to represent approximately the desired pattern $F(z)$. Superpositions of Bessel beams with $\beta_m = 2\pi m/L$ have actually been used in some work to obtain a large set of *transverse* amplitude profiles [26]. However, for our purposes, this choice is not appropriate, for two principal reasons: (1) It yields negative values for β_m (when $m < 0$), which implies backward-propagating components (since $\omega_0 > 0$); and (2) when $L >> \lambda_0$, which is of interest here, the main terms of the series would correspond to very small values of β_m, which results in a very short depth of field of the corresponding Bessel beams (when generated by finite apertures), preventing creation of the desired envelopes far from the source.

Therefore, we need to make a choice for the values of β_m that permits forward propagation components only and a good depth of field. This problem can be solved by putting

$$\beta_m = Q + \frac{2\pi}{L} m, \tag{2.70}$$

where $Q > 0$ is a value to be chosen (as we shall see) according to the given experimental situation and the desired degree of transverse field localization. Due to Eq. (2.68), we get

$$0 \leq Q \pm \frac{2\pi}{L} N \leq \frac{\omega_0}{c}. \tag{2.71}$$

Inequality (2.71) can be used to determine the maximum value of m, which we call $N_{\max}$, once Q, L, and ω_0 have been chosen.

As a consequence, to get a longitudinal intensity pattern approximately equal to the one desired, $|F(z)|^2$, in the interval $0 \leq z \leq L$, Eq. (2.67) should be rewritten as

$$\Psi(\rho = 0, z, t) = e^{-i\omega_0 t}\, e^{iQz} \sum_{m=-N}^{N} A_m\, e^{i(2\pi/L)mz}, \tag{2.72}$$

with

$$A_m = \frac{1}{L} \int_0^L F(z)\, e^{-i(2\pi/L)mz}\, dz\,. \tag{2.73}$$

Obviously, one obtains only an approximation to the desired longitudinal pattern, because the trigonometric series (2.72) is necessarily truncated ($N \leq N_{\max}$). Its total number of terms, let us repeat, will be fixed once the values of Q, L, and ω_0 are chosen.

When $\rho \neq 0$, the wave field $\Psi(\rho, z, t)$ becomes

$$\Psi(\rho, z, t) = e^{-i\omega_0 t}\, e^{iQz} \sum_{m=-N}^{N} A_m\, J_0(k_{\rho\, m}\, \rho)\, e^{i\,(2\pi/L)m\, z}\,, \tag{2.74}$$

with

$$k_{\rho m}^2 = \omega_0^2 - \left(Q + \frac{2\pi m}{L}\right)^2 . \tag{2.75}$$

The coefficients A_m will yield the amplitudes and relative phases of each Bessel beam in the superposition.

Because we are adding together zero-order Bessel functions, we can expect a high field concentration around $\rho = 0$. Moreover, due to the known nondiffractive behavior of Bessel beams, we expect that the resulting wave field will preserve its transverse pattern in the entire interval $0 \le z \le L$.

The methodology developed here deals with longitudinal intensity pattern control. Obviously, we cannot get total three-dimensional control, due to the fact that the field must obey the wave equation. However, to have some control also over the transverse behavior, we can use two methods. The first is through the parameter Q of Eq. (2.70). Actually, we have some freedom in the choice of this parameter, and frozen waves representing the same longitudinal intensity pattern can possess different values of Q. The important point is that in superposition (2.74), using a smaller value of Q makes the Bessel beams possess a higher transverse concentration (because decreasing the value of Q increases the value of the Bessel beams transverse wave numbers), and this will be reflected in the resulting field, which will present a narrower central transverse spot. The second way to control the transverse intensity pattern is using higher-order Bessel beams, and we show this later.

Next, we look at a few examples of our methodology.

Example 1 Let us imagine that we have an optical wave field with $\lambda_0 = 0.632$ μm (i.e., with $\omega_0 = 2.98 \times 10^{15}$ Hz), whose longitudinal pattern (along its z-axis) in the range $0 \le z \le L$ is given by the function

$$F(z) = \begin{cases} -4\,\dfrac{(z-l_1)(z-l_2)}{(l_2-l_1)^2} & \text{for } l_1 \le z \le l_2 \\ 1 & \text{for } l_3 \le z \le l_4 \\ -4\,\dfrac{(z-l_5)(z-l_6)}{(l_6-l_5)^2} & \text{for } l_5 \le z \le l_6 \\ 0 & \text{elsewhere,} \end{cases} \tag{2.76}$$

where $l_1 = L/5 - \Delta z_{12}$ and $l_2 = L/5 + \Delta z_{12}$ with $\Delta z_{12} = L/50$, $l_3 = L/2 - \Delta z_{34}$ and $l_4 = L/2 + \Delta z_{34}$ with $\Delta z_{34} = L/10$, and $l_5 = 4L/5 - \Delta z_{56}$ and $l_6 = 4L/5 + \Delta z_{56}$ with $\Delta z_{56} = L/50$. In other words, the longitudinal shape desired, in the range $0 \le z \le L$, is a parabolic function for $l_1 \le z \le l_2$, a unitary step function for $l_3 \le z \le l_4$, and a parabola again in the interval $l_5 \le z \le l_6$, being zero elsewhere (within the interval $0 \le z \le L$, as we said). In this example, let us put $L = 0.2$ m.

We can then easily calculate the coefficients A_m in the superposition (2.74), by inserting (2.76) into (2.73). Let us choose, for instance, $Q = 0.999\,\omega_0/c$. This choice permits the maximum value $N_{\max} = 316$ for m, as one can infer from Eq. (2.71). Let

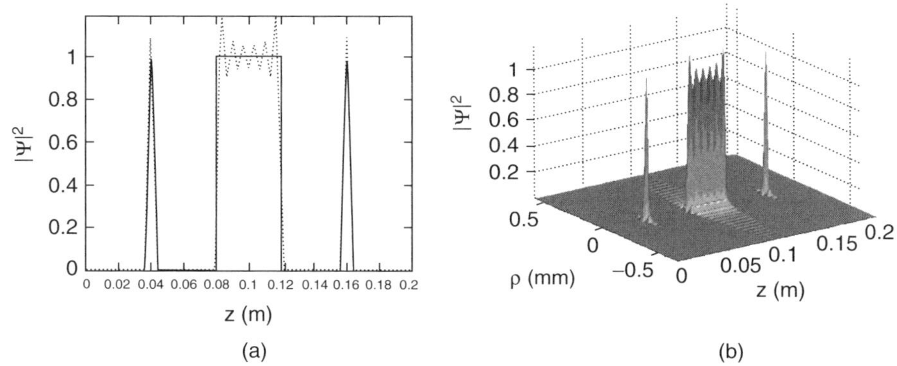

FIGURE 2.4 (a) Comparison of the intensity of the desired longitudinal function $F(z)$ and that of our frozen wave (FW), $\Psi(\rho = 0, z, t)$, obtained from Eq. (2.72). The solid line represents the function $F(z)$, and the dotted line our FW. (b) Three-dimensional plot of the field intensity of the FW chosen by us.

us emphasize that one is not compelled to use only $N = 316$, but can adopt for N any values *smaller* than 316: more generally, any value smaller than that calculated via inequality (2.71). Of course, using the maximum value allowed for N, one gets a better result.

In the present case, let us adopt the value $N = 30$. In Fig. 2.4a we compare the intensity of the longitudinal function $F(z)$ desired with that of the frozen wave, $\Psi(\rho = 0, z, t)$, obtained from Eq. (2.72) by adopting the value $N = 30$. One can verify that good agreement between the longitudinal behavior desired and our approximate FW is obtained with $N = 30$; the use of higher values only improves the approximation. Figure 2.4b shows the three-dimensional intensity of our FW, given by Eq. (2.74). One can observe that this field possesses the longitdinal pattern desired while being endowed with good transverse localization.

Example 2 We use this example to address an important concern. We can expect that for a desired longitudinal field intensity pattern, by choosing smaller values of the parameter Q, one will get FWs with a narrower transverse width [for the same number of terms in the series entering Eq. (2.74)], because the Bessel beams in Eq. (2.74) will possess larger transverse wave numbers, and consequently, higher transverse concentrations. We can verify this expectation by considering inside the usual range $0 \leq z \leq L$ the longitudinal pattern represented by the function

$$F(z) = \begin{cases} -4\,\dfrac{(z - l_1)(z - l_2)}{(l_2 - l_1)^2} & \text{for } l_1 \leq z \leq l_2 \\ 0 & \text{elsewhere,} \end{cases} \tag{2.77}$$

with $l_1 = L/2 - \Delta z$ and $l_2 = L/2 + \Delta z$. Such a function has a parabolic shape with its peak centered at $L/2$ and with longitudinal width $2\Delta z/\sqrt{2}$. By adopting

$\lambda_0 = 0.632\ \mu$m (i.e., $\omega_0 = 2.98 \times 10^{15}$ Hz), let us use superposition (2.74) with two different values of Q: We shall obtain two different FWs that despite having the same longitudinal intensity pattern, possess different transverse localizations. Namely, let us consider $L = 0.06$ m and $\Delta z = L/100$ and the two values $Q = 0.999\,\omega_0/c$ and $Q = 0.995\,\omega_0/c$. In both cases the coefficients A_m will be the same, calculated from Eq. (2.73), this time using the value $N = 45$ in superposition (2.74). The results are shown in Fig. 2.5a and b. Both FWs have the same longitudinal intensity pattern, but the one with the smaller Q is endowed with a narrower transverse width.

With this, we can get some control on the transverse spot size through the parameter Q. Actually, Eq. (2.74), which defines our FW, is a superposition of zero-order Bessel beams, and due to this fact, the resulting field is expected to possess a transverse localization around $\rho = 0$. Each Bessel beam in superposition (2.74) is associated with a central spot with transverse size, or width, $\Delta\rho_m \approx 2.4/k_{\rho\,m}$. On the basis of the convergence of series (2.74) expected, we can estimate the width of the transverse spot of the resulting beam as being

$$\Delta\rho \approx \frac{2.4}{k_{\rho\,m=0}} = \frac{2.4}{\sqrt{\omega_0^2/c^2 - Q^2}}, \tag{2.78}$$

which is the same value as that for the transverse spot of the Bessel beam with $m = 0$ in superposition (2.74). Relation (2.78) can be useful: Once we have chosen the longitudinal intensity pattern desired, *we can even choose the size of the transverse spot and use relation (2.78) to evaluate the needed corresponding value of parameter Q.*

For more detailed analysis of the spatial resolution and residual intensity of frozen waves, we refer readers to [17].

Increasing Control of the Transverse Shape Using Higher-Order Bessel Beams Here we argue that it is possible to increase our control of the transverse shape even more by using higher-order Bessel beams in our fundamental superposition (2.74). This new approach can be understood and accepted on the basis of simple and intuitive arguments, which are presented in [17]. A brief description of that approach follows.

The basic idea is to obtain the desired longitudinal intensity pattern not along the axis $\rho = 0$ but on a cylindrical surface corresponding to $\rho = \rho' > 0$. To do this, we proceed as before: Once we have chosen the longitudinal intensity pattern $F(z)$ desired, within the interval $0 \le z \le L$, we calculate the coefficients A_m as before [i.e., $A_m = (1/L)\int_0^L F(z)\exp(-i2\pi mz/L)\,dz$] and $k_{\rho m} = \sqrt{\omega_0^2 - (Q + 2\pi m/L)^2}$. Then we replace the zero-order Bessel beams $J_0(k_{\rho\,m}\rho)$, in superposition (2.74) with higher-order Bessel beams, $J_\mu(k_{\rho\,m}\rho)$, to get

$$\Psi(\rho, z, t) = e^{-i\omega_0 t}\, e^{iQz} \sum_{m=-N}^{N} A_m\, J_\mu(k_{\rho m}\rho)\, e^{i(2\pi/L)mz}. \tag{2.79}$$

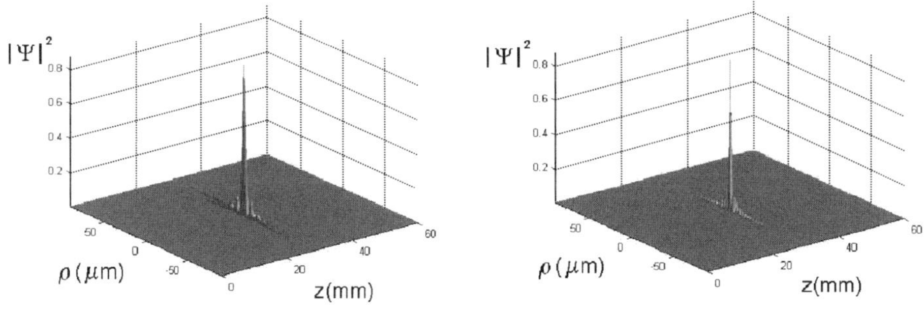

(a) (b)

FIGURE 2.5 (a) Frozen wave with $Q = 0.999\omega_0/c$ and $N = 45$, approximately reproducing the longitudinal pattern chosen, represented by Eq. (2.77). (b) A different frozen wave, now with $Q = 0.995\omega_0/c$ (but still with $N = 45$), forwarding the same longitudinal pattern. We can observe that in this case (with a lower value for Q) a higher transverse localization is obtained.

With this, and based on intuitive arguments [17], we can expect that the longitudinal intensity pattern desired, constructed initially for $\rho = 0$, will shift approximately to $\rho = \rho'$, where ρ' represents the position of the first maximum of the Bessel function [i.e., the first positive root of the equation $(dJ_\mu(k_{\rho\,m=0}\,\rho)/d\rho)|_{\rho'} = 0$]. Using such a procedure, one can obtain very interesting stationary configurations of field intensity, as "donuts," cylindrical surfaces, and much more.

In the following example, we show how to obtain, for example, a cylindrical surface of stationary light. To get it within the interval $0 \le z \le L$, let us first select the longitudinal intensity pattern given by Eq. (2.77), with $l_1 = L/2 - \Delta z$ and $l_2 = L/2 + \Delta z$, and with $\Delta z = L/300$. Moreover, let us choose $L = 0.05$ m, $Q = 0.998\,\omega_0/c$, and use $N = 150$.

Then, after calculating the coefficients A_m by Eq. (2.73), we have recourse to superposition (2.79). In this case we choose $\mu = 4$. According to the previous discussion, one can expect the desired longitudinal intensity pattern to appear shifted to $\rho' \approx 5.318/k_{\rho\,m=0} = 8.47$ μm, where 5.318 is the value of $k_{\rho\,m=0}\,\rho$ for which the Bessel function $J_4(k_{\rho\,m=0}\,\rho)$ assumes its maximum value, with $k_{\rho\,m=0} = \sqrt{\omega_0^2 - Q^2}$. Figure 2.6 shows the resulting intensity field. In Fig. 2.6a the transverse section of the resulting beam for $z = L/2$ is shown. The transverse peak intensity is located at $\rho = 7.75$ μm, with a 8.5% difference with respect to the predicted value of 8.47 μm. Figure 2.6b shows the orthogonal projection of the resulting field, which corresponds to a cylindrical surface of stationary light (or other fields).

We can see that the desired longitudinal intensity pattern has been obtained approximately, but, as desired, shifted from $\rho = 0$ to $\rho = 7.75$ μm, with the resulting field resembling a cylindrical surface of stationary light with radius 7.75 μm and length 238 μm. Donut-like configurations of light (or sound) are also possible.

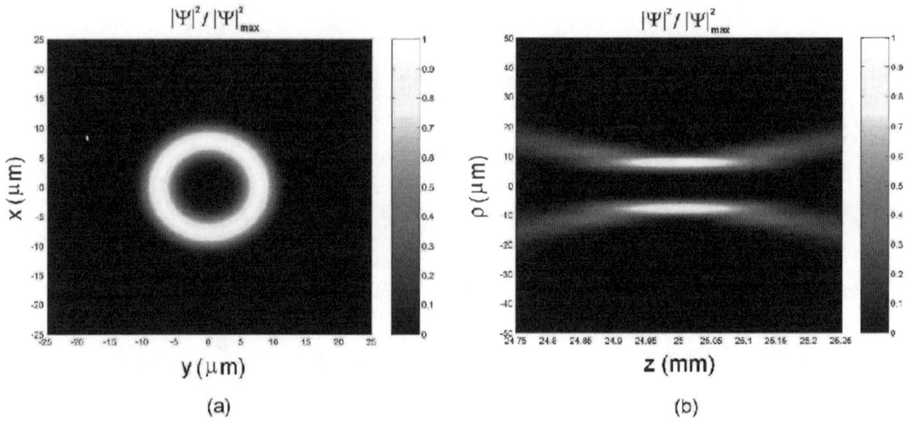

FIGURE 2.6 (a) Transverse section at $z = L/2$ of the higher-order FW. (b) Orthogonal projection of the three-dimensional intensity pattern of the same higher-order FW.

2.5.2 Stationary Wave Fields with Arbitrary Longitudinal Shape in Absorbing Media: Extending the Method

When propagating in a nonabsorbing medium, nondiffracting waves maintain their spatial shape for long distances. However, the situation is not the same when dealing with absorbing media. In this case, both ordinary and nondiffracting beams (and pulses) will suffer the same effect: an exponential attenuation along the propagation axis.

Here, we are going to extend [10] the method given above to show that through suitable superpositioning of equal-frequency Bessel beams, it is possible to obtain nondiffracting beams, in *absorbing media*, whose longitudinal intensity pattern can assume any shape desired within a chosen interval $0 \leq z \leq L$ of the propagation axis z. As a particular example, we obtain new nondiffracting beams capable of resisting the loss effects, maintaining the amplitude and spot size of their central core for long distances.

It is important to stress that this new method involves no active participation from the material medium. Actually, the energy absorption by the medium continues to occur normally, the difference being that these new beams have an initial transverse field distribution such that even in the presence of absorption, their central cores can be reconstructed for distances considerably longer than the penetration depths of ordinary (nondiffracting or diffracting) beams. In this sense, the present method can be regarded as extending, for absorbing media, the self-reconstruction properties [27] that normal localized waves are known to possess in lossless media.

In the same way as for lossless media, we construct a Bessel beam with angular frequency ω and axicon angle θ in the absorbing materials by superposing plane waves that have the same angular frequency ω and whose wave vectors lie on the surface of a cone with vertex angle θ. The refractive index of the medium can be written as

$n(\omega) = n_R(\omega) + in_I(\omega)$, the quantity n_R being the real part of the complex refraction index and n_I the imaginary part responsible for the absorbing effects. With a plane wave, the penetration depth δ for the frequency ω is given by $\delta = 1/\alpha = c/2\omega n_I$, where α is the absorption coefficient.

In this way, a zero-order Bessel beam in dissipative media can be written as $\psi = J_0(k_\rho \rho) \exp(i\beta z) \exp(-i\omega t)$ with $\beta = n(\omega)\omega \cos\theta/c = n_R\omega\cos\theta/c + in_I\omega\cos\theta/c \equiv \beta_R + i\beta_I$ and $k_\rho = n_R\omega \sin\theta/c + in_I\omega\sin\theta/c \equiv k_{\rho R} + ik_{\rho I}$, so $k_\rho^2 = n^2\omega^2/c^2 - \beta^2$. In this way, $\psi = J_0((k_{\rho R} + ik_{\rho I})\rho) \exp(i\beta_R z) \exp(-i\omega t) \exp(-\beta_I z)$, where β_R and $k_{\rho R}$ are the real parts of the longitudinal and transverse wavenumbers, and β_I and $k_{\rho I}$ are the imaginary parts. The absorption coefficient of a Bessel beam with an axicon angle θ is given by $\alpha_\theta = 2\beta_I = 2n_I\omega\cos\theta/c$, its penetration depth being $\delta_\theta = 1/\alpha_\theta = c/2\omega n_I \cos\theta$.

Due to the fact that k_ρ is complex, the amplitude of the Bessel function $J_0(k_\rho \rho)$ starts decreasing from $\rho = 0$ up to the transverse distance $\rho = 1/2k_{\rho I}$, and afterward it starts growing exponentially. This behavior is not physically acceptable, but one must remember that it occurs only because of the fact that an ideal Bessel beam requires an infinite aperture in order to be generated. However, in any real situation, when a Bessel beam is generated by finite apertures, that exponential growth in the transverse direction, starting after $\rho = 1/2k_{\rho I}$, will *not* occur indefinitely, stopping at a given value of ρ. Moreover, we emphasize that when generated by a finite aperture of radius R, the truncated Bessel beam [17] possesses a depth of field $Z = R/\tan\theta$ and can be described approximately by the solution given in the preceding paragraph for $\rho < R$ and $z < Z$.

Experimentally, to guarantee that exponential growth in the transverse direction does not even start, so as to generate only decreasing transverse intensity, the radius R of the aperture used to generate the Bessel beam should be $R \leq 1/2k_{\rho I}$. However, as noted by Durnin et al. [3], the same aperture also has to satisfy the relation $R \geq 2\pi/k_{\rho R}$. From these two conditions, we can infer that in an absorbing medium, a Bessel beam with decreasing transverse intensity can be generated only when the absorption coefficient is $\alpha < 2/\lambda$ (i.e., if the penetration depth is $\delta > \lambda/2$). The method developed in this subsection does refer to these cases; that is, we can always choose a suitable finite aperture size such that the truncated versions of all solutions presented here, including the general solution given by Eq. (2.80), will present no nonphysical behavior. Let us now present our method.

Consider an absorbing medium with the complex refraction index $n(\omega) = n_R(\omega) + in_I(\omega)$ and the following superposition of $2N + 1$ Bessel beams with the same frequency ω:

$$\Psi(\rho, z, t) = \sum_{m=-N}^{N} A_m J_0\left((k_{\rho R_m} + ik_{\rho I_m})\rho\right) e^{i\beta_{R_m} z} e^{-i\omega t} e^{-\beta_{I_m} z}, \qquad (2.80)$$

where m represents integer numbers, A_m constant (yet unknown) coefficients, and β_{R_m} and $k_{\rho R_m}$ (β_{I_m} and $k_{\rho I_m}$) the real parts (the imaginary parts) of the complex longitudinal and transverse wave numbers of the mth Bessel beam in superposition (2.80), with the following relations being satisfied:

$$k_{\rho_m}^2 = n^2 \frac{\omega^2}{c^2} - \beta_m^2 \tag{2.81}$$

$$\frac{\beta_{R_m}}{\beta_{I_m}} = \frac{n_R}{n_I}, \tag{2.82}$$

where $\beta_m = \beta_{R_m} + i\beta_{I_m}$ and $k_{\rho_m} = k_{\rho R_m} + ik_{\rho I_m}$, with $k_{\rho R_m}/k_{\rho I_m} = n_R/n_I$. Our goal now is to find the values of the longitudinal wave numbers β_m and the coefficients A_m in order to reproduce approximately, inside the interval $0 \leq z \leq L$ (on the axis $\rho = 0$), a freely chosen longitudinal intensity pattern that we call $|F(z)|^2$.

The problem for the particular case of lossless media [16,17] (i.e., when $n_I = 0 \rightarrow \beta_{I_m} = 0$) was solved earlier. For those cases it was shown that the choice $\beta = Q + 2\pi m/L$, with $A_m = \int_0^L F(z) \exp(-i2\pi mz/L)/L \, dz$, can be used to provide approximately the desired longitudinal intensity pattern $|F(z)|^2$ on the propagation axis within the interval $0 \leq z \leq L$, and at same time to regulate the spot size of the resulting beam by means of the parameter Q, which can also be used to obtain large depths of field and to validate the linear polarization approximation to the electric field in a transverse electromagnetic wave (see details in [16,17]).

However, when dealing with absorbing media, the procedure just described does not work, due to the presence of the functions $\exp(-\beta_{I_m} z)$ in the superposition (2.80), since in this case that series does not became a Fourier series when $\rho = 0$. To attempt to overcome this limitation, let us write the real part of the longitudinal wave number, in superposition (2.80), as

$$\beta_{R_m} = Q + \frac{2\pi m}{L}, \tag{2.83}$$

with

$$0 \leq Q + \frac{2\pi m}{L} \leq n_R \frac{\omega}{c}. \tag{2.84}$$

This inequality guarantees only forward propagation, with no evanescent waves. The superposition (2.80) can now be written as

$$\Psi(\rho, z, t) = e^{-i\omega t} e^{iQz} \sum_{m=-N}^{N} A_m J_0 \left((k_{\rho R_m} + ik_{\rho I_m})\rho \right) e^{i(2\pi m/L)z} e^{-\beta_{Im} z}, \tag{2.85}$$

where, by using (2.82), we have $\beta_{I_m} = (Q + 2\pi m/L)n_I/n_R$, and $k_{\rho_m} = k_{\rho R_m} + ik_{\rho I_m}$ is given by (2.81). Obviously, the discrete superposition (2.85) could be written as a continuous superposition (i.e., as an integral over β_{R_m}) by taking $L \rightarrow \infty$, but we prefer the discrete sum due to the difficulty of obtaining closed-form solutions to the integral form.

Now, let us examine the imaginary part of longitudinal wave numbers. The minimum and maximum values among the β_{I_m} are $(\beta_I)_{\text{min}} = (Q - 2\pi N/L)n_I/n_R$ and $(\beta_I)_{\text{max}} = (Q + 2\pi N/L)n_I/n_R$, the central value being given by $\overline{\beta}_I \equiv (\beta_I)_{m=0} = Qn_I/n_R$. With this in mind, let us evaluate the ratio $\Delta = [(\beta_I)_{\text{max}} - (\beta_I)_{\text{min}}]/\overline{\beta}_I = 4\pi N/LQ$.

Thus, when $\Delta << 1$, there are no substantial differences among the various β_{I_m}, since $\beta_{I_m} \approx \overline{\beta}_I$ holds for all m. In the same way, there are no substantial differences among the exponential attenuation factors, since $\exp(-\beta_{I_m} z) \approx \exp(-\overline{\beta}_I z)$. So when $\rho = 0$, the series in the right-hand side of Eq. (2.85) can be approximately considered a truncated Fourier series multiplied by the function $\exp(-\overline{\beta}_I z)$ and therefore superposition (2.85) can be used to reproduce approximately the desired longitudinal intensity pattern $|F(z)|^2$ (on $\rho = 0$), within $0 \le z \le L$, when the coefficients A_m are given by

$$A_m = \frac{1}{L} \int_0^L F(z)\, e^{\overline{\beta}_I z} e^{-i\,(2\pi m/L)z}\, dz, \tag{2.86}$$

the presence of the factor $\exp(\overline{\beta}_I z)$ in the integrand being necessary to compensate for the factors $\exp(-\beta_{I_m} z)$ in superposition (2.85). Since we are adding together zero-order Bessel functions, we can expect a good field concentration around $\rho = 0$.

In short, we have shown in this section how, in an absorbing medium, one can get a stationary wave-field with a good transverse concentration, whose longitudinal intensity pattern (on $\rho = 0$) can assume approximately any desired shape $|F(z)|^2$ within the predetermined interval $0 \le z \le L$. The method is a generalization of a previous method [16,17] and consists in the superposition of Bessel beams in (2.85), the real and imaginary parts of their longitudinal wave numbers being given by Eqs. (2.83) and (2.82) while their complex transverse wave numbers are given by Eq. (2.81) and the coefficients of the superposition are given by Eq. (2.86). The method is justified, since $4\pi N/LQ << 1$; happily enough, this condition is satisfied in many situations.

Regarding the generation of these new beams, given apparatus capable of generating a single Bessel beam, we can use an array of such apparatus to generate a group of beams with the appropriate longitudinal wave numbers and amplitudes/phases (as required by the method), thus producing the desired beam. For instance, we can use [16,17] a laser to illuminate an array of concentric annular apertures (located at the focus of a convergent lens), with the appropriate radii and transfer functions, able to yield both the correct longitudinal wave numbers (once a value for Q has been chosen) and the coefficients A_n of the fundamental superposition (2.85).

Some Examples For generality's sake, let us consider a hypothetical medium in which a typical XeCl excimer laser ($\lambda = 308$ nm $\rightarrow \omega = 6.12 \times 10^{15}$ Hz) has a penetration depth of 5 cm (i.e., an absorption coefficient $\alpha = 20\ \text{m}^{-1}$, and therefore $n_I = 0.49 \times 10^{-6}$). Also let us suppose that the real part of the refraction index for this wavelength is $n_R = 1.5$ and therefore that $n = n_R + in_I = 1.5 + i\,0.49 \times 10^{-6}$.

Note that the value of the real part of the refractive index is not very important since we are dealing with monochromatic wave fields.

A Bessel beam with $\omega = 6.12 \times 10^{15}$ Hz and an axicon angle $\theta = 0.0141$ rad (thus with a transverse spot of radius 8.4 μm), when generated by an aperture, say, of radius $R = 3.5$ mm, can propagate in vacuum a distance (its depth of field) equal to $Z = R/\tan\theta = 25$ cm while resisting diffraction effects. However, in the material medium considered here, the penetration depth of this Bessel beam would be only $z_p = 5$ cm. Next, we give two interesting applications of this method.

Example 1 Almost Undistorted Beams in Absorbing Media We can use the extended method to obtain, in the same medium and for the same wavelength, an almost undistorted beam capable of preserving its spot size and the intensity of its central core for a distance many times larger than the typical penetration depth of an ordinary beam (nondiffracting or not). Let us suppose that for this material medium, we wish to have a beam (with $\omega = 6.12 \times 10^{15}$ Hz) that maintains the amplitude and spot size of its central core for a distance of 25 cm (i.e., a distance five times greater than the penetration depth of an ordinary beam of the same frequency). We can model this beam by choosing the desired longitudinal intensity pattern $|F(z)|^2$ (on $\rho = 0$) within $0 \leq z \leq L$:

$$F(z) = \begin{cases} 1 & \text{for } 0 \leq z \leq Z \\ 0 & \text{elsewhere} \end{cases} \tag{2.87}$$

and by setting $Z = 25$ cm, with, for example, $L = 33$ cm. Now, Bessel beam superposition (2.85) can be used to reproduce this intensity pattern approximately. Let us choose $Q = 0.9999\omega/c$ for the β_{R_m} in (2.83), and $N = 20$ [note that according to inequality (2.84), N could assume a maximum value of 158]. Once we have chosen the values of Q, L, and N, the values of the complex longitudinal and transverse Bessel beams wave numbers happen to be defined by relations (2.83), (2.82), and (2.81). Eventually, we can use Eq. (2.86) and find the coefficients A_m of the fundamental superposition (2.85) that defines the resulting stationary wave field. Note that the condition $4\pi N/LQ << 1$ is perfectly satisfied in this case.

In Fig. 2.7a we see the three-dimensional field intensity of the resulting beam. One can see that the field possesses good transverse localization (with a spot size smaller than 10 μm) and is capable of maintaining the spot size and intensity of its central core up to the desired distance (a better result could be reached by using a higher value of N). It is interesting to note that at this distance (25 cm), an ordinary beam would have its initial field-intensity attenuated 148 times.

As we have said before, energy absorption by the medium continues to occur normally; the difference is that these new beams have an initial transverse field distribution that is sufficiently sophisticated to be able to reconstruct (even in the presence of absorption) their central cores up to a certain distance. For a better view of this field intensity distribution and of the energy flux, Fig. 2.7b shows the resulting beam in an orthogonal projection and logarithmic scale. It is clear that the energy comes from the lateral regions in order to reconstruct the central core of the beam. On the plane $z = 0$,

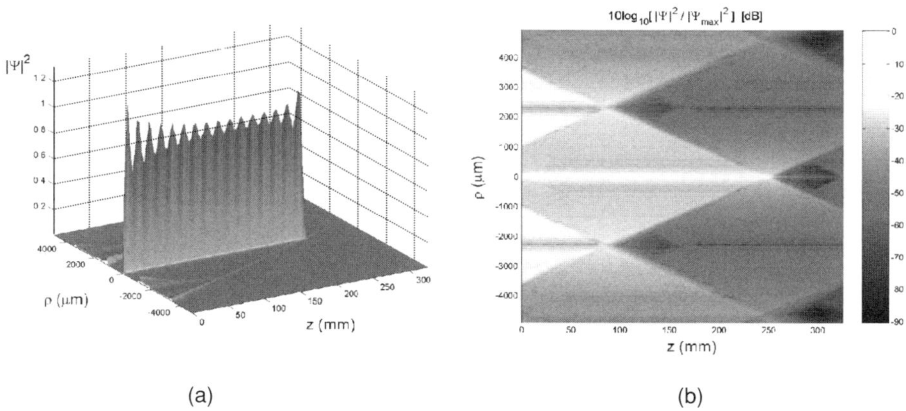

FIGURE 2.7 (a) Three-dimensional field intensity of the resulting beam. (b) The resulting beam, in an orthogonal projection and in logarithmic scale.

within the region $\rho \leq R = 3.5$ mm, there is a uncommon field-intensity distribution, being very disperse instead of concentrated. This uncommon initial field-intensity distribution is responsible for constructing the central core of the resulting beam and for its reconstruction up to the distance $z = 25$ cm. Due to the absorption, the beam (total) energy flowing through different z-planes is not constant, but the energy flowing through the beam spot area, and the beam spot size itself, are conserved up to (in this case) $z = 25$ cm.

Example 2 Beams in Absorbing Media with a Growing Longitudinal Field Intensity. Again considering a hypothetical medium in which an ordinary Bessel beam with $\theta = 0.0141$ rad and $\omega = 6.12 \times 10^{15}$ Hz has a penetration depth of 5 cm, we aim next to construct a beam that instead of possessing a constant core intensity up to $z = 25$ cm presents moderate exponential growth of its intensity up to that distance.

Let us assume that we wish to get a longitudinal intensity pattern $|F(z)|^2$ in the interval $0 < z < L$:

$$F(z) = \begin{cases} \exp(z/Z) & \text{for } 0 \leq z \leq Z \\ 0 & \text{elsewhere} \end{cases} \tag{2.88}$$

with $Z = 25$ cm and $L = 33$ cm. Again using $Q = 0.9999\omega/c$, $N = 20$, we proceed as in example 1, calculating the complex longitudinal and transverse Bessel beams wave numbers and finally, the coefficients A_m of the fundamental superposition (2.85). In Fig. 2.8 we see the three-dimensional field intensity of the resulting beam. One can see that the field presents the desired longitudinal intensity pattern with good transverse localization (a spot size smaller than 10 μm).

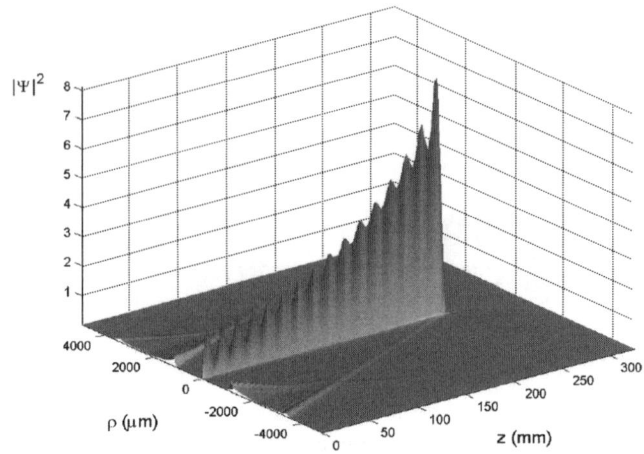

FIGURE 2.8 Three-dimensional field intensity of the resulting beam, in the absorbing medium, with a growing longitudinal field intensity.

Obviously, the amount of energy necessary to construct these new beams is greater than that necessary to generate an ordinary beam in a nonabsorbing medium. It is also clear that there is a limitation on the depth of field of these new beams. In example 1, for distances longer than 10 times the penetration depth of an ordinary beam, in addition to a greater energy demand, the field intensity in the lateral regions would be even higher than that of the core, and the field would lose the usual beam characteristics (i.e., transverse field concentration).

REFERENCES

1. A much more complete list of references may be found in Chapter 1.
2. J. N. Brittingham, Focus wave modes in homogeneous Maxwell's equations: transverse electric mode, *J. Appl. Phys.* **54**, 1179–1189 (1983).
3. J. Durnin, J. J. Miceli, and J. H. Eberly, Diffraction-free beams, *Phys. Rev. Lett.* **58**, 1499–1501 (1987).
4. J.-Y. Lu and J. F. Greenleaf, Nondiffracting X-waves: exact solutions to free-space wave equation and their finite aperture realizations, *IEEE Trans. Ultrason. Ferroelectr. Freq. Control* **39**, 19–31 (1992), and refs. therein.
5. H. Sõnajalg and P. Saari, Suppression of temporal spread of ultrashort pulses in dispersive media by Bessel beam generators, *Opt. Lett.* **21**, 1162–1164 (1996).
6. M. Zamboni-Rached, K. Z. Nóbrega, H. E. Hernández-Figueroa, and E. Recami, Localized superluminal solutions to the wave equation in (vacuum or) dispersive media, for arbitrary frequencies and with adjustable bandwidth, *Opt. Commun.* **226**, 15–23 (2003).
7. M. A. Porras, R. Borghi, and M. Santarsiero, Suppression of dispersion broadening of light pulses with Bessel–Gauss beams, *Opt Commun.* **206**, 235–241 (2003).

8. C. Conti, S. Trillo, P. Di Trapani, G.Valiulis, A. Piskarskas, O. Jedrkieicz, and J. Trull, Nonlinear electromagnetic X-waves, *Phys. Rev. Lett.* **90**, 170406 (May 2003), and refs. therein.
9. J. Salo, J. Fagerholm, A. T. Friberg, and M. M. Salomaa, Nondiffracting bulk-acoustic X waves in crystals, *Phys. Rev. Lett.* **83**, 1171–1174 (1999), and refs. therein.
10. M. Zamboni-Rached, Diffraction–attenuation resistant beams in absorbing media, *Opt. Express* **14**, 1804–1809 (2006).
11. I. M. Besieris, A. M. Shaarawi, and R. W. Ziolkowski, A bidirectional traveling plane wave representation of exact solutions of the scalar wave equation, *J. Math. Phys.* **30**, 1254–1269 (1989).
12. M. Zamboni-Rached, Localized waves: structure and applications, M.Sc. thesis, Physics Department, Unicamp, São Paulo, Brazil, 1999.
13. M. Zamboni-Rached, E. Recami, and H. E. Hernández-Figueroa, New localized superluminal solutions to the wave equations—with finite total energies and arbitrary frequencies, *Euro. Phys. J. D* **21**, 217–228 (2002).
14. M. Zamboni-Rached, Localized waves in diffractive/dispersive media, Ph.D. dissertation, DMO/FEEC Universidade Estadual de Campinas, Campinas, Brazil, Aug. 2004; http://libdigi.unicamp.br/document/?code=vtls000337794.
15. S. Longhi, Localized subluminal envelope pulses in dispersive media, *Opt. Lett.* **29**, 147–149 (2004), and refs. therein.
16. M. Zamboni-Rached, Stationary optical wave fields with arbitrary longitudinal shape by superposing equal frequency Bessel beams: frozen waves, *Opt. Expres* **12**, 4001–4006 (2004).
17. M. Zamboni-Rached, E. Recami, and H. E. Hernández-Figueroa, Theory of frozen waves: modelling the shape of stationary wave fields, *J. Opt. Soc. Am. A* **22**, 2465–2475.
18. S. V. Kukhlevsky and M. Mechler, Diffraction-free subwavelength-beam optics at nanometer scale, *Opt. Commun.* **231**, 35–43 (2004).
19. I. S. Gradshteyn and I. M. Ryzhik, *Integrals, Series and Products*, 4th ed., Academic Press, New York, 1965.
20. A. M. Shaarawi, I. M. Besieri, and T. M. Said, Temporal focusing by use of composite X-waves, *J. Opt. Soc. Am. A* **20**, 1658–1665, (Aug. 2003).
21. M. Zamboni-Rached, A. Shaarawi, and E. Recami, Focused X-shaped pulses, *J. Opt. Soc. Am. A* **21**, 1564–1574 (2004).
22. M. Zamboni-Rached, H. E. Hernández-Figueroa, and E. Recami, Chirped optical X-type pulses, *J. Opt. Soc. Am. A* **21**, 2455–2463 (2004).
23. G. Agrawal, *Nonlinear Fiber Optics*, 4th ed., Academic Press, San Diego, CA, 2006.
24. Z. Bouchal and M. Olivik, Non-diffractive vector Bessel beams, *J. Mod. Opt.* **42**, 1555–1566 (1995).
25. E.Recami, On localized X-shaped superluminal solutions to Maxwell equations, *Physica A*, **252**, 586–610 (1998).
26. Z. Bouchal, Controlled spatial shaping of nondiffracting patterns and arrays, *Opt. Lett.* **27**, 1376–1378 (2002).
27. R. Grunwald, U. Griebner, U. Neumann, and V. Kebbel, Self-reconstruction of ultrashort-pulse Bessel-like X-waves, presented at CLEO/QELS 2004, San Francisco, CA, May 16–21, 2004, *Conference Digest* (CD-ROM), paper CMQ7.

CHAPTER THREE

Two Hybrid Spectral Representations and Their Applications to the Derivations of Finite-Energy Localized Waves and Pulsed Beams

IOANNIS M. BESIERIS
Virginia Polytechnic Institute and State University, Blacksburg, Virginia

AMR M. SHAARAWI
The American University of Cairo, Cairo, Egypt

3.1 INTRODUCTION

Two spectral superposition methods are presented that allow smooth transitions between two seemingly distinct classes of localized waves: luminal or focus wave modes (FWMs), and superluminal or X-waves. First, a novel superluminal hybrid spectral representation is introduced which is based on the superposition of products of forward plane waves moving at a fixed speed $v > c$ and backward plane waves moving at the speed of light in a vacuum, c. This technique is used to construct a large class of finite-energy superluminal-type X-shaped localized waves. The latter are characterized by arbitrarily high-frequency bands and are suitable for applications in the microwave and optical regime. In the limiting case $v \to c$, one recaptures the well-known luminal FWM-type localized wave solutions. Analogously, a modified hybrid

Localized Waves, Edited by Hugo E. Hernández-Figueroa, Michel Zamboni-Rached, and Erasmo Recami

spectral representation is introduced which is based on the superposition of products of forward plane waves moving at a fixed speed c and backward plane waves moving at the speed $v > c$. In the limiting case $v \to c$, one recaptures variants of the well-known FWM-type localized wave solutions. In this framework, the modified hybrid spectral representation becomes an optimal setting for the derivation of a large class of finite-energy focused pulsed beams. The latter have properties closely resembling those arising in recent experiments involving focused terahertz pulsed beams.

The possibility of exact solutions to the scalar wave equation, Maxwell's equations, as well as other types of equations (e.g., Klein–Gordon, Schrödinger, Dirac) that describe ultrawideband, spatially localized, slowly decaying transmission of energy in space–time, has been suggested by several groups in recent years (see [1–8] for a review of the analytical literature and [9–13] for experimental verifications). Due to the aforementioned properties, such solutions may have potential applications in various research areas (e.g., remote sensing, directed energy transfer and secure communications, nondestructive evaluation, ultrafast microscopy, high-resolution imaging, tissue characterization, and photodynamic therapy). Of these solutions, the general classes of luminal or FWM-type and superluminal or X-shaped solutions, termed *localized waves* (LWs), can be obtained by means of, among other methods, two fundamental spectral superposition schemes: the bidirectional [2] and the superluminal [4] representations. Unfortunately, one cannot use the superluminal spectral representation to examine the limiting condition $v \to c$, due to a built-in singularity. Our specific aim in this chapter is to introduce two novel spectral representations involving the fixed speeds v and c, which allow a smooth transition between the seemingly disparate luminal (FWM-type) and X-shaped localized waves.

The chapter is structured as follows: A brief overview of the bidirectional and superluminal spectral representations is provided in Section 3.2, the superluminal hybrid and modified superluminal hybrid spectral methods are discussed in Sections 3.3 and 3.4, respectively, and concluding remarks are made in Section 3.5.

3.2 OVERVIEW OF BIDIRECTIONAL AND SUPERLUMINAL SPECTRAL REPRESENTATIONS

Consider, for simplicity, an azimuthally symmetric (with respect to the z-axis) homogeneous $(3+1)$-dimensional scalar wave equation in free space:

$$\frac{1}{\rho}\frac{\partial}{\partial\rho}\left[\rho\frac{\partial}{\partial\rho}\psi(\rho,z,t)\right]+\frac{\partial^2}{\partial z^2}\psi(\rho,z,t)-\frac{1}{c^2}\frac{\partial^2}{\partial t^2}\psi(\rho,z,t)=0, \tag{3.1}$$

where c denotes the speed of light in vacuo. In this section we review two fixed-speed spectral representations leading naturally to two seemingly distinct classes of localized pulse solutions to the scalar wave equation.

3.2.1 Bidirectional Spectral Representation

A general solution to the scalar wave equation (3.1) can be expressed in terms of the bidirectional spectral representation:

$$\psi(\rho, z, t) = \int_0^\infty d\alpha \int_0^\infty d\beta \int_0^\infty d\kappa\kappa \tilde{\psi}(\kappa, \alpha, \beta)\delta(\kappa^2 - 4\alpha\beta) J_0(\kappa\rho) e^{-i\alpha(z-ct)} e^{i\beta(z+ct)} \tag{3.2}$$

The wave function is given as a superposition of Bessel beams $J_0(\kappa\rho)\exp[-i\alpha(z - ct)]$ moving in the positive z-direction with speed c, modulated by plane waves $\exp[i\beta(z + ct)]$ moving in the negative z-direction with speed c.

The bidirectional spectral representation provides the most natural basis for synthesizing localized wave solutions of the focus wave mode type. A very broad class of axisymmetric luminal LW solutions is given as follows:

$$\psi(\rho, z, t) = \frac{1}{a_1 + i(z - ct)} f[\Theta(\rho, z, t)],$$
$$\Theta(\rho, z, t) \equiv a_2 - i(z + ct) + \frac{\rho^2}{a_1 + i(z - ct)}. \tag{3.3}$$

Here $a_{1,2}$ are free real positive parameters and $f(\cdot)$ is an arbitrary function. Two observations are in order regarding the class of solutions incorporated into Eq. (3.3):

1. The wave functions are clearly *bidirectional*, due to the presence of the two characteristic variables $z - ct$ and $z + ct$ of the $(1 + 1)$-dimensional scalar wave equation. However, the parameters $a_{1,2}$, as well as additional free parameters entering through the function $f(\cdot)$ in Eq. (3.3), can be "tweaked" so that the wave packets can propagate primarily in the positive z-direction. Very close replicas of such LWs can be launched causally from apertures constructed on the basis of the Huygens principle [14–16].

2. Depending on the choice of the function $f(\cdot)$ in Eq. (3.3), the resulting luminal LWs may have either infinite or finite energy content. The former propagate along the positive z-direction with only local deformations; more precisely, they regenerate periodically. The latter, on the other hand, can have extended ranges of localization in the near-to-far regions. In general, the axial localization near the pulse center is algebraic and the transverse localization is exponential. However, Kiselev and Perel [17] have shown that certain wave packets may possess both axial and transverse exponential localization.

The simplest luminal LW is the "pure" FWM

$$\psi_{\text{FWM}}(\rho, z, t) = \frac{1}{a_1 + i(z - ct)} \exp[i\beta(z + ct)] \exp\left[-\beta \frac{\rho^2}{a_1 + i(z - ct)}\right], \tag{3.4}$$

introduced by Brittingham [18] in 1983 (see also [19]). It follows from Eq. (3.3) by choosing $f(\cdot) = \exp[-\beta(\cdot)]$. In addition to a_1, the free real parameter β in Eq. (3.4) is also positive. The wave function ψ_{FWM} consists of an envelope traveling along the positive z-direction with speed c, modulated by a plane wave moving in the negative z-direction with speed c. The entire wave packet sustains only local deformations; more precisely, it regenerates for values of ct equal to $n\pi/\beta$, where n is an integer. The FWM is physically unrealizable because it contains infinite energy.

The choice of the complex analytic signal $f(\cdot) = \exp[-b(\cdot)][a_2 + (\cdot)]^{-q-1}$ in Eq. (3.3) leads to the finite-energy modified power spectrum (MPS) pulse

$$\psi_{\text{MPS}}(\rho, z, t) = \frac{1}{a_1 + i(z - ct)} \frac{\exp\left[-b\left(\dfrac{\rho^2}{a_1 + i(z - ct)} - i(z + ct)\right)\right]}{\left[a_2 + \left(\dfrac{\rho^2}{a_1 + i(z - ct)} - i(z + ct)\right)\right]^{q+1}}, \tag{3.5}$$

derived originally by Ziolkowski [20]. In addition to a_1 and a_2, the free real parameters b, p, and q are positive.

A variant of the solution in Eq. (3.3) is the following:

$$\begin{aligned} \psi(\rho, z, t) &= \frac{1}{a_2 - i(z + ct)} f\left[\bar{\Theta}(\rho, z, t)\right], \\ \bar{\Theta}(\rho, z, t) &\equiv a_1 + i(z - ct) + \frac{\rho^2}{a_2 - i(z + ct)}. \end{aligned} \tag{3.6}$$

An interesting aspect of this solution is that with the formal substitution $z + ct \to 2z$, the function

$$\begin{aligned} \psi_{\text{PB}}(\rho, z, t) &= \frac{1}{a_2 - i2z} f\left[\hat{\Theta}(\rho, z, t)\right], \\ \hat{\Theta}(\rho, z, t) &\equiv a_1 + i(z - ct) + \frac{\rho^2}{a_2 - i2z} \end{aligned} \tag{3.7}$$

obeys the pulsed beam equation [21,22]

$$\left(\frac{\partial^2}{\partial \rho^2} + \frac{1}{\rho}\frac{\partial}{\partial \rho} + 2\frac{\partial^2}{\partial \varsigma \partial z}\right) \psi_{\text{PB}}(\rho, z, \varsigma) = 0, \qquad \varsigma \equiv z - ct, \tag{3.8}$$

which, in turn, is the paraxial approximation of the scalar wave equation (3.1).

One can obtain from Eq. (3.6) the following variant FWM and MPS solutions:

$$\psi_{\text{VFWM}}(\rho, z, t) = \frac{1}{a_2 - i(z + ct)} \exp[-i\alpha(z - ct)] \exp\left[-\alpha \frac{\rho^2}{a_2 - i(z + ct)}\right] \tag{3.9}$$

$$\psi_{\text{VMPS}}(\rho, z, t) = \frac{1}{a_2 - i(z + ct)} \frac{\exp\left[-b\left(\dfrac{\rho^2}{a_2 - i(z + ct)} + i(z - ct)\right)\right]}{\left[a_1 + \left(\dfrac{\rho^2}{a_2 - i(z + ct)} + i(z - ct)\right)\right]^{q+1}}. \tag{3.10}$$

For $f(\cdot) = (\cdot)^{-q'}$, we obtain

$$\psi_{\mathrm{SP}}(\rho, z, t) = \frac{1}{a_2 - i(z + ct)} \left[a_1 + i(z - ct) + \frac{\rho^2}{a_2 - i(z + ct)} \right]^{-q'}. \quad (3.11)$$

This solution, following from the VMPS pulse under the restriction $b = 0$, is a variant of the $(q')^{\mathrm{th}}$-order *splash mode* introduced by Ziolkowski [20]. Recently, the solution given in Eq. (3.11) with $q' = 1$ has been used as a Hertz potential in an extensive study of the spatiotemporal evolution of focused single-cycle terahertz electromagnetic pulses [23]. Special attention has been paid to the limiting case $a_1 << a_2$, corresponding to the paraxial regime. The resulting pulsed beam solutions are natural spatiotemporal modes of an open electromagnetic cavity. As such, they can be generated in the laboratory using a curved mirror waveguide or by propagation along a lens waveguide.

3.2.2 Superluminal Spectral Representation

A general solution to the (3 + 1)-dimensional scalar wave equation (3.1) can be expressed in terms of the superluminal spectral representation [4]

$$\begin{aligned} \psi(\rho, z, t) = & \int_{-\infty}^{\infty} d\lambda \int_0^{\infty} d\kappa \kappa \tilde{\psi}(\kappa, \lambda) J_0(\kappa\rho) \exp\left[i\lambda\gamma \frac{v}{c} \left(z - \frac{c^2}{v} t \right) \right] \\ & \times \exp\left[-i\gamma(z - vt)\sqrt{\kappa^2 + \lambda^2} \right], \qquad \gamma = \frac{1}{\sqrt{(v/c)^2 - 1}}, \qquad v > c \end{aligned} \quad (3.12)$$

which is based on the invariance of the scalar wave equation under a generalized Lorentz transformation. This representation consists of superpositions of products of forward Bessel beams moving at a fixed speed $v > c$ and backward plane waves moving at the subluminal speed c^2/v.

The superluminal spectral representation is the most natural setting for deriving all X-shaped localized solutions to the scalar wave equation. The simplest superluminal solution is the zero-order X-wave:

$$\psi_{\mathrm{XW}}(\rho, z, t) = \frac{1}{\sqrt{\rho^2 + [\bar{a}_1 + i\gamma(z - vt)]^2}}, \qquad \bar{a}_1 > 0, \quad (3.13)$$

which was introduced independently by Lu and Greenleaf [24] and Ziolkowski et al. [14]. (In the latter reference, this solution was referred to as a *slingshot superluminal pulse.*) It is an infinite-energy LW pulse propagating without distortion along the z-direction with superluminal speed v.

Another infinite-energy X-shaped solution is given as follows:

$$\begin{aligned} \psi_{\mathrm{FXW}}(\rho, z, t) = & \frac{1}{\sqrt{\rho^2 + [\bar{a}_1 + i\gamma(z - vt)]^2}} \exp\left[i\beta\gamma \frac{v}{c} \left(z - \frac{c^2}{v} t \right) \right] \\ & \times \exp\left(-\beta\sqrt{\rho^2 + [\bar{a}_1 + i\gamma(z - vt)]^2} \right). \end{aligned} \quad (3.14)$$

This wave packet combines features present in both the zero-order X-wave [see Eq. (3.13)] and the focus wave mode (FWM) [see Eq. (3.4)]. For this reason it has been called a *focused X-wave* (FXW) [4]. It resembles a zero-order X-wave except that its highly focused central region has a tight exponential localization, in contrast to the loose algebraic transverse dependence of the zero-order X-wave.

A finite-energy superluminal LW is given as [4]

$$\psi_{\mathrm{MFXW}}(\rho, z, t) = \exp\left[ib\gamma\frac{v}{c}\left(z - \frac{c^2}{v}t\right)\right]\exp\left(-b\sqrt{\rho^2 + [\bar{a}_1 + i\gamma(z - vt)]^2}\right) \times\left[\bar{a}_2 - i\left(\gamma\frac{v}{c}\left(z - \frac{c^2}{v}t\right) + i\sqrt{\rho^2 + [\bar{a}_1 + i\gamma(z - vt)]^2}\right)\right]^{-q-1}, \tag{3.15}$$

where $\bar{a}_1, \bar{a}_2,\ b$, and q are positive real parameters. Due to its resemblance to the MPS pulse [cf. Eq. (3.5), it is referred to as the *modified focus X-wave* (MFXW).

3.3 HYBRID SPECTRAL REPRESENTATION AND ITS APPLICATION TO THE DERIVATION OF FINITE-ENERGY X-SHAPED LOCALIZED WAVES

Due to the presence of γ in the superluminal spectral representation in Eq. (3.12), one cannot examine the limiting condition $v \to c$. It seems, therefore, that the luminal solutions of the $(3+1)$-dimensional scalar wave derived from the bidirectional spectral representation and the superluminal solutions derived by means of the superluminal spectral representation are two different classes of spatiotemporally localized waves. In this section we show that this is not true by introducing a novel hybrid spectral representation involving the fixed speeds $v > c$ and c.

3.3.1 Hybrid Spectral Representation

In Eq. (3.12), the following transformations are made from the wave numbers (λ, κ) to the wave numbers (α, β):

$$\begin{aligned} -\gamma\sqrt{\lambda^2 + \kappa^2} + \gamma\lambda(c/v) &= \alpha + \beta(c/v), \\ -\gamma\sqrt{\lambda^2 + \kappa^2} + \gamma\lambda(v/c) &= \alpha - \beta. \end{aligned} \tag{3.16}$$

We then obtain

$$\psi(\rho, z, t) = \int_{-\infty}^{\infty} d\alpha \int_{-\infty}^{\infty} d\beta \int_{0}^{\infty} d\kappa\kappa\tilde{\psi}(\kappa, \alpha, \beta)\delta\left[-\kappa^2 + \frac{1}{\gamma^2}\alpha^2 + 2\left(1 + \frac{v}{c}\right)\alpha\beta\right] \times J_0(\kappa\rho)\exp[-i\alpha(z - vt)]\exp[i\beta(z + ct)]. \tag{3.17}$$

This *hybrid spectral representation* is based on superpositions of products of forward Bessel beams moving at a fixed speed $v > c$ and backward plane waves moving at the speed c. In the limiting case $v \to c$, one recaptures the bidirectional spectral representation given in Eq. (3.2), which leads to well-known FWM-type localized wave solutions.

3.3.2 (3 + 1)-Dimensional Focus X-Wave

It is convenient to assume that $\alpha, \beta > 0$ and integrate first with respect to κ in Eq. (3.17). One then obtains

$$\psi(\rho, z, t) = \int_0^\infty d\alpha \int_0^\infty d\beta \tilde{\psi}_1(\alpha, \beta) J_0 \left[\frac{\rho}{\gamma} \sqrt{\alpha^2 + 2\left(1 + \frac{v}{c}\right) \gamma^2 \alpha\beta} \right] \times \exp[-i\alpha(z - vt)] \exp[i\beta(z + ct)]. \quad (3.18)$$

The spectrum $\tilde{\psi}_1(\alpha, \beta)$ is chosen, next, as follows:

$$\tilde{\psi}_1(\alpha, \beta) = \delta(\beta - \beta') \exp(-a_1 \alpha). \quad (3.19)$$

Here $\delta(\cdot)$ denotes the Dirac delta function and a_1 and β' are arbitrary positive parameters. With the introduction of this spectrum, integration over β can be carried out easily and the remaining integration over α yields (see [25], p. 249) a variant of the FXW given in Eq. (3.14); specifically,

$$\begin{aligned} &\psi_{\mathrm{VFXW}}(\rho, z, t; \beta') \\ &= \frac{\exp[i\beta'(z + ct)]}{\sqrt{(\rho/\gamma)^2 + [a_1 + i(z - vt)]^2}} \exp\left[\gamma^2 \left(1 + \frac{v}{c}\right) \beta' \left[a_1 + i(z - vt)\right]\right] \\ &\quad \times \exp\left[-\gamma^2 \left(1 + \frac{v}{c}\right) \beta' \sqrt{(\rho/\gamma)^2 + [a_1 + i(z - vt)]^2}\right]. \end{aligned} \quad (3.20)$$

For $v > c$, this solution is an X-shaped transversely and axially localized pulse, except that its highly focused central region has a tight exponential localization, analogous to that of a FWM. For $\beta' = 0$, one obtains the zero-order X-wave given in Eq. (3.13) with $\bar{a}_1 = \gamma a_1$. Carrying out the limit $v \to c$ yields the FWM in Eq. (3.4) modulo a constant multiplier term. The main purpose of the hybrid form of $\psi_{\mathrm{VFXW}}(\rho, z, t; \beta')$ in Eq. (3.20) is to allow for such a smooth transition. An additional advantage of the hybrid form is that it obviates the presence of backward wave components propagating at the luminal speed c. To make this point clearer, the first two exponential terms in Eq. (3.20) can be combined to yield the following alternative form of the (3 + 1)-dimensional VFXW pulse [compare with Eq. (3.14)]:

$$\begin{aligned} \psi_{\mathrm{VFXW}}(\rho, z, t) = &\frac{\exp[\gamma^2(1 + v/c)\beta' a_1]}{\sqrt{(\rho/\gamma)^2 + [a_1 + i(z - vt)]^2}} \exp\left[i\beta' \left(1 - \frac{c}{v}\right)^{-1} \left(z - \frac{c^2}{v} t\right)\right] \\ &\times \exp\left[-\gamma^2 \left(1 + \frac{v}{c}\right) \beta' \sqrt{(\rho/\gamma)^2 + [a_1 + i(z - vt)]^2}\right]. \end{aligned} \quad (3.21)$$

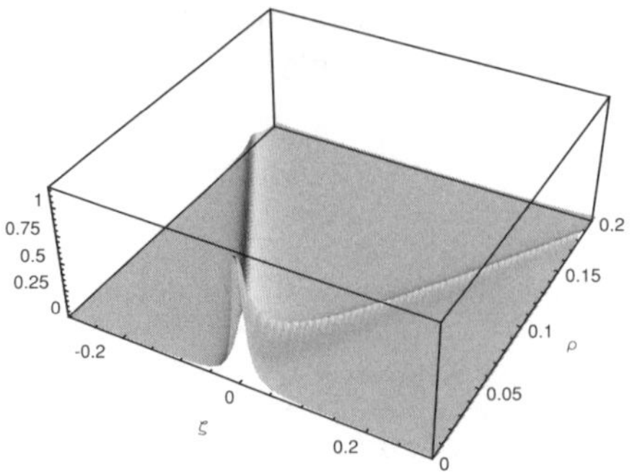

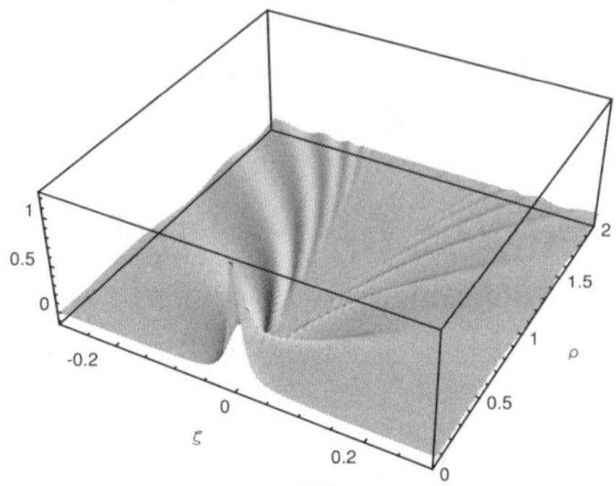

FIGURE 3.1 (a) (3 + 1)-Dimensional focus X-wave: $\mathrm{Re}[\psi_{\mathrm{VFXW}}(\rho, \varsigma; vt, \beta')]$ versus ρ, ς, with parameter values $\beta' = 1\,\mathrm{m}^{-1}$, $a_1 = 10^{-2}\,\mathrm{m}$, $v = 2c$, and $vt = 0$. (b) Transition to a (3 + 1)-dimensional focus wave mode: $\mathrm{Re}[\psi_{\mathrm{VFXW}}(\rho, \varsigma; vt, \beta')]$ versus ρ, ς with parameter values $a_1 = 10^{-2}\,\mathrm{m}$, $\beta' = 1\,\mathrm{m}^{-1}$, $vt = 0$, and $v = 1.0000001c$.

This expression seems to be unidirectional in the sense that it consists of an envelope traveling in the positive z-direction with superluminal speed v, multiplied by a plane wave also moving in the positive z-direction but at subluminal speed c^2/v. Of course, the unidirectionality of the wave packet is only apparent, as is made clear in the hybrid form.

The modulus of the (3 + 1)-dimensional VFXW, in either the hybrid form given in Eq. (3.20) or in the superluminal form given in Eq. (3.21), depends only on ρ and $\varsigma = z - vt$. As a consequence, it moves rigidly along the positive z-direction with superluminal speed v without sustaining distortion. This means that $\psi_{\mathrm{VFXW}}(\rho, z, t; \beta')$ contains infinite energy. The real part of this function regenerates periodically along the z-direction. The regeneration period equals $vt = 2\pi n/[\beta'(1 + (c/v))]$, where n is an integer. Figure 3.1a shows a surface plot of $\mathrm{Re}[\psi_{\mathrm{VFXW}}(\rho, \varsigma; vt, \beta')]$ for the parameter values $\beta' = 1\,\mathrm{m}^{-1}$, $a_1 = 10^{-2}\,\mathrm{m}$ and $v = 2c$. The X-shape of the (3 + 1)-dimensional VFXW is clearly evident. The transition of the wave packet into a (3 + 1)-dimensional FWM is depicted in Fig. 3.1b, where the values of β' and a_1 have been kept the same but $v = 1.0000001c$.

3.3.3 (3 + 1)-Dimensional Finite-Energy X-Shaped Localized Waves

Let $\omega' = \beta'/c$ (rad/s) and consider the superposition

$$\psi(\rho, z, t) = \frac{1}{\pi}\int_0^\infty d\omega'\, \psi_{\mathrm{VFXW}}(\rho, z, t; \omega')\tilde{G}(\omega'). \tag{3.22}$$

If $\hat{g}\,(t)$ denotes the complex analytic signal corresponding to the temporal spectrum $\tilde{G}(\omega')$, one obtains the general finite-energy (3 + 1)-dimensional X-shaped localized wave solution

$$\psi(\rho, z, t) = \hat{g}\left[\left(t + \frac{z}{c}\right) - i\frac{1}{c}\gamma^2\left(1 + \frac{v}{c}\right)[a_1 + i(z - vt)]\right.$$
$$\left.\times\sqrt{(\rho/\gamma)^2 + [a_1 + i(z - vt)]^2}\right]\frac{1}{\sqrt{(\rho/\gamma)^2 + [a_1 + i(z - vt)]^2}}. \tag{3.23}$$

As a specific illustrative example, consider the spectrum

$$\tilde{G}(\omega') = (\omega' - \omega_0)^q \exp[-a_3(\omega' - \omega_0)]H(\omega' - \omega_0) \tag{3.24}$$

with $a_3, \omega_0 > 0,\ q > -1$, and $H(\cdot)$ denoting the Heaviside unit step function. The complex analytic signal associated with this spectrum is given explicitly as follows ([25], p. 137):

$$\hat{g}(t) = \Gamma(q + 1)\exp[-\omega_0(a_3 - it)](a_3 - it)^{-q-1}. \tag{3.25}$$

When this result is used in conjunction with Eq. (3.23), one obtains a variant of the MFXW pulse given in Eq. (3.15) specifically,

$$\psi_{\text{VMFXW}}(\rho, z, t) = \Gamma(q + 1)\frac{1}{\sqrt{(\rho/\gamma)^2 + [a_1 + i(z - vt)]^2}}\frac{\exp(-\omega_0 p)}{p^{q+1}};$$
$$p = a_3 - i\left(t + \frac{z}{c}\right) - \frac{1}{c}\gamma^2\left(1 + \frac{v}{c}\right)$$
$$\times\left\{a_1 + i(z - vt) - \sqrt{(\rho/\gamma)^2 + [a_1 + i(z - vt)]^2}\right\}. \tag{3.26}$$

It should be noted that carefully carrying out the limit $v \to c$ yields the luminal MPS pulse [see Eq. (3.5)] modulo a constant term, with $b = \omega_0/c$ and $a_2 = ca_3$. For $v > c$, $\psi_{\text{VMFXW}}(\rho, z, t)$ is a finite-energy X-shaped transversely and axially localized wave characterized by a tight exponential localization of its highly focused central region, analogous to that of a MPS pulse. The main purpose of the hybrid form of $\psi_{\text{VMFXW}}(\rho, z, t)$ is to allow for such a smooth transition. An additional advantage of the hybrid form is that it obviates the presence of backward wave components propagating at the luminal speed c.

It should be noted that the presence of a superluminal speed in the finite-energy MFXW solution in Eq. (3.26) does not contradict relativity. If the parameters are chosen so that the solution contains mostly forward propagating wave components, the pulse moves superluminally with almost no distortion up to a certain distance z_d, and then it slows down to a luminal speed c, with significant accompanying distortion. Although the peak of the pulse does move superluminally up to z_d, it is not causally related at two distinct ranges $z_1, z_2 \in [0, z_d)$. The physical significance of the $(3 + 1)$-dimensional MFXW is due to its spatiotemporal localization.

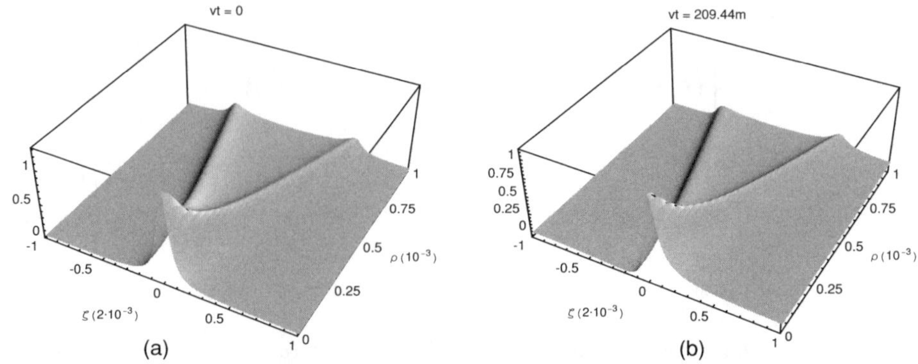

FIGURE 3.2 (3 + 1)-dimensional modified focus X-wave: $\mathrm{Re}[\psi_{\mathrm{MVFXW}}(\rho,\ \varsigma;\ vt)]$ versus $\rho,\ \varsigma$, with $v = 1.5c,\ a_1 = 10^{-4}\gamma^{-1}(\mathrm{m}),\ q = 0,\ a_3 = 10^3 c^{-1}(\mathrm{s})$, and $\omega_0 = 60c$ (rad/s). (a) $n = 0\,(vt = 0)$; (b) $n = 5000\,(vt \approx 209.44\,\mathrm{m})$.

Figure 3.2 shows surface plots of the real part of the (3 + 1)-dimensional VMFXW for the parameter values $v = 1.5c$, $a_1 = 10^{-4}/\gamma$ (m), $q = 0$, $a_3 = 10^3/c$ (s), $\omega_0 = 60c$ (rad/s), and two values of $vt = 2\pi n/\,[(\omega_0/c)(1 + (c/v)]$ at which the wave packet is expected to partially regenerate. The X-shape of the (3 + 1)-dimensional VMFXW is clearly evident. The finite energy of the pulse causes a distortion of the pulse shape at the range $vt \approx 209.44$ m for the parameters given. The transition of the wave packet into a (3 + 1)-dimensional MPS pulse is depicted in Fig. 3.3 for the parameter values $a_1 = 10^{-2}/\gamma$ (m), $a_3 = 10^3/c$ (s) , $q = 0$, $\omega_0 = 3c$ (rad/s), and $v = 1.00001c$, again for different ranges, $vt = 2\pi n/\,[(\omega_0/c)(1 + (c/v)]$, at which the wave packet is expected to partially regenerate.

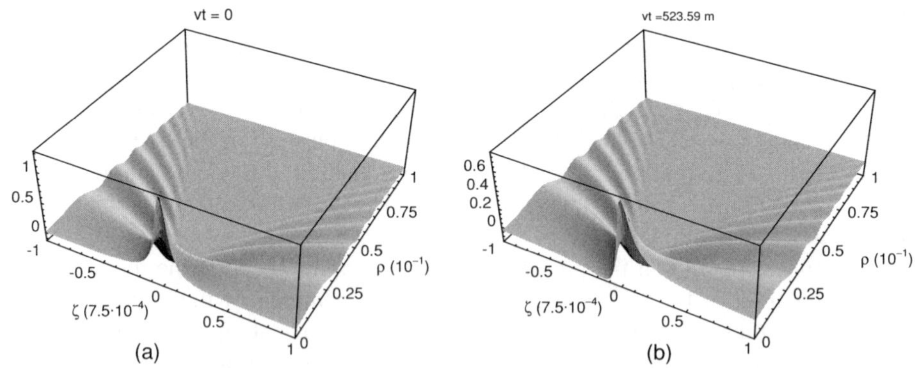

FIGURE 3.3 Transition to the (3 + 1)-dimensional modified power spectrum pulse: $\mathrm{Re}[\psi_{\mathrm{MVFXW}}(\rho,\ \varsigma;\ vt)]$ versus $\rho,\ \varsigma$, with $q = 0,\ v = 1.0000c,\ \ a_1 = 10^{-2}\gamma^{-1}(\mathrm{m})$, and $\omega_0 = 3c$ (rad/s). (a) $n = 0\,(vt = 0)$; (b) $n = 500\,(vt \approx 523.59\,\mathrm{m})$.

3.4 MODIFIED HYBRID SPECTRAL REPRESENTATION AND ITS APPLICATION TO THE DERIVATION OF FINITE-ENERGY PULSED BEAMS

3.4.1 Modified Hybrid Spectral Representation

In the superluminal spectral representation given in Eq. (3.12), the following transformations are made from the wave numbers (λ, κ) to the new wave numbers (α, β):

$$\begin{aligned} -\gamma\sqrt{\kappa^2+\lambda^2}+\lambda\gamma(c/v) &= \alpha(c/v)+\beta, \\ -\gamma\sqrt{\kappa^2+\lambda^2}+\lambda\gamma(v/c) &= \alpha-\beta. \end{aligned} \tag{3.27}$$

We then obtain

$$\psi(\rho, z, t) = \int_{-\infty}^{\infty} d\alpha \int_{-\infty}^{\infty} d\beta \int_0^{\infty} d\kappa\kappa\tilde{\psi}(\kappa, \alpha, \beta)\delta\left(-\kappa^2+\frac{1}{\gamma^2}\beta^2+2\left(1+\frac{v}{c}\right)\alpha\beta\right) \times J_0(\kappa\rho)\exp[-i\alpha(z-ct)]\exp[i\beta(z+vt)]. \tag{3.28}$$

This modified hybrid spectral representation is based on superpositions of products of forward Bessel beams moving at a fixed speed c and backward plane waves moving at the speed $v > c$. It will be shown that this representation allows a smooth transition from superluminal localized waves to pulsed beams.

3.4.2 (3 + 1)-Dimensional Splash Modes and Focused Pulsed Beams

Proceeding along the lines followed in Section 3.3, it can be established that a broad class of (3 + 1)-dimensional superluminal-type solutions based on the modified spectral representation is given by

$$\begin{aligned} \psi(\rho, z, t) &= \frac{1}{\sqrt{(\rho/\gamma)^2+[a_2-i(z+vt)]^2}} f[\theta(\rho, z, t)]; \\ \theta(\rho, z, t) &= -i(z-ct)+\left(\frac{v}{c}+1\right)\gamma^2 \\ &\quad \times\left\{a_2-i(z+vt)-\sqrt{(\rho/\gamma)^2+[a_2-i(z+vt)]^2}\right\}, \\ &\qquad \gamma = \frac{1}{\sqrt{(v/c)^2-1}}, \quad v > c. \end{aligned} \tag{3.29}$$

Here $a_2 > 0$ and $f(\cdot)$ is essentially an arbitrary function. $\theta(\rho, z, t)$ is an exact solution to the nonlinear characteristic equation

$$\left(\frac{\partial\theta}{\partial x}\right)^2+\left(\frac{\partial\theta}{\partial y}\right)^2+\left(\frac{\partial\theta}{\partial z}\right)^2-\frac{1}{c^2}\left(\frac{\partial\theta}{\partial t}\right)^2=0 \tag{3.30}$$

associated with the scalar wave equation (3.1). In this sense, the class of solutions embodied in Eq. (3.29) conforms to the Courant–Hilbert *ansatz* [26–28].

The choice $f(\cdot) = \exp(\cdot)$ in Eq. (3.29) leads to the specific solution

$$\begin{aligned}\psi_{\mathrm{BFXW}}(\rho, z, t; \alpha) = &\frac{1}{\sqrt{(\rho/\gamma)^2 + [a_2 - i(z + vt)]^2}} \\ &\times \exp\left[-i\alpha(z - ct) + \alpha\gamma^2\left(1 + \frac{v}{c}\right)\right. \\ &\left.\times\left\{a_2 - i(z + vt) - \sqrt{(\rho/\gamma)^2 + [a_2 - i(z + vt)]^2}\right\}\right]. \end{aligned} \quad (3.31)$$

In the limit $v \to c$ it is reduced to $\psi_{\mathrm{VFWM}}(\rho, z, t; \alpha)$[cf. Eq. (3.9)] modulo a constant multiplier term. Superposition over VFWMs, that is,

$$\psi_{\mathrm{SM/VFWM}}(\rho, z - ct, z + ct) = \int_0^\infty d\alpha\, \tilde{G}(\alpha)\psi_{\mathrm{VFWM}}(\rho, z - ct, z + ct; \alpha), \quad (3.32)$$

results in *luminal splash modes* [20]. A particular solution of this type is the qth-order splash mode $\psi_{\mathrm{VSM}}(\rho, z, t)$ given in Eq. (3.11). Finite-energy *luminal pulsed beam* solutions within the paraxial approximation can be obtained as follows:

$$\psi_{\mathrm{PB/VFWM}}(\rho, z - ct, z) = \int_0^\infty d\alpha\, \tilde{G}(\alpha)\psi_{\mathrm{VFWM}}(\rho, z - ct, z + ct \to 2z). \quad (3.33)$$

They obey the pulsed beam equation (3.8).

Analogously to Eq. (3.32), a new class of exact finite-energy (3 + 1)-dimensional splash modes can be obtained by means of the spectral synthesis

$$\psi_{\mathrm{SM/BFXW}}(\rho, z, t) = \int_0^\infty d\alpha\, \tilde{G}(\alpha)\psi_{\mathrm{BFXW}}(\rho, z, t; \alpha). \quad (3.34)$$

For an illustrative example, consider a spectrum analogous to that given in Eq. (3.24): that is,

$$\tilde{G}(\alpha) = (\alpha - b)^q \exp[-a_1(\alpha - b)]H(\alpha - b), \qquad a_1, b > 0. \quad (3.35)$$

The integration in Eq. (3.34) can be carried out explicitly, yielding the splash mode solution

$$\begin{aligned}\psi_{\mathrm{SM/BFXW}}(\rho, z, t) &= \Gamma(q + 1)\frac{1}{\sqrt{(\rho/\gamma)^2 + [a_2 - i(z + vt)]^2}}\frac{\exp(-bP)}{P^{q+1}}; \\ P &\equiv a_1 + i(z - ct) - \gamma^2\left(1 + \frac{v}{c}\right) \\ &\quad\times\left\{a_2 - i(z + vt) - \sqrt{(\rho/\gamma)^2 + [a_2 - i(z + vt)]^2}\right\}. \end{aligned} \quad (3.36)$$

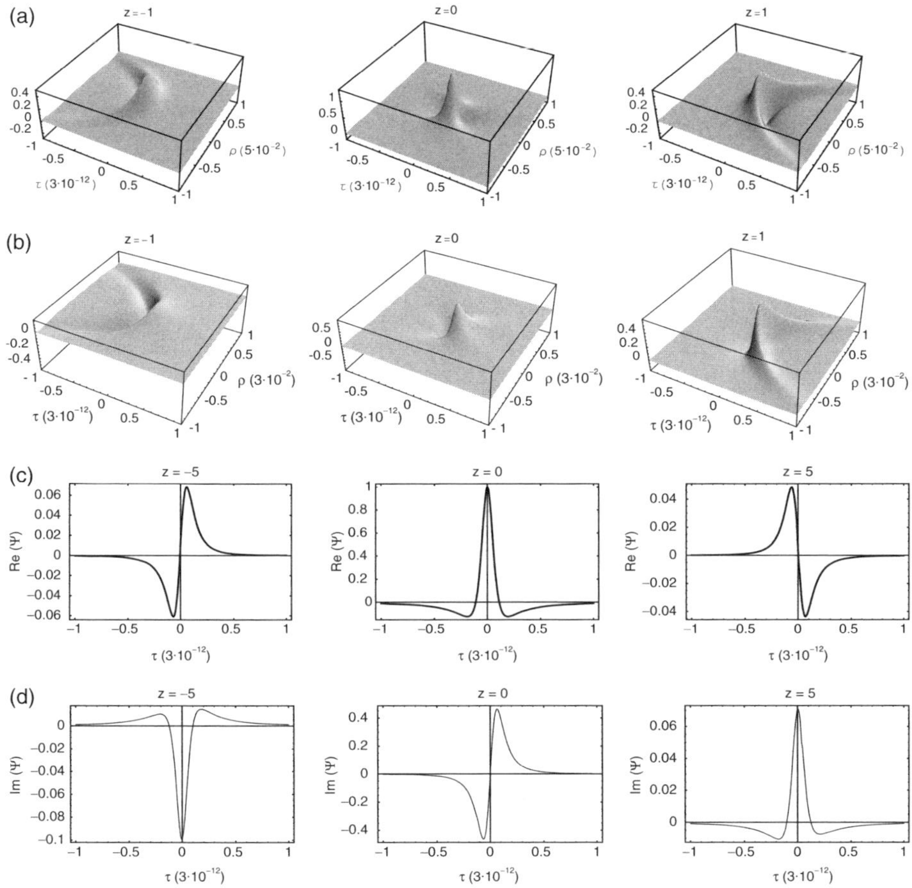

FIGURE 3.4 Spatiotemporal evolution of a focused pulsed beam from $z = -1$ through the focus ($z = 0$) to $z = 1$ m: (a) $\mathrm{Re}[\psi_{\mathrm{FBB}}(\rho, \tau; z)]$ versus ρ, τ; (b) $\mathrm{Im}[\psi_{\mathrm{FBB}}(\rho, \tau; z)]$ versus ρ, τ; $a_1 = 10^{-4}$ m, $a_2 = 1$ m, $q = 1$, $b = 2\,\mathrm{m}^{-1}$, and $v = 1.0000001c$. Temporal reshaping of a focused pulse beam from $z = -5$ m through the focus ($z = 0$) to $z = 5$ m: (c) $\mathrm{Re}[\psi_{\mathrm{FBB}}(0, \tau; z)]$ versus τ; (d) $\mathrm{Im}[\psi_{\mathrm{FBB}}(0, \tau; z)]$ versus τ, $a_1 = 10^{-4}$ m, $a_2 = 1$ m, $q = 1$, $b = 2\,\mathrm{m}^{-1}$, and $v = 1.0000001c$.

In the limit $v \to c$, one obtains $\psi_{\mathrm{VMPS}}(\rho, z, t)$ [see Eq. (3.10)] modulo a constant term. If in the latter $z + ct$ is formally replaced by $2z$, one obtains the paraxial pulsed beam solution

$$\psi_{\mathrm{PB/VMPS}}(\rho, z - ct, z) = \frac{\exp[-ib(z - ct)]}{a_2 - i \cdot 2z} \exp\left(-b \frac{\rho^2}{a_2 - i \cdot 2z}\right) \times \left[a_1 + i(z - ct) + \frac{\rho^2}{a_2 - i \cdot 2z}\right]^{-q-1}, \quad (3.37)$$

obeying the paraxial pulsed beam [Eq. (3.8)].

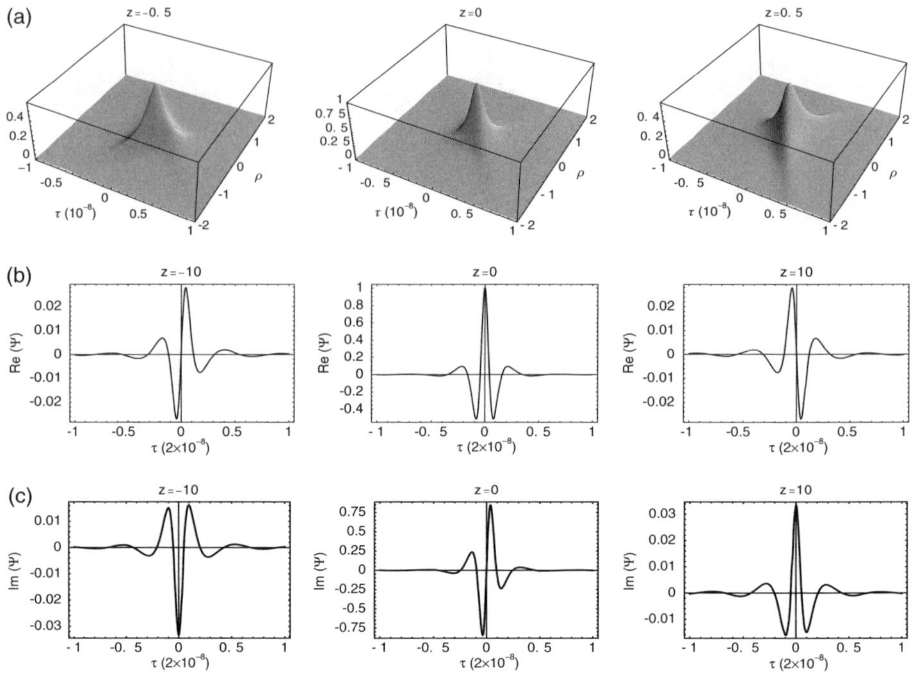

FIGURE 3.5 Spatiotemporal evolution of a nonparaxial splash mode from $z = -0.5$ through the focus ($z = 0$) to $z = 0.5\,\text{m}$: (a) $\text{Re}[\psi_{\text{SM/VFXW}}(\rho, \tau; z)]$ versus ρ, τ; (b) $\text{Im}[\psi_{\text{SM/VFXW}}(\rho, \tau; z)]$ versus ρ, τ; $a_1 = 0.3\,\text{m}$, $a_2 = 1\,\text{m}$, $q = 1$, $b = 4\,\text{m}^{-1}$, and $v = 2c$. Temporal reshaping of a nonparaxial splash mode from $z = -10\,\text{m}$ through the focus ($z = 0$) to $z = 10\,\text{m}$: (c) $\text{Re}[\psi_{\text{SM/VFXW}}(0, \tau; z)]$ versus τ; (d) $\text{Im}[\psi_{\text{SM/VFXW}}(0, \tau; z)]$ versus τ, $a_1 = 0.3\,\text{m}$, $a_2 = 1\,\text{m}$, $q = 1$, $b = 4\,\text{m}^{-1}$, and $v = 2c$.

For values of v very close to c and under the restriction $a_1 \ll a_2$, the splash mode solution $\psi_{\text{SM/BFXW}}(\rho, z, t)$ in Eq. (3.36) is indistinguishable from the corresponding paraxial focused pulsed beam obeying Eq. (3.8). The notation $\psi_{\text{FPB}}(\rho, z, t)$ will be used under these conditions. Figure 3.4a and b show, respectively, the spatiotemporal evolution of the real and imaginary parts of the solution $\psi_{\text{FPB}}(\rho, \tau \equiv t - z/c; z)$ from the plane $z = -1\,\text{m}$, passing through the focus ($z = 0$) to the plane $z = 1\,\text{m}$, for the parameter values $a_1 = 10^{-4}\,\text{m}$, $a_2 = 1\,\text{m}$, $q = 1$, $b = 2\,\text{m}^{-1}$, and $v = 1.0000001c$. One clearly observes the curved phase fronts, the polarity reversal and temporal reshaping as the wave packet evolves through the focus. Figure 3.4c and d show, respectively, plots of $\text{Re}[\psi_{\text{FPB}}(0, \tau; z)]$ versus $\tau \equiv t - z/c$ and $\text{Im}[\psi_{\text{FPB}}(0, \tau; z)]$ versus τ for the same parameter values and three ranges z. It should be noted, in particular, that the symmetric real solution at $z = 0$ evolves in the far field into an inverted version of the antisymmetric imaginary solution at $z = 0$. Furthermore, the antisymmetric imaginary solution at $z = 0$ evolves in the far field into the symmetric real solution. Such a transformation of the pulse temporal profile has been observed in terahertz experiments [29,30].

The spatiotemporal evolution of the modulus of the splash mode solution $\psi_{SM/BFXW}(\rho, z, t)$ given in Eq. (3.36) is depicted in Fig. 3.5a in a nonparaxial framework: specifically, for the parameter values $a_1 = 0.3\,\text{m}$, $a_2 = 1\,\text{m}$, $q = 1$, $b = 4\,\text{m}^{-1}$, and $v = 2c$. Figure 3.5b and c are plots of $\text{Re}[\psi_{SM/VFXW}(0, \tau; z)]$ versus τ and $\text{Im}[\psi_{SM/VFXW}(0, \tau; z)]$ versus τ, respectively, for the same parameter values.

3.5 CONCLUSIONS

A hybrid spectral representation method has been presented that allows a smooth transition between superluminal (X-shaped) and luminal (FWM-type) pulsed waves. This technique has been used to obtain both infinite-energy (FXW) and finite-energy (MFXW) transversely and axially localized pulse solutions to the $(3 + 1)$-dimensional scalar wave equation. Transverse electric and magnetic FXW and MFXW solutions to Maxwell's equations can be derived readily using the scalar-valued solutions obtained in this chapter as the z-components of electric and magnetic Hertz potentials. The hybrid spectral representation method can be used to obtain FXW and MFXW localized pulse solutions to the (n + 1)-dimentional, $n \geq 2$, scalar wave equation. Also, it can be used to derive analogous solutions to hyperbolic equations (e.g., the Klein–Gordon equation) governing wave propagation in media characterized by temporal dispersion.

A modified hybrid spectral representation method has also been presented, which allows a smooth transition from superluminal localized waves to exact splash modes. Within the framework of a certain parametrization, the latter are rendered indistinguishable from the paraxial luminal pulsed beam solutions governed by the pulsed beam equation (3.8). The specific example given in Section 3.4 illustrates the diffraction-induced temporal reshaping and polarity reversal of extremely short (one-cycle) focused pulses.

REFERENCES

1. R. W. Ziolkowski, Localized transmission of electromagnetic energy, *Phys. Rev. A* **39**, 2005–2033 (1989).
2. I. M. Besieris, A. M. Shaarawi, and R. W. Ziolkowski, A bidirectional traveling plane wave representation of exact solutions of the scalar wave equation, *J. Math. Phys.* **30**, 1254–1269 (1989).
3. R. W. Ziolkowski, I. M. Besieris, and A. M. Shaarawi, Localized wave representations of acoustic and electromagnetic radiation, *Proc. IEEE* **79**, 1371–1378 (1991).
4. I. M. Besieris, M. Abdel-Rahman, A. M. Shaarawi, and A. Chatzipetros, Two fundamental representations of localized pulse solutions to the scalar wave equation, *Prog. Electromagn. Res.* **19**, 1–48 (1998).
5. E. Recami, On localized "X-shaped" superluminal solutions to Maxwell's equations, *Physica A* **252**, 586–610 (1998).

6. J. Salo, J. Fagerholm, A. T. Friberg, and M. M. Salomaa, Unified description of nondiffracting X and Y waves, *Phys. Rev. E* **62**, 4261–4275 (2000).
7. M. Zamboni-Rached and E. Recami, A set of new localized superluminal solutions to Maxwell's equations, *Ann. Fond. Louis. de Broglie* **27**, 187–216 (2002).
8. P. Saari and K. Reivelt, Generation and classification of localized waves by Lorentz transformations in Fourier space, *Phys. Rev. E* **68,** 036612, 1–12 (2004).
9. R. W. Ziolkowski, D. K. Lewis, and B. D. Cook, Experimental verification of the localized wave transmission effect, *Phys. Rev. Lett.* **62**, 147–150 (1989).
10. R. W. Ziolkowski and D. K. Lewis, Verification of the localized wave transmission effect, *J. Appl. Phys.* **68**, 6083–6086 (1990).
11. P. Saari and K. Reivelt, Evidence of X-shaped propagation-invariant localized light waves, *Phys. Rev. Lett.* **79**, 4135–4137 (1997).
12. K. Reivelt and P. Saari, Experimental demonstration of realizability of optical focus wave modes, *Phys. Rev. E* **66**, 056611-1-9 (2002).
13. R. Grunwald, V. Kebbel, U. Neumann, A. Kummrow, M. Rini, E. T. Nibbering, M. Piche, G. Rousseau, and M. Fortin, Generation and characterization of spatially and temporally localized few-cycle optical wave packets, *Phys. Rev. A* **67**, 063820, 1–5 (2003).
14. R. W. Ziolkowski, I. M. Besieris, and A. M. Shaarawi, Aperture realizations of the exact solutions to homogeneous wave equations, *J. Opt. Soc. Am. A* **10**, 75–87 (1993).
15. R. W. Ziolkowski, I. M. Besieris, and A. M. Shaarawi, Aperture realizations of the exact solutions to homogeneous wave equations, *J. Opt. Soc. Am. A* **10**, 75–87 (1993).
16. A. M. Shaarawi, I. M. Besieris, R. W. Ziolkowski, and S. M. Sedky, Generation of approximate focus wave mode pulses from wideband dynamic Gaussian apertures, *J. Opt. Soc. Am. A* **12**, 1954–1964 (1995).
17. A. P. Kiselev and M. V. Perel, Highly localized solutions of the wave equations, *J. Math. Phys.* **41**, 1934–1955 (2000).
18. J. N. Brittingham, Focus wave modes in homogeneous Maxwell equations: Transverse electric mode, *J. Appl. Phys.* **54**, 1179–1189 (1983).
19. A. P. Kiselev, Modulated Gaussian beams, *Radio Phys. Quantum. Electron.* **26**, 1014 (1983).
20. R. W. Ziolkowski, Exact solutions of the wave equation with complex source locations, *J. Math. Phys.* **26**, 861–863 (1984).
21. E. Heyman, Pulsed beam propagation in an inhomogeneous medium, *IEEE Trans. Antennas Propag.* **42**, 311–319 (1994).
22. M. A. Porras, Ultrashort pulsed Gaussian light beams, *Phys. Rev.* E **58**, 1086–1093 (1998).
23. S. Feng, H. G. Winful, and R. W. Hellwarth, Spatio-temporal evolution of single-cycle electromagnetic pulses, *Phys. Rev. E* **59**, 4630–4649 (1999).
24. J. Y. Lu and J. F. Greenleaf, Nondiffracting X waves: exact solutions to the free space scalar wave equation and their finite aperture realization, *IEEE Trans. Ultrason. Ferroelectr. Freq. Control.* **39**, 19–31 (1992).
25. A. Erdelyi, Ed., *Tables of Integral Transforms*, Vol. I, McGraw-Hill, New York, 1954.
26. R. Courant and D. Hilbert, *Methods of Mathematical Physics*, Vol. II, Interscience, New York, 1962.
27. P. Hillion, Courant–Hilbert solutions of the wave equation, *J. Math Phys.* **33**, 34–45 (1992).

28. I. M. Besieris and A. M. Shaarawi, Three classes of Courant–Hilbert progressive solutions to the scalar wave equation, *J. Electromagn. Waves Appl.* **16**, 1047–1060 (2002).
29. E. Budiarto, N.-W. Pu, S. Jeong, and J. Bokor, Near-field propagation of terahertz pulses from a large-aperture antenna, *Opt. Lett.* **23**, 213–215 (1998).
30. M. T. Reiten, S. A. Harmon, and R. A. Cheville, Terahertz beam propagation measured through three-dimensional amplitude profile determination, *J. Opt. Soc. Am. B* **20**, 2215–2225 (2003).

CHAPTER FOUR

Ultrasonic Imaging with Limited-Diffraction Beams

JIAN-YU LU
The University of Toledo, Toledo, Ohio

4.1 INTRODUCTION

One type of limited-diffraction beam was first described by Stratton in 1941as *undistorted progressive waves* (UPWs) [1]. In 1987, without referring to Stratton's work, Durnin et al. studied UPWs both by computer simulation and as an optical experiment [2–4]. Because the UPWs in Stratton's book have a Bessel transverse beam profile, they are termed *Bessel beams*. Durnin et al. named the Bessel beams *nondiffracting* or *diffraction-free beams* [2–4]. Because Durnin's terminology is controversial in the scientific community, these beams are commonly termed *limited-diffraction beams* [5], since all practical beams or waves will eventually diffract. Bessel beams are localized in the transverse direction and may have potential applications [6–14]. In acoustics, the first Bessel annular array transducer was designed and constructed in 1990 [15–16] and patented in 1992 [17]. Applications of Bessel beams in acoustics have been studied extensively [18–30].

Localized waves (LWs) were developed by Brittingham in 1983 and termed *focus wave modes* [31]. LWs have properties similar to those of Bessel beams in terms of transverse localization. In addition, LWs contain multiple frequencies and may be localized in the axial direction. LWs have been studied by many investigators

Localized Waves, Edited by Hugo E. Hernández-Figueroa, Michel Zamboni-Rached, and Erasmo Recami

[32–40]. However, LWs are not propagation invariant; that is, they do not meet the propagation-invariant condition as defined in the following: If one travels with the wave at a speed c_1, he or she sees a wave packet, $\Phi(\vec{r}, t) = \Phi(x, y, z - c_1 t)$, that is unchanged for $z - c_1 t =$ constant, where z is the axial axis along the direction of wave propagation, $\vec{r} = (x, y, z)$ is a point in space, and t is the time.

To find multiple-frequency waves that are propagation invariant [i.e., $\Phi(\vec{r}, t) = \Phi(x, y, z - c_1 t)$], in 1991, X-waves were developed [41–43] and were studied subsequently [44–54]. The name X-waves was used because the beam profile in the axial cross section (a plane through the beam axis) resembles the letter "X." Due to the interest in X-waves for nonlinear optics and other applications, X-waves were introduced in 2004 in the "Search and Discovery" column of *Physics Today* [55]. The two 1992 X-wave papers [42–43] were awarded by the Ultrasonics, Ferroelectrics, and Frequency Control (UFFC) Society of the Institute of Electrical and Electronics Engineers (IEEE) in 1993. Later, an X-wave experiment in optics was performed by Saari and Reivelt and published in 1997 in *Physical Review Letters* [56]. To generalize X-waves, a transformation that is used to obtain limited-diffraction beams (including X-waves) in an N-dimensional space from any solutions to an $(N - 1)$-dimensional isotropic/homogeneous wave equation was developed in 1995 [44], where $N \geq 2$ is an integer. This formula has been related to part of the Lorentz transformation [57–58], and was used and demonstrated by other researchers [59–60]. Furthermore, an X-wave transform that is a transformation pair was developed in 2000 for any physically realizable waves using the orthogonal property of X-waves [46–47]. The orthogonal property of X-waves was studied further by Salo et al. in 2001 [61]. The transformation pair allows one to decompose an arbitrary physically realizable wave into X-waves (inverse X-wave transform) and determine the coefficients (forward X-wave transform) of the decomposition. Based on X-wave theory, a method and its extension that are capable of ultrahigh frame rate (HFR) two- or three-dimensional imaging were developed in 1997 [62–87]. Due to the importance of this method, it was noted as one of the predictions of the twenty-first century medical ultrasonics in 2000 [88]. After the introduction of X-waves in 1991 [41–43], these waves have been studied extensively by many investigators [56,58–60,89–123]. There are also some review papers on X-waves and their applications [124–131].

In this chapter, fundamentals of limited-diffraction beams are reviewed and studies of these beams are put into a unified theoretical framework. The theory of limited-diffraction beams is developed further. New limited-diffraction solutions to the Klein–Gordon and Schrödinger equations as well as limited-diffraction solutions to these equations in confined spaces are obtained. The relationship between the transformation that converts any solutions to an $(N - 1)$-dimensional wave equation to limited-diffraction solutions of an N-dimensional equation and the Lorentz transformation is clarified and extended. The transformation is also applied to the Klein–Gordon equation. In addition, some applications of limited-diffraction beams are summarized.

4.2 FUNDAMENTALS OF LIMITED-DIFFRACTION BEAMS

4.2.1 Bessel Beams

An N-dimensional isotropic/homogeneous wave equation is given by

$$\left(\sum_{j=1}^{N} \frac{\partial^2}{\partial x_j^2} - \frac{1}{c^2}\frac{\partial^2}{\partial t^2}\right)\Phi(\vec{r}, t) = 0, \tag{4.1}$$

where $x_j (j = 1, 2, \ldots, N)$ represents rectangular coordinates in an N-dimensional space, $N \geq 1$ is an integer, $\Phi(\vec{r}, t)$ is a scalar function (sound pressure, velocity potential, or Hertz potential in electromagnetics) of spatial variables, $\vec{r} = (x_1, x_2, \ldots, x_N)$, t is time, and c is the speed of light in vacuum or the speed of sound in a medium.

In three–dimensional space, we have

$$\left(\nabla^2 - \frac{1}{c^2}\frac{\partial^2}{\partial t^2}\right)\Phi(\vec{r}, t) = 0, \tag{4.2}$$

where ∇^2 is the Laplace operator. In cylindrical coordinates, the wave equation is given by

$$\left[\frac{1}{r}\frac{\partial}{\partial r}\left(r\frac{\partial}{\partial r}\right) + \frac{1}{r^2}\frac{\partial^2}{\partial \phi^2} + \frac{\partial^2}{\partial z^2} - \frac{1}{c^2}\frac{\partial^2}{\partial t^2}\right]\Phi(\vec{r}, t) = 0, \tag{4.3}$$

where $r = \sqrt{x^2 + y^2}$ is the radial distance, $\phi = \tan^{-1}(y/x)$ is the polar angle, and z is the axial axis.

One generalized solution to the N-dimensional wave equation (4.1) is given by [42,124]

$$\Phi(x_1, x_2, \ldots, x_N; t) = f(s), \tag{4.4}$$

where

$$s = \sum_{j=1}^{N-1} D_j x_j + D_N(x_N \pm c_1 t), \qquad N \geq 1 \tag{4.5}$$

and where D_j are complex coefficients, $f(s)$ is any well-behaved complex function of s, and

$$c_1 = c\sqrt{\frac{1 + \sum_{j=1}^{N-1} D_j^2}{D_N^2}}. \tag{4.6}$$

If c_1 is real, $f(s)$ and its linear superposition represent limited-diffraction solutions to the N-dimensional wave equation (4.1). For example, if $N = 3$, $x_1 = x$, $x_2 = y$, $x_3 = z$, $D_1 = \alpha_0(k, \zeta)\cos\theta$, $D_2 = \alpha_0(k, \zeta)\sin\theta$, and $D_3 = b(k, \zeta)$, with cylindrical coordinates, we obtain families of solutions to (4.3) [42,124]:

$$\Phi_\zeta(s) = \int_0^\infty T(k)\left[\frac{1}{2\pi}\int_{-\pi}^{\pi} A(\theta) f(s)\, d\theta\right] dk \tag{4.7}$$

and

$$\Phi_K(s) = \int_{-\pi}^{\pi} D(\zeta)\left[\frac{1}{2\pi}\int_{-\pi}^{\pi} A(\theta) f(s)\, d\theta\right] d\zeta, \tag{4.8}$$

where

$$s = \alpha_0(k, \zeta) r\cos(\phi - \theta) + b(k, \zeta)\left[z \pm c_1(k, \zeta)t\right], \tag{4.9}$$

$$c_1(k, \zeta) = c\sqrt{1 + [\alpha_0(k, \zeta)/b(k, \zeta)]^2}, \tag{4.10}$$

$\alpha_0(k, \zeta)$, $b(k, \zeta)$, $A(\theta)$, $T(k)$, and $D(\zeta)$ are well-behaved functions, and θ, k, and ζ are free parameters. If $c_1(k, \zeta)$ is real and is not a function of k and ζ, respectively, $\Phi_\zeta(s)$ and $\Phi_K(s)$ are families of limited-diffraction solutions to the wave equation (4.3).

The following function is also a family of limited-diffraction solution to the wave equation [42,124]:

$$\Phi_L(r, \phi, z - ct) = \Phi_1(r, \phi)\Phi_2(z - ct), \tag{4.11}$$

where $\Phi_2(z - ct)$ is any well-behaved function of $z - ct$ and $\Phi_1(r, \phi)$ is a solution to the transverse Laplace equation:

$$\left[\frac{1}{r}\frac{\partial}{\partial r}\left(r\frac{\partial}{\partial r}\right) + \frac{1}{r^2}\frac{\partial^2}{\partial \phi^2}\right]\Phi_1(r, \phi) = 0. \tag{4.12}$$

If $T(k) = \delta(k - k')$, $f(s) = e^s$, $\alpha_0(k, \zeta) = -i\alpha$, and $b(k, \zeta) = i\beta$ in (4.7) and (4.9), we have

$$\Phi_\zeta(s) = \left[\frac{1}{2\pi}\int_{-\pi}^{\pi} A(\theta) e^{-i\alpha r\cos(\phi - \theta)}\, d\theta\right] e^{i(\beta z - \omega t)}, \tag{4.13}$$

where $\beta = \sqrt{k'^2 - \alpha^2}$ is the propagation parameter, $\delta(k - k')$ is the Dirac delta function, $k' = \omega/c > 0$ is the wave number, and ω is the angular frequency. If $A(\theta) = i^n e^{in\theta}$, one obtains an nth-order Bessel beam [2–4,15–17]:

$$\begin{aligned}\Phi_{B_n}(\vec{r}, t) &= \Phi_{B_n}(r, \phi, z - c_1 t)\\ &= e^{in\phi} J_n(\alpha r) e^{i(\beta z - \omega t)}, \qquad n = 0, 1, 2, \ldots,\end{aligned} \tag{4.14}$$

where the subscript B_n represents an nth-order Bessel beam, α is a scaling parameter, $J_n(\cdot)$ is an nth-order Bessel function of the first kind, and $c_1 = \omega/\beta$ is the phase velocity of the wave. It is clear that Bessel beams are single-frequency waves and are localized in the transverse direction. The scaling parameter, α, determines the degree of localization. Because of this property, Bessel beams can be applied to medical ultrasonic imaging [15–21]. Bessel beams are studied further [22–30] along with the studies of acoustic transducers and ultrasound waves [132–135].

4.2.2 Nonlinear Bessel Beams

In medical imaging, nonlinear properties are important to provide additional information on diseased tissues. Harmonics of Bessel beams due to the tissue nonlinearity are useful to obtain higher-quality images by combining the localized properties of limited-diffraction beams [22,23].

4.2.3 Frozen Waves

It is clear from (4.14) that single-frequency Bessel beams have two free parameters. One is the order of the Bessel function, and the other is the scaling parameter that changes the phase velocity of the Bessel beams. The order of the Bessel beams, n, in (4.14) has been exploited to produce various limited-diffraction beams of different transverse beam profiles since 1995 [29,30]. Another parameter, the scaling parameter, α, in (4.14), has also been used for a linear superposition of Bessel beams to produce a beam of a desired axial profile [24–27] for zeroth-order Bessel beams. Although an annular array was used in the production of superposed Bessel beams in these studies, the number of annuli and the width of each ring are free to change. When the number of annuli approaches infinity and the width of each ring shrinks to zero with a given circular aperture, the field distribution at the surface of the annular array is in fact a continuous function. In a more general way, one could use X-wave transform [28,46,47] to produce a wave whose shape would be close to the shape desired under conditions such as the least-squares criterion [136] by changing both the order of the beams and the scaling parameter.

In 2004, Zamboni-Rached developed an analytical relationship between the scaling parameter of Bessel beams and the axial beam profile along the beam axis ($r = 0$) for the zeroth-order Bessel beams. The resulting linear superposition of Bessel beams of different scaling parameter, α, was called *frozen waves* [137]. The method was extended to include superposition over both the scaling parameter and the order of the Bessel beams [138] to better control the transverse beam profile of the frozen waves. These studies not only provide computationally efficient ways for beam designs but may also have applications in optical tweezers [139].

4.2.4 X-Waves

If $T(k) = B(k)e^{-a_0 k}$, $A(\theta) = i^n e^{in\theta}$, $\alpha_0(k, \zeta) = -ik \sin\zeta$, $b(k, \zeta) = ik \cos\zeta$, and $f(s) = e^s$, one obtains an nth-order X-wave [41–53], which is a superposition of limited-diffraction portion of axicon beams [140,141]:

$$\begin{aligned}\Phi_{X_n}(\vec{r}, t) &= \Phi_{X_n}(r, \phi, z - c_1 t) \\ &= e^{in\phi} \int_0^\infty B(k) J_n(kr \sin\zeta) e^{-k[a_0 - i\cos\zeta(z - c_1 t)]}\, dk, \qquad n = 0, 1, 2, \ldots,\end{aligned} \tag{4.15}$$

where the subscript X_n represents an nth-order X-wave, $c_1 = c/\cos\zeta \geq c$ is both the phase and group velocity of the wave, $|\zeta| < \pi/2$ is the axicon angle [141,142], a_0 is a positive free parameter that determines the decaying speed of the high-frequency components of the wave, and $B(k)$ is an arbitrary well-behaved transfer function of a device (acoustic transducer or electromagnetic antenna) that produces the wave. Comparing (4.15) with (4.14), it is easy to see the similarity and difference between a Bessel beam and an X-wave. X-waves are multiple-frequency waves, whereas Bessel beams have a single frequency. However, both waves have the same limited-diffraction property (i.e., they are propagation invariant). Because of multiple frequencies, X-waves can be localized in both transverse space and time to form a tight wave packet. They can propagate in free space or isotropic/homogeneous media without spreading or dispersion. Choosing specific $B(k)$, one can obtain analytical X-wave solutions [41–43]. One example is the zeroth-order [$n = 0$ and $B(k) = a_0$] X-wave [42]:

$$\begin{aligned}\Phi_{X_0}(\vec{r}, t) &= \Phi_{X_0}(r, \phi, z - c_1 t) \\ &= \int_0^\infty a_0 J_0(kr \sin\zeta) e^{-k[a_0 - i\cos\zeta(z - c_1 t)]}\, dk \\ &= \frac{a_0}{\sqrt{(r\sin\zeta)^2 + [a_0 - i\cos\zeta(z - c_1 t)]^2}}.\end{aligned} \tag{4.16}$$

4.2.5 Obtaining Limited-Diffraction Beams with Variable Transformation

If $\Phi_{N-1}(\vec{r}_{N-1}, t)$ is a solution to the $(N-1)$-dimensional isotropic/homogeneous wave equation, one can always obtain a limited-diffraction solution, $\Phi_N(\vec{r}_N, t)$, to the N-dimensional wave equation [see (4.1)] with the following variable substitutions [44]:

$$\begin{aligned}&\vec{r}_{N-1}\sin\zeta \to \vec{r}_{N-1} && &&\vec{r}_{N-1}\sin\zeta \to \vec{r}_{N-1} \\ &\frac{x_N\cos\zeta}{c} - t \to t && \text{or} && t - \frac{x_N\cos\zeta}{c} \to t,\end{aligned} \tag{4.17}$$

where $\vec{r}_{N-1} = (x_1, x_2, \ldots, x_{N-1})$, $\vec{r}_N = (x_1, x_2, \ldots, x_N)$, $N \geq 2$ is an integer, and $|\zeta| < \pi/2$ is the axicon angle [141,142] [for $N = 1$, $\Phi_{N-1}(\vec{r}_{N-1}, t) = \Phi_0(t)$ is a vibration and not a wave; in this case, (4.17) and the procedure above work only when $\zeta = 0$]. Because $x_N\cos\zeta/c - t$ appears as a single variable in the equation

$$\Phi_N(\vec{r}_N, t) = \Phi_N(\vec{r}_{N-1}, x_N - c_1 t) = \Phi_{N-1}(\vec{r}_{N-1}\sin\zeta, x_N\cos\zeta/c - t), \tag{4.18}$$

$\Phi_N(\vec{r}_N, t)$ is a limited-diffraction beam propagating along the axis, x_N. As shown in [57,58], (4.17) is related to part of the Lorentz transformation (missing the transformation on x_N) after dividing all variables by the same constant, $\sin\zeta$:

$$\vec{r}_{n-1} \to \vec{r}_{n-1}$$
$$\frac{t}{\sin\zeta} - \frac{x_n \cos\zeta}{c \sin\zeta} = \frac{1}{\sin\zeta}\left(t - \frac{x_n \cos\zeta}{c}\right) = \gamma\left(t - \frac{\beta}{c}x_n\right) \to t, \tag{4.19}$$

where $\beta = \cos\zeta = v/c$ and $\gamma = 1/\sin\zeta = 1/\sqrt{1-\beta^2}$, and where $0 \le v < c$ is the velocity of the moving coordinates (observer) along the axis, x_N. Contrary to the report in [57,58], if $\Phi_{N-1}(\vec{r}_{N-1}, t)$ is a solution to the $(N-1)$-dimensional isotropic/homogeneous wave equation, $\Phi_N(\vec{r}_N, t)$ will not be a solution to the N-dimensional wave equation (4.1) with the partial Lorentz transformation (4.19). Equation (17) has also been used in [59,60] to derive limited-diffraction beams in waveguides.

4.2.6 Limited-Diffraction Solutions to the Klein–Gordon Equation

An N-dimensional Klein–Gordon equation for a free relativistic particle is given by [143]:

$$\left(\sum_{j=1}^{N} \frac{\partial^2}{\partial x_j^2} - \frac{1}{c^2}\frac{\partial^2}{\partial t^2} - \frac{m^2c^2}{\hbar^2}\right)\Phi_N(\vec{r}_N, t) = 0, \tag{4.20}$$

where $x_j (j = 1, 2, \ldots, N)$ represents rectangular coordinates in an N-dimensional space; $N \ge 1$ is an integer; $\Phi_N(\vec{r}_N, t)$ is a scalar wave function of spatial variables $\vec{r}_N = (x_1, x_2, \ldots, x_N)$ and time t; c is the speed of light in vacuum; $\hbar = h/2\pi$, where h is the Planck constant; $m = m' \sin\zeta$ is the mass of the particle at rest, where m' is a mass-related constant; and $|\zeta| < \pi/2$ is the axicon angle [141,142].

Assuming that $\Phi_{N-1}(\vec{r}_{N-1}, t)$ is a solution to the following $(N-1)$-dimensional Klein–Gordon equation with a mass m' [143]:

$$\left(\sum_{j=1}^{N-1} \frac{\partial^2}{\partial x_j^2} - \frac{1}{c^2}\frac{\partial^2}{\partial t^2} - \frac{m'^2c^2}{\hbar^2}\right)\Phi_{N-1}(\vec{r}_{N-1}, t) = 0, \tag{4.21}$$

where $\vec{r}_{N-1} = (x_1, x_2, \ldots, x_{N-1})$; (4.18) is a solution to (4.20) after the variable substitution (4.17). This can be proved easily in a manner similar to that described in [44]. Using (4.18) and (4.21), we have

$$\left(\sum_{j=1}^{N-1} \frac{\partial^2}{\partial x_j^2}\right)\Phi_{N-1}\left(\vec{r}_{N-1}\sin\zeta, \frac{x_N\cos\zeta}{c} - t\right)$$
$$= \sin^2\zeta\left(\frac{1}{c^2}\frac{\partial^2}{\partial t^2} + \frac{m'^2c^2}{\hbar^2}\right)\Phi_{N-1}\left(\vec{r}_{N-1}\sin\zeta, \frac{x_N\cos\zeta}{c} - t\right) \tag{4.22}$$

and

$$\frac{\partial^2}{\partial x_N^2}\Phi_{N-1}\left(\vec{r}_{N-1}\sin\zeta, \frac{x_N\cos\zeta}{c} - t\right)$$
$$= \frac{\cos^2\zeta}{c^2}\frac{\partial^2}{\partial t^2}\Phi_{N-1}\left(\vec{r}_{N-1}\sin\zeta, \frac{x_N\cos\zeta}{c} - t\right). \tag{4.23}$$

Summing both the left- and right-hand sides of (4.22) and (4.23), and comparing the results with (4.20), it is clear that (4.18) is a solution to (4.20). Limited-diffraction solutions to the Klein–Gordon equation mean that a free relativistic particle may be accompanied by a rigidly propagating wave along the axis, x_N, at a velocity that is greater than the speed of light in vacuum in a manner similar to that of X-waves [41–43] (for $\zeta \neq 0$). If $|\zeta| \rightarrow \pi/2$, the wave speed $c_1 = c/\cos\zeta \rightarrow \infty$ and then one has $m' \rightarrow m$. For photons where $m = 0$, (4.22) and (4.23) are the same as those in [44]. It is worth noting that from the proofs in (4.22) and (4.23) and in [44], it is clear that the functions $\sin\zeta$ and $\cos\zeta$ in (4.17) can be other functions as long as the summation of the squares of those functions is equal to 1: $f_1^2(\zeta) + f_2^2(\zeta) \equiv 1$, where $f_1(\zeta)$ and $f_2(\zeta)$ are any well-behaved functions of ζ or other free parameters. This extends the transformation formula in (4.17).

In the following we obtain some localized limited-diffraction solutions to the Klein–Gordon equation. Assuming that $f(s) = e^s$ in (4.4), where s is given by (4.5), and inserting (4.4) into (4.20), one obtains the velocity of the wave:

$$c_1 = c\sqrt{\frac{\left(\sum_{j=1}^{N} D_j^2 - m^2c^2/\hbar^2\right)}{D_N^2}}. \tag{4.24}$$

If $N = 3$, $x_1 = x$, $x_2 = y$, and $x_3 = z$, (4.24) becomes

$$c_1 = c\sqrt{\frac{\left(D_1^2 + D_2^2 + D_3^2 - m^2c^2/\hbar^2\right)}{D_3^2}}. \tag{4.25}$$

Choosing $D_1 = \alpha_0\cos\theta$ and $D_2 = \alpha_0\sin\theta$, where $-\pi \leq \theta \leq \pi$ is a free parameter and α_0 is a well-behaved function of any free parameters, if $\alpha_0 = -imc/\hbar\sin\zeta$, one obtains

$$D_3 = i\frac{mc}{\hbar}\left(\frac{\sqrt{1-(\hbar/mc)^2\alpha_0^2}}{\sqrt{(c_1/c)^2 - 1}}\right) = i\frac{mc}{\hbar}\frac{\sqrt{1+\sin^2\zeta}}{\sqrt{(c_1/c)^2 - 1}}. \tag{4.26}$$

Since e^s in (4.4) is a solution to the Klein–Gordon equation (4.20), a linear superposition over the free parameter, θ, is still a solution:

$$\begin{aligned}\Phi_{B_n}^{\mathrm{KG}}(\vec{r},t) &= \frac{1}{2\pi}\int_{-\pi}^{\pi} A(\theta)e^s\,d\theta \\ &= e^{in\phi}J_n\left(\frac{mc}{\hbar}r\sin\zeta\right)\exp\left[i\frac{mc}{\hbar}\left(\frac{\sqrt{1+\sin^2\zeta}}{\sqrt{(c_1/c)^2-1}}\right)(z-c_1t)\right], \\ &\qquad (n=0,1,2,\ldots), \qquad (4.27)\end{aligned}$$

where the subscript B_n and the superscript KG represent an nth-order Bessel beam and the Klein–Gordon equation, respectively, and $i(mc/\hbar)(\sqrt{1+\sin^2\zeta}/\sqrt{(c_1/c)^2-1})$ is the propagation constant. Equation (4.27) is a localized solution to (4.20) and its localization increases with the mass, m. For electrons at rest, $m = 9.1\times10^{-31}$ kg, and thus $mc/\hbar = 2.6\times10^{12}\,\mathrm{m}^{-1}$ ($\hbar = 1.05\times10^{-34}$ J · s and $c = 3.0\times10^8$ m/s). The wave in (4.27) is localized in picometer scale if $\sin\zeta \approx 1$. There are other choices of α_0. If α_0 is a constant, a localized limited-diffraction solution that has a fixed transverse beam profile can be obtained. If $\alpha_0 = -i(\gamma m v/\hbar)\sin\zeta$, where $\gamma = 1/\sqrt{1-\beta^2}$ and $\beta = v/c$, and where v is the velocity of the particle, the transverse localization of the solutions will increase with the speed of the particle. In this case, the propagation constant is given by $i(mc/\hbar)(\sqrt{1+(\gamma v/c)^2\sin^2\zeta}/\sqrt{(c_1/c)^2-1})$.

Superposing $\Phi_{B_n}^{\mathrm{KG}}(\vec{r},t)$ in (4.27) over the mass, m, one obtains a composed wave function that is similar to the X-wave [41–53] but may not necessarily be a solution to (4.20) where m is a constant for a given particle (the physical meaning could be a group of independent particles of different masses traveling in space). Using (4.7) and (4.27), and letting $T(k) = B(k)e^{-a_0 k}$, where $k = mc/\hbar$, one obtains

$$\begin{aligned}\Phi_{X_n}^{\mathrm{KG}}(\vec{r},t) &= \Phi_{X_n}^{\mathrm{KG}}(r,\phi,z-c_1t) \\ &= \frac{c}{\hbar}e^{in\phi}\int_0^{\infty} B\left(\frac{mc}{\hbar}\right)J_n\left(\frac{mc}{\hbar}r\sin\zeta\right) \\ &\quad\times\exp\left\{\frac{mc}{\hbar}\left[a_0 - i\left(\sqrt{1+\sin^2\zeta}/\sqrt{(c_1/c)^2-1}\right)(z-c_1t)\right]\right\}dm, \\ &\qquad n=0,1,2,\ldots, \qquad (4.28)\end{aligned}$$

where the subscript X_n represents an nth-order X-wave and the superscript KG the Klein–Gordon equation, a_0 is a positive free parameter, and $B(k)$ is an arbitrary well-behaved transfer function. If $n = 0$ and $B(k) = a_0$, from (4.28) and (4.16) one has (where c_1 is a constant) [42]

$$\begin{aligned}\Phi_{X_0}^{\mathrm{KG}}(\vec{r},t) &= \Phi_{X_0}^{\mathrm{KG}}(r,\phi,z-c_1t) \\ &= \frac{a_0}{\sqrt{(r\sin\zeta)^2+\left[a_0 - i\left(\sqrt{1+\sin^2\zeta}/\sqrt{(c_1/c)^2-1}\right)(z-c_1t)\right]^2}}. \qquad (4.29)\end{aligned}$$

It is clear from (4.26)–(4.29) that if $c_1 < c$, the solutions or functions are no longer waves. If $c_1 = c$, D_3 in (4.26) is infinity. For $c_1 > c$, one obtains rigidly propagating superluminal waves or functions, as in the case of X-waves [42]. One example is to assume that $c_1 = c/\cos\zeta$, as given in (4.17) [44]. A superposition that is similar to (4.28) can also be done over the velocity, v, instead of the mass, m, of a particle if, say, $\alpha_0 = -i(\gamma m v/\hbar)\sin\zeta$. In this case, the superposition is a limited-diffraction solution to the Klein–Gordon equation (4.20).

4.2.7 Limited-Diffraction Solutions to the Schrödinger Equation

The general nonrelativistic, time-dependent, and three-dimensional Schrödinger wave equation for multiple particles is given by (see, e.g., [144])

$$-\sum_{j=1}^{M} \frac{\hbar^2}{2m_j} \nabla_j^2 \Phi + V\Phi = i\hbar \frac{\partial \Phi}{\partial t}, \tag{4.30}$$

where $\Phi = \Phi(x_1, x_2, x_3; \ldots; x_{3M-2}, x_{3M-1}, x_{3M}; t)$ is the wave function (related to the probability of finding particles in space and time) and $V = V(x_1, x_2, x_3; \ldots; x_{3M-2}, x_{3M-1}, x_{3M}; t)$ is the potential of the system. Φ and V are determined by all the particles and their interactions. ∇_j^2 is the Laplace in terms of the position of the jth particle in space, $\vec{r}_j = (x_{3j-2}, x_{3j-1}, x_{3j})$, where $j = 1, 2, \ldots, M$ (M is an integer) and m_j is the mass at rest of the jth particle. Assuming that V is not a function of spatial variables and time, and $\Phi(s) = e^s$, where s is given by (4.5), one obtains [54]

$$c_1 = \frac{\sum_{j=1}^{M} (-\hbar^2/2m_j)\left(D_{3j-2}^2 + D_{3j-1}^2 + D_{3j}^2\right) + V}{-i\hbar D_{3M}}. \tag{4.31}$$

If $M = 1, x_1 = x, x_2 = y, x_3 = z, m_1 = m, D_1 = \alpha_0 \cos\theta$, and $D_2 = \alpha_0 \sin\theta$, where $|\zeta| < \pi/2$ is an axicon angle, $-\pi \le \theta \le \pi$ is a free parameter, and α_0 is a well-behaved function of any free parameters, (4.31) is simplified [54] as

$$c_1 = \frac{(-\hbar^2/2m)\left(\alpha_0^2 + D_3^2\right) + V}{-i\hbar D_3}. \tag{4.32}$$

If $V = 0$ and $\alpha_0 = -i\,(mc/\hbar)\sin\zeta$, one has

$$D_3 = \begin{cases} i\dfrac{mc}{\hbar}\left(\dfrac{c_1}{c} \pm \sqrt{\left(\dfrac{c_1}{c}\right)^2 + \dfrac{\hbar^2}{m^2c^2}\alpha_0^2}\right) = i\dfrac{mc}{\hbar}\left(\dfrac{c_1}{c} \pm \sqrt{\left(\dfrac{c_1}{c}\right)^2 - \sin^2\zeta}\right), & \alpha_0 \ne 0 \\ i \cdot 2\dfrac{mc}{\hbar}\dfrac{c_1}{c}, & \alpha_0 = 0. \end{cases} \tag{4.33}$$

Following the steps to obtain (4.27), one obtains a localized solution to the Schrödinger equation in (4.30) under the conditions leading to (4.33) [54]:

$$\Phi_{B_n}^S(\vec{r}, t) = \frac{1}{2\pi}\int_{-\pi}^{\pi} A(\theta)e^s \, d\theta$$
$$= e^{in\phi} J_n\left(\frac{mc}{\hbar} r \sin\zeta\right) \exp\left[i\frac{mc}{\hbar}\left(c_1/c \pm \sqrt{(c_1/c)^2 - \sin^2\zeta}\right)(z - c_1 t)\right],$$
$$n = 0, 1, 2, \ldots, \qquad (4.34)$$

where the subscript B_n and the superscript S represent an nth-order Bessel beam and the Schrödinger equation, respectively, and $i(mc/\hbar)(c_1/c \pm \sqrt{(c_1/c)^2 - \sin^2\zeta})$ is the propagation constant. Similar to the Klein–Gordon equation [see the text below (4.27)], one can select $\alpha_0 = $ constant, $\alpha_0 = -i(\gamma m v/\hbar)\sin\zeta$, or other functions to obtain more limited-diffraction beams [the corresponding D_3 can be obtained easily by inserting different α_0 into (4.33)].

Following the derivations of (4.28) and substituting $(mc/\hbar)(\sqrt{1 + \sin^2\zeta}/\sqrt{(c_1/c)^2 - 1})$ with $(mc/\hbar)(c_1/c \pm \sqrt{(c_1/c)^2 - \sin^2\zeta})$, one obtains a function that is similar to the X-wave [41–53] but may not necessarily be a solution to (4.30) (the physical meaning could be a group of independent particles of different masses traveling in space):

$$\Phi_{X_n}^S(\vec{r}, t) = \Phi_{X_n}^S(r, \phi, z - c_1 t)$$
$$= \frac{c}{\hbar} e^{in\phi} \int_0^\infty B\left(\frac{mc}{\hbar}\right) J_n\left(\frac{mc}{\hbar} r \sin\zeta\right)$$
$$\times \exp\left\{-\frac{mc}{\hbar}\left[a_0 - i\left(c_1/c \pm \sqrt{(c_1/c)^2 - \sin^2\zeta}\right)(z - c_1 t)\right]\right\} dm,$$
$$n = 0, 1, 2, \ldots, \qquad (4.35)$$

where the subscript X_n represents an nth-order X-wave and the superscript S represents the Schrödinger equation, a_0 is a positive free parameter, and $B(k)$ is an arbitrary well-behaved transfer function. If $n = 0$ and $B(k) = a_0$, from (4.35) and (4.16), one obtains (where c_1 is a constant) [42]

$$\Phi_{X_0}^S(\vec{r}, t) = \Phi_{X_0}^S(r, \phi, z - c_1 t)$$
$$= \frac{a_0}{\sqrt{(r\sin\zeta)^2 + \left[a_0 - i\left(c_1/c \pm \sqrt{(c_1/c)^2 - \sin^2\zeta}\right)(z - c_1 t)\right]^2}}. \qquad (4.36)$$

In (4.33–4.36), if $(c_1/c)^2 - \sin^2\zeta < 0$, the solutions or functions are unbounded for some z or t and may not be of interest. If $(c_1/c)^2 - \sin^2\zeta \geq 0$, one obtains limited-diffraction solutions or functions [42]. One example is to assume that $c_1 = c/\cos\zeta$, as given in (4.17) [44]. A superposition that is similar to (4.35) can also be done over the velocity, v, instead of the mass, m, of a particle if, say, $\alpha_0 = -i(\gamma m v/\hbar)\sin\zeta$.

In this case, the superposition is a limited-diffraction solution to the Schrödinger equation (4.30).

4.2.8 Electromagnetic X-Waves

The free-space Maxwell's equations are given by [145]

$$\begin{aligned} \nabla \times \vec{E} &= -\mu_0 \frac{\partial \vec{H}}{\partial t} \qquad \nabla \cdot \vec{E} = 0 \\ \nabla \times \vec{H} &= \varepsilon_0 \frac{\partial \vec{E}}{\partial t} \qquad \nabla \cdot \vec{H} = 0, \end{aligned} \tag{4.37}$$

where $\vec{E}$ is the electric field strength, $\vec{H}$ is the magnetic field strength, ε_0 is the dielectric constant of free space ($\varepsilon_0 \approx \pi/36 \times 10^{-9} F/m$), μ_0 is the magnetic permeability of free space ($\mu_0 = 4\pi \times 10^{-7}$H/m), and t is the time.

Because of the third equation ($\nabla \cdot \vec{E} = 0$) of (4.37), the electric field strength can be written [54,146]

$$\vec{E} = -\mu_0 \frac{\partial}{\partial t} \nabla \times \vec{\prod}_m, \tag{4.38}$$

where $\vec{\prod}_m = \Phi \vec{n}^0$ is a magnetic Hertz vector potential with transverse electrical (TE) polarization, where Φ is a scalar function and $\vec{n}^0$ represents a unit vector. Inserting (4.38) into the first equation of (4.37), one obtains

$$\vec{H} = \nabla \times \left(\nabla \times \vec{\prod}_m \right). \tag{4.39}$$

From (4.37–4.39), one obtains the vector wave equation

$$\nabla^2 \vec{\prod}_m - \frac{1}{c^2} \frac{\partial^2 \vec{\prod}_m}{\partial t^2} = 0. \tag{4.40}$$

Letting $\vec{n}^0 = \vec{z}^0$, where $\vec{z}^0$ is a unit vector along the z-axis, and using cylindrical coordinates from (4.38) and (4.39) we obtain

$$\vec{E} = -\mu_0 \frac{1}{r} \frac{\partial^2 \Phi}{\partial t \, \partial \phi} \vec{r}^0 + \mu_0 \frac{\partial^2 \Phi}{\partial t \, \partial r} \vec{\phi}^0 \tag{4.41}$$

and

$$\vec{H} = \frac{\partial^2 \Phi}{\partial r \, \partial z} \vec{r}^0 + \frac{1}{r} \frac{\partial^2 \Phi}{\partial \phi \, \partial z} \vec{\phi}^0 + \left(\frac{\partial^2 \Phi}{\partial z^2} - \frac{1}{c^2} \frac{\partial^2 \Phi}{\partial t^2} \right) \vec{z}^0, \tag{4.42}$$

respectively, where Φ is a solution to the free-space scalar wave equation (4.2), and where $\vec{r}^0$ and $\vec{\phi}^0$ are the unit vectors along the variables r and ϕ, respectively. Once a solution to (4.2) is found, the electrical field strength, $\vec{E}$, and the magnetic field strength, $\vec{H}$, can be obtained from Eqs. (4.41) and (4.42), respectively.

If Φ is an nth-order broadband X-wave solution or a general X-wave solution [see (4.15)] to (4.2), the components of $\vec{E}$ and $\vec{H}$ are also X-wave functions [54]. From the $\vec{E}$ and $\vec{H}$ expressions, the Poynting energy flux vector and the energy density can be derived [54]. Solutions to $\vec{E}$ and $\vec{H}$ obtained this way will be limited-diffraction solutions to Maxwell's equations in (4.37) [54].

4.2.9 Limited-Diffraction Beams in Confined Spaces

Limited-diffraction beams in confined spaces are of interest [59,60,147]. Previously, Shaarawi et al. [148] and Ziolkowski et al. [149] have shown that localized waves such as focused wave modes and modified power spectrum pulses can also propagate in waveguides for an extended propagation depth. In the following, theoretical results of X-waves propagating in a confined space such as a waveguide are developed for acoustics, electromagnetics, and quantum mechanics [147].

1. *Acoustic waves.* Assuming that Φ in (4.2) represents acoustic pressure in an infinitely long cylindrical acoustical waveguide (radius a), which is filled with an isotropic/homogeneous lossless fluid medium enclosed in an infinitely rigid boundary, the normal vibration velocity of the medium at the wall of the cylindrical waveguide is zero for all the frequency components of the X-waves [i.e., $\partial\Phi_{X_n}(\vec{r}, t; \omega)/\partial r \equiv 0$, $\forall\omega \geq 0$ at $r = a$, where $\Phi_{X_n}(\vec{r}, t; \omega)$ is the X-wave component at angular frequency ω; see (4.15)]. To meet this boundary condition, the parameter k in (4.15) is quantized:

$$k_{nj} = \frac{\mu_{nj}}{a \sin\zeta}, \qquad n, j = 0, 1, 2, \ldots, \tag{4.43}$$

where μ_{nj} are the roots of the equations

$$\begin{aligned} J_1(x) &= 0, & n &= 0 \\ J_{n-1}(x) &= J_{n+1}(x), & n &= 1, 2, \ldots. \end{aligned} \tag{4.44}$$

Thus, the integral in (4.15) can be changed to a series representing frequency-quantized X-waves [147]:

$$\Phi_{X_n}(\vec{r}, t) = e^{in\phi} \sum_{j=0}^{\infty} \Delta k_{nj} B(k_{nj}) J_n(k_{nj} r \sin\zeta) e^{-k_{nj}[a_0 - i\cos\zeta(z - c_1 t)]}, \qquad r \leq a, \qquad n = 0, 1, 2, \ldots, \tag{4.45}$$

where $\Delta k_{n0} = k_{n1}$ and $\Delta k_{nj} = k_{nj+1} - k_{nj}$ $(j = 1, 2, 3, \ldots)$. Unlike conventional guided waves, frequency-quantized X-waves contain multiple frequencies and propagate through waveguides at the speed of c_1 without dispersion, and similar results can be obtained for waveguides of other homogeneous boundary conditions. For an

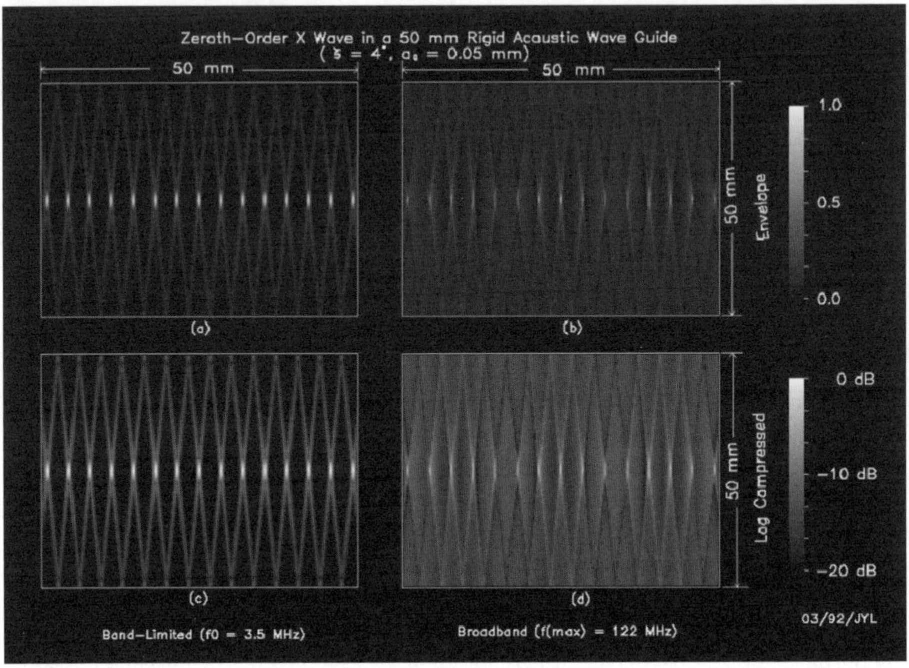

FIGURE 4.1 Envelope-detected zeroth-order X-wave in a 50 mm-diameter rigid acoustic waveguide. The waves shown has an axicon angle of 4° and $a_0 = 0.05$ mm. (a) and (c) Bandlimited version with a Blackman window function centered at 3.5 MHz with about 81% of fractional −6 dB bandwidth. (b) and (d) are broadband versions. The images in the top row are on a linear scale and those in the bottom row are of log scale, to show the sidelobes.

infinitely long cylindrical acoustical waveguide consisting of isotropic/homogeneous lossless media in a free space (vacuum) with radius a, the acoustical pressure is zero at the boundary of the waveguide, $r = a$ [i.e., $\mu_{nj}(n, j = 1, 2, 3, \ldots)$ in (4.43) are roots of $J_n(x) = 0\,(j = 1, 2, \ldots)$]. See Figs. 4.1 to 4.3 for examples of X-waves in an acoustic waveguide [147].

It is clear that if $n = 0$, (4.45) represents an axially symmetric frequency-quantized X-wave. If $a \to \infty$, $\Delta k_{nj} \to 0$ and the summation in (4.45) becomes an integration that represents the X-waves in (4.15). On the other hand, if $a \to 0$, both k_{nj} and $\Delta k_{nj} \to \infty$ $(n, j = 0, 1, 2, \ldots)$. This means that for a small waveguide, only high-frequency-quantized X-waves can propagate through it.

2. *Electromagnetic waves.* The free-space vector wave equations from the free-space Maxwell's equations (4.37) are given by [150]

$$\nabla^2 \vec{E} - \frac{1}{c^2}\frac{\partial^2 \vec{E}}{\partial t^2} = 0 \tag{4.46}$$

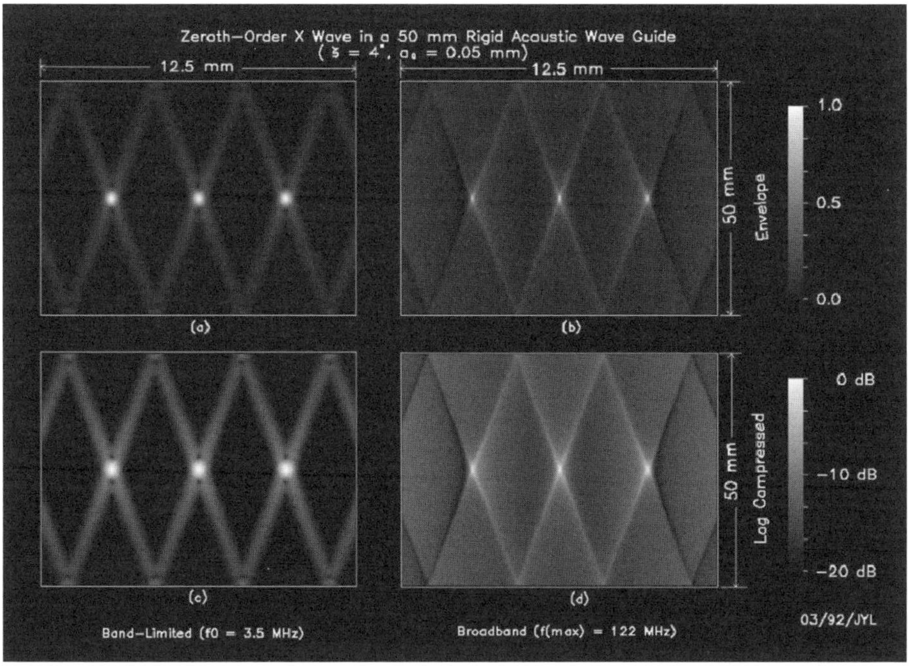

FIGURE 4.2 The same as those in Fig. 4.1 except that the images are zoomed horizontally around the center.

and

$$\nabla^2 \vec{H} - \frac{1}{c^2} \frac{\partial^2 \vec{H}}{\partial t^2} = 0. \tag{4.47}$$

A solution to (4.46) can be written as

$$\vec{E}(\vec{r}, t) = \vec{E}_{\perp}(r, \phi) e^{\gamma z - i\omega t}, \tag{4.48}$$

where $\gamma = i\beta$ is a propagation constant, $\beta = \sqrt{k^2 - k_c^2} > 0$ (for propagation waves), $k = \omega / c$ is the wave number, and $\vec{E}_{\perp}(r, \phi)$ is a solution of the transverse vector Helmholtz equation:

$$\nabla_{\perp}^2 \vec{E}_{\perp}(r, \phi) + k_c^2 \vec{E}_{\perp}(r, \phi) = 0, \tag{4.49}$$

where $\nabla_{\perp}^2$ is the transverse Laplace operator and k_c is a parameter that is independent of r, ϕ, z, and t. For transverse magnetic (TM) waves, $\vec{E}_{\perp}(r, \phi) = E_z(r, \phi)\vec{z}^0$ and (4.49) becomes a scalar Helmholtz equation of $E_z(r, \phi)$, where $\vec{z}^0$ is a unit vector along the z-axis.

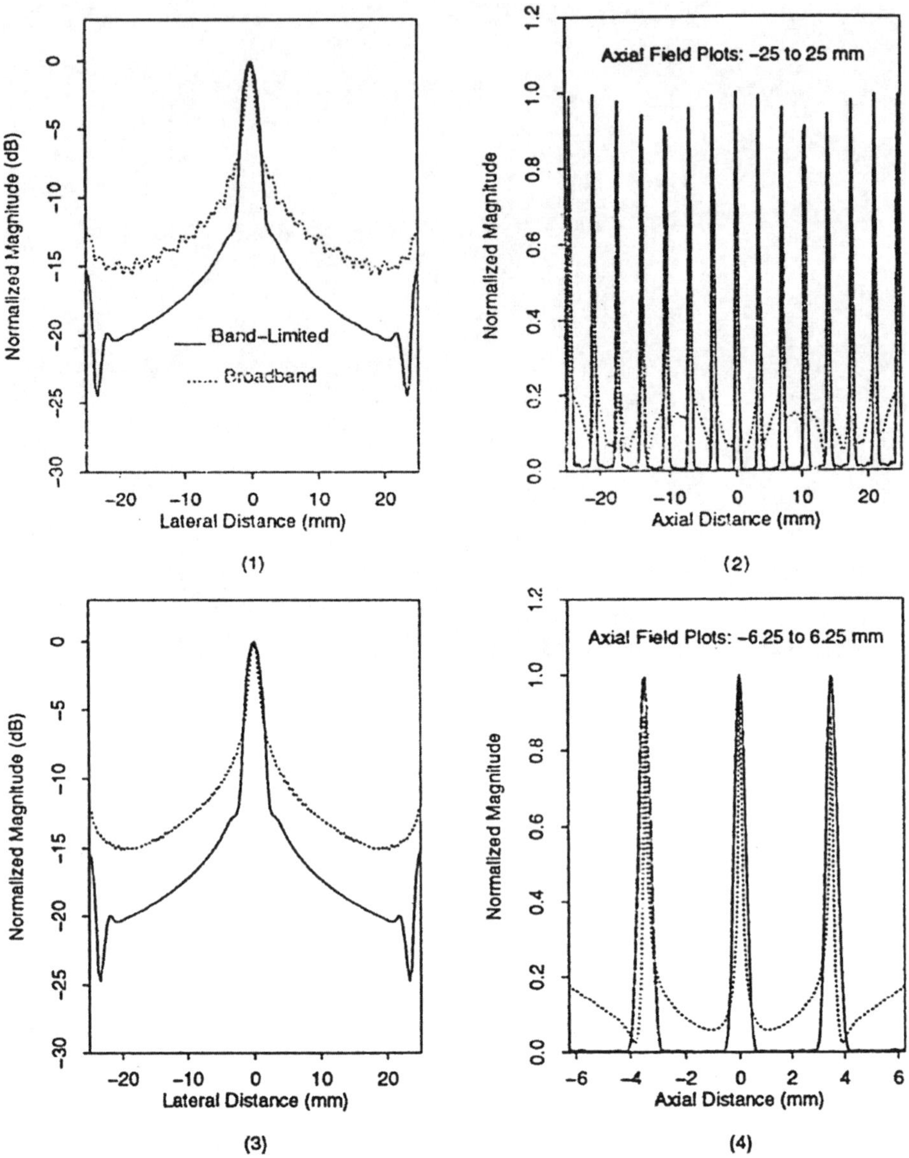

FIGURE 4.3 Transverse [(1) and (3)] and axial [(2) and (4)] sidelobe plots of the images in Fig. 4.1 [(1) and (2)] and Fig. 4.2 [(3) and (4)], respectively. Solid and dotted lines are for bandlimited and broadband cases, respectively.

If $k_c = k \sin\zeta$, where $|\zeta| < \pi/2$ is a constant, after taking into consideration the exponential term in (4.48) and integrating the solution of (4.49) from 0 to ∞ over k, one obtains an nth-order X-wave solution [replace the symbol, $\Phi_{X_n}(\vec{r}, t)$, in (4.15) with $E_{z_{n\mathrm{X}}}(\vec{r}, t)$, where the subscript X indicates an X-wave]. Assuming that electromagnetic X-waves travel in vacuum in a totally conductive cylindrical waveguide of a radius, a [i.e., $E_{z_{n\mathrm{X}}}(\vec{r}, t) \equiv 0$ at $r = a$], similar to the frequency quantization procedure of the acoustic case (4.45), one obtains [147]

$$E_{z_{n\mathrm{X}}}(\vec{r}, t) = e^{in\phi} \sum_{j=0}^{\infty} \Delta k_{nj} B(k_{nj}) J_n(k_{nj} r \sin\zeta) e^{-k_{nj}[a_0 - i\cos\zeta(z - c_1 t)]}, \qquad r \le a,$$
$$n = 0, 1, 2, \ldots, \qquad (4.50)$$

where the k_{nj} $(n, j = 0, 1, 2, \ldots)$ are given by (4.43) and the μ_{nj} $(n, j = 0, 1, 2, \ldots)$ in (4.43) are roots of $J_n(x) = 0\,(n = 0, 1, 2, \ldots)$. Other components of $\vec{E}$ and $\vec{H}$ can be derived from $E_z(\vec{r}, t)$ using the free-space Maxwell's equations (4.37). They will have the same speed, c_1, as E_z. For transverse electric (TE) waves, the results are similar.

3. *DeBroglie waves.* With a finite transverse spatial extension (such as a free particle passing through a hole with a finite aperture), the function $\Phi_{X_n}^{\mathrm{KG}}(\vec{r}, t)$ in (4.28) or $\Phi_{X_n}^{\mathrm{S}}(\vec{r}, t)$ in (4.35) would change (spread or diffract) after certain distance behind the hole. However, in cases such as particles passing through a pipe, $\Phi_{X_n}^{\mathrm{KG}}(\vec{r}, t)$ and $\Phi_{X_n}^{\mathrm{S}}(\vec{r}, t)$ need to meet the boundary conditions that they are zero on the wall of the pipe. This gives the following quantized X-wave functions corresponding to (4.28) and (4.35), respectively [147]:

$$\Phi_{X_n}^{\mathrm{KG}}(\vec{r}, t) = e^{in\phi} \sum_{j=0}^{\infty} \Delta k_{nj} B(k_{nj}) J_n(k_{nj} r \sin\zeta) e^{-k_{nj}\left[a_0 - i\left(\sqrt{1+\sin^2\zeta}/\sqrt{(c_1/c)^2 - 1}\right)(z - c_1 t)\right]},$$
$$r \le a, \quad n = 0, 1, 2, \ldots, \qquad (4.51)$$

and

$$\Phi_{X_n}^{\mathrm{S}}(\vec{r}, t) = e^{in\phi} \sum_{j=0}^{\infty} \Delta k_{nj} B(k_{nj}) J_n(k_{nj} r \sin\zeta) e^{-k_{nj}\left[a_0 - i\left(c_1/c \pm \sqrt{(c_1/c)^2 - \sin^2\zeta}\right)(z - c_1 t)\right]},$$
$$r \le a, \quad n = 0, 1, 2, \ldots, \qquad (4.52)$$

where $k_{nj} = m_{nj} c/\hbar\ (n, j = 0, 1, 2, \ldots)$ are given by (4.43) and μ_{nj} $(n, j = 0, 1, 2, \ldots)$ in (4.43) are roots of $J_n(x) = 0\,(n = 0, 1, 2, \ldots)$. Equations (4.51) and (4.52) represent particles in a confined space with their quantized de Broglie's waves. The quantization may only allow particles of a certain mass to pass through the pipe (waveguide). As mentioned in the text below (4.27) and (4.34), the free parameter α_0 can be chosen differently. If $\alpha_0 = -i(\gamma m v/\hbar)\sin\zeta$, the quantization in (4.51) and (4.52) may be modified for summation over the velocity v, instead of m, of the particles. In this case, only particles with certain velocities are allowed to pass through the pipe or a small nanotube.

There are other implications of the studies above. As we know, light in free space behaves like a wave but acts as particles (photons) when interacting with materials. Some microscopic structures of materials could be considered as optical waveguides within which the light waves are confined. From our discussion above of X-waves in confined spaces, it is understood that only light waves that have a higher energy (or frequency) can penetrate these materials or cause interactions.

4.2.10 X-Wave Transformation

Because X-waves are orthogonal [61], similar to plane waves, any physically realizable waves or well-behaved solutions to the wave equation can be expressed as a linear superposition of X-waves (inverse X-wave transform), and the coefficients of the superposition can be determined (forward X-wave transform), [46,47]. The inverse X-wave transform is given by (Eq. (4.15) of [46])

$$\begin{aligned}\Phi(\vec{r}, t) &= \sum_{n=-\infty}^{\infty} \int_0^{\pi/2} d\zeta \int_0^{\infty} dk T_{n,\zeta}(k) \Phi_{A_{n,k,\zeta}}(r, \phi, z - c_1 t) \\ &= \sum_{n=-\infty}^{\infty} \int_0^{\pi/2} \left[e^{in\phi} \int_0^{\infty} T_{n,\zeta}(k) J_n(kr \sin\zeta) e^{ik\cos\zeta(z - c_1 t)}\, dk \right] d\zeta \\ &= \sum_{n=-\infty}^{\infty} \int_0^{\pi/2} \Phi_{X_{n,\zeta}}(r, \phi, z - c_1 t)\, d\zeta, \end{aligned} \tag{4.53}$$

where

$$T_{n,\zeta}(k) = B_{n,\zeta}(k) e^{-ka_0} \tag{4.54}$$

and

$$\Phi_{A_{n,k,\zeta}}(r, \phi, z - c_1 t) = e^{in\phi} J_n(kr \sin\zeta) e^{ik\cos\zeta(z - c_1 t)}, \tag{4.55}$$

where $c_1 = c/\cos\zeta$ and $|\zeta| < \pi/2$.

The forward X-wave transform can be used to determine the coefficients (Eq. (4.26) of [46]):

$$\begin{aligned}T_{n,\zeta}(k) = &\frac{k^2 c \sin\zeta \cos\zeta\, H(k)}{(2\pi)^2} \\ &\times \int_0^{\infty} r dr \int_{-\pi}^{\pi} d\phi \int_{-\infty}^{\infty} dt \Phi(r, \phi, z, t) \Phi^*_{A_{n,k,\zeta}}(r, \phi, z - c_1 t), \end{aligned} \tag{4.56}$$

where

$$\Phi^*_{A_{n,k,\zeta}}(r, \phi, z - c_1 t) = e^{-in\phi} J_n(kr \sin \zeta) e^{-ik \cos \zeta (z - c_1 t)} \tag{4.57}$$

is a complex conjugate of $\Phi_{A_{n,k,\zeta}}(r, \phi, z - c_1 t)$ and $H(k)$ is the Heaviside step function [151]:

$$H(k) = \begin{cases} 1, & k \geq 0 \\ 0, & \text{otherwise.} \end{cases} \tag{4.58}$$

$H(k)$ is used to indicate that k is positive and thus can be placed on either side of (56).

4.2.11 Bowtie Limited-Diffraction Beams

If $\Phi_N(\vec{r}_N, t) = \Phi_N(\vec{r}_{N-1}, x_N - c_1 t)$ is a limited-diffraction solution to the isotropic/homogeneous wave equation (4.1), the Klein–Gordon equation (4.20), or the Schrödinger equation (4.30) (assuming that V is not a function of the corresponding component of $\vec{r}_{N-1}$), where $\vec{r}_N = (x_1, x_2, \ldots, x_N)$, $\vec{r}_{N-1} = (x_1, x_2, \ldots, x_{N-1})$, N is an integer, and c_1 is the speed of the wave, any partial derivatives of $\Phi_N(\vec{r}_{N-1}, x_N - c_1 t)$ along any component of $\vec{r}_{N-1}$ are still limited-diffraction solutions to these equations [152–156]. These solutions are called *bowtie beams* because their transverse beam shapes are similar to the shape of a bowtie. These beams may have applications in medical imaging of a lower sidelobe because one part of the sidelobe of a transmission beam may be used to cancel the other part of the sidelobe of a reception beam [152–156]. [Note that the following properties are also true. Any partial derivatives of a limited-diffraction solution, $\Phi_N(\vec{r}_{N-1}, x_N - c_1 t)$, in terms of the time t will also be a limited-diffraction solution to (4.1), (4.20), and (4.30), respectively. One example is the second derivative X-wave in terms of time given in [44]. Replacing t with $-t$ in $\Phi_N(\vec{r}_{N-1}, x_N - c_1 t)$, one obtains a time-reversal mirror limited-diffraction wave propagating in a backward direction along x_N.]

4.2.12 Limited-Diffraction Array Beams

If the partial derivatives are carried out on more than one component of $\vec{r}_{N-1} = (x_1, x_2, \ldots, x_{N-1})$ for $\Phi_N(\vec{r}_N, t) = \Phi_N(\vec{r}_{N-1}, x_N - c_1 t)$, limited-diffraction grid or layered array beams may be produced for equations (4.1), (4.20), and (4.30) (assuming that V is not a function of the corresponding components of $\vec{r}_{N-1}$) [157–160]. Array beams may have applications to three-dimensional imaging [157], blood-flow velocity measurements [158], and high-frame-rate imaging [62,63,79–85].

4.2.13 Computation with Limited-Diffraction Beams

Efficient computation of limited-diffraction beams produced by a finite aperture is important for understanding the properties of these beams. A Fourier–Bessel method

[24–28] has been used to calculate arbitrary waves of axial symmetry. Limited-diffraction array beams [157–160] have been used for efficient computation of waves produced by a two-dimensional array transducer [159,160]. Angular spectrum decomposition has been used for the study [161], and various methods have been investigated [162].

4.3 APPLICATIONS OF LIMITED-DIFFRACTION BEAMS

4.3.1 Medical Ultrasound Imaging

Limited-diffraction beams are localized waves and are, in theory, propagation invariant. In practice, because the dimension of wave sources is always finite, these waves will eventually diffract. However, these waves have a large depth of field, meaning that they will propagate over a large distance without spreading. This property is useful in medical ultrasound imaging, where an extended depth of focus is needed to provide clear images over the entire depth of interest within the thickness of the human body. Studies on this subject have been reported in the literature (e.g., [15–17, 163–168]).

4.3.2 Tissue Characterization (Identification)

Due to the large depth of field of limited-diffraction beams, these beams may be used for tissue characterization (identification) [169–171]. For example, different tissues have different attenuations on ultrasound waves. If the waves diffract as they propagate, such as conventional focused waves, one has to compensate for the diffraction effects of the waves in the estimation of tissue attenuation. The compensation process could be computationally intensive and tedious. An example of tissue characterization with limited-diffraction beams is given in [171].

4.3.3 High-Frame-Rate Imaging

High-frame-rate two- and three-dimensional ultrasound imaging is important for visualizing fast-moving objects such as the heart. Based on our previous studies of ultrasound diffraction tomography [172–176] and limited-diffraction beams such as X-waves [41–53], we have developed the high-frame-rate imaging method [62–88]. Recently, the method has been extended to include steered plane wave and limited-diffraction array-beam transmissions [79–85].

4.3.4 Two-Way Dynamic Focusing

A two-way dynamic focusing method was developed by transmitting limited-diffraction array beams and receiving ultrasound echo signals with array beam weightings of the same parameters. This method increases the image field of view and image resolution due to enlarged coverage of spatial Fourier domain [177].

4.3.5 Medical Blood-Flow Measurements

Blood-flow velocity measurements and imaging are important for medical diagnoses [178–179]. However, with the conventional Doppler method, only flow velocity that is along the ultrasound beam can be measured. To measure the velocity vector, velocity components along and transverse to the beam are both needed. Limited-diffraction beams may help to measure the transverse component of the velocity more accurately, due to their spatial modulation properties [158,180,181].

4.3.6 Nondestructive Evaluation of Materials

Nondestructive evaluation (NDE) is important for many applications, such as finding defects in aircraft engines with ultrasound without slicing them apart or destroying them. Similar to medical imaging, limited-diffraction beams can also be applied to NDE on various industrial materials by getting images of a large depth of field [182,183].

4.3.7 Optical Coherent Tomography

Optical coherent tomography (OCT) uses the same principle of conventional ultrasound pulse-echo imaging. It is able to obtain microscopic images of a cross section along an optical beam. Similar to ultrasound imaging, limited-diffraction beams can be used to increase the depth of field of OCT [184].

4.3.8 Optical Communications

Limited-diffraction beams such as X-waves [41–43] are orthogonal in space. Because of this property, signals such as television programs in different channels can be sent over the same space from the same channel (carrier frequency). Limited-diffraction beams have been exploited to increase the capacity in communications using the property of their spatial orthogonality [185–186].

4.3.9 Reduction of Sidelobes in Medical Imaging

Limited-diffraction beams can maintain high resolution in medical imaging over a large depth of field. However, compared to focused beams at their focuses, limited-diffraction beams have higher sidelobes. Sidelobes may lower image contrast in ultrasound imaging, making the differentiation between benign and malignant tissues difficult. Various methods have been developed to reduce sidelobes of limited-diffraction beam in medical imaging [5,187–190].

4.4 CONCLUSIONS

Limited-diffraction beams are a class of waves that may be localized in both space and time and can propagate rigidly in free space or confined spaces to an infinite distance in theory at superluminal speed. Because of the localized property and the fact that they are solutions to various wave equations, limited-diffraction beams may provide insight into various physical phenomena and may have theoretical significance. In addition, limited-diffraction beams can be produced approximately with a finite aperture and energy over a large depth of field, meaning that they can keep a small beam width over a large distance. This and other properties of limited-diffraction beams make them suitable for various applications, such as medical imaging, tissue characterization, blood-flow measurement, nondestructive evaluation of materials, and optical communications.

Acknowledgments

This work was supported in part by grant HL60301 from the National Institutes of Health.

REFERENCES

1. J. A. Stratton, *Electromagnetic Theory,* McGraw-Hill, New York, 1941, p. 356.
2. J. Durnin, Exact solutions for nondiffracting beams: I. The scalar theory, *J. Opt. Soc. Am. A* **4**(4), 651–654 (1987).
3. J. Durnin, J. J. Miceli, Jr., and J. H. Eberly, Diffraction-free beams, *Phys. Rev. Lett.* **58**(15), 1499–1501 (Apr. 13, 1987).
4. J. Durnin, J. J. Miceli, Jr., and J. H. Eberly, *Experiments with Nondiffracting Needle Beams*, Optical Society of America, Washington, DC, 1987, p. 208; available from IEEE Service Center, Piscataway, NJ (cat. no. 87CH2391–1).
5. J.-y. Lu and J. F. Greenleaf, Sidelobe reduction for limited diffraction pulse-echo systems, *IEEE Trans. Ultrason. Ferroelectr. Freq. Control* **40**(6), 735–746 (Nov. 1993).
6. G. Indebetow, Nondiffracting optical fields: some remarks on their analysis and synthesis, *J. Opt. Soc. Am. A* **6**(1), 150–152 (Jan. 1989).
7. F. Gori, G. Guattari, and C. Padovani, Model expansion for J0-correlated Schell-model sources, *Opt. Commun.* **64**(4), 311–316 (Nov. 15, 1987).
8. K. Uehara and H. Kikuchi, Generation of near diffraction-free laser beams, *Appl. Phys. B* **48**, 125–129 (1989).
9. L. Vicari, Truncation of nondiffracting beams, *Opt. Commun.* **70**(4), 263–266 (Mar. 15, 1989).
10. M. Zahid and M. S. Zubairy, Directionally of partially coherent Bessel–Gauss beams, *Opt. Commun.* **70**(5), 361–364 (Apr. 1, 1989).
11. S. Y. Cai, A. Bhattacharjee, and T. C. Marshall, "Diffraction-free" optical beams in inverse free electron laser acceleration, *Nucl. Instrum. Methods Phys. Res. A* **272**(1–2), 481–484 (Oct. 1988).
12. J. Ojeda-Castaneda and A. Noyola-lglesias, Nondiffracting wavefields in grin and free-space, *Microwave Opt. Technol. Lett.* **3**(12), 430–433 (Dec. 1990).
13. D. K. Hsu, F. J. Margetan, and D. O. Thompson, Bessel beam ultrasonic transducer: fabrication method and experimental results, *Appl. Phys. Lett.* **55**(20), 2066–2068 (Nov. 13, 1989).

14. J. A. Campbell and S. Soloway, Generation of a nondiffracting beam with frequency independent beam width, *J. Acoust. Soc. Am.* **88**(5), 2467–2477 (Nov. 1990).
15. J.-y. Lu and J. F. Greenleaf, Ultrasonic nondiffracting transducer for medical imaging, *IEEE Trans. Ultrasonics, Ferroelectr. Freq. Control* **37**(5), 438–447, (Sept. 1990).
16. J.-y. Lu and J. F. Greenleaf, Pulse-echo imaging using a nondiffracting beam transducer, *Ultrasound Med. Biol.* **17**(3), 265–281 (May 1991).
17. J.-y. Lu and J. F. Greenleaf, Ultrasonic nondiffracting transducer, U.S. patent 5,081,995, Jan. 21, 1992.
18. J.-y. Lu and J. F. Greenleaf, A computational and experimental study of nondiffracting transducer for medical ultrasound, *Ultrason. Imag.* **12**(2), 146–147 (Apr. 1990).
19. J.-y. Lu and J. F. Greenleaf, Effect on nondiffracting beam of deleting central elements of annular array transducer, *Ultrason. Imag.* **13**(2), 203 (Apr. 1991).
20. J.-y. Lu and J. F. Greenleaf, Simulation of imaging contrast of nondiffracting beam transducer, *J. Ultrasound Med.* **10**(3) (suppl.), S4 (Mar. 1991).
21. J.-y. Lu, Theoretical study of medical ultrasonic nondiffracting transducers, presented at the Whitaker Foundation's 3rd Annual Conference, July 30–Aug. 1, 1993.
22. D. Ding and J.-y. Lu, Second harmonic generation of the *n*th-order Bessel beams, *Phys. Rev. E* **61**(2), 2038–2041 (Feb. 2000).
23. D. Ding and J.-y. Lu, Higher-order harmonics of limited diffraction Bessel beams, *J. Acoust. Soc. Am.* **107**(3), 1212–1214 (Mar. 2000).
24. P. Fox, J. Cheng, and J.-y. Lu, Fourier–Bessel field calculation and tuning of a CW annular array, *IEEE Trans. Ultrason. Ferroelectr. Freq. Control* **49**(9), 1179–1190 (Sept. 2002).
25. P. D. Fox, J. Cheng, and J.-y. Lu, Theory and experiment of Fourier–Bessel field calculation and tuning of a PW annular array, *J. Acoust. Soc. Am.* **113**(5), 2412–2423 (May 2003).
26. P. D. Fox, J. Cheng, and J.-y. Lu, Ultrasound pulse modeling and tuning with Fourier–Bessel method for an annular array, *Ultrason. Imag.* **22**(4), 257 (Oct. 2000).
27. P. Fox and J.-y. Lu, Modelling of pulsed annular arrays using limited diffraction Bessel beams, *Ultrasound Med.* **20**(3) (suppl.), S66 (Mar. 2001).
28. P. Fox, J.-y. Lu, S. Holm, and F. Tranquart, Connection between X-waves, Fourier–Bessel series and optimal modeling aperture for circular symmetric arrays, *2005 IEEE Ultrasonics Symposium Proceedings*, 05CH37716C, Vol. 2, 2005, pp. 1644–1647.
29. J.-y. Lu, Construction of limited diffraction beams with Bessel bases, *1995 IEEE Ultrasonics Symposium Proceedings*, 95CH35844, Vol. 2, 1995, pp. 1393–1397.
30. J.-y. Lu, Designing limited diffraction beams, *IEEE Trans. Ultrason. Ferroelectr. Freq. Control* **44**(1), 181–193 (Jan. 1997).
31. J. N. Brittingham, Focus wave modes in homogeneous Maxwell's equations: transverse electric mode, *J. Appl. Phys.* **54**(3), 1179–1189 (1983).
32. R. W. Ziolkowski, Exact solutions of the wave equation with complex source locations, *J. Math. Phys.* **26**(4), 861–863 (Apr. 1985).
33. R. W. Ziolkowski, D. K. Lewis, and B. D. Cook, Evidence of localized wave transmission, *Phys. Rev. Lett.* **62**(2), 147–150 (Jan. 9, 1989).
34. A. M. Shaarawi, I. M. Besieris, and R. W. Ziolkowski, Localized energy pulse train launched from an open, semi-infinite, circular waveguide, *J. Appl. Phys.* **65**(2), 805–813 (1989).

35. I. M. Besieris, A. M. Shaarawi, and R. W. Ziolkowski, A bidirectional traveling plane wave representation of exact solutions of the scalar wave equation, *J. Math. Phys.* **30**(6), 1254–1269 (1989).
36. E. Heyman, B. Z. Steinberg, and L. B. Felsen, Spectral analysis of focus wave modes, *J. Opt. Soc. Am. A* **4**(11), 2081–2091 (Nov. 1987).
37. R. W. Ziolkowski, Localized transmission of electromagnetic energy, *Phys. Rev. A* **39**(4), 2005–2033 (Feb. 15, 1989).
38. J. V. Candy, R. W. Ziolkowski, and D. K. Lewis, Transient waves: reconstruction and processing, *J. Acoust. Soc. Am.* **88**(5), 2248–2258 (Nov. 1990).
39. J. V. Candy, R. W. Ziolkowski, and D. K. Lewis, Transient wave estimation: a multichannel deconvolution application, *J. Acoust. Soc. Am.* **88**(5), 2235–2247 (Nov. 1990).
40. R. W. Ziolkowski and D. K. Lewis, Verification of the localized wave transmission effect, *J. Appl. Phys.* **68**(12), 6083–6086, (Dec. 15, 1990).
41. J.-y. Lu and J. F. Greenleaf, Theory and acoustic experiments of nondiffracting X-waves, *1991 IEEE Ultrasonics Symposium Proceedings*, 91CH3079-1, Vol. 2, 1991, pp. 1155–1159.
42. J.-y. Lu and J. F. Greenleaf, Nondiffracting X-waves: exact solutions to free-space scalar wave equation and their finite aperture realizations, *IEEE Trans. Ultrason. Ferroelectr. Freq. Control* **39**(1), 19–31(Jan. 1992).
43. J.-y. Lu and J. F. Greenleaf, Experimental verification of nondiffracting X-waves, *IEEE Trans. Ultrason. Ferroelectr. Freq. Control* **39**(3), 441–446 (May 1992).
44. J.-y. Lu, H. Zou, and J. F. Greenleaf, A new approach to obtain limited diffraction beams, *IEEE Trans. Ultrason. Ferroelectr. Freq. Control* **42**(5), 850–853 (Sept. 1995).
45. H. Zou, J.-y. Lu, and J. F. Greenleaf, Obtaining limited diffraction beams with the wavelet transform, *1993 IEEE Ultrasonics Symposium Proceedings*, 93CH3301-9, Vol. 2, 1993, pp. 1087–1090.
46. J.-y. Lu and A. Liu, An X-wave transform, *IEEE Trans. Ultrason. Ferroelectr. Freq. Control* **47**(6), 1472–1481 (Nov. 2000).
47. J.-y. Lu and A. Liu, Representation of solutions to wave equation with X-waves, *Ultrason. Imag.* **21**(1), 70 (Jan. 1999).
48. W. A. Rodrigues, Jr., and J.-y. Lu, On the existence of undistorted progressive waves (UPWs) of arbitrary speeds $0 < v < \infty$ in nature, *Found. Phys.* **27**(3), 435–508 (Mar. 1997).
49. Tai K. Song, J.-y. Lu, and J. F. Greenleaf, Modified X-waves with improved field properties, *Ultrason. Imag.* **15**(1), 36–47 (Jan. 1993).
50. J.-y. Lu, and J. F. Greenleaf, Formation and propagation of limited diffraction beams, in Yu Wei and Ben-li Gu, Eds; *Acoustical Imaging*, Vol. 20, Kluwer Academic/Plenum, NY, 1993, pp. 331–343.
51. J.-y. Lu and J. F. Greenleaf, Comparison of sidelobes of limited diffraction beams and localized waves, in J. P. Jones, Ed., *Acoustical Imaging*, Vol. 21, Kluwer Academic/Plenum, NY, 1995, pp. 145–152.
52. J.-y. Lu and J. F. Greenleaf, Steering of limited diffraction beams with a two-dimensional array transducer, *1992 IEEE Ultrasonics Symposium Proceedings*, 92CH3118-7, Vol. 1, 1992, pp. 603–607.

53. J.-y. Lu and J. F. Greenleaf, A study of two-dimensional array transducers for limited diffraction beams, *IEEE Trans. Ultrason. Ferroelectr. Frequency Control* **41**(5), 724–739 (Sep. 1994).

54. J.-y. Lu, J. F. Greenleaf, and E. Recami, Limited diffraction solutions to Maxwell and Schrödinger equations, LANL archives physics/9610012, Report INFN/FM-96/01, Instituto Nagionole de Fisica Nucleare, Frascati, Italy, Oct. 1996.

55. C. Day, Intense X-shaped pulses of light propagate without spreading in water and other dispersive media, "Search and Discovery" column of *Phys. Today* **57**(10), 25–26 (Oct. 2004).

56. P. Saari and K. Reivelt, Evidence of X-shaped propagation-invariant localized light waves, *Phys. Rev. Lett.* **79**, 4135–4138 (1997).

57. I. Besieris, M. Abdel-Rahman, A. Shaarawi, and A. Chatzipetros, Two fundamental representations of localized pulse solutions to the scalar wave equation, *Prog. Electromagn. Res.* **19**, 1–48 (1998).

58. P. Saari and K. Reivelt, Generation and classification of localized waves by Lorentz transformations in Fourier space, *Phys. Rev. E* **69**, 036612 (2004).

59. M. Zamboni-Rached, K. Z. Nóbrega, E. Recami, and H. E. Hernández-Figueroa, Superluminal X-shaped beams propagating without distortion along a coaxial guide, *Phys. Rev. E* **66**, 046617 (2002).

60. M. Zamboni-Rached, E. Recami, and F. Fontana, Superluminal localized solutions to Maxwell equations propagating along a normal-sized waveguide, *Phys. Rev. E* **64**, 066603 (2001).

61. J. Salo, A. T. Friberg, and M. M. Salomaa, Orthogonal X-waves, *J. Phys. A Math. Gen.* **34**, 9319–9327 (2001).

62. J.-y. Lu, 2D and 3D high frame rate imaging with limited diffraction beams, *IEEE Trans. Ultrason. Ferroelectr. Freq. Control* **44**(4), 839–856 (July 1997).

63. J.-y. Lu, Experimental study of high frame rate imaging with limited diffraction beams, *IEEE Trans. Ultrason. Ferroelectr. Freq. Control* **45**(1), 84–97 (Jan. 1998).

64. J.-y. Lu, Limited diffraction beams for high frame rate 2D and 3D pulse-echo imaging, *J. Ultrasound Med.* **16**(3) (suppl.), S10 (Mar. 1997).

65. J.-y. Lu and J. F. Greenleaf, High frame rate imaging with limited diffraction beams, U. S. patent 5,720,708, Feb. 24, 1998.

66. H. Peng and J.-y. Lu, High frame rate 2D and 3D imaging with a curved or cylindrical array, *2002 IEEE Ultrasonics Symposium Proceedings*, 02CH37388, Vol. 2, 2002, pp. 1725–1728.

67. H. Peng and J.-y. Lu, Construction of high frame rate images with Fourier transform, *J. Acoust. Soc. Am.* **111**(5), pt. 2, 2385–2386 (2002).

68. J.-y. Lu and A. Liu, Comparison of high-frame rate and delay-and-sum imaging methods, *J. Acoust. Soc. Am.* **106**(4), pt. 2, 2136 (Oct. 1999).

69. J.-y. Lu and S. He, High frame rate imaging with a small number of array elements, *IEEE Trans. Ultrason. Ferroelectr. Freq. Control* **46**(6), 1416–1421 (Nov. 1999).

70. J.-y. Lu and S. He, Reducing number of elements of transducer arrays in Fourier image construction method, *1998 IEEE Ultrasonics Symposium Proceedings*, 98CH36102, Vol. 2, 1998, pp. 1619–1622.

71. J.-y. Lu and S. He, Effects of phase aberration on high frame rate imaging, *Ultrasound Med. Bio.* **26**(1), 143–152 (2000).
72. J.-y. Lu, Assessment of phase aberration effects on high frame rate imaging, *Ultrason. Imag.* **19**(1), 53 (Jan. 1997).
73. J.-y. Lu, A study of signal-to-noise ratio of the Fourier method for construction of high frame rate images, in Hua Lee, Ed., *Acoustical Imaging*, vol. **24**, Kluwer Academic/Plenum, NY, 2000, pp. 401–405.
74. J.-y. Lu, J. Cheng, and J. Wang, High frame rate imaging system for limited diffraction array beam imaging with square-wave aperture weightings, *IEEE Trans. Ultrason. Ferroelectr. Freq. Control* **53**(10), 1796–1812 (October 2006).
75. J.-y. Lu and J. L. Waugaman, Development of a linear power amplifier for high frame rate imaging system, *2004 IEEE Ultrasonics Symposium Proceedings*, 04CH37553C, Vol. 2, 2004, pp. 1413–1416.
76. J.-y. Lu, A multimedia example, *IEEE Trans. Ultrason. Ferroelectr. Freq. Control* **50**(9), 1078 (Sep. 2003).
77. A. Liu, J. Wang, and J.-y. Lu, Logic design and implementation of FPGA for a high frame rate ultrasound imaging system, *J. Acoust. Soc. Am.* **111**(5), pt. 2, 2484 (2002).
78. J. Wang, J. L. Waugaman, A. Liu, and J.-y. Lu, Design and construction of a high frame rate imaging system, *Acoust. Soc. Am.* **111**(5), pt. 2, 2484 (2002).
79. J. Cheng and J.-y. Lu, Extended high frame rate imaging method with limited diffraction beams, *IEEE Trans. Ultrason. Ferroelectr. Freq. Control* **53**(5), 880–899 (May 2006).
80. J. Cheng and J.-y. Lu, Fourier based imaging method with steered plane waves and limited-diffraction array beams, *2005 IEEE Ultrasonics Symposium Proceedings*, 05CH37716C, Vol. 2, 2005, pp. 1976–1979.
81. J. Wang and J.-y. Lu, A study of motion artifacts of Fourier-based image construction, *2005 IEEE Ultrasonics Symposium Proceedings*, 05CH37716C, Vol. 2, 2005, pp. 1439–1442.
82. J. Wang and J.-y. Lu, Motion artifacts of extended high frame rate imaging, *IEEE Trans. Ultrason. Ferroelectr. Freq. Control* **54**(7), 1303–1315 (July 2007).
83. J.-y. Lu and J. Wang, Square-wave aperture weightings for reception beam forming in high frame rate imaging, *2006 IEEE Ultrasonics Symposium Proceedings* 06CH37777, Vol. 1, 2006, pp. 124–127.
84. J.-y. Lu, Z. Wang, and S. Kwon, Blood flow velocity vector imaging with high frame rate imaging methods, *2006 IEEE Ultrasonics Symposium Proceedings* 06CH37777, Vol. 2, 2006, pp. 963–966.
85. J. Wang and J.-y. Lu, Effects of phase aberration and noise on extended high frame rate imaging, *Ultrasonic Imaging* **29**(2), 105–121 (April 2007).
86. J.-y. Lu and S. He, Increasing field of view of high frame rate ultrasonic imaging, *J. Acoust. Soc. Am.* **107**(5), pt. 2, 2779 (May 2000).
87. J.-y. Lu, Nonlinear processing for high frame rate imaging, *J. Ultrasound Med.* 18(3), (suppl.), S50 (Mar. 1999).
88. G. Wade, Human uses of ultrasound: ancient and modern, *Ultrasonics*, **38**, 1–5 (2000).
89. M. A. Porras, A. Parola, D. Faccio, A. Dubietis, and P. Di Trapani, Nonlinear unbalanced Bessel beams: stationary conical waves supported by nonlinear losses, *Phys. Rev. Lett.* **93**, 153902 (2004).

90. M. Kolesik, E. M. Wright, and J. V. Moloney, Dynamic nonlinear X-waves for femtosecond pulse propagation in water, *Phys. Rev. Lett.* **92**, 253901 (2004).
91. C. Conti and S. Trillo, Nonspreading wave packets in three dimensions formed by an ultracold Bose gas in an optical lattice, *Phys. Rev. Lett.* **92**, 120404 (2004).
92. P. Di Trapani, G. Valiulis, A. Piskarskas, O. Jedrkiewicz, J. Trull, C. Conti, and S. Trillo, Spontaneously generated X-shaped light bullets, *Phys. Rev. Lett.* **91**, 093904 (2003).
93. C. Conti, S. Trillo, P. Di Trapani, G. Valiulis, A. Piskarskas, O. Jedrkiewicz, and J. Trull, Nonlinear electromagnetic X-waves, *Phys. Rev. Lett.* **90**, 170406 (2003).
94. J. Salo, J. Fagerholm, A. T. Friberg, and M. M. Salomaa, Nondiffracting bulk-acoustic X-waves in crystals, *Phys. Rev. Lett.* **83**, 1171–1174 (1999).
95. T. Wulle and S. Herminghaus, Nonlinear optics of Bessel beams, *Phys. Rev. Lett.* **70**, 1401–1404 (1993).
96. S. Longhi and D. Janner, X-shaped waves in photonic crystals, *Phys. Rev. B* **70**, 235123 (2004).
97. S. Longhi, Localized and nonspreading spatiotemporal Wannier wave packets in photonic crystals, *Phys. Rev. E* **71**, 016603 (2005).
98. A. Ciattoni and P. Di Porto, Electromagnetic nondiffracting pulses in lossless isotropic plasmalike media, *Phys. Rev. E* **70**, 035601(R) (2004).
99. M. A. Porras, and P. Di Trapani, Localized and stationary light wave modes in dispersive media, *Phys. Rev. E* **69**, 066606 (2004).
100. A. Ciattoni and P. Di Porto, One-dimensional nondiffracting pulses, *Phys. Rev. E* **69**, 056611 (2004).
101. A. Ciattoni, C. Conti, and P. Di Porto, Vector electromagnetic X waves, *Phys. Rev. E* **69**, 036608 (2004).
102. E. Recami, M. Zamboni-Rached, and C. A. Dartora, Localized X-shaped field generated by a superluminal electric charge, *Phys. Rev. E* **69**, 027602 (2004).
103. J. Trull, O. Jedrkiewicz, P. Di Trapani, A. Matijosius, A. Varanavicius, G. Valiulis, R. Danielius, E. Kucinskas, A. Piskarskas, and S. Trillo, Spatiotemporal three-dimensional mapping of nonlinear X waves, *Phys. Rev. E* **69**, 026607 (2004).
104. S. Longhi, Parametric amplification of spatiotemporal localized envelope waves, *Phys. Rev. E* **69**, 016606 (2004).
105. S. Longhi, Spatial-temporal Gauss–Laguerre waves in dispersive media, *Phys. Rev. E* **68**, 066612 (2003).
106. R. W. Ziolkowski and Ching-Ying Cheng, Existence and design of trans-vacuum-speed metamaterials, *Phys. Rev. E* **68**, 026612 (2003).
107. O. Jedrkiewicz, J. Trull, G. Valiulis, A. Piskarskas, C. Conti, S. Trillo, and P. Di Trapani, Nonlinear X waves in second-harmonic generation: experimental results, *Phys. Rev. E* **68**, 026610 (2003).
108. M. A. Porras, G. Valiulis, and P. Di Trapani, Unified description of Bessel X waves with cone dispersion and tilted pulses, *Phys. Rev. E* **68**, 016613 (2003).
109. C. Conti, X wave-mediated instability of plane waves in Kerr media, *Phys. Rev. E* **68**, 016606 (2003).
110. M. A. Porras, I. Gonzalo, and A. Mondello, Pulsed light beams in vacuum with superluminal and negative group velocities, *Phys. Rev. E* **67**, 066604 (2003).

111. J. Salo and M. M. Salomaa, Nondiffracting waves in anisotropic media, *Phys. Rev. E* **67**, 056609 (2003).

112. M. Zamboni-Rached, F. Fontana, and E. Recami, Superluminal localized solutions to Maxwell equations propagating along a waveguide: the finite-energy case, *Phys. Rev. E* **67**, 036620 (2003).

113. K. Reivelt and P. Saari, Experimental demonstration of realizability of optical focus wave modes, *Phys. Rev. E* **66**, 056611 (2002).

114. A. M. Shaarawi, B. H. Tawfik, and I. M. Besieris, Superluminal advanced transmission of X waves undergoing frustrated total internal reflection: the evanescent fields and the Goos–Hänchen effect, *Phys. Rev. E* **66**, 046626 (2002).

115. K. Reivelt and P. Saari, Optically realizable localized wave solutions of the homogeneous scalar wave equation, *Phys. Rev. E* **65**, 046622 (2002).

116. M. A. Porras, Diffraction effects in few-cycle optical pulses, *Phys. Rev. E* **65**, 026606 (2002).

117. R. W. Ziolkowski, Superluminal transmission of information through an electromagnetic metamaterial, *Phys. Rev. E* **63**, 046604 (2001).

118. Z. L. Horváth and Z. Bor, Diffraction of short pulses with boundary diffraction wave theory, *Phys. Rev. E* **63**, 026601 (2001).

119. M. A. Porras, R. Borghi, and M. Santarsiero, Few-optical-cycle Bessel–Gauss pulsed beams in free space, *Phys. Rev. E* **62**, 5729–5737 (2000).

120. J. Salo, J. Fagerholm, A. T. Friberg, and M. M. Salomaa, Unified description of non-diffracting X and Y waves, *Phys. Rev. E* **62**, 4261–4275 (2000).

121. A. Ranfagni and D. Mugnai, Possibility of superluminal behaviors for X-like and Zenneck waves, *Phys. Rev. E* **58**, 6742–6745 (1998).

122. J. Fagerholm, A. T. Friberg, J. Huttunen, D. P. Morgan, and M. M. Salomaa, Angular-spectrum representation of nondiffracting X waves, *Phys. Rev. E* **54**, 4347–4352 (1996).

123. D. Power and R. Donnelly, Spherical scattering of superpositions of localized waves, *Phys. Rev. E* **48**, 1410–1417 (1993).

124. J.-y. Lu, H. Zou, and J. F. Greenleaf, Biomedical ultrasound beam forming, *Ultrasound Med. Bio.* **20**(5), 403–428 (July 1994).

125. J.-y. Lu and J. F. Greenleaf, Diffraction-limited beams and their applications for ultrasonic imaging and tissue characterization, in F. L. Lizzi, Ed., *New Developments in Ultrasonic Transducers and Transducer Systems*, *Proc. SPIE* **1733**, 92–119 (1992).

126. J.-y. Lu, Principle and applications of limited diffraction beams, *Acta Phys. Sin.* **8** (suppl.) 22–26, (Aug. 1999).

127. J.-y. Lu, Principle and applications of limited diffraction beams, *Proceedings of the 7th International Workshop on Modern Acoustics—Ultrasonics*, 1998, p. 8.

128. J.-y. Lu, New development in beam propagation, *J. Ultrasound Med.* **12**(3) (suppl.), S29 (Mar. 1993).

129. J.-y. Lu, Engineering applications of limited diffraction beams, *J. Acoust. Soc. Am.* **117**(4), pt. 2, 2446 (2005).

130. J.-y. Lu, Limited diffraction beams and their applications on signal processing, *J. Acoust. Soc. Am.* **106**(4), pt. 2, 2125 (Oct. 1999).

131. J.-y. Lu, Limited diffraction beams: their principles and potential applications, presented at the Ultrasonic Transducer Engineering Workshop, Aug. 16–18, (1995).

132. J.-y. Lu, T. Kinter, and J. F. Greenleaf, Measurement and dynamic display of acoustic wave pulses, *1989 IEEE Ultrasonics Symposium Proceedings*, 89CH2791-2, Vol. 1, 1989, pp. 673–676.

133. J.-y. Lu, R. R. Kinnick, J. F. Greenleaf, and C. M. Sehgal, Multi-electrical excitations of a transducer for ultrasonic imaging, in H. Lee and G. Wade, Eds., *Acoustical Imaging*, Vol. 18, Kluwer Academic/Plenum, NY, 1990, pp. 511–519.

134. J.-y. Lu and J. F. Greenleaf, Evaluation of transducers with near-field scanning of their surfaces, *1994 IEEE Ultrasonics Symposium Proceedings*, 94CH3468-6, Vol. 2, 1994, pp. 1163–1167.

135. T. A. Pitts, J. F. Greenleaf, J.-y. Lu, and R. R. Kinnick, Tomographic Schlieren imaging for measurement of beam pressure and intensity, *1994 IEEE Ultrasonics Symposium Proceedings*, 94CH3468-6, Vol. 3, 1994, pp. 1665–1668.

136. W. H. Press, B. P. Flannery, S. A. Teukolsky, and W. T. Vetterling, *Numerical Recipes in C*, Cambridge University Press, Cambridge, 1989, Chaps. 2 and 14.

137. M. Zamboni-Rached, Stationary optical wave fields with arbitrary longitudinal shape by superposing equal frequency Bessel beams: frozen waves, *Opt. Express* **12**(17), 4001–4006 (Aug. 23, 2004).

138. M. Zamboni-Rached, E. Recami, and H. E. Hernández-Figueroa, Theory of frozen waves: modeling the shape of stationary wave fields, *J. Opt. Soc. Am. A* **22**(11), 2465–2475 (Nov. 2005).

139. V. Garcés-Chávez, D. McGloin, H. Melville, W. Sibbett, and K. Dholakia, Simultaneous micromanipulation in multiple planes using a self-reconstructing light beam, *Nature* **419**, 145–147 (Sept 12, 2002).

140. J. A. Campbell and S. Soloway, Generation of a nondiffracting beam with frequency independent beam width, *J. Acoust. Soc. Am.* **88**(5), 2467–2477 (Nov. 1990).

141. M. S. Patterson and F. S. Foster, Acoustic fields of conical radiators, *IEEE Trans. Son. Ultrason.* **SU-29**(2), 83–92 (Mar. 1982).

142. J. H. Mcleod, The axicon: a new type of optical element, *J. Opt. Soc. Am.* **44**, 592 (1954).

143. Fayyazuddin and Riazuddin, *Quantum Mechanics*, World Scientific, Singapore, 1990, Chap. 18.

144. T. R. Sandin, *Essentials of Modern Physics*, Addison-Wesley, Reading, MA, 1989, p. 146.

145. J. R. Meyer-Arendt, *Introduction to Classical and Modern Optics*, Prentice-Hall, Englewood Cliffs, NJ, 1972, Chap. 6 .

146. R. W. Ziolkowski, Localized transmission of electromagnetic energy, *Phys. Rev. A* **39** 2005–2033 (1989).

147. J.-y. Lu, Frequency quantized nondiffracting X waves in a confined space, prepared on July 25, 1991 and revised on March 7, 1992 for the *Journal of Acoustical Society of America*. However, it is unpublished but was acknowledged on page 5 in M. Zamboni-Rached, E. Recami, and F. Fontana, Superluminal localized solutions to Maxwell equations propagating along a normal-sized waveguide, *Phys. Rev. E* **64**, 066603 (2001).

148. A. M. Shaarawi, I. M. Besieris, and R. W. Ziolkowski, Localized energy pulse trains launched from an open, semi-infinite, circular wave guide, *J. Appl. Phys.* **65**(2), 805–813 (Jan. 15, 1989).

149. R. W. Ziolkowski, I. M. Besieris, and A. M. Shaarawi, Localized wave representations of acoustic and electromagnetic radiation, *Proc. IEEE* **79**(10), 1371–1377, (Oct. 1991).

150. J. R. Meyer-Arendt, *Introduction to Classical and Modern Optics*, Prentice-Hall, Englewood Cliffs, NJ, 1972, Chap. 6.
151. R. Bracewell, *The Fourier Transform and Its Applications*, McGraw-Hill, New York, 1965, Chaps. 4 and 6.
152. J.-y. Lu, Bowtie limited diffraction beams for low-sidelobe and large depth of field imaging, *IEEE Trans. Ultrason. Ferroelectr. Freq. Control* **42**(6), 1050–1063 (Nov. 1995).
153. J.-y. Lu, Producing bowtie limited diffraction beams with synthetic array experiment, *IEEE Trans. Ultrason. Ferroelectr. Freq. Control* **43**(5), 893–900 (Sept. 1996).
154. J.-y. Lu, Method for reducing sidelobes of limited diffraction pulse-echo images, U.S. patent 5,492,121, Feb. 20, 1996.
155. J.-y. Lu and J. F. Greenleaf, Unsymmetrical limited diffraction beams for sidelobe reduction, *Ultrasound Med.* **14**(3) (supp.), S36 (Mar. 1995).
156. J.-y. Lu, Experiment with bowtie limited diffraction beams, *Ultrason. Imag.* **18**(1), 41 (Jan. 1996).
157. J.-y. Lu, Limited diffraction array beams, *Int. J. Imag. Sys. Technol.* **8**(1), 126–136 (Jan. 1997).
158. J.-y. Lu, Improving accuracy of transverse velocity measurement with a new limited diffraction beam, *1996 IEEE Ultrasonics Symposium Proceedings*, 96CH35993, Vol. 2, 1996, pp. 1255–1260.
159. J.-y. Lu and J. Cheng, Field computation for two-dimensional array transducers with limited diffraction array beams, *Ultrason. Imag.* **27**(4), 237–255 (Oct. 2005).
160. J.-y. Lu and J. Cheng, Efficient computation of field of 2D array with limited diffraction array beams, *J. Acoust. Soc. Am.* **109**(5), pt. 2, 2397–2398 (2001).
161. J. Cheng and J.-y. Lu, A study of angular spectrum and limited diffraction beams for calculation of field of array transducers, *J. Acoust. Soc. Am.* **111**(5), pt. 2, 2483 (2002).
162. H. Peng, Z. Zhang, and J.-y. Lu, A study of simulation methods of limited diffraction beams, *J. Acoust. Soc. Am.* **108**(5), pt. 2, 2595 (Nov. 2000).
163. J.-y. Lu, Tai K. Song, R. R. Kinnick, and J. F. Greenleaf, In vitro and in vivo real-time imaging with ultrasonic limited diffraction beams, *IEEE Trans. Med. Imag.* **12**(4), 819–829 (Dec. 1993).
164. J.-y. Lu, T.-K. Song, R. R. Kinnick, and J. F. Greenleaf, Real-time imaging with limited diffraction beams, *Ultrason. Imag.* **15**(2), 176 (Apr. 1993).
165. J.-y. Lu, T.-K. Song, R. R. Kinnick, J. F. Greenleaf, J. E. Meredith, and J. W. Charboneau, A general-purpose real-time scanner for evaluation of beam geometry in medical imaging, *J. Ultrasound Med.* **13**(3) (suppl.), S40 (Mar. 1994).
166. J.-y. Lu, M. Fatemi, and J. F. Greenleaf, Pulse-echo imaging with X wave, in P. Tortoli and L. Masotti, Eds., *Acoustical Imaging*, Vol. 22, Kluwer Academic/Plenum, NY, 1996, pp. 191–196.
167. J.-y. Lu and J. F. Greenleaf, Experiment of imaging contrast of Bessel nondiffracting beam transducer, *J. Ultrasound Med.* **11**(3) (suppl.), S43 (Mar. 1992).
168. J.-y. Lu, J. Wang, H. Peng, and J. Cheng, Increasing frame rate of limited diffraction imaging system with code excitations, in R. G. Maev, Ed., *Acoustical Imaging*, Vol. 26, Kluwer Academic/Plenum, NY, 2002, pp. 467–475.
169. K. F. Dajani, S. Salles-Cunha, H. G. Beebe, and J.-y. Lu, Ultrasonographic volumetry of atherosclerotic plaques, *J. Vasc. Technol.* **26**(2), 89–97 (June 2002).

170. K. F. Dajani, R. K. Vincent, S. Salles-Cunha, J.-y. Lu, and H. G. Beebe, Multi-frequency ultrasound imaging for carotid plaque characterization: a feasibility study, *Ultrason. Imag.* **21**(4), 292–293 (Oct. 1999).

171. J.-y. Lu and J. F. Greenleaf, Evaluation of a nondiffracting transducer for tissue characterization, *1990 IEEE Ultrasonics Symposium Proceedings*, 90CH2938-9, Vol. 2, 1990, pp. 795–798.

172. J.-y. Lu, A computational study for synthetic aperture diffraction tomography: interpolation versus interpolation-free, in L. W. Kessler, Ed., *Acoustical Imaging*, Vol. 16, Kluwer Academic/Plenum, NY, 1988, pp. 421–443.

173. J.-y. Lu and Y. Wei, An improvement of the Fourier-domain interpolation reconstruction algorithms for synthetic aperture diffraction tomography, *Appl. Acoust.* (Chinese quarterly) **8**(1), 1–5 (1989).

174. J.-y. Lu and Y. Wei, A study of the Fourier-domain interpolation reconstruction algorithms for synthetic aperture diffraction tomography, *Chin. J. Med. Imag. Technol.* **4**(1), 47 (1988).

175. J.-y. Lu and Y. Wei, A new method for quantitative reflection imaging, in Hiroshi Shimizu, Noriyoshi Chubachi, and Jun-ichi Kushibiki, Eds., *Acoustical Imaging*, Vol. 17, 1989, pp. 533–541.

176. J.-y. Lu and Y. Wei, Experiments for a new quantitative reflection imaging method, *Phys. Med. Bio.* **33** (suppl. 1), 291 (1988).

177. J.-y. Lu, Transmit–receive dynamic focusing with limited diffraction beams, *1997 IEEE Ultrasonics Symposium Proceedings*, 97CH36118, Vol. 2, 1997, pp. 1543–1546.

178. S. He, K. F. Dajani, and J.-y. Lu, Doppler beam steering for blood flow velocity vector imaging, in M. Halliwell and P. N. T. Wells, Eds., *Acoustical Imaging*, Vol. 25, Kluwer Academic/Plenum, NY, 2000, pp. 443–448.

179. J.-y. Lu, S. He, and K. F. Dajani, Ultrasound storage correlator arrays for real-time blood flow velocity vector imaging, in M. Halliwell and P. N. T. Wells, Eds., *Acoustical Imaging*, Vol. 25, 2000, pp. 427–436.

180. J.-y. Lu, X.-L. Xu, H. Zou, and J. F. Greenleaf, Application of Bessel beam for Doppler velocity estimation, *IEEE Trans. Ultrason. Ferroelectr. Freq. Control* **42**(4), 649–662 (July 1995).

181. J.-y. Lu, X.-L. Xu, H. Zou, and J. F. Greenleaf, Modeling and experiment of Doppler shift and its spectral broadening for a Bessel beam, *Ultrason. Imag.* **16**(1), 53–54 (Jan. 1994).

182. J.-y. Lu and J. F. Greenleaf, Producing deep depth of field and depth-independent resolution in NDE with limited diffraction beams, *Ultrason. Imag.* **15**(2), 134–149 (Apr. 1993).

183. J.-y. Lu and J. F. Greenleaf, Nondestructive evaluation of materials with a Bessel transducer, *Ultrason. Imag.* **14**(2), 203–204 (Apr. 1992).

184. J.-y. Lu, J. Cheng, and B. Cameron, Low sidelobe limited diffraction optical coherence tomography, *Proc. SPIE* **4619**, 300–311 (2002).

185. J.-y. Lu and S. He, Optical X waves communications, *Opt. Commun.* **161**, 187–192 (Mar. 15, 1999).

186. J.-y. Lu, High-speed transmissions of images with limited diffraction beams, in S. Lees and L. A. Ferrari, Eds., *Acoustical Imaging*, Vol. 23, Kluwer Academic/Plenum, NY, 1997, pp. 249–254.

187. J.-y. Lu and J. F. Greenleaf, A study of sidelobe reduction for limited diffraction beams, *1993 IEEE Ultrasonics Symposium Proceedings*, 93CH3301-9, Vol. 2, 1993, pp. 1077–1082.

188. S. He and J.-y. Lu, Sidelobe reduction of limited diffraction beams with Chebyshev aperture apodization, *J. Acoust. Soc. Am.* **107**(6), 3556–3559 (June 2000).

189. J.-y. Lu, J. Cheng, and H. Peng, Sidelobe reduction of images with coded limited diffraction beams, *2001 IEEE Ultrasonics Symposium Proceedings*, 01CH37263, Vol. 2, 2001, pp. 1565–1568.

190. J.-y. Lu and J. F. Greenleaf, Sidelobe reduction of nondiffracting pulse-echo images by deconvolution, *Ultrason. Imag.* **14**(2), 203 (Apr. 1992).

CHAPTER FIVE

Propagation-Invariant Fields: Rotationally Periodic and Anisotropic Nondiffracting Waves

JANNE SALO
Helsinki University of Technology, Espoo, Finland

ARI T. FRIBERG
Royal Institute of Technology, Kista, Sweden

5.1 INTRODUCTION

Diffraction is a natural phenomenon that influences wave propagation even in homogeneous media. Any deviation from transverse uniformity will eventually distort the field profile, leading to wave components that move in different directions. The strength of diffraction depends on the scale of spatial variations of the field compared with the wavelength. The best known examples are the spreading of the Gaussian laser beam and the classic Airy pattern formed by a circular aperture.

It was generally thought that only a uniform plane wave is completely devoid of diffractive effects. Hence it came as somewhat of a surprise when Durnin [1987] predicted, and shortly afterward, Durnin et al. [1987a] demonstrated experimentally the feasibility of a narrow diffraction-free light beam (see also DeBeer et al. [1987],

To the memory of our colleague, Professor Martti M. Salomaa (1949–2004).

Localized Waves, Edited by Hugo E. Hernández-Figueroa, Michel Zamboni-Rached, and Erasmo Recami

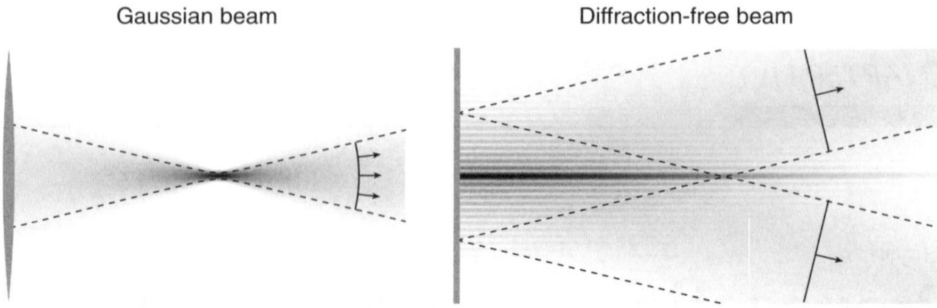

FIGURE 5.1 Gaussian beam and finite diffraction-free beam. The central maximum of the diffraction-free beam has the same width as the waist of the Gaussian beam. The gray-scale shading denotes the absolute value of the wave amplitude, and the arrows indicate the direction of local energy propagation.

reply by Durnin et al. [1987b], Sprangle and Hafizi [1991], reply by Durnin et al. [1991], as well as Hafizi and Sprangle [1991] and Lapointe [1992]). Transversely, the field has a Bessel function profile. It was called a *diffraction-free* (or *nondiffracting*) *beam*, due to the fact that the central peak appears to defy diffraction, despite the spot size being on the order of the wavelength. It was quickly noted, however, that a strictly diffraction-free beam has to be of infinite extent. Even though the field is concentrated on the axis, an infinite number of sidelobes of decreasing brightness are needed to support the main beam. Limiting their number will unavoidably also reduce the length of the diffraction-free beam, making it once again subject to diffraction.

In Fig. 5.1 we illustrate the differences between an ordinary Gaussian laser beam and an apertured diffraction-free field (represented here by a Bessel–Gauss beam [Gori et al., 1987]). Although a Gaussian beam can easily be focused into a bright focal spot, it will rapidly dissolve after the focus. A diffraction-free beam, on the other hand, supports a long axial focal line together with additional sidelobes. The total energy carried by the diffraction-free field is, however, much higher if equal intensity is to be achieved along the axis.

5.1.1 Brief Overview of Propagation-Invariant Fields

After the experiment by Durnin et al. [1987a], the properties, extensions, and potential applications of the diffraction-free beam have been studied extensively. However, the same field distribution had already been discovered by [Stratton, 1941, pp. 354–357] as a "wave function of the circular cylinder." The transverse form of this field is described by Bessel functions or other circular cylinder functions, depending on the radial boundary conditions. The same solution was later encountered in rigorous electromagnetic calculations by van Nie [1964]. The elementary nondiffracting beams are often referred to simply as *Bessel beams*.

Independently, Bessel-like beams had actually been produced some 50 years ago by McLeod [1954] with the use of conical axicons (see also McLeod [1960] and

Jaroszewicz et al. [2005]). These beams were not considered nondiffracting at the time, although they were recognized as axial line images of pointlike sources. Kelly et al. [1963] was able to show that their radial pattern is described by the zeroth-order Bessel function. Apparently, nearly unbeknownst to the Western world, in the 1980's scientists in Moscow had used an axicon to generate "continuous laser sparks" from high-power Gaussian laser beams [Bunkin et al., 1983; Korobkin et al., 1986] (see also [Polonskiy, 2005]). Indebetouw [1989] suggested, based on the McCutchen theorem [McCutchen, 1964], that an axicon could be used to form a Bessel beam, and this was later realized experimentally by Scott and McArdle [1992]. The first to use diffractive optical elements (i.e., computer-generated holograms, diffractive axicons) to produce diffraction-free beams were Turunen et al. [1988]. The advantage of this technique is that even higher-order Bessel beams could be realized with ease [Vasara et al., 1989; Paterson and Smith, 1996] (see also Dholakia [2000]). Later, diffraction-free beams were produced by a wide variety of methods, including laser resonators [Uehara and Kikuchi, 1989; Khilo et al., 2001; Hakola et al., 2004], spherically aberrated lenses [Herman and Wiggins, 2001; Burvall et al., 2004], dual-element systems [Iftekharuddin et al., 1993; Sedukhin, 1998; Honkanen and Turunen, 1998], interferometers [Cox and Dibble, 1992; Horváth et al., 1997], spatial light modulators [Davis et al., 1996; Chattrapiban et al., 2003], and nonlinear liquid crystals [Hakola et al., 2006], to name just a few. For general overviews of axicon fields and Bessel beams, we refer to Soroko [1996], Jaroszewicz [1997], and McGloin et al. [2005].

Ideal diffraction-free waves are characterized by the property that the field, which satisfies the source-free Helmholtz equation, factors in its transverse and longitudinal parts, with the longitudinal part being in intensity independent of the propagation distance. An almost limitless variety of diffraction-free patterns and arrays can be produced in the Cartesian coordinate system [Bouchal, 2002], while the Bessel beams and other more general rotating and spiraling fields containing, for example, Neumann and Hankel functions are examples of propagation-invariant waves in circular cylindrical coordinates [Chávez-Cerda et al., 1996; Chávez-Cerda, 1999]. The Helmholtz equation also separates in appropriate transverse and longitudinal parts in both elliptical and parabolical cylindrical coordinates. Examples of nondiffracting fields in the former coordinate system are the propagation-invariant Mathieu beams [Gutiérrez-Vega et al., 2000, 2001; Dartora et al., 2003], whereas in the latter system one obtains propagation-invariant parabolic beams [Bandrés et al., 2004; López-Mariscal et al., 2005]. Higher-order doughnut (dark, hollow) beams exhibit helical wavefronts, which correspond to optical vortices (e.g., phase singularities, dislocations). Nondiffracting vortex beams exert radiation pressure and carry orbital angular momentum; they can be used for microparticle trapping (optical tweezers) [He et al., 1995] and rotation (optical spanners) [Volke-Sepulveda et al., 2002; Chávez-Cerda et al., 2002].

Another defining feature of nondiffracting beams is that the plane waves in their angular-spectrum representation are confined to lying on the surface of a cone; for this reason, such fields are sometimes referred to as *conical waves*. A general description for (spatially) partially coherent propagation-invariant fields was put forward early by Turunen et al. [1991] (see also Friberg et al. [1991] and Kowarz and Agarwal [1995]).

The plane-wave components that make up a partially coherent nondiffracting field are radially uncorrelated, while the correlations in the azimuthal variable may be arbitrary. The invariant propagation of partially coherent uniform-intensity (infinite) fields of the Schell-model type was studied by Turunen [2002]. On allowing the plane waves to be coherent beams of arbitrary form, Shchegrov and Wolf [2000] introduced a new model of partially coherent conical fields of finite transverse extent that propagate over large distances without changing their intensity profiles or coherence properties. Other recent work in this area deals with the generation and properties of white-light propagation-invariant beams [Fischer et al., 2005].

Additionally, the nondiffracting beams and fields originally analyzed in scalar theory were naturally soon generalized to vector waves that satisfy Maxwell's equations rigorously [Mishra, 1991; Turunen and Friberg, 1993; Bouchal and Olivík, 1995; Horák et al., 1997], spectrally even in the nonstationary case [Bouchal et al., 1998a]. A classification of various propagation-invariant vector fields on the basis of the complex Poynting theorem, as well as experimental demonstrations, were presented by Bouchal et al. [1996]. Use of polarization-grating axicons to produce electromagnetic propagation-invariant fields has been suggested [Tervo and Turunen, 2001a]. Propagation-invariant azimuthally and radially polarized fields have been examined [Tervo et al., 2002], and a general vectorial decomposition of the electric field into two orthogonal components was recently applied to study propagation-invariant, rotating, and self-imaging vector fields [Pääkkönen et al., 2002].

Pulse-like fields, formed as wave packets consisting of a range of wavelengths, are naturally equally subject to diffraction (and dispersion) effects. The acoustics community showed interest in Bessel beams and the first (narrowband) ultrasonic transducer was fabricated by Hsu et al. [1989]. The application potential of nondiffracting waves was soon recognized in high-resolution medical imaging [Lu and Greenleaf, 1990]. However, an entirely new line of research was initiated by the seminal work of [Lu and Greenleaf, 1992] with the realization that Bessel beams could be generalized into nondiffracting pulses, called *X-waves*. Despite the simple principle that all frequency components of the field propagate at the same angle to the beam axis [Fagerholm et al., 1996], such pulses are difficult to produce in optics, where simultaneous, coherent control of a wide range of frequencies is complicated. This was, however, later achieved in a ground-breaking experiment by [Saari and Reivelt, 1997]. With acoustic transducers, production of nondiffracting pulses was hardly more difficult than the generation of continuous-wave Bessel beams. The localized (nondiffracting and nondispersive) pulses were also generalized to dispersive media [Sõnajalg et al., 1997; Porras, 2001; Zamboni-Rached et al., 2003].

Following the discovery of X-waves, localized nondiffracting pulses were studied extensively. Again, however, these wave solutions had their predecessors. Earlier, Brittingham [1983] had published the finding of *focus wave modes*, pulselike waves propagating in three-dimensional space at the speed of light. They have the peculiar property of remaining focused at all times (i.e., their spatial envelope does not change under propagation, although the wave field within evolves periodically). Thus, they were considered nondispersive. The original, fairly cumbersome formulation of focus wave modes was generalized by Sezginer [1985], who also showed that no focus wave modes of finite energy can exist in free space without sources. Although focus wave modes and X-waves clearly belong to different subclasses of waves, they exhibit similar spectral properties, as pointed out by Shaarawi [1997]. Experimental realization of optical focus wave modes was recently reported by Reivelt and Saari [2002] (see also [Valtna et al., 2006]). We note further that the localized solutions have been extended from scalar waves [Besieris et al., 1998; Salo et al., 2000; Zamboni-Rached et al., 2002] to electromagnetic fields (Maxwell's equations) [Recami, 1998; Recami et al., 2003; Mugnai and Mochi, 2005] and to other field equations in physics [Rodrigues and Lu, 1997]. Within the microwave regime, pulsed X-waves were experimentally observed by Mugnai et al. [2000].

Many theoretical and practical questions associated with Bessel beams, nondiffracting fields, and X-waves have been addressed. Mathematically ideal fields inevitably have properties that are nonphysical; for instance, all these wave types require an infinite aperture and an infinite amount of energy. Strangely, X-waves also appeared to propagate with a superluminal velocity [Lu and Greenleaf, 1992], which has also been verified experimentally [Alexeev et al., 2002]. The latter contradiction was resolved by understanding that energy (or information) carried by the X-wave is not, in fact, transported by the central pulse; it is, instead, launched at an earlier time from the outer regions of the aperture, thus having exactly sufficient time to propagate at the speed of light (see also [Recami, 2001]). The issue of the infinite radial extent has, by now, turned into an engineering type of question of optimal aperture design to avoid unnecessary ripples or edge diffractions within the desired finite propagation length. In this area, *frozen waves* have attracted interest [Zamboni-Rached et al., 2005]; they are transversely highly localized fields that can have an arbitrary longitudinal profile, even in an absorbing medium [Zamboni-Rached, 2006]. Other recent topics include paraxial envelope X-waves [Porras et al., 2003] and spatiotemporal linear optical bullets [Ponomarenko and Agrawal, 2006]. Nondiffracting fields have the self-reconstructing property in the linear [Bouchal et al., 1998b] and nonlinear regimes [Butkus et al., 2002]. After the seminal work of Wulle and Herminghaus [1993], spatiotemporally localized optical waves have been studied and applied extensively in nonlinear optics. As examples, we mention second-harmonic generation of X-waves [Conti and Trillo, 2003] and spontaneous formation of X-shaped light bullets [Di Trapani et al., 2003]. A valuable recent review of nondiffracting optical waves, including properties, experiments, and applications, is that of Bouchal [2003].

5.1.2 Scope of This Chapter

Continuous-wave self-imaging, propagation-invariant, and rotating wave fields have attracted much attention in recent years in optics [Pääkkönen et al., 1998; Piestun and Shamir, 1998; Tervo and Turunen, 2001b; Saastamoinen et al., 2004]. In this chapter we consider a broader class of fields, called *rotationally periodic waves* (RPWs) [Salo and Salomaa, 2001; Salo, 2003], which, ideally, avoid diffractive deterioration under propagation. This class encompasses several types of waves usually treated separately, such as monochromatic propagation-invariant beams (e.g., Bessel beams, self-imaging fields) and various pulsed wave fields (nondiffracting pulses and focus wave modes). While purely nondiffracting waves are limited to superluminal velocities and focus wave modes to the speed of light, rotationally periodic waves allow for synchronous pulsed waves propagating in a uniform medium at velocities that are entirely independent of the speed of light (or sound). As a further extension we also discuss nondiffracting waves that propagate in anisotropic crystalline solids in the ultrasonic regime [Salo et al., 1999; Salo and Salomaa, 2003], and analyze their features that emerge due specifically to the anisotropy. We additionally develop an asymptotic representation that allows a physical interpretation in terms of transverse energy propagation, phonon focusing, caustics, and internal diffraction, all characteristics of strongly anisotropic wave propagation.

5.2 ROTATIONALLY PERIODIC WAVES

We consider scalar waves that propagate in an infinite source-free medium and satisfy the linear, nondispersive wave equation

$$\left(\frac{\partial^2}{\partial x^2}+\frac{\partial^2}{\partial y^2}+\frac{\partial^2}{\partial z^2}\right)\Phi(x,y,z;t)=\frac{1}{c^2}\frac{\partial^2}{\partial t^2}\Phi(x,y,z;t), \tag{5.1}$$

where c is the phase velocity, assumed independent of frequency, direction of propagation, and field strength. We look for wave solutions $\Phi(x, y, z; t)$ that retain their spatial shape on propagation (i.e., which do not spread, or if they do, will return repeatedly to their original form). Expressing an arbitrary wave field using spatial cylindrical coordinates $\mathbf{r} = (r_\perp, \varphi, z)$ and time t, we require that it be periodic in the following sense:

$$\Phi(r_\perp, \varphi+\gamma, z+v\tau; t+\tau) = \Phi(r_\perp, \varphi, z; t). \tag{5.2}$$

This condition may be interpreted as follows: After a certain time τ, the field has propagated along z a distance $\xi = v\tau$, and it has also rotated through an angle γ. We refer to v as the velocity of propagation—to be distinguished from the phase velocity c characteristic to the medium. No assumptions are made regarding the behavior of the field between the instants t and $t+\tau$; for instance, it is not required to rotate uniformly, but rather, it may evolve in an arbitrarily complicated manner. The evolution is, nonetheless, periodic along the corkscrew curve $[r_\perp, \varphi+(\gamma/\tau)t, z+vt]$ and hence the waves are called *rotationally periodic waves*. The cylindrical coordinate representation also introduces a further periodicity condition,

$$\Phi(r_\perp, \varphi+2\pi, z; t) = \Phi(r_\perp, \varphi, z; t), \tag{5.3}$$

due to the requirement that the wave solution be uniquely-valued.

5.2.1 Fourier Representation of General RPWs

An arbitrary wave field in cylindrical coordinates may be expressed formally using the Fourier–Bessel representation as

$$\Phi(r_\perp, \varphi, z; t) = \sum_n \int_0^\infty k_\perp dk_\perp \int_{-\infty}^\infty dk_z \int_{-\infty}^\infty d\omega\, a_n(k_\perp, k_z, \omega)\, e^{in\varphi} \times J_n(r_\perp k_\perp) e^{i(k_z z - \omega t)}, \quad (5.4)$$

where $k_\perp$ is the radial wave number and J_n is the Bessel function of the first kind and of order n. However, the wave equation (5.1) and the two periodicity requirements (5.2) and (5.3) imply constraints in the integrations above. They lead to the ordinary wave dispersion relation

$$k_z^2 + k_\perp^2 = \frac{\omega^2}{c^2} \quad (5.5)$$

and to an additional *RPW dispersion relation*,

$$k_z = \frac{\omega}{v} + \frac{2\pi l - n\gamma}{\xi} \equiv \frac{\omega}{v} + \mu, \quad (5.6)$$

where the *mode indices* l and n are arbitrary integers. Hence, the radial wave number is also determined uniquely as

$$k_\perp = \sqrt{\frac{\omega^2}{c^2} - k_z^2} \quad (5.7)$$

and the general rotationally periodic waves can be written in the form [Salo and Salomaa, 2001]

$$\Phi(r_\perp, \varphi, z; t) = \sum_l \sum_n e^{i[n\varphi + (2\pi l - \gamma n)z/\xi]} \int a_{ln}(\omega) J_n(r_\perp k_\perp) e^{i\omega(z/v - t)}\, d\omega, \quad (5.8)$$

where the ω integral is evaluated over such frequencies for which the radial wave number has a positive real value. The integral now represents an envelope field that does not depend on z and t separately but only through $z - vt$, and therefore propagates with velocity v. The actual solution is a sum of such waves, modulated by exponentials that guarantee the periodicity requirements.

5.2.2 Special Propagation Symmetries

Several special cases of general RPWs given by Eq. (5.8) are of interest on their own, and we describe some of them briefly. Various subclasses are summarized in Table 5.1.

TABLE 5.1 Classification of Various Rotationally Periodic Waves

Wave Type	γ	n[a]	l
General RPWs	free	free	free
Uniformly rotating waves	free	free	$l = 0$
Nondiffracting waves	$\gamma = 0$	free	$l = 0$
Self-imaging waves	$\gamma = 0$	free	free
Rotationally invariant waves	—	$n = 0$	free

Source: Salo and Salomaa, 2001; Salo, 2003.
[a] If $n = 0$, the rotation parameter γ plays no role. It is therefore omitted from the rotationally invariant waves.

Uniformly Rotating Waves General RPWs may show complex behavior within each period. Thus, we first identify uniformly rotating waves that have trajectories of constant value on propagation. That is, we require that the periodicity condition (5.2) hold for all $\xi = v\tau$ and $\gamma = \nu\tau$, where τ is real and positive. Hence, the constant-value trajectories are given by $[r_\perp \cos(\varphi + \nu\tau), r_\perp \sin(\varphi + \nu\tau), z + v\tau]$ for all $r_\perp$, φ, and z, which corresponds to $l = 0$ in the general RPW expression (5.8).

Nondiffracting Waves Nondiffracting waves (NDWs) are defined as fields that propagate uniformly at a fixed velocity; hence, they satisfy $\Phi(r_\perp, \varphi, z, t) = \Phi(r_\perp, \varphi, z - vt)$. Clearly, for these fields $\gamma = 0$ and they also belong to the class of uniformly rotating waves. Nondiffracting waves are often called *X-waves* due to their X-shaped (or conical) cross-sectional form [Lu and Greenleaf, 1992; Salo et al., 2000]. A pulsed NDW with $n = 0$ is known as the fundamental X-wave, and those with $n \neq 0$ are called higher-order X-waves.

Self-Imaging Waves Self-imaging waves are nonrotating periodic fields that reproduce their amplitude pattern with the period τ at locations $z + v\tau$ [i.e., $\Phi(r_\perp, \varphi, z + v\tau, t + \tau) = \Phi(r_\perp, \varphi, z, t)$]. They have $\gamma = 0$, but they may contain different l and n components.

Rotationally Invariant Waves Rotationally invariant waves are fields with no azimuthal dependence, which leads directly to the requirement that $n = 0$. On the other hand, while $n = 0$, the field becomes independent of the γ parameter, which may thus be set to zero. However, rotationally invariant waves can evolve on propagation; that is, their crosssections are not required to remain constant, and therefore different l-modes are permitted, which serves to make them a subclass of self-imaging waves.

5.2.3 Monochromatic Waves

If all spectral coefficients $a_{ln}(\omega)$ are proportional to $\delta(\omega - \omega_0)$, with common ω_0 for all n and l, the ensuing monochromatic solution may be written in the form

$$\Phi(r_\perp, \varphi, z; t) = \sum_l \sum_n a_{ln} e^{i[n\varphi + (2\pi l - n\gamma)z/\xi]} J_n(r_\perp k_\perp) e^{i\omega_0(z/v - t)}. \tag{5.9}$$

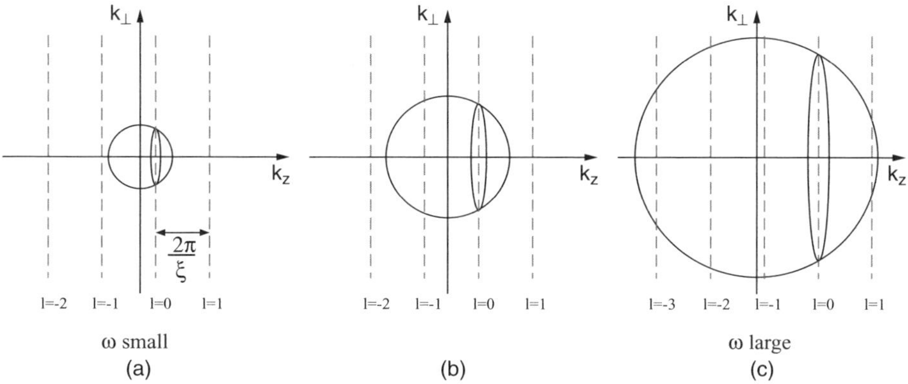

FIGURE 5.2 Values of k_z allowed for a given ω. The circle denotes the dispersion relation $k_\perp^2 + k_z^2 = \omega^2/c^2$, which all Fourier components of the field must satisfy. The fundamental mode ($n = l = 0$) is always possible, provided that $|v| \geq c$, as it scales with frequency ω. Since different l modes have spacings of $2\pi/\xi$, higher modes can exist only if ω is large enough. In this schematic representation we have chosen $n = 0$; otherwise, the n spacing is given by γ/ξ. In (a) only the fundamental mode satisfies the dispersion relation; in (b) the mode $l = -1$ is also permitted; in (c) several higher modes are allowed.

Originally, both l and n were allowed to have all integer values, but now the summation is more limited. Since the spacings of the different l and n terms are $2\pi/\xi$ and γ/ξ, respectively, only a finite number of modes can satisfy the dispersion relation $|k_z| \leq \omega_0/c$; see Fig. 5.2. From the requirement that the radial wave number be real and positive, we observe that for a fixed frequency $\omega_0 > 0$ only such modes are permitted that fulfill

$$-\frac{c+v}{cv}\,\omega_0 \leq \mu \leq -\frac{c-v}{cv}\,\omega_0. \tag{5.10}$$

This limitation also holds for polychromatic fields: When integrating over ω, the Fourier weight of a given frequency must vanish for those l and n modes which are not permitted for that frequency.

Each term in Eq. (5.9) has the form of a Bessel beam,

$$\begin{aligned}\Phi(r_\perp, \varphi, z; t) &= e^{i(n\varphi + (2\pi l - n\gamma)z/\xi)} J_n(r_\perp k_\perp) e^{i\omega_0(z/v - t)} \\ &= e^{in\varphi} J_n(r_\perp k_\perp) e^{i\omega_0[(\mu/\omega_0 + 1/v)z - t]}\end{aligned} \tag{5.11}$$

with an axial wave number $k_z = \omega_0/v + \mu$, as required by Eq. (5.6). Hence, the Bessel beam depends on l and γ only through the parameter μ, which may be further removed by redefining the propagation velocity as $v' = (\mu/\omega_0 + 1/v)^{-1}$. The individual Bessel beams thus have only two independent parameters, the velocity v' and azimuthal order n.

Since $k_\perp$ is real and positive, $|k_z| = \sqrt{\omega_0^2/c^2 - k_\perp^2} \leq |\omega_0/c|$ and the (axial) phase velocity $v_z = \omega_0/k_z$ is necessarily superluminal. Hence, each individual Bessel field always has a superluminal velocity of propagation along the beam axis, regardless of the parameters v and μ. This ambiguity is a consequence of the definition of the rotational periodicity of a wave: The field pattern is repeated after a time interval τ and spatial spacing ξ, which together define the velocity of propagation $v = \xi/\tau$: $\Phi(z + \xi, t + \tau) = \Phi(z, t)$. If, however, the field also satisfies $\Phi(z + \xi, t + \tau/2) = \Phi(z, t)$, we may argue that the velocity is, in fact, $v' = 2\xi/\tau = 2v$, which is equally consistent. It is therefore only the superpositions of the Bessel beams, given by Eq. (5.9), or their spectral generalizations (see pulsed single-mode waves below), whose velocity of propagation is determined directly by the parameter v.

5.2.4 Pulsed Single-Mode Waves

We have shown in Eq. (5.8) that all rotationally periodic waves (RPWs) can be decomposed into separate fields with fixed indices l and n. We now discuss the expressions of such individual fields and show that the physical nature of these fields depends essentially on the velocity of propagation of the wave.

A pulsed (time-dependent) single-mode wave is expressed in its most general form as

$$\Phi(r_\perp, \varphi, z; t) = e^{in\varphi} \int a(\omega) J_n(r_\perp k_\perp) e^{i(k_z z - \omega t)}\, d\omega \tag{5.12}$$

with the axial and radial wave numbers given by

$$k_z = \frac{\omega}{v} + \mu, \qquad k_\perp = \sqrt{\frac{\omega^2}{c^2} - \left(\frac{\omega}{v} + \mu\right)^2}, \tag{5.13}$$

where $\mu = (2\pi l - n\gamma)/\xi$. The dispersion relation dictates that $|k_z| \le |\omega|/c$, which introduces different types of conditions for the cases of superluminal, subluminal, and luminal waves.

Superluminal Single-Mode Wave Let us consider a pulsed single-mode wave with a superluminal propagation velocity $v > c$ and omit the case $v < -c$, which has exactly the same physical properties, apart from an inverted direction of propagation. Care has to be exercised in the ω-integration in Eq. (5.12), since only a range of frequencies can support waves for given l and n. If $\mu = 0$, all frequencies ω are allowed; in this case the solution is actually a nondiffracting wave of index n. Clearly, the wave will also satisfy the periodicity conditions specified by the values of γ, l, and n. The general expression for the field is

$$\Phi(r_\perp, \varphi, z; t) = e^{in\varphi} \int_{-\infty}^{\infty} a(\omega) J_n \left(r_\perp |\omega| \sqrt{\frac{v^2 - c^2}{v^2 c^2}} \right) e^{i\omega(z/v - t)} d\omega, \tag{5.14}$$

where $a(\omega)$ is an arbitrary spectral function of the pulse.

When $\mu \neq 0$, the radial wave number $k_\perp$ is given by Eq. (5.13), and the limits of integration are obtained from the dispersion relation, which requires that

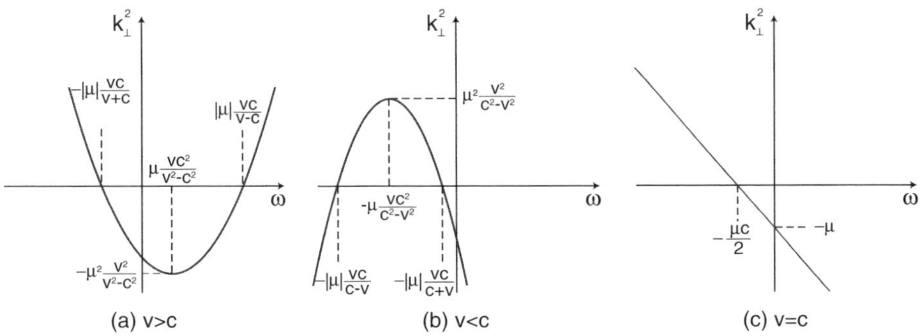

FIGURE 5.3 Frequencies allowed for (a) superluminal, (b) subluminal, and (c) luminal RPWs. The RPW dispersion relation and the wave equation specify the transverse wavenumber as a function of frequency, and the permitted range is determined by the condition $k_\perp^2 \geq 0$. In these schematic graphs, $\mu > 0$.

$|k_z| = |\omega/v + \mu| \leq |\omega|/c$. On the other hand, this condition is equivalent to the requirement that $k_\perp$ be real and positive (i.e., that $k_\perp^2 > 0$), as illustrated in Fig. 5.3. This leads to

$$\omega \geq |\mu|\frac{vc}{v-c} \quad \text{or} \quad \omega \leq -|\mu|\frac{vc}{v+c}. \tag{5.15}$$

Changing the integration variable as $\omega = \mu[v^2c/(v^2-c^2)](\chi + c/v)$ gives

$$k_z = \frac{\mu vc}{v^2-c^2}\chi + \frac{\mu v^2}{v^2-c^2}, \qquad k_\perp = \sqrt{\frac{\mu^2 v^2}{v^2-c^2}}\sqrt{\chi^2-1} \equiv \beta\sqrt{\chi^2-1}, \tag{5.16}$$

and the integration limits now correspond directly to the condition $|\chi| \geq 1$. We incorporate the altered integration measure in the weight function according to $a(\omega)\,d\omega \equiv a(\chi)\,d\chi$, which leads to

$$\Phi(r_\perp, \varphi, z; t) = e^{in\varphi}e^{i\mu\{[v^2/(v^2-c^2)][z-(c^2/v)t]\}} \times \left[\int_{-\infty}^{-1} + \int_1^{\infty}\right] a(\chi) J_n(r_\perp\beta\sqrt{\chi^2-1})\, e^{i\mu\{[vc/(v^2-c^2)](z-vt)\chi\}}\, d\chi, \tag{5.17}$$

where $\beta = |\mu|\sqrt{v^2/(v^2-c^2)}$. This is the most general formula for a rotationally periodic, superluminal single-mode wave with $\mu \neq 0$ [Salo and Salomaa, 2001]. If $\mu < 0$, the direction of integration is reversed, since then, $\omega \propto -\chi$. This is seen in the exponential terms, where μ still has a negative value whereas β depends only on the absolute value of μ.

We observe that the wave is composed of a pulse envelope propagating at velocity v and a modulating plane wave at velocity c^2/v. Hence, the velocity of pulse propagation is superluminal, whereas the modulating phase is subluminal. Their geometric average

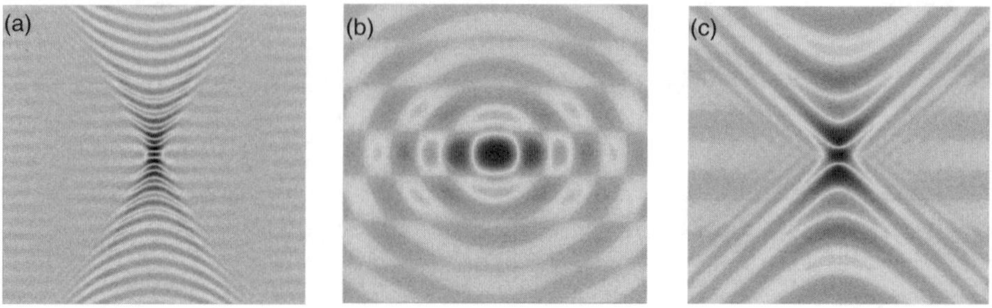

FIGURE 5.4 (a) Superluminal, $v = 2c$; (b) subluminal, $v = c/2$; and (c) luminal, $v = c$, propagation-invariant pulses. The pulse core propagates upward and the parameters are $\xi = 1$, $n = 0$, and $l = 1$. The spectra are given by (a) $a(\chi) = \sqrt{\chi^2 - 1}\, e^{-2\chi}$, (b) $a(\chi) = \sqrt{1 - \chi^2}$, and (c) $a(\chi) = e^{-\chi}$.

equals, however, exactly the ordinary phase velocity c, and the superluminality of the pulse is compensated with a correspondingly slower phase. We emphasize that neither the propagation velocity of the pulse nor the velocity of the phase depends on the mode (i.e., on the mode parameter μ). Nevertheless, the wavelength of the modulation depends on μ.

In Fig. 5.4 we illustrate a superluminal propagation-invariant pulse together with its subluminal and luminal counterparts. Observe that the superluminal wave has an X-wave-like form. This is due to the fact that for high frequencies k_z becomes proportional to frequency, which is the case for nondiffracting waves. This high-frequency limit fixes the ratio $k_\perp / k_z$ and leads to the cone of propagation, typical of nondiffracting waves and X-waves, in particular.

Subluminal Single-Mode Wave We next consider pulsed subluminal single-mode waves that satisfy $0 < v < c$. The frequency is again restricted to values that correspond to real radial wave numbers $k_\perp$. First we point out that for $\mu = 0$, we have $k_\perp^2 < 0$ for all frequencies, and thus no fundamental mode can exist for subluminal waves. This is also known as the statement that all nondiffracting waves necessarily are superluminal [Vasara et al., 1989]. For $\mu \neq 0$, we obtain the following bounds on the frequency:

$$\begin{aligned} -\mu\frac{vc}{c-v} &\leq \omega \leq -\mu\frac{vc}{c+v}, &&\text{when} \quad \mu > 0, \\ -\mu\frac{vc}{c+v} &\leq \omega \leq -\mu\frac{vc}{c-v}, &&\text{when} \quad \mu < 0. \end{aligned} \tag{5.18}$$

On changing the integration variable, now as $\omega = \mu[v^2c/(c^2 - v^2)](\chi - c/v)$, we find that

$$k_z = \frac{\mu vc}{c^2 - v^2}\chi - \frac{\mu v^2}{v^2 - c^2}, \qquad k_\perp = \sqrt{\frac{\mu^2 v^2}{c^2 - v^2}}\sqrt{\chi^2 - 1} \equiv \delta\sqrt{1 - \chi^2}, \tag{5.19}$$

and the subluminal single-mode waves, in general, are then given by [Salo and Salomaa, 2001]

$$\Phi(r_\perp, \varphi, z; t) = e^{in\varphi} e^{-i\mu\{[v^2/(c^2-v^2)][z-(c^2/v)t]\}} \times \int_{-1}^{1} a(\chi) J_n(r_\perp \delta \sqrt{1-\chi^2}) e^{i\mu\{[vc/(c^2-v^2)](z-vt)\chi\}} \, d\chi. \quad (5.20)$$

Hence, the subluminal waves necessarily have a limited bandwidth. Similar to the superluminal case, the wave is composed of a propagating envelope and a modulating phase, but now the velocity of the envelope is subluminal and that of the phase is superluminal. The geometric average velocity still equals the ordinary phase velocity c.

A subluminal wave for $v = c/2$ is illustrated in Fig. 5.4b. Note that the wave is far less localized than the similar superluminal wave, which is due to the fact that the superluminal solution may contain arbitrarily high frequencies, while the frequency band of the subluminal wave is strictly bounded from above. Also, no conical structure appears in the case of subluminal pulses, since no asymptotic high-frequency value exists for $k_\perp / k_z$.

Luminal Single-Mode Wave Finally, we consider the case of pulsed luminal waves with $v = c$. The radial wave number, from Eq. (5.13), now is

$$k_\perp = \sqrt{-2\frac{\omega}{c}\mu - \mu^2}. \quad (5.21)$$

If $\mu = 0$, the transverse wave number vanishes identically and the field becomes independent of $r_\perp$. Since all other Bessel functions except $J_0(r_\perp k_\perp)$ vanish at zero argument, only the $n = 0$ term remains and the field simply is composed of parallel plane waves.

For $\mu \neq 0$, the radial wave number is real only if $\omega \leq -\mu c/2$ for $\mu > 0$ or $\omega \geq |\mu| c/2$ for $\mu < 0$. Hence, we set $\omega = -(\mu c/2)(\chi + 1)$, which leads to

$$k_z = -\frac{\mu}{2}(\chi - 1), \qquad k_\perp = |\mu| \sqrt{\chi}, \quad (5.22)$$

and the single-mode wave in this case is represented by [Salo and Salomaa, 2001]

$$\Phi(r_\perp, \varphi, z; t) = e^{in\varphi} e^{i(\mu/2)(z+ct)} \int_0^\infty a(\chi) J_n(r_\perp |\mu| \sqrt{\chi}) \, e^{-i(\mu/2)(z-ct)\chi} \, d\chi. \quad (5.23)$$

Luminal RWP pulses thus factor in two terms, both moving with the speed of light: The pulse envelope advances along the *positive* z-axis carrying the wave front, but it is modulated with a *counter-propagating* plane wave of wavelength $\lambda = 4\pi/|\mu|$ that moves along the *negative* z-axis. The same property, in fact, already appears in the focus wave modes, which are inherently luminal since they are expressed in terms of coordinates $z - ct$ and $z + ct$, the first being associated with the envelope and the second with the modulation Sezginer [1985]. A luminal wave is shown in Fig. 5.4c; it shares the high-frequency limit of superluminal waves, since the bandwidth is not limited from above, but the characteristic ratio $k_\perp / k_z$ for high frequencies tends to zero.

5.2.5 Discussion

The classical wave equation contains a parameter, the phase velocity, which under certain conditions in a natural way describes the velocity of wave propagation. If it is strictly constant, it also sets boundaries to causality. In some media, the phase velocity may depend on frequency (dispersion), direction (anisotropy or birefringence), and/or field strength (nonlinearity), in which cases several new characteristic velocities appear, such as the group velocity, energy velocity, and signal velocity (see, e.g., Oughstun et al. [1994]).

The velocity of rotationally periodic waves is not inherent to the medium, but rather, to the construction of the wavefront. Its formation is a synchronous process: The wave is generated in such a way that its spatial shape is recreated periodically along the line of propagation. The main body of the beam or the pulse itself does not contain the necessary energy or information for the reconstruction, as is the case for solitons. Therefore, in empty space, an infinite amount of energy is needed because the energy does not propagate along the focal line but radiates away from it, and "new" energy must be transmitted to the axis to maintain the wave pattern.

Generation of rotationally periodic waves poses challenges similar to those encountered with X-waves and focus wave modes. The innocent-looking Fourier representation of the fields, in fact, also raises problems of causality—if not fundamental, at least practical [Salo and Salomaa, 2001]. The RPWs may contain wave components that carry energy in the direction opposite to pulse propagation, which occurs whenever k_z and ω have different signs. These components must be launched from the receiving end of wave transmission rather than from the emitting part. On the other hand, exclusion of these components does not destroy the periodicity property. Nonetheless, one must be careful in designing suitable transducers, and generally, large apertures are needed to feed energy for long focal lines. This severely complicates the generation of rotationally periodic waves and to some extent impedes their use in energy and signal transmission.

5.3 NONDIFFRACTING WAVES IN ANISOTROPIC CRYSTALS

In recent years, remarkable progress has been made in experimental studies of ballistic heat flow and phonon propagation in anisotropic nonmetallic crystals. Wavefronts associated with heat pulses can be measured in crystalline media at a few kelvin with phonon imaging techniques [Wolfe, 1998], and coherent phonon wave packets can be observed in real time using ultrafast optical excitation and detection [Sugawara et al., 2002]. Continuous (monochromatic) wave detection on crystal surface can be achieved using scanning homodyne Michelson interferometers [Knuuttila et al., 2000], while improved heterodyne techniques may be used to obtain both amplitude and phase information of the wave patterns [Kokkonen et al., 2003]. At ultrasonic frequencies (up to tens of gigahertz), acoustic wave devices based on anisotropic and piezoelectric crystals are also regularly employed in various signal-processing applications, such as resonators and filters. In view of the potential applications in these fields, we consider nondiffracting wave modes in strongly anisotropic media [Salo et al., 1999], and briefly discuss methods for their generation and detection [Salo, 2003; Salo and Salomaa, 2003].

5.3.1 Representation of Anisotropic Nondiffracting Waves

The propagation of acoustic vibrations in elastic and piezoelectric solids is governed by elastic equations of motion, which reflect the underlying periodic crystal structure. In the presence of anisotropy, the phase velocity varies as a function of propagation direction, which gives rise to several new effects:

1. The group velocity may have a transverse component relative to the wave vector, which makes the beam pattern move at a finite angle compared with the wave vectors, an effect called *beam steering*.
2. Isotropic energy distribution in Fourier space no longer implies isotropy in real space, where *phonon focusing* causes direction-dependent variations in the energy flow.
3. In the case of strong anisotropy, the energy flow may even diverge along some crystal directions, causing *caustics* to form in the wavefront.
4. Several wave components may occasionally transport energy along the same (spatial) directions, and their interference creates an effect called *internal diffraction*.

Both low-frequency acoustic waves and high-frequency heat flow phonons obey the elastic equations of motion, characterized by the stiffness tensor of the crystal. These equations are nondispersive in the long-wavelength limit, where the wavelength λ is much larger than the average interatomic separation. For the general theory of elasticity and piezoelectricity, see, for instance, Landau and Lifshitz [1986]. The wave equation for a plane wave,

$$\mathbf{u}(\mathbf{r}; t) = \mathbf{U}\mathrm{e}^{\mathrm{i}(\mathbf{k}\cdot\mathbf{r}-\omega t)}, \tag{5.24}$$

where $\mathbf{r} = (x, y, z)$, leads to the Christoffel (eigenvalue) equation for the amplitudes,

$$\sum_{l,m,n=1}^{3} \tilde{c}_{klmn} k_l k_n U_m = \rho\, \omega^2 U_k; \tag{5.25}$$

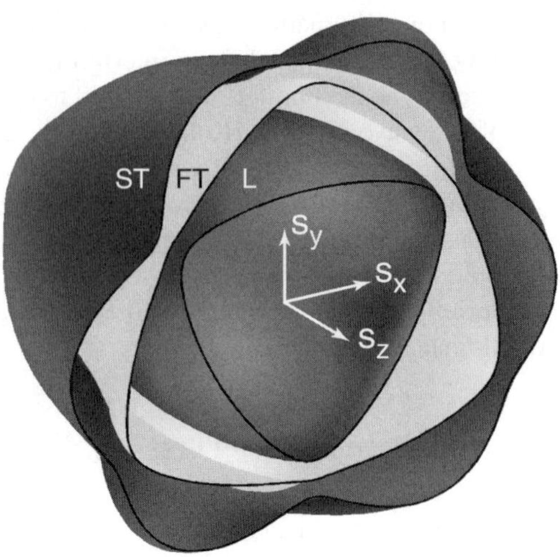

FIGURE 5.5 Three nondiffracting wave modes in quartz. The slowness surfaces are cut at $s_z = 70 \times 10^{-6}$ s/m, corresponding to phase velocity $v = 14{,}286$ m/s. The innermost surface represents the longitudinal (L) mode, the middle surface, the fast transverse (FT) mode, and the outermost surface, the slow transverse (ST) mode. The cutting lines also correspond to the three slowness curves used to produce the nondiffracting waves.

here U_k are the spatial components of the displacement vector, $\tilde{c}_{klmn}$ is the tensor of (piezoelectrically stiffened) elastic moduli, and ρ is the medium density [Auld, 1973].

In elastic solids, three different plane-wave modes exist with collinear wave vectors; they have orthogonal polarizations (directions of spatial displacement) $\mathbf{U}$ and often different phase velocities. These plane-wave modes are characterized by their slowness vectors $\mathbf{s} = \mathbf{k}/\omega$, with amplitudes given by the corresponding inverse phase velocities. In the nondispersive regime, the slowness vectors are independent of frequency and trace three closed slowness surfaces, characteristic to the medium in question, as illustrated in Fig. 5.5. The direction of energy propagation is determined by the associated group velocity $\mathbf{V}_g$ that is normal to the slowness surface and satisfies (in the nondispersive limit) $\mathbf{s} \cdot \mathbf{V}_g = 1$.

The nondiffracting waves in anisotropic media are defined as in Section 5.2 as fields that propagate uniformly along the z-axis with a constant velocity v,

$$\mathbf{u}(\mathbf{r}; t) = \mathbf{u}(x, y, z - vt), \tag{5.26}$$

and similarly, their spatiotemporal Fourier transform proves to be proportional to $\delta(k_z - \omega/v)$. Hence, they can be represented as superpositions of plane waves that all satisfy the common condition $k_z = \omega/v$ (i.e., $s_z = 1/v$),

$$\mathbf{u}(\mathbf{r}; t) = \iint A(\omega, \theta)\mathbf{U}(\theta)e^{i\omega[\mathbf{s}(\theta)\cdot\mathbf{r} - t]}\, d\theta\, d\omega, \tag{5.27}$$

where the integration path is given by the intersection of the slowness surfaces with the ($s_z = 1/v$)-plane, as shown in Fig. 5.5, and θ is the arc length parametrization of one (or all) slowness curves [Salo et al., 1999]. The polarizations $\mathbf{U}(\theta)$ are defined by the Christoffel equation only up to a complex constant factor, and they must be chosen continuous along the path; all other contributions are attributed to the spectral weight function $A(\omega, \theta)$. For a fixed propagation velocity v, at most three slowness curves may exist, and they characterize the longitudinal (L), fast transverse (FT), and slow transverse (ST) nondiffracting waves. Note that the orientation of the z-axis refers only to the desired direction of nondiffracting wave propagation and is independent of the crystal axes.

We consider two specific cases of nondiffracting waves: the monochromatic (continuous wave) beam, given by

$$\mathbf{v}(\mathbf{r}, t; \omega) = \int \beta(\theta)\mathbf{U}(\theta)e^{i\omega[s_x(\theta)x+s_y(\theta)y+s_z z-t]}\, d\theta, \tag{5.28}$$

and the X-wave pulse defined by spectral function $A(\omega, \theta) = \omega^m e^{-\alpha\omega}\beta(\theta)$ with the maximum at frequency $\omega = m/\alpha$. Using

$$\int_0^\infty \omega^m e^{-\alpha\omega} e^{i\omega[\mathbf{s}(\theta)\cdot\mathbf{r}-t]}\, d\omega = \frac{m!}{\{\alpha - \mathrm{i}[\mathbf{s}(\theta)\cdot\mathbf{r} - t]\}^{m+1}}, \tag{5.29}$$

the entire pulse is

$$\mathbf{u}(\mathrm{r}; t) = \int \frac{m!\,\beta(\theta)\mathbf{U}(\theta)}{\{\alpha - i[\mathbf{s}(\theta)\cdot\mathbf{r} - t]\}^{m+1}}\, d\theta. \tag{5.30}$$

These two expressions then depend only on integration over the slowness curve, with the integral weight determined by $\beta(\theta)$. We refer to the special case of $\beta(\theta) \equiv 1$ as the fundamental mode. Fundamental L and ST beam modes are illustrated in Fig. 5.6, and the corresponding pulses, in Fig 5.7.

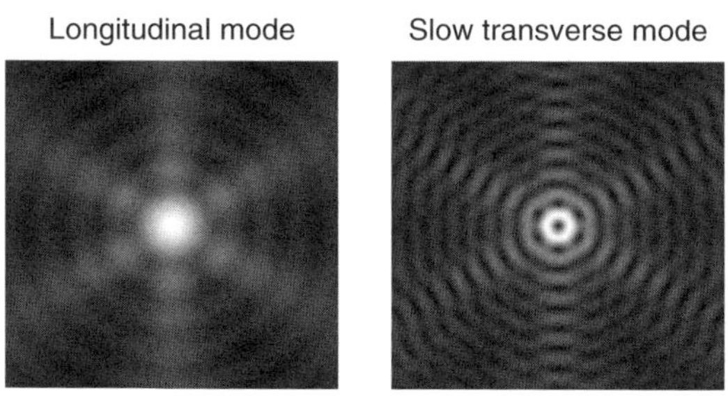

FIGURE 5.6 Beam cross sections of nondiffracting L and ST waves in quartz corresponding to the slowness cuts in Fig. 5.5. The L mode has only direction-dependent variation in the amplitude, while the nonconvexity of the ST slowness curve causes displacements of the field minima.

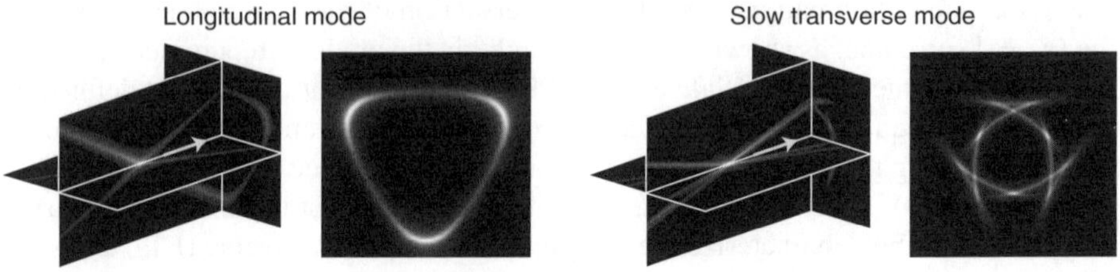

FIGURE 5.7 Three-dimensional illustrations of L and ST nondiffracting pulses in quartz. The strong anisotropy of ST mode creates a highly folded cone of propagation. The pulses correspond to the slowness cuts in Fig. 5.5, and the white arrows indicate the propagation direction of the pulse pattern.

5.3.2 Effects Due to Anisotropy

We gain physical insight into the nature of anisotropic nondiffracting waves by considering approximations of Eqs. (5.28) and (5.27) in the asymptotic region away from the beam/pulse center. In cylindrical coordinates, $\mathbf{r} = (r_\perp, \varphi, z)$, the nondiffracting beam assumes the form

$$\mathbf{v}(\mathbf{r}, t; \omega) = e^{i\omega(s_z z - t)} \int \beta(\theta)\mathbf{U}(\theta)e^{i\omega r_\perp \hat{\mathbf{r}}_\perp \cdot \mathbf{s}(\theta)}\, d\theta, \tag{5.31}$$

where $\hat{\mathbf{r}}_\perp = \hat{\mathbf{i}}\cos\varphi + \hat{\mathbf{j}}\sin\varphi$ is the radial unit vector in the transverse plane. For large enough $\omega r_\perp$, the exponential term in the integrand of Eq. (5.31) oscillates rapidly, and the integral averages out to zero. In the asymptotic evaluation of the integral, we look for the leading terms that decay comparatively slowest for large r.

The main contribution to the integral for a fixed azimuthal direction φ arises from those points on the slowness curve for which the normal of the curve is collinear with the direction chosen, and the asymptotic contributions to the nondiffracting wave are expressed as [Salo and Salomaa, 2003]

$$\mathbf{v}(\mathbf{r}, t; \omega) \approx \sum_{k:\, \mathbf{n}_k || \mathbf{r}_\perp} \sqrt{\frac{2\pi}{\omega r_\perp \kappa(\theta_k)}}\, \beta(\theta_k)\mathbf{U}(\theta_k)\, e^{\pm i\pi/4} e^{i\omega[\mathbf{s}(\theta_k)\cdot\mathbf{r} - t]}. \tag{5.32}$$

The summation is taken over the points specified above, $\kappa(\theta) = |\mathbf{s}''(\theta)|$ is the curvature of the slowness curve, and $\pm$ refers to the sign of $\mathbf{r}_\perp \cdot \mathbf{s}''_\perp(\theta_k)$ in the exponent. Each term in the wave expression is either a plane wave arriving at the beam axis from the direction φ or moving away from it, and together, they constitute a generalized conical wave. In the case of an isotropic medium, the curvature is constant and the conventional Bessel beams are obtained by setting $\beta(\theta) = e^{in\theta}$. This approximation then results in the asymptotic representation of the Bessel functions as $J_n(x) \approx (2/\pi x)^{1/2}\cos(x - \pi/4 - n\pi/2)$.

The asymptotic form for the nondiffracting pulse is obtained by a frequency integral over Eq. (5.32). Choosing an X-wave spectrum $f(\omega) = \omega^m e^{-\alpha\omega}$ and using the result

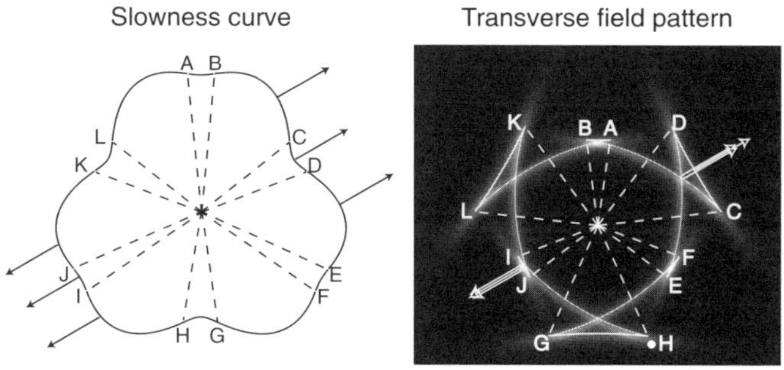

FIGURE 5.8 Slowness curve and asymptotic transverse field pattern of fundamental ST mode pulse. Capital letters indicate the zero-curvature points on the slowness surface (left), which imply caustics as turning points of the folds in the cross section of the wavefront (right). The six collinear arrows represent wave components that all propagate along the same line. If they overlap, their mutual interference gives rise to internal diffraction. Here the radial shape curve, Eq. (5.35), is superposed on the cross section of the field pattern.

[Gradshteyn and Ryzhik, 1994]

$$\int_0^{\infty} \frac{\omega^m}{\sqrt{\omega}} e^{-\alpha\omega}\, d\omega = \frac{\Gamma(m+1/2)}{\alpha^{m+1/2}}, \tag{5.33}$$

where $\Gamma(n)$ is the gamma function, the wave assumes the analytical form

$$\mathbf{u}_m(\mathbf{r}, t) \approx \sum_k \sqrt{\frac{2\pi}{r_\perp \kappa(\theta_k)}}\, \beta(\theta_k)\mathbf{U}(\theta_k) e^{\pm i\pi/4} \frac{\Gamma\,(m+1/2)}{\{\alpha - i[\mathbf{s}(\theta_k)\cdot\mathbf{r} - t]\}^{m+1/2}}. \tag{5.34}$$

The transverse wave pattern may be readily understood with the use of this approximation: The maximal amplitude of the individual wave components occurs at the minimum of the denominator in Eq. (5.34). The radial vectors pointing to the maxima are given by

$$\mathbf{R}_\perp = \frac{t - s_z z}{\mathbf{s}_\perp \cdot \mathbf{n}_\perp}\mathbf{n}_\perp = \frac{t - z/v}{\mathbf{s}_\perp \cdot \mathbf{n}_\perp}\mathbf{n}_\perp \tag{5.35}$$

for fixed z and t. Here, $\mathbf{s}_\perp$ are radial slowness vectors and $\mathbf{n}_\perp$ are (radial) normals to the slowness curve. Allowing $\mathbf{s}_\perp$ to trace the entire slowness curve in the $(s_z = 1/v)$-plane, $\mathbf{R}_\perp$ forms the *radial shape curve* that characterizes the conical shape of the propagating pulse, as shown in Fig. 5.8.

The beam amplitude of the asymptotic expression, Eq. (5.32), is proportional to $|\beta(\theta)|/[\kappa(\theta)]^{1/2}$, where the shape function $\beta(\theta)$ describes the strength of the excitation in the Fourier domain, while the curvature $\kappa(\theta)$ is purely characteristic of the medium and the direction and velocity of propagation (which together serve to define the slowness curve). A small curvature of the slowness (i.e., large radius of curvature) implies an elevated level of wave amplitude in a specified real-space direction, an effect called *phonon focusing*. If the slowness curve becomes locally flat, that is, its curvature approaches zero, the asymptotic approximation fails since the denominator vanishes. Although such singularities occur only in the asymptotic expression (the original integral does not diverge), they imply existence of a caustic associated with a flat point in the slowness surface.

Another consequence of a vanishing curvature is the appearance of folds in the propagation cone. In the case of weak anisotropy and a convex slowness curve, only two waves propagate along each direction, one approaching the axis and one emanating from it. If, however, the anisotropy is strong enough to render the slowness curve nonconvex, several waves may contribute: If they overlap spatially, the interference is observed as internal diffraction; otherwise, they are seen as folds in the wavefronts (see Fig. 5.8). This is strongly associated with the formation of caustics since they appear for wave modes for which the curvature κ vanishes and may change sign (i.e., the slowness curve becomes flat), a requirement for the slowness curve's being nonconvex.

5.3.3 Acoustic generation of NDWs

In this subsection we discuss briefly a scheme for producing longitudinal nondiffracting waves in piezoelectric crystals [Salo, 2003; Salo and Salomaa, 2003]. The method is based on a piezoelectric transducer similar to the computer-generated holograms used in optics (see, e.g., Turunen et al. [1988]). A planar transducer is located on one surface of crystal wafer while the nondiffracting beam patters is to be observed on the opposite surface (see Fig. 5.9).

According to the asymptotic expansion, Eq. (5.32), the nondiffracting beam essentially consists of just a few plane waves along each radial direction. If the wave mode has merely moderate anisotropy, such that the slowness curve remains convex

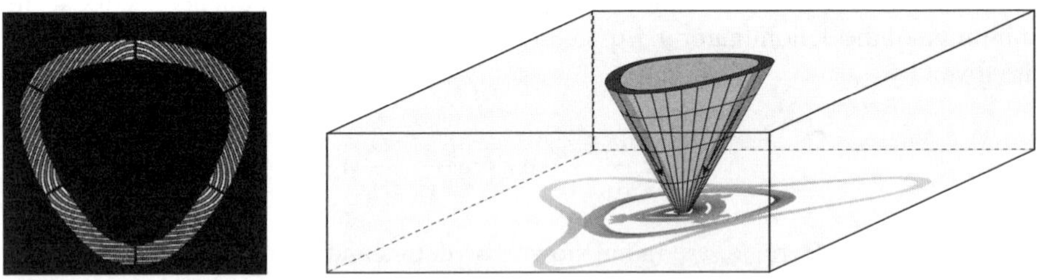

FIGURE 5.9 *Left*: Sketch of a transducer for exciting an L-mode nondiffracting beam in quartz. White denotes grounded electrodes, gray driving electrodes, and black areas are free substrate surface. *Right*: Entire radiation pattern of the transducer as observable on the opposite crystal surface. The cone represents the longitudinal (L) mode, and the actual nondiffracting beam cross section is formed on the cone waist, where both inward- and outward-propagating components overlap. The transducer also produces spurious FT and ST modes, displayed on the background surface. The boundary diffraction effects due to finite aperture are discarded.

as for the L and FT modes in Fig 5.5, only two plane waves appear: one propagating toward the beam axis and one propagating away from it. To produce the focal line of the beam, it is only necessary to generate the first, with the wave vector (along the surface) given by $(k_x, k_y) = \omega \mathbf{s}_\perp$. Therefore, the transducer must have the same periodicity, which is most easily realized with interdigital-type transducers (IDTs) [White and Volmer, 1965; Milsom et al., 1977].

The presence of different wave modes, however, causes trouble that is not encountered when using optical holograms: An active transducer produces not only the desired longitudinal wave mode but also unwanted transverse modes [Peach, 1995]. Similarly, a periodic transducer also generates different diffraction orders [Petit, 1980]. Since each mode generated has its own group velocity, they all transport energy in different directions (apart from the evanescent modes, which do not carry energy into the crystal), and a finite-sized transducer can be designed such that the modes separate spatially and only one of them constitutes the nondiffracting beam itself. The even diffraction orders can also be made to vanish using a locally inversion symmetric transducer layout. In simple transducer schemes, a locally periodic metallic interdigital structure is driven with a time-oscillating voltage, a technology widely used in surface-acoustic wave (SAW) devices.

The transducer shown in Fig. 5.9 consists of an annular element that produces a longitudinal nondiffracting beam in quartz. The aperture is parametric in the sense that it may be scaled arbitrarily, depending on the thickness of the quartz substrate, and the frequency used only affects the scaling of the local periodicity of the transducer structure. If the frequency is high enough so that boundary diffractions can be ignored, the various wave modes and diffraction orders separate and a pure longitudinal beam pattern is produced on the rear surface of the wafer, where it can be measured, for instance, using interferometric scanning [Knuuttila et al., 2000, 2001; Kokkonen et al., 2003].

5.3.4 Discussion

Nondiffracting waves in anistropic media propagate uniformly with velocity v [see Eq. (5.26)], and they obey the same NDW dispersion relation $k_z = \omega/v$ as in isotropic media. Similarly, they focus (acoustic) energy onto the focal line but also require a large aperture and sidelobes to maintain the central beam. However, the presence of a possibly strong anisotropy affects the properties of the wave, and beam steering, in particular, may complicate their use: occasionally, wave components with $k_z > 0$ may still carry energy along the negative z-axis, and such waves cannot be created with a causal transducer. Even if this is not the case, the energy of some wave components may move nearly in the transverse (xy) plane, which would require an immensely large aperture. On the other hand, transducers can be designed such that wave components are properly focused to form a nondiffracting wave pattern within a limited volume in the vicinity of the back surface of the wafer. In particular, the transducer and detection scheme presented above can be used to confirm the formation of a longitudinal nondiffracting wave.

Inside the elastic crystal, nondiffracting waves focus acoustic energy onto the propagation axis, where other physical phenomena may take place. If the crystal features nonlinear behavior, such effects should only take place in regions of high amplitude; nondiffracting waves would thus allow the investigation of nonlinear acoustic effects, such as harmonic generation or soliton formation [Hao and Maris, 2001]. They may also be potentially useful for studying novel effects in piezoelastic crystals, such as acoustic memory [McPherson et al., 2002] or dispersively backward-propagating waves [Alippi et al., 2001]. High acoustic power may even cause irreversible structural changes on the propagation axis, while formation of the focal line is only weakly affected by the nonlinearity, since the wave amplitude away from the axis is much lower. The entire beam pattern within a transparent crystal can be measured by a frequency-shift holographic method [Togami and Chiba, 1978]: The crystal is illuminated through a side surface and the light is scattered from the beam due to the photoelastic effect, which is particularly strong near the beam focal line. Holographic recording of the transmitted light then yields tomographic cuts of the beam pattern within the crystal bulk.

5.4 CONCLUSIONS

Nondiffracting waves feature two characteristic properties that distinguish them from other modes of wave propagation:

1. The wave pattern remains invariant on propagation within the region defined by the geometry of the transducer aperture. This is due to the fact that all wave components share a common phase velocity along the propagation axis, thus keeping their mutual interference pattern unchanged.
2. The wave energy is focused onto the axis of propagation, not merely into a single spot as for Gaussian-type waves but along an entire focal line, whose length is determined by the transducer.

Nondiffracting beams and nondispersive pulselike fields such as X-waves have shown application potential in several fields. The long, narrow focal line of Bessel beams, which suggests good lateral resolution in imaging applications, has been the driving motivation for limited-diffraction-wave research in medical imaging (see, e.g., [Lu, 1998] and references therein). The depth of focus and high transverse resolution of the focal line have proven advantageous in optical microlithography [Erdélyi et al., 1997], and Bessel beams have been considered for improving heterodyne detection efficiency [Iftekharuddin and Karim, 1992]. Especially strong has been the interest in utilizing Bessel beams in second- and higher-order harmonic generation in nonlinear optics [Nisoli et al., 2002]. This is due to the ease of tuning the axial wavelength of the Bessel beam to meet the phase-matching criteria [Glushko et al., 1993]. Several other fields of applications have recently also been considered, such as manipulation of microscopic particles [Garcés-Chávez et al., 2002].

At present, the field of nondiffracting classical waves appears to have reached a state of certain maturity: the novel and amazing properties of Bessel beams and pulsed X-waves are well understood, and the main emphasis of research has moved toward practical scientific and industrial applications. However, due to the rapid progress of nano-optics, it is safe to predict that the nonlinear and quantum-mechanical interactions of diffraction-free beams ("Bessel photons" [Jáuregui and Hacyan, 2005]) with atomic systems, metamaterials, and other subwavelength structures will still have several surprises in store in many fields of physics and engineering sciences.

REFERENCES

Alexeev, I., K. Y. Kim, and H. M. Milchberg, 2002, Measurement of the superluminal group velocity of an ultrashort Bessel beam pulse, *Phys. Rev. Lett.* **88**, 073901.

Alippi, A., A. Bettucci and M. Germano, 2001, Conditions for anomalous acousto-optical diffraction by backward propagating acoustic waves, *Phys. Rev. E* **63**, 026604.

Artl, J. and K. Dholakia, 2000, Generation of high-order Bessel beams by use of an axicon, *Opt. Commun.* **177**, 297–301.

Auld, M. A., 1973, *Acoustic Fields and Waves in Solids*, Vol. 1, Wiley, New York.

Bandrés, M. A., J. C. Gutiérrez-Vega, and S. Chávez-Cerda, 2004, Parabolic nondiffracting optical wavefields, *Opt. Lett.* **29**, 44–46.

Besieris, I., M. Abdel-Rahman, A. Shaarawi, and A. Chatzipetros, 1998, Two fundamental representations of localized pulse solutions to the scalar wave equation, *Prog. Electromagn. Res.* **19**, 1–48.

Bouchal, Z., 2002, Controlled spatial shaping of nondiffracting patterns and arrays, *Opt. Lett.* **27**, 1376–1378.

Bouchal, Z., 2003, Nondiffracting optical beams: physical properties, experiments, and applications, *Czech. J. Phys.* **53**, 537–578.

Bouchal, Z., and M. Olivík, 1995, Non-diffractive vector Bessel beams, *J. Mod. Opt.* **42**, 1555–1566.

Bouchal, Z., R. Horák, and J. Wagner, 1996, Propagation-invariant electromagnetic fields: theory and experiment, *J. Mod. Opt.* **423**, 1905–1920.

Bouchal, Z., J. Bajer, and M. Bertolotti, 1998a, Vectorial spectral analysis of the nonstationary electromagnetic field, *J. Opt. Soc. Am. A* **15**, 2172–2181.

Bouchal, Z., M. Wagner, and M. Chlup, 1998b, Self-reconstruction of a distorted nondiffracting beam, *Opt. Commun.* **151**, 207–211.

Brittingham, J. N., 1983, Focus wave modes in homogeneous Maxwell's equations: transverse electric mode, *J. Appl. Phys.* **54**, 1179–1189.

Bunkin, F. V., V. V. Korobkin, Yu. A. Kurinyi, L. Ya. Polonskii, and L. N. Pyatnitskii, 1983, Laser spark with a continuous channel in air, *Sov. J. Quantum Electron.* **13**, 254–255.

Burvall, A., K. Kolacz, Z. Jaroszewicz, and A. T. Friberg, 2004, Simple lens axicon, *Appl. Opt.* **43**, 4838–4844.

Butkus, R., R. Gadonas, J. Janušonis, A. Piskarskas, K. Regelskis, V. Smilgevičius, and A. Stabinis, 2002, Nonlinear self-reconstruction of truncated Bessel beam, *Opt. Commun.* **206**, 201–209.

Chattrapiban, N., E. A. Rogers, D. Cofield, W. T. Hill III, and R. Roy, 2003, Generation of nondiffracting Bessel beams using a spatial light modulator, *Opt. Lett.* **28**, 2183–2185.

Chávez-Cerda, S., 1999, A new approach to Bessel beams, *J. Mod. Opt.* **46**, 923–930.

Chávez-Cerda, S., G. S. McDonald, and G. H. C. New, 1996, Nondiffracting beams: travelling, standing, rotating and spiral waves, *Opt. Commun.* **123**, 225–233.

Chávez-Cerda, S., M. J. Padgett, I. Allison, G. H. C. New, J. C. Gutiérrez-Vega, A. T. O'Neil, I. MacVicar, and J. Courtial, 2002, Holographic generation and orbital angular momentum of high-order Mathieu beams, *J. Opt. B Quantum Semiclass. Opt.* **4**, S52–S57.

Conti, C., and S. Trillo, 2003, X-waves generated at second harmonic, *Opt. Lett.* **28**, 1251–1253.

Cox, A. J., and D. C. Dibble, 1992, Nondiffracting beams from a spatially filtered Fabry–Perot resonator, *J. Opt. Soc. Am. A* **9**, 282–286.

Dartora, C. A., M. Zamboni-Rached, K. Z. Nóbrega, E. Recami, and H. E. Hernández-Figueroa, 2003, General formulation for the analysis of scalar diffraction-free beams using angular modulation: Mathieu and Bessel beams, *Opt. Commun.* **222**, 75–80.

Davis, J. A., E. Carcole, and D. M. Cottrell, 1996, Nondiffracting interference patterns generated with programmable spatial light modulators, *Appl. Opt.* **35**, 599–602.

DeBeer, D., S. R. Hartmann, and R. Friedberg, 1987, Comment on "Diffraction-free beams," *Phys. Rev. Lett.* **59**, 2611.

Di Trapani, P., G. Valiulis, A. Piskarskas, O. Jedrkiewicz, J. Trull, C. Conti, and S. Trillo, 2003, Spontaneously generated X-shaped light bullets, *Phys. Rev. Lett.* **91**, 093904.

Durnin, J., 1987, Exact solutions for nondiffracting beams: I. The scalar theory, *J. Opt. Soc. Am. A* **4**, 651–654.

Durnin, J., J. J. Miceli, Jr., and J. H. Eberly, 1987a, Diffraction-free beams, *Phys. Rev. Lett.* **58**, 1499–1501.

Durnin, J., J. J. Miceli, Jr., and J. H. Eberly, 1987b, Durnin, Miceli and Eberly reply, *Phys. Rev. Lett.* **59**, 2612.

Durnin, J., J. J. Miceli, Jr., and J. H. Eberly, 1991, Durnin, Miceli and Eberly reply, *Phys. Rev. Lett.* **66**, 838.

Erdélyi, M., Z. L. Horváth, G. Szabo, Z. Bor, F. K. Tittel, J. R. Cavallaro, and M. C. Smayling, 1997, Generation of diffraction-free beams for applications in microlithography, *J. Vac. Sci. Technol. B* **15**, 287–292.

Fagerholm, J., A. T. Friberg, J. Huttunen, D. P. Morgan, and M. M. Salomaa, 1996, Angular-spectrum representation of nondiffracting X waves, *Phys. Rev. E* **54**, 4347–4352.

Fischer, P., C. T. A. Brown, J. E. Morris, C. López-Mariscal, E. M. Wright, W. Sibbett, and K. Dholakia, 2005, White light propagation invariant beams, *Opt. Express* **13**, 6657–6666.

Friberg, A. T., A. Vasara, and J. Turunen, 1991, Partially coherent propagation-invariant beams: passage through paraxial optical systems, *Phys. Rev. A* **43**, 7079–7082.

Garcés-Chávez, V., D. McGloin, H. Melville, W. Sibbett, and K. Dholakia, 2002, Simultaneous micromanipulation in multiple planes using a self-reconstructing light beam, *Nature* **419**, 145–147.

Glushko, B., B. Kryzhanovsky, and D. Sarkisyan, 1993, Self-phase-matching mechanism for efficient harmonic generation in a ring pump beam geometry, *Phys. Rev. Lett.* **71**, 243–246.

Gori, F., G. Guattari, and C. Pavadoni, 1987, Bessel–Gauss beams, *Opt. Commun.* **64**, 491–495.

Gradshteyn, I. S., and I. M. Ryzhik, 1994, *Table of Integrals, Series and Products*, 5th ed., Academic Press, San Diego, CA.

Gutiérrez-Vega, J. C., M. D. Iturbe-Castillo, and S. Chávez-Cerda, 2000, Alternative formulation for invariant optical fields: Mathieu beams, *Opt. Lett.* **25**, 1493–1495.

Gutiérrez-Vega, J. C., M. D. Iturbe-Castillo, G. A. Ramirez, S. Chávez-Cerda, and G. H. C. New, 2001, Experimental demonstration of optical Mathieu beams, *Opt. Commun.* **195**, 35–40.

Hafizi, B., and P. Sprangle, 1991, Diffraction effects in directed radiation beams, *J. Opt. Soc. Am. A* **8**, 705–717.

Hakola, A., S. C. Buchter, T. Kajava, H. Elfström, J. Simonen, P. Pääkkönen, and J. Turunen, 2004, Bessel–Gauss output beam from diode-pumped Nd: YAG laser, *Opt. Commun.* **238**, 335–340.

Hakola, A., A. Shevchenko, S. C. Buchter, M. Kaivola, and N. V. Tabiryan, 2006, Creation of a narrow Bessel-like laser beam using a nematic liquid crystal, *J. Opt. Soc. Am. B* **23**, 637–641.

Hao, H.-Y., and H. J. Maris, 2001, Experiments with acoustic solitons in crystalline solids, *Phys. Rev. B* **64**, 064302.

He, H., N. R. Heckenberg, and H. Rubinstein-Dunlop, 1995, Optical particle trapping with higher-order doughnut beams produced using high efficiency computer generated holograms, *J. Mod. Opt.* **42**, 217–223.

Herman, R. M., and T. A. Wiggins, 2001, Production and uses of diffractionless beams, *J. Opt. Soc. Am. A* **8**, 932–942.

Honkanen, M., and J. Turunen, 1998, Tandem systems for efficient generation of uniform-axial-intensity Bessel fields, *Opt. Commun.* **154**, 368–375.

Horák, R., Z. Bouchal, and J. Bajer, 1997, Nondiffracting stationary electromagnetic field, *Opt. Commun.* **133**, 315–327.

Horváth, Z. L., M. Erdélyi, G. Szabó, Z. Bor, F. K. Tittel, and J. R. Cavallaro, 1997, Generation of nearly nondiffracting Bessel beams with a Fabry–Perot interferometer, *J. Opt. Soc. Am. A* **14**, 3009–3013.

Hsu, D. K., F. J. Margetan, and D. O. Thompson, 1989, Bessel beam ultrasonic transducer: fabrication method and experimental results, *Appl. Phys. Lett.* **55**, 2066–2068.

Iftekharuddin, K. M., and M. A. Karim, 1992, Heterodyne detection by using a diffraction-free beam: tilt and offset effects, *Appl. Opt.* **31**, 4853–4856.

Iftekharuddin, K. M., A. A. S. Awwal, and M. A. Karim, 1993, Gaussian-to-Bessel beam transformation using a split refracting system, *Appl. Opt.* **32**, 2251–2256.

Indebetouw, G., 1989, Nondiffracting optical fields: some remarks on their analysis and synthesis, *J. Opt. Soc. Am. A* **6**, 150–152.

Jaroszewicz, Z., 1997, in *Axicons: Design and Propagation Properties, R&D Treatises*, Vol. 5, M. Pluta, Ed., SPIE Polish Chapter, Warsaw, Poland.

Jaroszewicz, Z., A. Burvall, and A. T. Friberg, 2005, Axicon: the most important optical element, *Opt. Photon. News* **16**, 34–39.

Jáuregui, R., and S. Hacyan, 2005, Quantum-mechanical properties of Bessel beams, *Phys. Rev. A* **71**, 033411.

Kelly, D. H., 1963, Axicons, annuli, and coherent light, *J. Opt. Soc. Am.* **53**, 525.

Khilo, A. N., E. G. Katranji, and A. A. Ryzhevich, 2001, Axicon based Bessel resonator: analytical description and environment, *J. Opt. Soc. Am. A* **18**, 1986–1992.

Knuuttila, J. V., P. T. Tikka, and M. M. Salomaa, 2000, Scanning Michelson interferometer for imaging surface acoustic wave fields, *Opt. Lett.* **25**, 613–615.

Knuuttila, J. V., J. Saarinen, C. S. Hartmann, V. P. Plessky, and M. M. Salomaa, 2001, Measurement of BAW radiation from low-loss LSAW resonators, *Electron. Lett.* **37**, 1055–1056.

Kokkonen, K., J. V. Knuuttila, V. P. Plessky, and M. M. Salomaa, 2003, Phase-sensitive absolute-amplitude measurements of surface waves using heterodyne interferometry, *Proceedings of the International IEEE Ultrasonics Symposium*, New York, pp. 1145–1148.

Korobkin, V. V., L. Y. Polonskii, V. P. Poponin, and L. N. Pyatnitskii, 1986, Focusing of Gaussian and super-Gaussian laser beams by axicon to obtain continuous laser sparks, *Sov. J. Quantum Electron.* **16**, 178–182.

Kowarz, M. W., and G. S. Agarwal, 1995, Bessel-beam representation for partially coherent fields, *J. Opt. Soc. Am. A* **12**, 1324–1330.

Landau, L. D., and E. M. Lifshitz, 1986, *Theory of Elasticity*, 3rd ed., Pergamon Press, London.

Lapointe, M. R., 1992, Review of nondiffracting Bessel beams experiments, *Opt. Laser Technol.* **24**, 315–321.

López-Mariscal, C., M. A. Bandrés, S. Chávez-Cerda, and J. C. Gutiérrez-Vega, 2005, Observation of parabolic nondiffracting wave fields, *Opt. Express* **13**, 2364–2369.

Lu, J.-Y., 1998, Experimental study1998 of high frame rate imaging with limited diffraction beams, *IEEE Trans. Ultrason. Ferroelectr. Freq. Control* **45**, 84–97.

Lu, J.-Y., and J. F. Greenleaf, 1990, Ultrasonic nondiffracting transducer for medical imaging, *IEEE Trans. Ultrason. Ferroelectr. Freq. Control* **37**, 438–447.

Lu, J.-Y., and J. F. Greenleaf, 1992, Nondiffracting X waves: exact solutions to free-space scalar wave equation and their finite aperture realizations, *IEEE Trans. Ultrason. Ferroelectr. Freq. Control* **39** 19–31.

McCutchen, C. W., 1964, Generalized aperture and three-dimensional diffraction image, *J. Opt. Soc. Am.* **54**, 240–244.

McGloin, D., and K. Dholakia 2005, Bessel beams: diffraction in a new light, *Contemp. Phys.* **46**, 15–28.

McLeod, J. H., 1954, The axicon: a new type of optical element, *J. Opt. Soc. Am.* **44**, 591–597.

McLeod, J. H., 1960, Axicons and their uses, *J. Opt. Soc. Am.* **50**, 166–169.

McPherson, M. S., I. Ostrovskii, and M. A. Breazeale, 2002, Observation of acoustical memory in $LiNbO_3$, *Phys. Rev. Lett.* **89**, 115506.

Milsom, R. F., N. H. C. Reilly, and M. Redwood, 1977, Analysis of generation and detection of surface and bulk acoustic waves by interdigital transducers, *IEEE Trans. Soni. Ultrason.* **SU-24**, 147–166.

Mishra, S. R., 1991, A vector wave analysis of a Bessel beam, *Opt. Commun.* **85**, 159–161.

Mugnai, D., and I. Mochi, 2005, Superluminal X-wave propoagation: energy localization and velocity, *Phys. Rev. E* **73**, 016606.

Mugnai, D., A. Ranfagni, and R. Ruggeri, 2000, Observation of superluminal behaviors in wave propagation, *Phys. Rev. Lett.* **84**, 4830–4833.

Nisoli, M., E. Priori, G. Sansone, S. Stagira, G. Cerullo, and S. De Silvestri, 2002, High-brightness high-order harmonic generation by truncated Bessel beams in the sub-10-fs regime, *Phys. Rev. Lett.* **88**, 033902.

Oughstun, K. E., and G. C. Sherman, 1994, *Electromagnetic Pulse Propagation in Causal Dielectrics*, Springer-Verlag, Berlin.

Pääkkönen, P., J. Lautanen, M. Honkanen, M. Kuittinen, S. N. Khonina, V. V. Kotlyar, V. A. Soifer, and A. T. Friberg, 1998, Rotating optival fields: experimental demonstration with diffractive optics, *J. Mod. Opt.* **46**, 2355–2369.

Pääkkönen, P., J. Tervo, P. Vahimaa, and J. Turunen, 2002, General vectorial decomposition of electromagnetic fields with application to propagation-invariant and rotating fields, *Opt. Express* **10**, 949–959.

Paterson, C., and R. Smith, 1996, Higher-order Bessel waves produced by axicon-type computer-generated holograms, *Opt. Commun.* **124**, 121–130.

Peach, R. C., 1995, A general Green function analysis of SAW devices, *Proceedings of the International IEEE Ultrasonics Symposium*, New York, pp. 221–225.

Petit, R., Ed., 1980, *Electromagnetic Theory of Gratings*, Springer-Verlag, Berlin.

Piestun, R., and J. Shamir, 1998, Generalized propagation-invariant wave fields, *J. Opt. Soc. Am. A* **15**, 3039–3044.

Polonskiy, L., 2005, Axicons: more to the story, *Opt. Photon. News* **16**, 7.

Ponomarenko, S. A. and G. P. Agrawal, 2006, Linear optical bullets, *Opt. Commun.* **261**, 1–4.

Porras, M. A., 2001, Diffraction-free and dispersion-free pulsed beam propagation in dispersive media, *Opt. Lett.* **17**, 1364–1366.

Porras, M. A., S. Trillo, C. Conti, and P. Di Trapani, 2003, Paraxial envelope X waves, *Opt. Lett.* **13**, 1090–1092.

Recami, E., 1998, On localized "X-shaped" superluminal solutions to Maxwell equations, 1998, *Physica A* **252**, 586–610.

Recami, E., 2001, Superluminal motions? Bird's-eye view of the experimental situation, *Found. Phys.* **31**, 1119–1135.

Recami, E., M. Zamboni-Rached, K. Z. Nóbrega, C. A. Dartora, and H. E. Hernández-Figueroa, 2003, On the localized superluminal solutions to the Maxwell equations, *IEEE J. Sel. Top. Quantum Electron.* **9**, 59–73.

Reivelt, K., and P. Saari, 2002, Experimental demonstration of realizability of optical focus wave modes, *Phys. Rev. E* **66**, 056611.

Rodrigues, W. A., and J.-Y. Lu, 1997, On the existence of undistorted progressive waves (UPWs) of arbitrary speeds $0 \leq v < \infty$ in nature, *Found. Phys.* **27**, 435–508.

Saari, P., and K. Reivelt, 1997, Evidence of X-shaped propagation-invariant localized light waves, *Phys. Rev. Lett.* **79**, 4135–4138.

Saastamoinen, T., J. Tervo, P. Vahimaa, and J. Turunen, 2004, Exact self-imaging and transversely periodic fields, *J. Opt. Soc. Am. A* **21**, 1424–1429.

Salo, J., 2003, Propagation-invariant waves in acoustic, optical, and radio-wave fields, Ph.D. dissertation, Helsinki University of Technology, Espoo, Finland.

Salo, J., and M. M. Salomaa, 2001, Diffraction-free pulses at arbitrary speeds, *J. Opt. A Pure Appl. Opt.* **3**, 366–373.

Salo, J., and M. M. Salomaa, 2003, Nondiffracting waves in anisotropic media, *Phys. Rev. E* **67**, 056609.

Salo, J., J. Fagerholm, A. T. Friberg, and M. M. Salomaa, 1999, Nondiffracting bulk X waves in crystals, *Phys. Rev. Lett.* **83**, 1171–1174.

Salo, J., J. Fagerholm, A. T. Friberg, and M. M. Salomaa, 2000, Unified description of nondiffracting X and Y waves, *Phys. Rev. E* **62**, 4261–4275.

Scott, G., and N. McArdle, 1992, Efficient generation of nearly diffraction-free beams using using an axicon, *Opt. Eng.* **31**, 2640–2643.

Sedukhin, A. G., 1998, Beam-preshaping axicon focusing, *J. Opt. Soc. Am. A* **15**, 3057–3066.

Sezginer, A., 1985, A general formulation of focus wave modes, *J. Appl. Phys.* **57**, 678–683.

Shaarawi, A. M., 1997, Comparison of two localized wave fields generated from dynamic apertures, *J. Opt. Soc. Am. A* **14**, 1804–1816.

Shchegrov, A. V., and E. Wolf, 2000, Partially coherent conical beams, *Opt. Lett.* **25**, 141–143.

Sõnajalg, H., M. Rätsep, and P. Saari, 1997, Demonstration of the Bessel-X pulse propagating with strong lateral and longitudinal localization in a dispersive medium, *Opt. Lett.* **22**, 310–312.

Soroko, L. M., 1996, *Meso-Optics: Foundations and Applications*, World Scientific, Singapore.

Sprangle, P., and B. Hafizi, 1991, Comment on nondiffracting beams, *Phys. Rev. Lett.* **66**, 837.

Stratton, J. A., 1941, *Electromagnetic Theory*, McGraw-Hill, New York.

Sugawara, Y., O. B. Wright, O. Matsuda, M. Takigahira, Y. Tanaka, S. Tamura, and V. E. Gusev, 2002, Watching ripples on crystals, *Phys. Rev. Lett.* **88**, 185504.

Tervo, J., and J. Turunen, 2001a, Generation of vectorial propagation-invariant fields by polarization-grating axicons, *Opt. Commun.* **192**, 13–18.

Tervo, J., and J. Turunen, 2001b, Rotating scale-invariant electromagnetic fields, *Opt. Express* **9**, 9–15.

Tervo, J., P. Vahimaa, and J. Turunen, 2002, On propagation-invariant and self-imaging intensity distributions of electromagnetic fields, *J. Mod. Opt.* **49**, 1537–1543.

Togami, Y., and T. Chiba, 1978, Observation of bulk waves and surface waves through side planes of several cuts of $LiNbO_3$ SAW devices, *J. Appl. Phys.* **49**, 3587–3589.

Turunen, J., 2002, Invariant propagation of uniform-intensity Schell-model fields, *J. Mod. Opt.* **49**, 1795–1799.

Turunen, J., and A. T. Friberg, 1993, Self-imaging and propagation-invariance in electromagnetic fields, *JEOS A Pure Appl. Opt.* **2**, 51–61.

Turunen, J., A. Vasara, and A. T. Friberg, 1988, Holographic generation of diffraction-free beams, *Appl. Opt.* **27**, 3959–3962.

Turunen, J., A. Vasara, and A. T. Friberg, 1991, Propagation invariance and self-imaging in variable-coherence optics, *J. Opt. Soc. Am. A* **8**, 282–289.

Uehara, K., and H. Kikuchi, 1989, Generation of nearly diffraction-free laser beams, *Appl. Phys. B* **48**, 125–129.

Valtna, H., K. Reivelt, and P. Saari, 2006, Modifications of the focused X wave as suitable models of strongly localized waves for realization in the optical domain, *J. Opt. A Pure Appl. Opt.* **8**, 118–122.

van Nie, A. G., 1964, Rigorous calculation of the electromagnetic field of wave beams, *Philips Res. Rep.* **19**, 378–394.

Vasara, A., J. Turunen, and A. T. Friberg, 1989, Realization of general nondiffracting beams with computer-generated holograms, *J. Opt. Soc. Am. A* **6**, 1748–1754.

Volke-Sepulveda, K., V. Garcés-Chávez, S. Chávez-Cerda, J. Artl, and K. Dholakia, 2002, Orbital angular momentum of a high-order Bessel light beam, *J. Opt. B Quantum Semiclass. Opt.* **4**, S82–S89.

White, R. M., and F. W. Volmer, 1965, Direct piezoelectric coupling to surface elastic waves, *Appl. Phys. Lett.* **7**, 314–316.

Wolfe, J. P., 1998, *Imaging Phonons: Acoustic Wave Propagation in Solids*, Cambridge University Press, Cambridge.

Wulle, T., and S. Herminghaus, 1993, Nonlinear optics of Bessel beams, *Phys. Rev. Lett.* **70**, 1401–1404.

Zamboni-Rached, M., 2002, Diffraction–attenuation resistant beams in absorbing media, *Opt. Express* **14**, 1804–1809.

Zamboni-Rached, M., E. Recami, and H. E. Hernández-Figueroa, 2002, New localized superluminal solutions to the wave equations with finite total energies and arbitrary frequencies, *Eur. Phys. J. D* **21**, 217–228.

Zamboni-Rached, M., K. Z. Nóbrega, H. E. Hernández-Figueroa, and E. Recami, 2003, Localized superluminal solutions to the wave equation in (vacuum or) dispersive media, for arbitrary frequencies and with adjustable bandwidth, *Opt. Commun.* **226**, 15–23.

Zamboni-Rached, M., E. Recami, and H. E. Hernández-Figueroa, 2005, Theory of "frozen waves": modelling the shape of stationary fields, *J. Opt. Soc. Am. A* **22**, 2465–2475.

CHAPTER SIX

Bessel X-Wave Propagation

DANIELA MUGNAI AND IACOPO MOCHI
Nello Carrara Institute of Applied Physics–CNR, Florence Research Area, Sesto Fiorentino, Italy

6.1 INTRODUCTION

The propagation of a Bessel X-wave is of great interest in physics. This interest is due to the unusual features of these types of waves, which are nondiffracting [1–7] and show superluminal behavior in both phase and group velocities [8–13]. Because of these two characteristics, the interest that has arisen regarding Bessel X-waves has led to several theoretical and experimental investigations, none of which has yet been able to provide definite information on energy transfer and its velocity.

As is well known, the scalar field of a Bessel X-wave propagating along the z-axis of a cylindrical coordinate system $(\rho,\ \psi,\ z)$ is given by [14]

$$u(\rho, \psi, z) = 2\pi A J_0(k_0 n\rho \sin\theta_0) \exp(ik_0 nz \cos\theta_0) \exp(-i\omega t), \qquad (6.1)$$

where A is an amplitude factor, θ_0 the parameter that characterizes the aperture of the beam (the axicon angle), J_0 the zero-order Bessel function of the first kind [15], n the refractive index of the medium where the beam propagates, and k_0 the wave number in the vacuum.† The field as given by Eq. (6.1) can be obtained from the superposition of infinite plane waves of the same amplitude, each propagating in a different direction and forming the same angle θ_0 with the z-axis.

Equation (6.1) is known as a Bessel X-beam. To be more precise, the Bessel X-wave does not properly describe a beam, since it is not limited by a caustic surface,

† The field (6.1) is rotationally symmetric and thus independent of the angular coordinate ψ.

Localized Waves, Edited by Hugo E. Hernández-Figueroa, Michel Zamboni-Rached, and Erasmo Recami

inasmuch as J_0 oscillates when its argument tends to infinity. However, the definition of beam is widely accepted because of its characteristic of having well-localized energy. Looking at Eq. (6.1) we see immediately that the phase propagates in the z-direction with a velocity $v = c/(n \cos\theta_0)$ greater than c/n, and that it does not changes its shape during propagation since the amplitude is independent of z.†

6.2 OPTICAL TUNNELING: FRUSTRATED TOTAL REFLECTION

The propagation of electromagnetic waves through classically forbidden regions is of great interest mainly because of its implications with regard to the topic of superluminality. From an experimental point of view, many papers provide evidence of superluminality in various physical systems [16–21].

From a theoretical point of view, the problem of a passage through a forbidden gap has been treated by considering both plane waves and wave packets impinging, at oblique incidence, into a plane parallel layer of a medium with a refractive index that is smaller than the index of the surrounding medium [22–25].

Bessel X-wave propagation at normal and oblique incidence into a layer, in the presence of frustrated total reflection, has been analyzed extensively. In the case of normal incidence, the analysis has been performed in both scalar [14,26] and vectorial approximation [14], while for oblique incidence only a scalar approach was considered [27]. In passing through the tunneling region, even if the shape of the Bessel beam can suffer deformation, the particular symmetry of this type of beam enables it to travel through the gap with no displacement in its direction of propagation. This behavior seems to be a peculiarity of the Bessel beam, since in passing through a classically forbidden region, a plane wave undergoes a shift in its direction of propagation [25].

6.2.1 Bessel Beam Propagation into a Layer: Normal Incidence

Scalar Treatment The expansion of a Bessel beam in plane waves, whose directions of propagation cover a conical surface of semiaperture θ_0, indicates immediately what happens when the Bessel beam impinges (at $z = 0$) into a plane surface that separates two media of different refractive indexes n and n' (see Fig. 6.1). Each plane wave forms the same incidence angle θ_0 and hence the same refraction angle θ'. Since the refracted waves have the same amplitude, their superposition u once again originates a Bessel beam, which is now characterized by the parameter θ' such that

$$n' \sin\theta' = n \sin\theta_0 . \tag{6.2}$$

†The situation is reminescent of what occurs when only two plane waves interfere, the only difference being that the two-wave interference pattern occupies the entire space, while the field (6.1) is practically limited to a restricted zone of space, that is, between the first zeros of the function J_0.

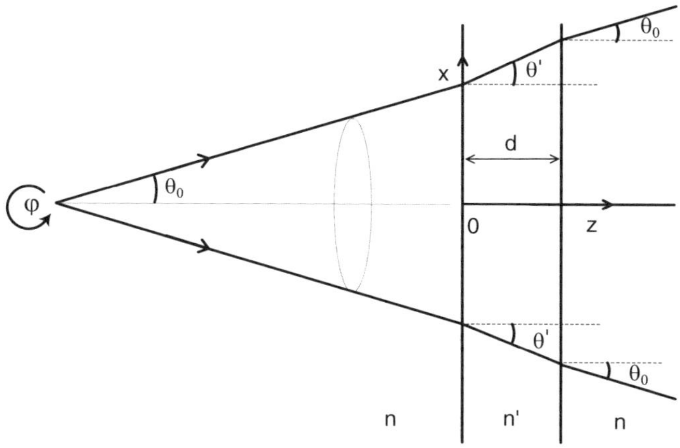

FIGURE 6.1 Schematic representation of the layer and the impinging Bessel beam characterized by the axicon angle θ_0; d is the width of the layer, and n and n' are the refractive indexes of the two media.

Thus, inside the slab we have

$$u(\rho, \psi, z) = A' J_0(k_0 n' \rho \sin\theta') \exp(i k_0 n' z \cos\theta'), \tag{6.3}$$

where the temporal factor $\exp(-i\omega t)$ is omitted since it is of no importance in the present treatment.

If $\cos\theta'$ is real, the phase of the Bessel beam for $z > 0$ propagates in the z-direction: normally, therefore, to the boundary. The phase velocity is found to be $c/(n' \cos\theta')$ (i.e., greater than the velocity light in the n' medium). If $\cos\theta'$ is purely imaginary [i.e., for θ_0 larger than the limit angle $i_0 = \arcsin(n/n')$] the single plane waves composing the incident Bessel beam are in total reflection and therefore give rise to plane refracted waves whose phase propagates parallel to the boundary.

In the case of total reflection, namely

$$\sin\theta' = (n/n') \sin\theta_0 > 1, \qquad \cos\theta' = i\Gamma \quad (\Gamma > 0), \tag{6.4}$$

if the beam of Eq. (6.1) impinges normally into a layer of width d, it gives rise:

1. On the left of the first boundary ($z < 0$) to a reflected Bessel beam that propagates in the negative direction of the z-axis.
2. Inside the layer ($0 < z < d$) to two Bessel beams: one "progressive" and the other "regressive."
3. On the right of the second boundary ($z > d$) to one transmitted Bessel beam that propagates in the positive z-direction.

Using a standard procedure, we can obtain the reflected and transmitted fields as well as the field inside the layer. The analysis of the transmitted field shows that propagation through the layer does not modify the shape of the incident field, and the amplitude transmitted is still a Bessel function, even if narrower with respect to the incident field. Moreover, in traveling through the tunneling region the beam does not modify its direction of propagation [14].

The field inside the layer is given by the superposition of the progressive and regressive waves. From its knowledge we can evaluate the wavelength λ' and the phase velocity v' in the gap, which result in

$$\lambda' = \frac{\lambda_0}{n\cos\theta_0}\,\frac{\cosh^2[k_0 n'\Gamma(d-z)]}{\cos^2[\eta(z)]}$$

$$v' = \frac{c}{n\cos\theta_0}\,\frac{\cosh^2[k_0 n'\Gamma(d-z)]}{\cos^2[\eta(z)]}, \tag{6.5}$$

where $\lambda_0 = 2\pi/k_0$ denotes the free-space plane-wave wavelength, and $\eta(z)$ is such that

$$\tan\eta(z) = -\tanh[k_0 n'\Gamma(d-z)]\frac{n\cos\theta_0}{n'\Gamma}. \tag{6.6}$$

It is interesting to note that for $z \to d,\ v' \to c/(n\cos\theta_0)$, which is equal to the phase velocity of the transmitted ($z > d$) and incident ($z < 0$) fields.

Vectorial Approach A vectorial analysis can be performed by considering, for example, the function of Eq. (6.1) as the tangential component of the electric field **E**, and through Maxwell's equations we can determine the longitudinal component of **E** and the magnetic field **H** as well. The longitudinal component of **E** vanishes on the z-axis for $\rho = 0$, where the tangential component has its maximum: this field (**E**, **H**) may be named quasi-TE. If the field (**E**, **H**) impinges normally on the layer of Fig. 6.1, it gives rise as in the scalar approximation to a reflected Bessel beam for $z < 0$, to a transmitted beam for $z > d$, and to a progressive and a regressive Bessel beam for $0 < z < d$. The progressive and regressive beams are evanescent if Eq. (6.4) holds.

Analogously, we can consider the scalar field of Eq. (6.1) as the tangential component of a magnetic field $\mathbf{H}'$, and from Maxwell's equations we can then obtain the longitudinal component of $\mathbf{H}'$ and the associated electric field $\mathbf{E}'$. The field ($\mathbf{E}'$, $\mathbf{H}'$) may be termed *quasi-TM*.

The (complex) amplitudes of the aforesaid beams may be determined by imposing, at the two boundaries, the continuity conditions for the tangential component of both the total electric and magnetic fields. In this way we can determine the total field inside the forbidden region as well as its wavelength and phase velocity. Due to the fact that the treatment is a little cumbersome, we limit ourselves to reporting that different wavelength and different phase velocity are found for the quasi-transverse electric (TE) and quasi-transverse magnetic (TM) cases.

A simpler analysis can be performed in the TE and TM cases. Let us consider a plane TE wave with direction of propagation $\mathbf{s}^i$ given by

$$\mathbf{s}^i = \alpha\mathbf{i} + \beta\mathbf{j} + \gamma\mathbf{k} = \sin\theta_0 \cos\varphi\mathbf{i} + \sin\theta_0 \sin\varphi\mathbf{j} + \cos\theta_0\mathbf{k}. \tag{6.7}$$

For the incident field $\mathbf{E}^i = E_x\mathbf{i} + E_y\mathbf{j}$ $(E_z = 0)$, let us put

$$\begin{aligned} E_x &= A_x \exp[\,ik_0 n(x \sin\theta_0 \cos\varphi + y \sin\theta_0 \sin\varphi + z \cos\theta_0)\,]\,d\varphi \\ E_y &= A_y \exp[\,ik_0 n(x \sin\theta_0 \cos\varphi + y \sin\theta_0 \sin\varphi + z \cos\theta_0)\,]\,d\varphi\,. \end{aligned} \tag{6.8}$$

Since $\mathbf{E}^i$ is normal to $\mathbf{s}^i$ (i.e., $\mathbf{E}^i \cdot \mathbf{s}^i = 0$), we have

$$A_x = A \sin\varphi, \qquad A_y = -A\cos\varphi\,, \tag{6.9}$$

which shows that A_x and A_y depend on φ (A is a constant).

Insertion of Eqs. (6.9) into Eqs. (6.8), and integration with respect to φ yields

$$\begin{aligned} E_{x,\text{beam}} &= i\,a \sin\psi\, J_1(k_0 n\rho \sin\theta_0) \exp(ik_0 nz \cos\theta_0) \\ E_{y,\text{beam}} &= -i\,a \cos\psi\, J_1(k_0 n\rho \sin\theta_0) \exp(ik_0 nz \cos\theta_0), \end{aligned} \tag{6.10}$$

where $a = 2\pi A$ and J_1 is the Bessel function of the first order [15].

The magnetic field $\mathbf{H}^i = H_x\mathbf{i} + H_y\mathbf{j} + H_z\mathbf{k}$ related to the $\mathbf{E}^i$ field can be found easily, and after integration with respect to φ we obtain

$$\begin{aligned} H_{x,\text{beam}} &= \frac{i\,a}{Z} \cos\theta_0 \cos\psi\, J_1(k_0 n\rho \sin\theta_0) \exp(ik_0 nz \cos\theta_0) \\ H_{y,\text{beam}} &= \frac{i\,a}{Z} \cos\theta_0 \sin\psi\, J_1(k_0 n\rho \sin\theta_0) \exp(ik_0 nz \cos\theta_0) \\ H_{z,\text{beam}} &= -\frac{a}{Z} \sin\theta_0\, J_0(k_0 n\rho \sin\theta_0) \exp(ik_0 nz \cos\theta_0), \end{aligned} \tag{6.11}$$

where $Z = Z_0/n$ and Z_0 is the free-space impedance. A field described by Eqs. (6.10) and (6.11) can be denoted as a Bessel beam and has properties very similar to those Eq. (6.1). However, only the longitudinal component of the magnetic field is of the type Eq. (6.1), that is, picked on the axis at $\rho = 0$; all other components for $\rho = 0$ vanish. A TE plane wave like the one described above gives rise at the incidence on the layer to a reflected first-order Bessel beam and to two transmitted "evanescent" Bessel beams (one of these is progressive and the other regressive), whereas at the second boundary, a transmitted first-order Bessel beam forms. The wavelength λ_{TE} and the phase velocity v_{TE} inside the layer turn out to be [14]

$$\begin{aligned} \lambda_{\text{TE}} &= \frac{2\pi}{|\nabla\eta(z)|} = \frac{\lambda_0}{n\cos\theta_0} \frac{\cosh^2[k_0 n'\Gamma(d-z)]}{\cos^2[\eta(z)]} \\ v_{\text{TE}} &= \frac{\omega}{|\nabla\eta(z)|} = \frac{c}{n\cos\theta_0} \frac{\cosh^2[k_0 n'\Gamma(d-z)]}{\cos^2[\eta(z)]}. \end{aligned} \tag{6.12}$$

We can analyze the TM case analogously. The results obtained are similar to those of the TE case, the only difference being that n must be replaced by n', and vice versa, everywhere (with the exception of the argument of the hyperbolic functions, and Γ), analogously to what happens in the Fresnel formulas for the reflection and transmission coefficients of a real plane wave at a plane interface. By comparing the results obtained for the TE and TM cases, we can conclude that:

1. The phase shift of the TM beam transmitted at $z = d$ with respect to the incident beam at $z = 0$ is different from that of the TE case.
2. The transmission coefficient for the TM case is different in amplitude and phase from that of the TE case. Consequently, an incident Bessel beam formed by a TE component and a TM component gives rise to a transmitted Bessel beam with a different polarization. The amplitudes are slowly varying functions of θ_0 in both the TE and TM cases. On the contrary, the phase in the case of TE differs greatly with respect to the TM case, and both of them vary greatly from θ_0.
3. The wavelength λ_{TM} and the phase velocity v_{TM} inside the layer, in the TM case, are different from those of the TE case:

$$\begin{aligned}\lambda_{\text{TM}} &= \frac{n\lambda_0}{n'^2\cos\theta_0}\,\frac{\cosh^2[k_0 n'\Gamma(d-z)]}{\cos^2[\eta_{\text{TM}}(z)]}\\ v_{\text{TM}} &= \frac{nc}{n'^2\cos\theta_0}\,\frac{\cosh^2[k_0 n'\Gamma(d-z)]}{\cos^2[\eta_{\text{TM}}(z)]},\end{aligned}\tag{6.13}$$

where $\eta_{\text{TM}}(z)$ is such that

$$\tan[\eta_{\text{TM}}(z)] = -\tanh[k_0 n'\Gamma(d-z)]\,\frac{n'\cos\theta_0}{n\Gamma}.\tag{6.14}$$

Figure 6.2 shows the normalized phase velocities v_{TM}/c (which is equal to $\lambda_{\text{TM}}/\lambda_0$) and v_{TE}/c (equal to $\lambda_{\text{TE}}/\lambda_0$). We note that immediately after the first boundary the motion is extremely fast since the effect of the second boundary, which forms the regressive (or "antievanescent") beam, is negligible. We wish to recall that in the absence of the second boundary, the phase velocity is infinite. For $z \longrightarrow d$ the phase velocities v_{TE}/c and v_{TM}/c intersect and the velocity of the TM wave becomes greater than that of the TE wave (see the inset in Fig. 6.2).

6.2.2 Oblique Incidence

Next, we investigate if the beam also propagates by maintaining its localized wave characteristic in the case of oblique incidence. To this end, let us consider a Bessel beam that impinges at oblique incidence onto a layer in such a way that each plane-wave component impinges with an angle larger than the critical angle [27,28]. The

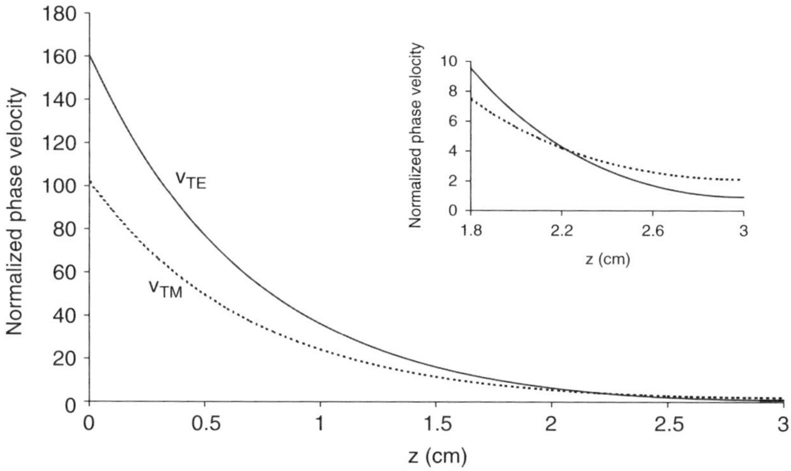

FIGURE 6.2 Normalized phase velocities v_{TE}/c (solid line) and v_{TM}/c (dotted line) as a function of z inside the gap for $d = 3$ cm. Other parameter values are $\omega = 60$ rad/s ($\simeq 9.5$ GHz), $n = 1.5$, $n' = 1$, $\theta_0 = 45°$.

main difference with respect to the normal incidence is that the axis of the beam is no longer coincident with the **k**-axis.

Denoting the axis of the Bessel beam as $\mathbf{k}_b$, for oblique incidence: we have $\mathbf{k}_b = \mathbf{k}' \neq \mathbf{k}$. In this case it is expedient to introduce a new reference system, S', whose axis $\mathbf{k}'$ is rotated in the x–z plane of a given angle Ω with respect to $\mathbf{k}$ (see Fig. 6.3). The Cartesian axes $\mathbf{i}', \mathbf{j}', \mathbf{k}'$ of S' are related to $\mathbf{i}, \mathbf{j}, \mathbf{k}$ by

$$\begin{aligned} \mathbf{i}' &= \gamma_k \mathbf{i} - \alpha_k \mathbf{k} \\ \mathbf{j}' &= \mathbf{j} \\ \mathbf{k}' &= \alpha_k \mathbf{i} + \gamma_k \mathbf{k}\,, \end{aligned} \tag{6.15}$$

where α_k and γ_k are

$$\alpha_k = \sin\Omega, \qquad \gamma_k = \cos\Omega. \tag{6.16}$$

The incident Bessel beam now consists of a set of plane waves, all forming the same angle θ_0 with $\mathbf{k}'$. This implies that the plane-wave component impinges onto the first interface of the slab with different angles of incidence. It turns out that the incidence angle varies from $\theta_0 - \Omega$ to $\theta_0 + \Omega$ when ϕ ranges from 0 to 2π. Denoting by ϕ' the azimuthal angle in the plane $x'y'$, the incidence u^i_{ob}, reflected u^r_{ob}, and transmitted u^t_{ob} fields are given by [27]

$$\begin{aligned} u^i_{ob} &= \exp\{ik_0 n[\alpha_{\phi'} x + \beta_{\phi'} y + \gamma_{\phi'} z]\} \\ u^r_{ob} &= R_{ob} \exp\big\{ik_0 n[\alpha_{\phi'} x + \beta_{\phi'} y - \gamma_{\phi'} z]\big\} \\ u^t_{ob} &= T_{ob} \exp\big\{ik_0 n[\alpha_{\phi'} x + \beta_{\phi'} y + \gamma_{\phi'}(z - d)]\big\}\,, \end{aligned} \tag{6.17}$$

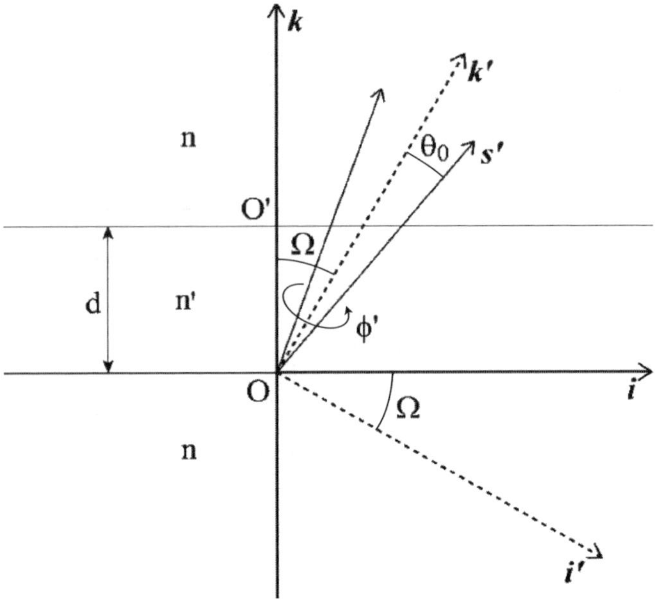

FIGURE 6.3 Slab of finite thickness d of a medium with refractive index n', surrounded on both sides by a different medium with refractive index $n > n'$. The reference system S of unit vectors **i**, **j**, **k** is shown together with the system S', rotated an angle Ω with respect to S. The vector $\mathbf{s}'$ represents the direction of propagation of a single plane wave. The contribution of all the plane waves, making the same angle θ_0 with the z'-axis, generates a Bessel beam characterized by the axicon angle θ_0 whose axis of symmetry is coincident with z'. The angle ϕ' is the azimuthal angle in the plane $x'y'$.

where (see Fig. 6.3)

$$\begin{aligned} \alpha_{\phi'} &= \gamma_k \sin\theta_0 \cos\phi' + \alpha_k \cos\theta_0 \\ \beta_{\phi'} &= \sin\theta_0 \sin\phi' \\ \gamma_{\phi'} &= \gamma_k \cos\theta_0 - \alpha_k \sin\theta_0 \cos\phi'. \end{aligned} \tag{6.18}$$

The transmitted T_{ob} and reflected R_{ob} coefficients can be found by imposing the continuity conditions at the interfaces.

To find the reflected and transmitted beams, we have to integrate Eqs. (6.17) over ϕ', between 0 and 2π. It can be verified that the amplitude transmitted ($z > d$) propagates rigidly in the $\mathbf{k}'$ direction (the axis of the beam), with no modification in its shape during propagation (an analogous property also holds for the reflected field). In Fig. 6.4 we show the behavior of the transmitted field as a function of the x-coordinate for three different values of the slab's thickness. We note that the dependence on d produces a strong variation in the amplitude value with no appreciable effect in the shape of the field: the maxima and minima positions remain unchanged. Moreover, the

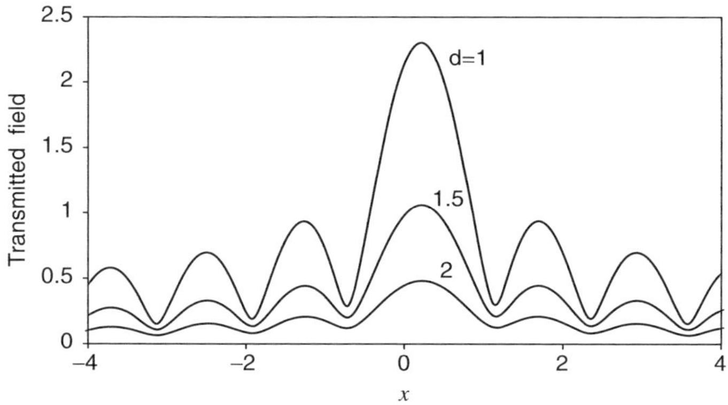

FIGURE 6.4 Transmitted field obtained by numerical integration of Eqs. (6.17) as a function of the x coordinate, for $\Omega = 6°$, $\omega = 60$ rad/s, $n = 1.5$, $n' = 1$, $\theta_0 = 60°$. With these values of the parameter, the amplitude of the field decreases by a factor of 5 by varying d from 1 to 2, but does not suffer appreciable modification in its shape. This kind of behavior is the same for greater values of Ω. However, Ω must not overcome its critical value: for $\theta_0 = 60°$, the value of $\Omega \simeq 18°$ represents the maximum angle possible in order to have total reflection at the first interface of all the waves forming the beam.

direction of propagation is independent of the width of the gap; that is, the direction of propagation depends only on angle Ω.

The propagation of the emerging Bessel beam can be followed by analyzing the field inside the slab (optical tunneling region). To this end, let us start again with a single plane wave. We recall that in the absence of the second half-space ($d = \infty$), the propagation after the first surface is due to evanescent waves that propagate parallel to the slab. The presence of the second boundary at $z = d$ forms antievanescent (or regressive) waves, and as is well known, the superposition of the two waves makes the Poynting vector different from zero perpendicular to the slab as well.

The progressive u^+ and regressive u^- waves within the slab can be written as

$$u^+_{ob} = p_{\phi'} \exp\{ik_0[n(\alpha_{\phi'}x + \beta_{\phi'}y) + in'\Gamma_{\phi'}z]\} \tag{6.19}$$

$$u^-_{ob} = r_{\phi'} \exp\{ik_0[n(\alpha_{\phi'}x + \beta_{\phi'}y) - in'\Gamma_{\phi'}z]\} \tag{6.20}$$

with

$$p_{\phi'} = \frac{e_{2\phi'}}{2n'\Gamma_{\phi'}}(n'\Gamma_{\phi'} - in\gamma_{\phi'})T_{\phi'} \tag{6.21}$$

$$r_{\phi'} = \frac{e_{1\phi'}}{2n'\Gamma_{\phi'}}(n'\Gamma_{\phi'} + in\gamma_{\phi'})T_{\phi'}. \tag{6.22}$$

Equations (6.21) and (6.22) were obtained from the continuity conditions for the tangential component of both the electric and magnetic fields across the two boundaries, at $z = 0$ and $z = d$ [24].

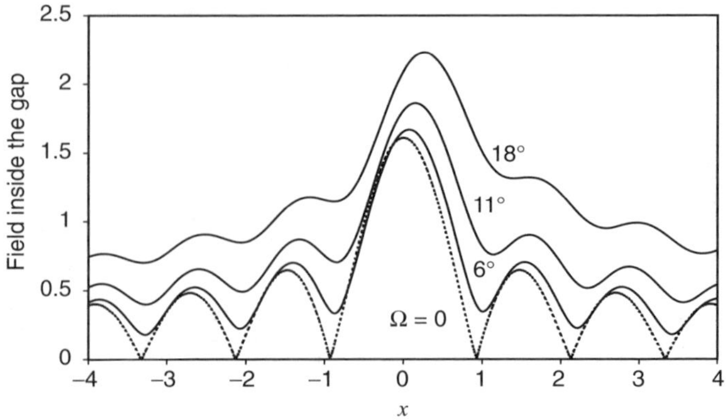

FIGURE 6.5 Total field inside the gap for three values of the parameter Ω: that is, for three different values of incidence angle. The field given by the superposition of progressive and regressive waves was obtained by numerical integration of Eq. (6.23) over ϕ' at $y = 0$ and $z = d/2$. Other parameter values are $\omega = 60$ rad/s, $n = 1.5$, $n' = 1$, $\theta_0 = 60°$, $d = 2$. The dashed line refers to normal incidence ($\Omega = 0$).

The field u^g_{tot} inside the gap is given by the superposition of progressive u^+ and regressive u^- waves of Eqs. (6.19) and (6.20) and can be expressed as

$$u^g_{\text{tot}} = \exp[i(k_0 n(\alpha_{\phi'} x + \beta_{\phi'} y)] \, |T_{\phi'}| e^{i\phi_T} \frac{1}{n'\Gamma_{\phi'}} \left[|A^g| e^{i\eta(z)} \right], \tag{6.23}$$

where $|A^g|$ and $\eta(z)$ are given by

$$|A^g| = \left\{ n'^2 \Gamma^2_{\phi'} + (n^2 - n'^2) \sinh^2[k_0 n' \Gamma_{\phi'}(z - d)] \right\}^{1/2}$$

and

$$\eta(z) = \arctan \left\{ \frac{n\gamma_{\phi'}}{n'\Gamma_{\phi'}} \tanh[k_0 n' \Gamma_{\phi'}(z - d)] \right\}. \tag{6.24}$$

To find the total Bessel beam, we have again to integrate the total field (6.23) over ϕ', from 0 to 2π. The amplitude of the Bessel beam inside the gap, at $z = d/2$ and $y = 0$, is shown in Fig. 6.5 as a function of the x-coordinate for three values of incidence angle Ω. The same behavior holds for the reflected and internal fields, the only difference lying in the fact that the variation in the amplitude is inappreciable for the reflected field, while the amplitude is halved in the field inside the slab. To obtain a clearer evidence of the delocalization effect due to the passage through the slab, a numerical analysis was performed for an axicon angle of 60°. For smaller axicon angles, the incidence angle must be even smaller in order to have total reflection, and the effect of deformation due to oblique incidence is very poor.

6.3 FREE PROPAGATION

In recent years, experimental measurements showed that Bessel beams also demonstrate group velocity greater than light velocity c in free propagation (propagation in vacuum or air) [8–10]. Superluminal group velocity in nondispersive media is a very new aspect of electromagnetic propagation, while a superluminal group velocity for pulses propagating in dispersive media has been predicted even since the beginning of the twentieth century [29].

It is well known that in the presence of anomalous dispersion, group velocity can largely overcome the speed of light c in vacuum and can even become negative. Anomalous dispersion can be achieved, for instance, in the presence of an absorption band. Even if measurements of delay time in the presence of an absorption band may be difficult due to attenuation (which also produces a distortion of the pulses), experimental measurements can be performed by confirming superluminal behavior in the group velocity [30,31]. However, as demonstrated by Sommerfeld and Brillouin, this superluminality cannot be extended to the signal velocity, which should always be limited to c [32].

In the absence of dispersion the situation is different, since the three velocities that characterize the propagation (phase, group, and signal velocities) tend to coincide. This statement is valid not only for waves but also for pulses, since all components at different frequencies propagate with the same velocity. In other words, in the absence of dispersion the pulse does not suffer reshaping, a process that is always present in dispersive systems.

In the case of the Bessel beam (6.1), the phase velocity v_{ph} along the z-axis is $(n = 1)$

$$v_{\text{ph}} = \frac{\omega}{\nabla(kz\cos\theta_0)} = \frac{c}{\cos\theta_0}, \tag{6.25}$$

and it is easy to demonstrate that the group velocity v_{gr} also has the same expression. In fact, by denoting the component of the wave vector along the direction of propagation z with $k_z = k\cos\theta$, and by recalling that $\omega = kc$ (this condition is required to satisfy the wave equation with Eq. (6.1)), the group velocity v_{gr} is expressed as

$$v_{\text{gr}} = \frac{d\omega}{dk_z} = \frac{d\omega}{dk}\frac{dk}{dk_z} = \frac{c}{\cos\theta}. \tag{6.26}$$

Once the superluminality in the group velocity is well established, the question as to whether this conclusion can be extended to the signal velocity naturally arises.

6.3.1 Phase, Group, and Signal Velocity: Scalar Approximation

In looking at Eq. (6.1), the first thing that we note is that the dependence on t and z occurs only through the quantity

$$\frac{z}{c}\cos\theta_0 - t, \tag{6.27}$$

and therefore the beam (6.1) takes the same value after a time dt equal to $(dz/c)\cos\theta_0$; that is, it propagates with a velocity

$$v = \frac{c}{\cos\theta_0}\,, \tag{6.28}$$

even greater than c. Moreover, if we insert in (6.1) a spectral function (depending only on the frequency), this presence does not modify the situation and the previous conclusion is also found to be true in this case. These considerations appear to be so widespread that we could also conclude that the signal propagates with a velocity greater that c.

However, great caution is required on this point, since the present analysis is related to a physical situation in which the product $v_{\mathrm{ph}}v_{\mathrm{gr}}$ is greater than c^2 even in the absence of dispersion. The latter represents a rather new field of investigation, and it is not clear whether the relation $v_{\mathrm{ph}}v_{\mathrm{gr}} = c^2$, which is valid for waves, can be extended to beams, or whether a new definition of signal velocity has to be sought.

According to Brillouin [29], a signal can be defined as a pulse of finite temporal extension, that is, of infinite extension in the frequency domain.† On the basis of this definition, let us construct a signal $U(\rho, z, t)$ by superimposing a set of Bessel beams that differ by the value of the frequency ω and have the same value of the parameter θ_0 and the same amplitude and phase at $t = 0$; that is,

$$U(\rho, z, t) = \int_{-\infty}^{\infty} J_0(k\rho\sin\theta_0)\exp(ikz\cos\theta_0)e^{-i\omega t}\,d\omega. \tag{6.29}$$

Apart from an inessential factor, the Bessel function J_0 can be written as [15]

$$J_0(x) = \int_0^{2\pi} \exp(ix\cos\varphi)\,d\varphi. \tag{6.30}$$

By introducing Eq. (6.30) into Eq. (6.29) and changing the integration order,‡ we obtain

$$\begin{aligned} U(\rho, z, t) &= \int_0^{2\pi} d\varphi \int_{-\infty}^{\infty} \exp\left[i\omega\left(\frac{\rho}{c}\sin\theta_0\cos\varphi + \frac{z}{c}\cos\theta_0 - t\right)\right] d\omega \\ &= \int_0^{2\pi} \delta\left(\frac{\rho}{c}\sin\theta_0\cos\varphi + \frac{z}{c}\cos\theta_0 - t\right) d\varphi\,, \end{aligned} \tag{6.31}$$

†The question is an extremely delicate one. As for the definition of *signal*, the scientific community has been discussing it since the beginning of the last century (and even before), but an analytical definition of this velocity is still lacking. Broadly speaking, a signal is a pulse (or wave packet) that supplies information. However, there may be a difference between *information* and *communication*. In fact, a single pulse (in a finite time interval) can supply information, but for a communication it is necessary to have a sequence of different pulses or, more generally, to know the boundary conditions.

‡This procedure is correct, since the function is continuous and converges in the integration interval.

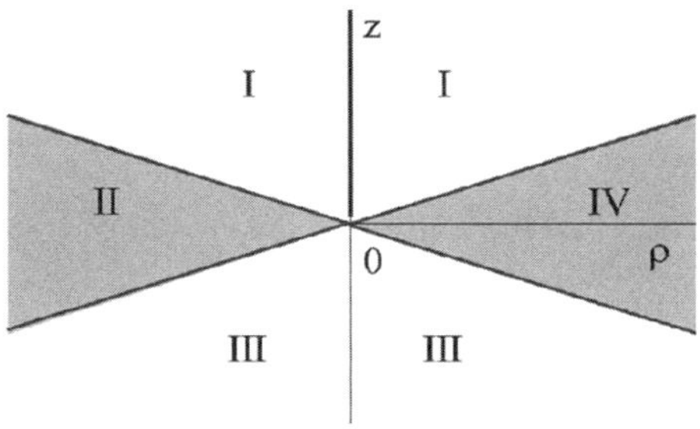

FIGURE 6.6 Representation of zones of existence of a Bessel pulse in the ρ–z plane for $\theta_0 = 20°$ and $t = 0$. In zones I and III the field is zero, while in II and IV its value is different from zero; that is, the field exists only inside the double cone delimited by the straight lines $\rho = |z \cot\theta_0 - (c/\sin\theta_0)t|$. For $\rho = 0$ the field is reduced to a δ-pulse.

where δ denotes the Dirac δ-function. It can immediately be seen that:

- For $\rho = 0$, the only solution that makes the integral different from zero is

$$\frac{z}{t} = \frac{c}{\cos\theta_0}, \quad \text{that is,} \quad v = \frac{c}{\cos\theta_0}.$$

- For $\rho \leq |z \cot\theta_0 - (c/\sin\theta_0)t|$ the field is zero, since the δ-function has no zeros in the integration interval.
- For $\rho \geq |z \cot\theta_0 - (c/\sin\theta_0)t|$, the field is different from zero and is given by

$$U \propto \left[\frac{\rho^2}{c^2}\sin^2\theta_0 - \left(\frac{z}{c}\cos\theta_0 - t\right)^2\right]^{-1/2}. \tag{6.32}$$

The plane z is therefore divided into four zones, which are shown in Fig. 6.6 for $t = 0$ (for $t > 0$, the diagram moves with velocity $v = c/\cos\theta_0$ in the direction of the z-axis). In zones I and III, U is zero, whereas in zones II and IV it is different from zero. Along the straight lines $\rho = |z \cot\theta_0 - (c/\sin\theta_0)t|$, the field is discontinuous, remaining equal to zero on one side and going to infinity on the other side.

In Eq. (6.31) we assumed a constant spectral function that may appear as a crude approximation for a real physical situation. A more realistic situation can be obtained by introducing in Eq. (6.29) a spectral function of the $(\sin\omega\tau)/\omega\tau$ type, which originates a rectangular pulse; that is,

$$U(\rho, z, t) = \int_{-\infty}^{\infty} \frac{\sin\omega\tau}{\omega\tau} J_0(k\rho\sin\theta_0)\exp(ikz\cos\theta_0)e^{-i\omega t}\,d\omega. \tag{6.33}$$

The result of the numerical integration of Eq. (6.33) is shown in Fig. 6.7. We note that the pulse travels without any deformation, and that the time spent by each point of the front, in traveling a distance $\Delta z = 10$ cm, is $\Delta t = 300$ ps. We therefore obtain, as

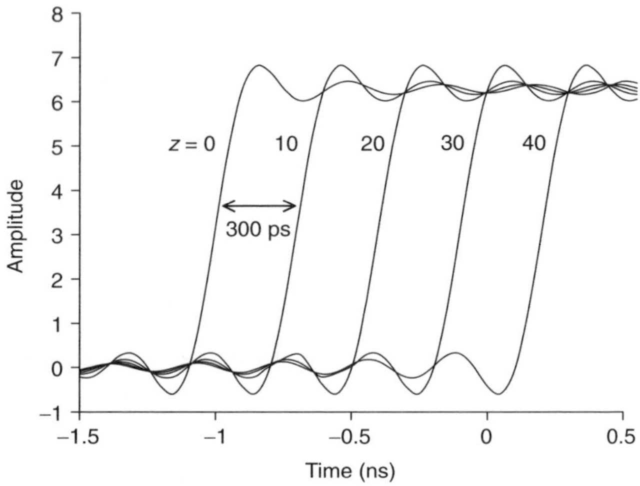

FIGURE 6.7 Results relative to the numerical integration of Eq. (6.33). Starting from the left, we show the rise fronts for $z = 0,\ 10,\ 20,\ 30,\ 40$ cm, respectively. The finite slope of the front, as well as the damped oscillations before and after the front (Gibbs effect), are due to the frequency truncation in the integration. Parameter values are $\tau = 1$ ns, $\omega = 60$ rad/ns, $\rho = 0,\ \theta_0 = 25°$. For ρ different from zero, the shape of the pulse is different, but the delay is unchanged in both the peak and the rise (or fall) front.

expected, $v = 33$ cm/ns $= c/\cos\theta_0$. In fact, independent of the shape of the pulse, the temporal duration is finite and the propagation velocity is equal in each point of the pulse since the entire system moves rigidly in the direction of the z-axis.

6.3.2 Energy Localization and Energy Velocity: A Vectorial Treatment

Previous analysis has shown that the propagation velocity of a Bessel pulse (or Bessel wave packet) on the z-direction is equal to $c/\cos\theta_0$, independent of its shape. However, the question as to whether it is possible to transport information at a velocity greater than c remains unresolved, mainly because the information transport is closely connected with the velocity of energy, and the scalar approximation cannot give indications with regard to this quantity.

To obtain information about the propagation of energy flux and the velocity of energy, let us analyze the electromagnetic propagation of a Bessel X-wave on the basis of a vectorial treatment. We wish to recall that knowledge of the electric and magnetic fields (and consequently, of the Poynting vector) is of great interest, since these quantities are connected not only with the topic of superluminality, but also with the production of localized electromagnetic energy.

First Approach Vectorial fields with an amplitude that is proportional to the zero-order Bessel function can be found in different ways. However, we are interested in deriving a vectorial field that has Eq. (6.1) as its scalar approximation, and to this

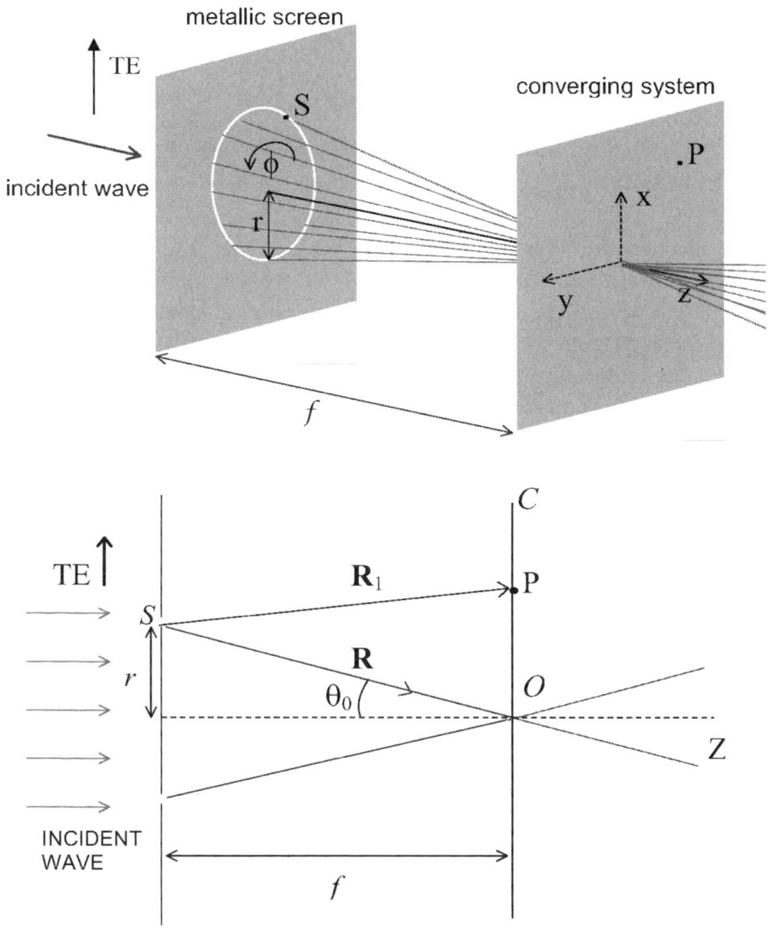

FIGURE 6.8 Scheme of the system considered for vectorial treatment. The metallic screen, over which the ring-shaped aperture is placed, and the converging system must be considered as being of infinite dimension. On the lower side, details of the reference system are shown.

end, we consider a realistic situation that is able to generate a field of this particular type [33].

Let us consider the system of Fig. 6.8, which consists of a ring-shaped aperture, of radius r, over a metallic screen on which a linearly polarized plane wave impinges in the $\mathbf{i}$ direction. The ring is placed on the focal plane of the converging system C with focal length $f \gg \lambda$ (λ being the wavelength) and with thickness d of the ring very small with respect to λ. Under these conditions, we may assume that the element dA of the ring at $S(r, \varphi, -f)$ behaves as an elementary dipole, parallel to $\mathbf{i}$, with amplitude proportional to $d\varphi$.†

† This position requires a suitable choice of transparency of the ring at S.

As a first vectorial treatment, let us consider that the associated field at the optical center O of C has the well-known characteristics of the far field radiated by an elementary dipole, that is, it is a spherical wave centered at S, with:

- The electric field **e** in the meridional plane of the dipole through O.
- The electric field **e** perpendicular to the direction **R** from S to O.

From these characteristics it is possible to evaluate the single-dipole electric and magnetic fields after the converging system. First, we assume that the only effect of the converging system is the transformation of impinging spherical wavefronts into plane fronts, directed from S to O, without changing the polarization of the impinging field. Then, after integration over φ, we obtain the total electric **E** and magnetic **H** fields arising from the contribution of all dipoles. Avoiding analytical details, we have

$$E_x = 2\pi\, e_0 e^{i\xi z} \left\{ J_0(\eta\rho) + \tan^2\theta_0 \left[\left(J_0(\eta\rho) - \frac{J_1(\eta\rho)}{\eta\rho} \right) - \cos^2\psi \left(J_0(\eta\rho) - \frac{2J_1(\eta\rho)}{\eta\rho} \right) \right] \right\} \tag{6.34}$$

$$E_y = -2\pi\, e_0 e^{i\xi z} \left[\frac{\sin 2\psi}{2} \tan^2\theta_0 \left(J_0(\eta\rho) - \frac{2J_1(\eta\rho)}{\eta\rho} \right) \right] \tag{6.35}$$

$$E_z = -2\pi i\, e_0 e^{i\xi z} \left[\tan\theta_0 \cos\psi\, J_1(\eta\rho) \right] \tag{6.36}$$

and

$$H_x = 0 \tag{6.37}$$

$$H_y = \frac{2\pi}{Z} e_0 e^{i\xi z} \frac{1}{\cos\theta_0} J_0(\eta\rho) \tag{6.38}$$

$$H_z = -i\frac{2\pi}{Z} e_0 e^{i\xi z} \sin\psi \frac{\sin\theta_0}{\cos^2\theta_0} J_1(\eta\rho), \tag{6.39}$$

where $\xi = k\cos\theta_0$, $\eta = k\sin\theta_0$, and J_1 denotes the first-order Bessel function of the first kind. The field E_x [Eq. (6.34)], which represents the main contribution to the electric field, describes a field that is different from the scalar field of Eq. (6.1), due to the presence of the term that depends on $\tan\theta_0$. However, for $\theta_0 \ll \pi/2$ $(r \ll f)$, as in the present case, this term is negligible. We note that the dependence on ψ, which is absent in the scalar approximation, is also negligible. This dependence may presumably be due to the fact that the amplitude of the field is not constant over a wavefront. In the present treatment we neglect this variation, which limits the applicability of our results to a small value of θ_0 and ρ.† In Fig. 6.9 we report the normalized value of E_x, E_y, and E_z versus ρ, for $\theta_0 = 10°$, together with the scalar

†This position is also justified because we need to avoid the zeros of the field. In fact, when the field is zero, we lose information regarding the direction of propagation, and the Poynting vector may have some anomalies.

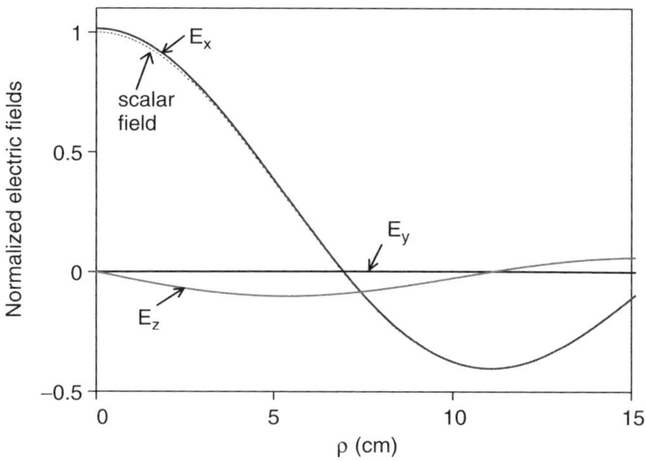

FIGURE 6.9 Electric fields E_x, E_y, and E_z normalized to $2\pi e_0 \exp(i\xi z)$ (continuous lines) versus ρ, as given by Eqs. (6.34)–(6.36), for $k = 2$, $\psi = 10°$, and $\theta_0 = 10°$. In E_z, phase factor $e^{i\pi/2}$ is disregarded. The dashed line very close to the E_x field represents the normalized scalar field (6.1) evaluated for the same parameter values.

field of Eq. (6.29). The scalar field is practically coincident with E_x, and we can therefore conclude that the vectorial field derived above has Eq. (6.29) as its scalar approximation, at least for $r \ll f$.

From a knowledge of the electric and magnetic fields, we are now in a position to evaluate the mean density of the energy flux, which is defined as being one-half of the real part of the complex Poynting vector [34,35]; that is, $\mathbf{S} = \mathrm{Re}\,(\mathbf{E} \times \mathbf{H}^\star)\,/2$.

For fields (6.34)–(6.36) and (6.37)–(6.39), the real part of the Poynting vector is found to be given by the k-component; that is, propagation of the energy flux occurs only in the z-direction, in accordance with the information given by the scalar field (6.1). Moreover, since the flux is independent of z, the energy propagates with no deformation. In Fig. 6.10, the behavior of the energy flux

$$S_z = \frac{1}{2}\left(E_x H_y^\star\right) \tag{6.40}$$

is shown as a function of ρ for several values of θ_0. For θ_0 very small (almost plane wave) the flux is almost independent of ρ, while when the beam forms, the flux increases by increasing θ_0 and tends to concentrate near $\rho = 0$, that is, along the z-axis. Thus, for small values of ρ (i.e., in the proximity of the z-axis), the energy supplied by a Bessel X-wave is always greater than the energy supplied by a plane wave. On the basis of the previous treatment, we are now in a position to evaluate the quantity of energy that several Bessel beams can supply.

When more than one beam impinges in the converging system, in order to have an increase in the energy (along the z-axis), each beam must be in phase with respect to the others. This condition is satisfied, for instance, if the optical path among the beams

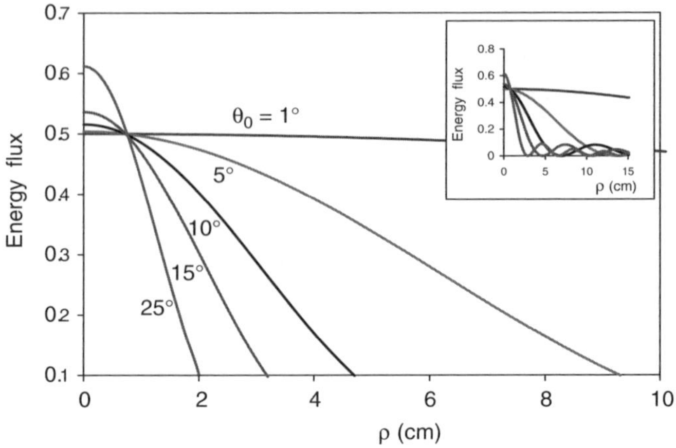

FIGURE 6.10 Energy flux S_z, as given by Eq. (6.40) versus ρ, for a few values of θ_0. For the sake of completeness, in the inset S_z is shown for larger values of ρ, even if only the region around the main peak is of physical interest. Parameter values are as in Fig. 6.9.

differs by one wavelength. Thus, it is easy to find the geometric characteristics that the rings must have in order to supply localized energy. Starting from the external ring-shaped aperture, let us denote the optical paths by $R_1, R_2, \ldots, R_n$, the radii by $r_1, r_2, \ldots, r_n$, and the axicon angles related to the n rings, by $\theta_1, \theta_2, \ldots, \theta_n$ respectively. We have

$$\begin{aligned} R_1 &= \frac{f}{\cos\theta_1} \\ R_2 &= R_1 - \lambda = \frac{f}{\cos\theta_2} \\ &\vdots \\ R_n &= R_1 - (n-1)\lambda = \frac{f}{\cos\theta_n}, \end{aligned} \tag{6.41}$$

where f is the focal length. Since the radius of the nth ring is $r_n = f\tan\theta_n$, it is easy to find the radius of rings such as Bessel beams have the same phase. In Fig. 6.11 we show the energy flux for a system consisting of three rings and for $\lambda = 3 \times 10^{-2}$ cm. As can be seen, the quantity of energy supplied by three rings is about nine times greater than the quantity due to a single ring (dotted line), as expected, due to the quadratic dependence of the Poynting vector with respect to the fields.

We note also that the energy maintains its localization, since the position of the first zero is unchanged. This effect is due to small values of the wavelength, which make the optical paths very close to each other. For higher values of the wavelength, the axicon angles differ substantially from each other, and consequently, the first zero of the Bessel function suffers a shift toward a higher value for ρ. In this situation, the amount of energy flux is about the same, but the flux tends to lose its localization.

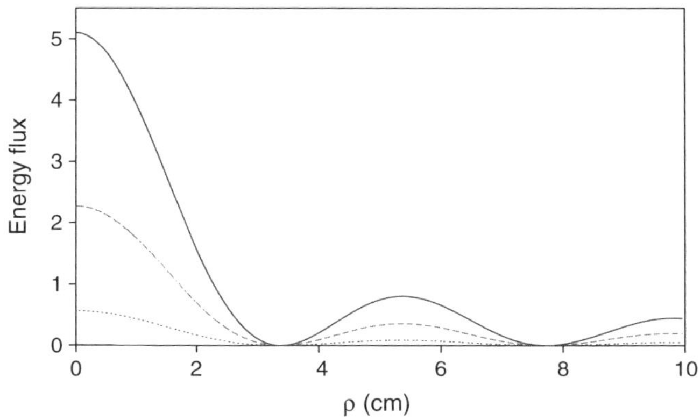

FIGURE 6.11 Energy flux in the presence of one (dotted line), two (dashed line), and three (solid line) ring-shaped apertures as a function of ρ. Parameter values are $f = 100$ cm, $\lambda = 0.3$ cm, $\theta_1 = 20°$.

As for the speed of energy flow (or mean-energy velocity) v_e, we can evaluate it as [32,34]

$$v_e = \frac{S_z}{\frac{1}{4}\left(\varepsilon E \cdot E^\star + \mu H \cdot H^\star\right)}, \tag{6.42}$$

where the quantity $(1/4)(\varepsilon E \cdot E^\star + \mu H \cdot H^\star)$ is the *time-averaged energy density* [see Eqs. (6.34)–(6.36) and (6.37)–(6.39)]. In Fig. 6.12 we report the normalized

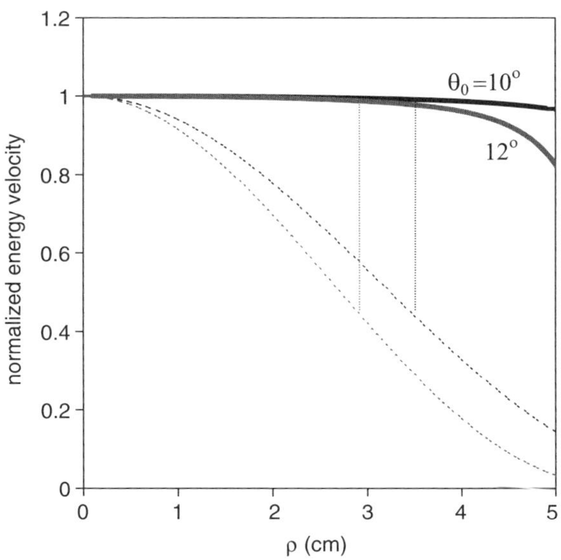

FIGURE 6.12 Energy velocity, normalized to the light velocity, as a function of ρ for $k = 2$, $\psi = 10°$, $\theta_0 = 10°$, and $\theta_0 = 12°$. Dashed lines represent the normalized energy flux relative to the same parameter values; vertical lines denote the half-height of the flux.

velocity of energy as a function of the radial coordinate ρ for two values of θ_0. The velocity is found to be equal to c from $\rho = 0$ up to near the first zero of the Bessel function: that is, the beam moves like an almost rigid system, despite its dependence on ρ and ψ. In proximity to the first zero of the Bessel function, the velocity decreases and tends to zero. Naturally, the zero in the velocity does not represent a stopping of the motion but, more simply, the absence of energy flux. In this situation, the concept of velocity has no physical meaning.

We would like to comment on this rather surprising result. In fact, we recall that for propagation in vacuum "if an energy density is associated with the magnitude of the wave ... the transport of energy occurs with the group velocity, since that is the rate at which the pulse travels along" [36]. If the definition of the energy velocity as given by Eq. (6.42) is also applicable to a Bessel beam (or, more generally, to localized waves), it is not clear what kind of physical mechanism makes the energy velocity different from the phase and group ones. The speed of energy flow as given by Eq. (6.42) refers to mean values of energy and time. Surely, Eq. (6.42) works well for waves that have a constant energy distribution, like plane waves for which phase, group, energy, and signal velocities coincide (and are equal to c in the absence of dispersion). On the contrary, the suspicion that the definition of Eq. (6.42) could not be suitable for localized waves is strong. This doubt is also supported by the fact that the product between phase and group velocities is different from c^2.

Other, More Rigorous Treatment of the Problem The preceding treatment, based on some approximations which even if they seem to work well from a fundamental point of view, may appear rather crude. Thus, to evaluate the correctness of the approximations involved previously, we wish to present a new, more rigorous approach to the problem.

Again with reference to Fig. 6.8, let us consider a point $P(x, y, z = f)$ on the converging system (hereafter referred to for simplicity as a lens), and let us denote with $\mathbf{R}_1$ the direction from S to P. The field $\mathbf{e}_o$ emerging from the lens is given by

$$\mathbf{e}_o = \mathbf{e}_i \, e^{i\Phi(x,y)}, \tag{6.43}$$

where $\Phi(x, y)$ is the phase introduced by the lens and $\mathbf{e}_i$ is the incoming electric dipole field, which in the far-field approximation is given by [34]

$$\mathbf{e}_i = \frac{k^2}{4\pi\epsilon_0}\left[\mathbf{s} \times (\mathbf{i} \times \mathbf{s}) \, \frac{e^{ikR_1}}{R_1}\right]. \tag{6.44}$$

In Eq. (6.44), $\mathbf{i}$ represents the electric dipole moment and $\mathbf{s}$ is the unit vector in the direction of $\mathbf{R}_1$; that is, $\mathbf{s} = \mathbf{R}_1/R_1$.

We have

$$\mathbf{R}_1 = (x - r\cos\phi)\mathbf{i} + (y - r\sin\phi)\mathbf{j} + f\mathbf{k}, \tag{6.45}$$

and therefore [by disregarding the constant factor $k^2/4\pi\epsilon_0$ in Eq. (6.44)]

$$\mathbf{e}_i = \frac{e^{ikR_1}}{R_1^3}(e_x\mathbf{i} + e_y\mathbf{j} + e_z\mathbf{k}), \tag{6.46}$$

with

$$\begin{aligned} e_x &= (y - r\sin\phi)^2 + f^2 \\ e_y &= -(x - r\cos\phi)(y - r\sin\phi) \\ e_z &= -f(x - r\cos\phi), \end{aligned} \tag{6.47}$$

and in the second-order approximation,

$$R_1 \simeq f\left[1 + \frac{1}{2f^2}\left((x - r\cos\phi)^2 + (y - r\sin\phi)^2\right)\right]. \tag{6.48}$$

As for the phase $\Phi(x, y)$ due to the lens, it is well known that

$$\Phi(x, y) \simeq -\frac{k}{2}\frac{x^2 + y^2}{f}. \tag{6.49}$$

By introducing Eqs. (6.46) and (6.49) into Eq. (6.43), we obtain [see also Eq. (6.48)]

$$\begin{aligned} \mathbf{e}_o = \frac{1}{f^3}\Big\{ & [(y - r\sin\phi)^2 + f^2]\mathbf{i} - (x - r\cos\phi)(y - r\sin\phi)\mathbf{j} \\ & - f(x - r\cos\phi)\mathbf{k}\Big\}\left[1 - A(x, y) + \frac{3r}{f^2}(x\cos\phi + y\sin\phi)\right] \\ & \times \exp\left[ik\left(f + \frac{r^2}{2f}\right)\right]\exp\left[-\frac{ikr}{f}(x\cos\phi + y\sin\phi)\right], \end{aligned} \tag{6.50}$$

where

$$A(x, y) = \frac{3(x^2 + y^2 + r^2)}{2f^2}. \tag{6.51}$$

The field $\mathbf{E}$ emerging from the lens is given by the superposition of the fields due to all dipoles. By integrating Eq. (6.50) over ϕ, we can evaluate the three components E_x, E_y, and E_z of $\mathbf{E}$.

At the moment, let us consider the field E_x, which should be the main contribution to the field $\mathbf{E}$. After integration we obtain the amplitude of the field, which (apart from inessential factors) is found to be equal to the first-order Bessel function $J_0(k\rho\tan\theta_0)$, very similar to the leading contribution of Eq. (6.34). The only difference lies in the dependence on the parameter θ_0. This means that the axicon angle of the Bessel beam as obtained in the present treatment does not coincide with the angle θ_0 under which the ring is seen from the optical center of the lens. However, in the Gaussian approximation, that is for small values of θ_0, the difference between $\tan\theta_0$ and $\sin\theta_0$ is inappreciable.

6.4 SPACE–TIME AND SUPERLUMINAL PROPAGATION

Before to conclude, we would like to analyze again the field of existence of a Bessel pulse as given by Eq. (6.31). In particular, let us consider the value of the time that makes the integral (6.31) different from zero [39]. For $\rho \longrightarrow 0$, the only value of time for which the field exists is $t = z\cos\theta_0/c$. However, since $0 < \theta_0 < \theta_{\max}$, the time interval in which the field is different from zero is

$$t_{\min}(\theta_0 = \theta_{\max}) < t < t_{\max}(\theta_0 = 0), \tag{6.52}$$

where $\theta_{\max} < \pi/2$ depends on the experimental setup. Since the Bessel pulse propagates along the z-axis, we can deduce that its temporal evolution in three-dimensional space–time (horizontal axes are chosen to be spatial dimensions, while the vertical axis is time) is within a conical surface similar to the light cone [40], where light velocity c is replaced by velocity $v_b = c/\cos\theta_0$ and t is a real quantity.

Thus, we can say that the propagation of a Bessel pulse in Euclidean space corresponds to a super-light cone in the pseudo-Euclidean space-time of Minkowski. In other words, by introducing a second spatial coordinate, for a given value of θ_0, we obtain a super-light cone that wraps itself around the well-known light cone. This situation is shown in Fig. 6.13, where straight line v_b, which depends on θ_0, is the pulse velocity. For $\theta_0 = \theta_{\max}$, v_b represents the borderline that determines the existence of the field; the Bessel pulse exists only in the gray zone. Inside this cone of existence, the past super-light cone, $t < 0$, represents the time interval prior to generation of the pulse. The pulse originates at $t = 0$ and for $t > 0$ (future super-light cone) propagates along the z-axis with velocity v_b. For $\theta_0 = 0$ the beam reduces to a plane wave, its velocity becomes equal to c, and the super-light cone becomes the light cone (black cone in Fig. 6.13).

Since Bessel beams are real quantities (they have been generated and measured experimentally), and since Eq. (6.1) is capable of describing the scalar field of the beam as being due to a specific experimental setup [33], we can conclude that the super-light cone places a new upper speed limit for all objects. Massless particles can travel not only along the light cone but also along the super-light cone in the region between the super-cone and the cone, while the world lines remain confined within the light cone. In substance, we can think that c is the velocity of light in its simplest manifestation (wave), while more complex electromagnetic phenomena, such as the interference among an infinite number of waves, may originate different velocities. The maximum value $\theta_{\max}$ of axicon angle θ_0 sets the maximum value of the Bessel pulse velocity. Since the field depth, that is, the spatial range in which the beam exists, is proportional to $\tan\theta_0^{-1}$, θ_0 can never reach the value of $\pi/2$. If it were possible to obtain values of θ_0 close to $\pi/2$, we should have almost immediate propagation in a nearly-zero space rather like an ultrafast shot destined to slow down immediately.

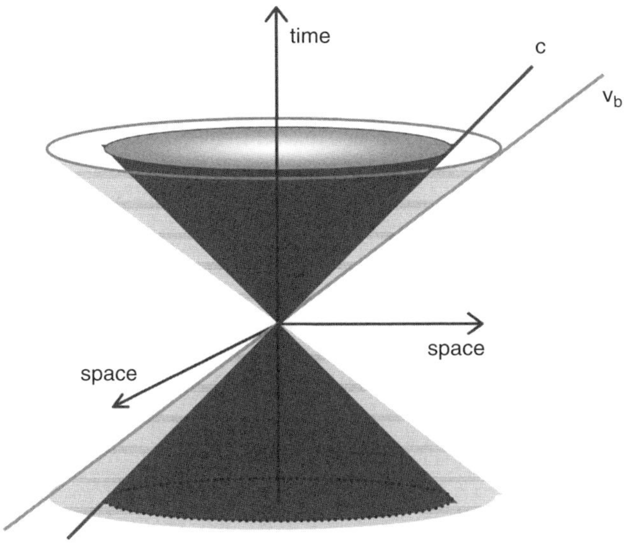

FIGURE 6.13 Schematic representation of the super-light cone in Minkowski space–time (pseudo-Euclidean space). The black zone represents the light cone, and the gray zone around it is the field of existence of the Bessel beam. Quantity v_b is the beam velocity for a given axicon angle, θ_0. For $\theta_0 = 0$, the beam is reduced to a plane wave, and its velocity then becomes equal to c. In this situation, the field of existence of the beam goes to zero and the super-light cone narrows and becomes equal to the light-cone.

The change in the upper limit of the light velocity does not modify the fundamental principles of relativity and the principle of causality, as demonstrated by recent theory dealing with the new geometrical structure of space–time [41]. The principle that "the speed of light is the same for all inertial observers, regardless of the motion of the source," remains unchanged, provided that the substitution $c \longrightarrow v_b\,(= c/\cos\theta_0)$ is made in the Lorentz transformations. In this way, the direction of the beam light does not depend on the motion of the source, and all observers measure the same speed (v_b) in all directions, independent of their motions.

Acknowledgments

Special thanks are due to Laura Ronchi Abbozzo for useful suggestions and discussions.

REFERENCES

1. J. Durnin, J. J. Miceli, Jr., and J. H. Eberly, Diffraction-free beams, *Phys. Rev. Lett.* **58**, 1499 (1987).
2. J. Durnin, Exact solution of nondiffracting beams, *J. Opt. Soc. Am. A* **4**, 651 (1987).
3. R. W. Ziolkowski, Localized transmission of electromagnetic energy, *Phys. Rev. A* **39**, 2005 (1989).

4. P. Sprangle and B. Hafizi, Comment on nondiffracting beams, *Phys. Rev. Lett.* **66**, 837 (1991).
5. J. Durnin, J. J. Miceli, Jr., and J. H. Eberly, Durnin, Miceli, and Eberly reply, *Phys. Rev. Lett.* **66**, 838 (1991).
6. K. Tanaka, M. Taguchi, and T. Tanaka, Quasi-diffraction-free beams, *J. Opt. Soc. Am. A* **18**, 1644 (2001).
7. K. Reivelt and P. Saari, Experimental demonstration of realizability of optical focus wave modes, *Phys. Rev. E* **66**, 056611 (2002).
8. P. Saari and K. Reivelt, Evidence of X-shaped propagation-invariant localized light waves, *Phys. Rev. Lett.* **79**, 4135 (1997).
9. D. Mugnai, A. Ranfagni, and R. Ruggeri, Observation of superluminal behavior in wave propagation, *Phys. Rev. Lett.* **84**, 4830 (2000).
10. I. Alexeev, K.Y. Kim, and H. M. Milchberg, Measurement of the superluminal group velocity of an ultrashort Bessel beam pulse, *Phys. Rev. Lett.* **88**, 073901 (2002).
11. I. M. Besieris and A. M. Shaarawi, Paraxial localized wave in free space, *Opt. Express* **12**, 3848 (2004).
12. M. Zamboni-Rached, A. M. Shaarawi, and E. Recami, Focused X-shaped pulses, *J. Opt. Soc. Am. A* **21**, 1564 (2004); Focused X-shaped pulses:errata, *J. Opt. Soc. Am. A* **22**, 2900 (2005).
13. P. Saari and K. Reivelt, Generation and classification of localized waves by Lorentz trasformation in Fourier space, *Phys. Rev. E* **69**, 036612 (2004)
14. D. Mugnai, Passage of a Bessel beam through a classically forbidden region, *Opt. Commun.* **188**, 17 (2001).
15. G.N. Watson, *Theory of Bessel Functions*, Cambridge University Press, Cambridge, 1922.
16. A. Ranfagni, D. Mugnai, P. Fabeni, and G. P. Pazzi, Semiclassical tunneling time in the presence of dissipation: an optical model, *Phys. Scri.* **42**, 508 (1990); A. Ranfagni, D. Mugnai, P. Fabeni, and G. P. Pazzi, Delay-time measurements in narrowed waveguides as a test of tunneling, *Appl. Phys. Lett.* **58**, 774 (1991).
17. A. Enders and G. Nimtz, On superluminal barrier traversal, *J. Phys. I (France)* **2**, 1693 (1992).
18. D. Mugnai, A. Ranfagni, and L. S. Schulman, Delay time measurements in a diffraction experiment: a case of optical tunneling, *Phys. Rev. E* **55**, 3593 (1997).
19. D. Mugnai, A. Ranfagni, and L. Ronchi, The question of tunneling time duration: a new experimental test at microwave scale, *Phys. Lett. A* **247**, 281 (1998).
20. A. M. Steinberg, P. G. Kwiat, and R. Y. Chiao, Measurement of the single-photon tunneling time, *Phys. Rev. Lett.* **71**, 708 (1993).
21. Ph. Balcou and L. Dutriaux, Dual optical tunneling times in frustrated total internal reflection, *Phys. Rev. Lett.* **78**, 851 (1997).
22. S. Bosanac, Propagation of electromagnetic wave packets in nondispersive dielectric media, *Phys. Rev. A* **28**, 577 (1983).
23. A. M. Steinberg and R. Y. Chiao, Tunneling delay times in one or two dimensions, *Phys. Rev. A* **49**, 3283 (1994).
24. D. Mugnai, A. Ranfagni, and L. Ronchi, Tunneling time in the case of frustrated total reflection, *Giorgio Ronchi Found.* **1**, 777 (1998).

25. D. Mugnai, The tunnel effect in electromagnetic propagation, *Opt. Commun.* **175**, 309 (2000).
26. A. M. Shaarawi and I. M. Besieris, Superluminal tunneling of an electromagnetic X wave through a planar slab, *Phys. Rev. E* **62**, 7415 (2000).
27. D. Mugnai, Bessel beam through a dielectric slab at oblique incidence: the case of total reflection, *Opt. Commun.* **207**, 95 (2002).
28. For propagation in a layered medium for normal and oblique incidence in the presence and absence of total reflection, respectively, see, e.g., A. M. Attiya, E. El-Diwany, A. M. Shaarawi, and I. M. Besieris, Reflection and transmission of X-waves in the presence of planarly layered media: the pulsed plane wave representation, *Prog. Electromagn. Res.* **30**, 191 (2001); A. M. Shaarawi, I. M. Besieris, A. M. Attiya, and E. El-Diwany, Reflection and transmission of an electromagnetic X-wave incident on a planar air–dielectric interface: spectral analysis, *Prog. Electromagn. Res.* **30**, 213 (2001).
29. L. Brillouin, *Wave Propagation and Group Velocity*, Academic Press, New York, 1960, p. 76, and refs. therein.
30. L. J. Wang, A. Kuzmich, and A. Dogariu, Gain-assisted superluminal light propagation, *Nature* **406**, 277 (2000).
31. D. Mugnai, A. Ranfagni, and R. Ruggeri, A note on microwave guided propagation in the case of anomalous dispersion, *IEEE Trans. Antennas Propag.* **51**, 914 (2003).
32. J. A. Stratton, *Electromagnetic Theory*, McGraw-Hill, New York, 1941, Sec. 5.18.
33. D. Mugnai and I. Mochi, Superluminal X-wave propagation: energy localization and velocity, *Phys. Rev. E* **73**, 016606 (2006).
34. J. D. Jackson, *Classical Electrodynamics*, Wiley, New York, 1999, Sec. 7.1.
35. See [32], p. 342.
36. See [34], Sec. 7.8.
37. A. P. Prudnikov, Yu. A. Brychkov, and O. I. Marichev, *Integrals and Series: Elementary Functions*, Vol. 1, Gordon and Breach, New York, 1986, p. 390.
38. M. Abramowitz and I. A. Stegun, Eds., *Handbook of Mathematical Functions with Formulas, Graphs, and Mathematical Tables*, Dover Publications, New York, 1972, p. 374; see also [15], p. 78.
39. D. Mugnai, *Phys. Lett. A* **364**, 435 (2007).
40. See ref. [34], Sec. 11.3.
41. F. Cardone and R. Mignani, *Energy and Geometry*, World Scientific Series in Contemporary Chemical Physics, Vol. 22, World Scientific, Singapore, 2004.

CHAPTER SEVEN

Linear-Optical Generation of Localized Waves

KAIDO REIVELT AND PEETER SAARI
University of Tartu, Tartu, Estonia

7.1 INTRODUCTION

Since this is the first collective monograph in a growing research field, the terminology is not yet well established. Therefore, we specify that we use the term *localized waves* as the generic name for spatially sharply peaked wideband wave packets propagating over large distances without spreading. Although the term is somewhat inexact, it is self-explanatory compared to its less frequently used synonyms: nondispersive, limited diffraction and undistorted progressive waves. Although questionable physically, the terms *nondiffracting* and *diffraction-free* have taken root in the massive literature on monochromatic Bessel beams, and we prefer to leave these terms for narrowband propagation-invariant wave fields, including many-cycle Bessel beam pulses. Finally, in a wider sense, the generic term *localized waves* (LWs) might also cover profoundly propagation-*variant* wideband waves such as sharply focused few-cycle and subcycle pulses.

During more than a decade after the pioneering theoretical paper of J. N. Brittingham in 1983 [1], the feasibility of electromagnetic LWs remained questionable for various reasons: from principal doubts up to understandable obstacles that one encounters, especially in the optical domain, due to the large bandwidth and spatiotemporal nonseparability inherent in LWs. Indeed, the ideas that had been proposed for generation of complicated LW solutions in the papers of that period of the new research field (see, e.g., [2–24] and references therein) are hardly realizable in optical domain.

Localized Waves, Edited by Hugo E. Hernández-Figueroa, Michel Zamboni-Rached, and Erasmo Recami

We entered into LW research in the mid-1990s with the idea that optical LWs must be realizable experimentally by making use of linear-optical elements: in particular, the wavelength dispersion of the cone angle (the axicon angle) of Bessel beam generators. We found that good approximations to the LW solutions can be generated by combinations of refractive elements (lenses and axicons) and circular diffraction gratings. Our approach, as well as the feasibility of optical LWs in general were verified in further studies, particularly, by the first successful experiments on the generation and recording of superluminal X-type and luminal focus-wave-mode-type light fields [25–32]. Having in mind the results of other groups [33–36], particularly on further development in the techniques of recording of ultrashort LWs and their generation in nonlinear media, we note that this research field has grown from an initial theoretical-only stage, as evidenced by the coverage of recent experimental studies by *Physics Today* in its editorial section "Search & Discovery" (*Physics Today*, vol. 57, October 2004).

In this chapter we propose a short overview of our work on experimental realizations of LWs. However, before we begin to describe our experiments, we introduce briefly a Fourier-space picture of LWs, in which our approach to the generation of various LWs is most comprehensible. Also, we discuss various approximations and deviations from an ideal LW, which since the latter is nonphysical in principle, one must adopt in order to design a realistic experiment. For this reason we also focus on understanding the physical nature of various LWs and peculiarities of their behaviour — all this is the backbone in which our approach relies. Later, when we review experiments, we briefly consider our generalization of the concept of LWs to partially coherent wave fields, which supplies us with a convenient method of recording the spatiotemporal behavior of wave fields. In the concluding section we make some remarks on the possible impact of studies of LWs in various branches of physics.

Due to limits on length, the list of references is far from complete even given the narrow range of this chapter. For a more comprehensive overview, particularly of theory of LWs, see [37–41] and the references therein. We trust that, together, the chapters of this book will supply readers with a complete bibliography on the rapidly growing field of localized waves and diffraction-free beams.

7.2 DEFINITION OF LOCALIZED WAVES

By now, it has been shown in many publications that there is a general physically transparent description that can be used to understand any propagation-invariant LW as an exact solution to the scalar free-space wave equation [5,6,28,37–39,41] and this "physical backbone" comes within the Fourier context. In those terms a general cylindrically symmetric solution of the free-space wave equation reads

$$\Psi(\rho, z, t) = \int_0^\infty dk \int_0^\pi d\theta \; A(k, \theta) J_0(k\rho \sin\theta) \exp[ik(z\cos\theta - ct)]. \qquad (7.1)$$

Here the field at a point (ρ, z) is represented as a superposition of monochromatic Bessel beams:

$$\Psi_B(\rho, z, t) = J_0(k\rho \sin\theta) \exp[ik(z\cos\theta - ct)], \tag{7.2}$$

each propagating both forward and backward along the z axis with a different cone angle θ and carrier wave number $k = \omega/c$ (c is the speed of light in vacuum). The zero-order Bessel function $J_0(k\rho \sin\theta)$ arises as the plane waves are integrated over the azimuthal angle with the wave number k propagating at angles θ relative to the z-axis.

An integral representation of LWs can be derived from the condition that the superposition of Bessel beams in Eq. (7.1) should form a nondispersing pulse propagating along the symmetry axis (z-axis) [28,29,37]. In terms of group-velocity dispersion of one-dimensional wave packets, this means that the group velocity $v_g = d\omega/dk_z$, where $k_z = k\cos\theta$, should be constant over the entire spectral range. This restriction allows nontrivial solutions only if we assume that the cone angle is a function of the wave number $\theta\,(k)$. Corresponding support of the angular spectrum of the plane-wave constituents of the pulse is a cylindrically symmetric surface in k-space, and the angular spectrum itself can be expressed by means of the Dirac delta function as

$$A(k, \theta) = \tilde{A}\,(k)\,\delta(\theta - \theta(k)). \tag{7.3}$$

The condition

$$v^g = \frac{c\,dk}{dk_z} = \left[\frac{1}{c}\frac{d}{dk}(k\cos\theta_{\mathrm{LW}}(k))\right]^{-1} = \frac{c}{\gamma}, \tag{7.4}$$

where the constant, γ, which determines the group velocity, yields

$$\cos\theta_{\mathrm{LW}}(k) = \frac{\gamma(k - 2\beta)}{k}; \tag{7.5}$$

see Fig. 7.1.

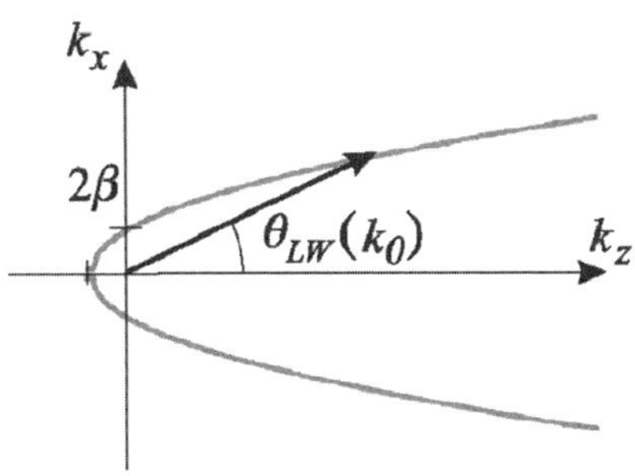

FIGURE 7.1 Geometrical interpretation of the parameter β of the supports of angular spectrum of plane waves of LWs (gray line).

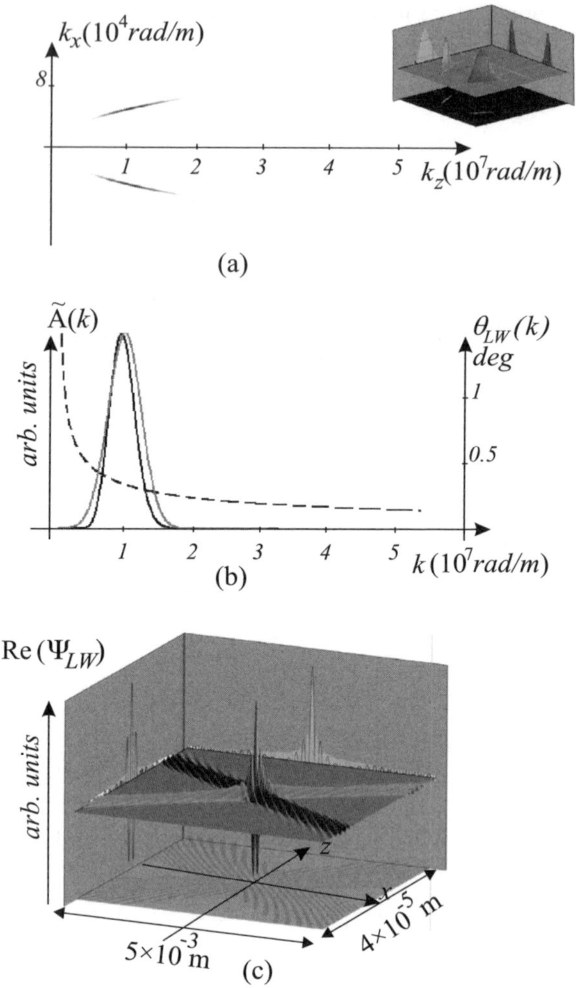

FIGURE 7.2 Numerical example of a LW optimized for optical generation with the parameters $\beta = 40$ rad/m, $\gamma = 1$: (a) angular spectrum of plane waves in two perspectives; (b) frequency spectrum of an optically feasible wave field (gray line) and Bessel–Gauss pulse (black line), and the angle $\theta_{\mathrm{LW}}(k)$ as the function of the wave number (dashed line); (c) spatial field distribution of the pulse.

The integration constant β in Eq. (7.5) is extracted from condition $\theta_{\mathrm{LW}}(2\beta) = 90^{\mathrm{o}}$. Thus, the angular spectrum of plane waves can be written as

$$A(k, \theta) = \tilde{A}(k)\delta\left(\cos\theta - \frac{\gamma(k - 2\beta)}{k}\right) \tag{7.6}$$

(see Fig. 7.2), which is a general expression for free-space LWs that move with group velocity v_g along the optical axis z. Substituting (7.6) into general solution (7.1) we get

$$\Psi_{\mathrm{LW}}(\rho, z, t) = \int_0^\infty dk \int_0^\pi d\theta\, \tilde{A}(k)\delta\left(\cos\theta - \frac{\gamma(k-2\beta)}{k}\right) \times J_0(k\rho\sin\theta)\exp[ik(z\cos\theta - ct)], \tag{7.7}$$

so that

$$\Psi_{\mathrm{LW}}(\rho, z, t) = \int_0^\infty dk\, \tilde{A}(k)\, J_0\left(k\rho\sin\theta_{\mathrm{LW}}(k)\right)\exp\left[ik\left(z\cos\theta_{\mathrm{LW}}(k) - ct\right)\right], \tag{7.8}$$

and this is the general integral expression for the LW solutions of the scalar wave equation.

An even more elegant approach is possible if we write the general axisymmetric expansion over the zeroth-order Bessel beams as

$$\Psi(\rho, z, t) = \int_{-\infty}^{\infty} dk_z \int_{|k_z|}^{\infty} dk\, \hat{A}(k_z, k) J_0\left(\sqrt{k^2 - k_z^2}\rho\right)\exp(ik_z z - ikct). \tag{7.9}$$

and note that for $|\Psi(\rho, z, t)|^2$ to be propagation-invariant (i.e., to depend on z and t through the propagation variable $z - v_g ct$, where v_g is a constant group velocity along the z-axis in units of c), the variables k and k_z must be bound linearly [5,6,39]:

$$k = v_g k_z + b, \tag{7.10}$$

where b is a constant. Hence, the spectrum has to be singular and may be factorized in the form

$$\hat{A}(k_z, k) = S(k)\,\delta(k - v_g k_z - b)\,\Theta(k^2 - k_z^2), \tag{7.11}$$

where $S(k)$ is any complex-valued function of one real positive variable, and the Heaviside unit step $\Theta(x)$ has been introduced as a factor to allow the k-integration in Eq. (7.9) to start from $k = 0$ instead of $k = |k_z|$. Thus, for an axisymmetric wave packet to be a propagation-invariant LW, its spectral support must be a line of intersection of the cone surface by a plane perpendicular to the plane (k_z, k) (see Fig. 7.3). The projection of the line onto the plane (k_z, k) is a straight line with the slope v_g (Fig. 7.3).

In the numerical example in Fig. 7.2 we have chosen the parameters of the LW (7.8) so that the resulting wave field is optically feasible; that is, (1) the frequency spectrum of the LW is in optical domain, and (2) the plane-wave spectrum of the LW should contain only plane waves that propagate at paraxial angles relative to the optical axis. The frequency spectrum of the LW is compared with that of the Bessel–Gauss pulse [9], which is apparently the best closed-form LW to simulate this type of LW [32].

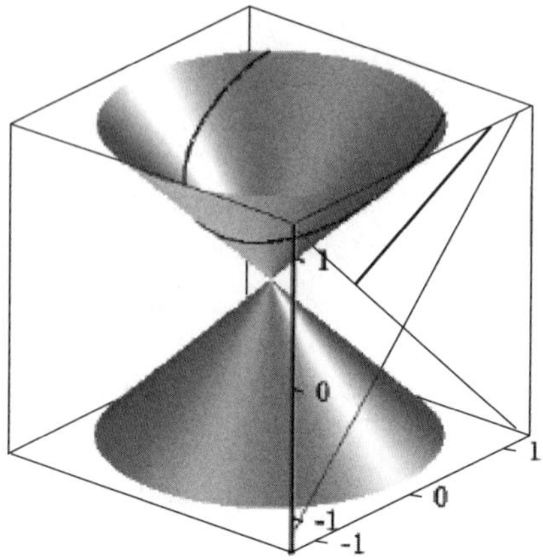

FIGURE 7.3 The conical surface in the Fourier space, which the support of any wave has to lie on. The frequency axis k (or ω/c) is vertical; the one with a numbered scale is the k_z axis. The third axis depicts one of the transverse components (for cylindrically symmetric waves—any transverse component) of the wave vector. Shown on the surface is the support line of a localized wave, which has subluminal velocity because, as shown, the projection of the line onto the $k_z - k$ plane is a straight line with slope less than 1. Since for solutions to the wave equation as analytic signals, the lower cone with negative frequencies is absent and for k_ρ representing the transverse components in the cylindrical system there are no negative values, henceforth we deal only with the rear-left half of the upper cone.

The simple derivations above provide some very essential and general consequences for LWs:

1. In many discussions, terms such as *localized waves* or *focus wave mode* have been used in connection with a specific frequency spectrum $\tilde{A}(k)$, which enables analytical integration of Eq. (7.8) (see, e.g., [7]). However, our approach implies that *all* the wave fields given by Eq. (7.8), propagate without any longitudinal or transversal dispersion, (i.e., they possess the characteristic property of LWs). Moreover, many of the frequency spectrums that have been assigned to LWs, are barely realizable experimentally, as the corresponding angular spectrums do not vanish for backward-propagating components ($k_z < 0$) within the support of an angular spectrum [37], and as their frequency spectrums cannot be optimized to approximate the frequency spectrum of real light sources. Therefore, in work by the authors of this chapter we have used the term *localized wave* (or *focus wave mode*) for all the wave fields, the support of angular spectrum of which satisfies condition (7.5), and the field distribution of which is given by Eq. (7.8)

2. As the plane-wave components in Eq. (7.8) can travel both forward and backward, this property was used in some early publications as a serious argument against the feasibility of LWs [3]. However, the simple arguments above show that in all practical situations, the causality of resulting wave fields is simply up to the proper

choice of frequency spectrum $\tilde{A}(k)$ and that acausal fields arise in situations where the unique analytical solutions of Eq. (7.8) are discussed.

3. As the angular spectrum of plane waves of LWs in Eq. (7.8) is singular, the energy content of the LWs is infinite. However, in the following sections we will see that the optical setups for finite-energy LWs can be constructed by means of the idealized theory above, as the finite-energy approximations arise simply due to the finite physical and time aperture of the system.

4. LW-type wave fields in paraxial approximation have been investigated extensively, and both monochromatic (see, e.g., [42,43] and references therein) and pulsed (see, e.g., [44–47] and references therein) wave fields with LW-like properties have been discovered and studied. In our discussion we work only in a Fourier representation; although both approaches provide means of evaluating wave propagation, in our opinion the characteristic properties of LWs can be better interpreted in a Fourier context.

Next we consider various aspects of LWs that are essential to an understanding of the optical realizations of LWs and discuss experiments that have been carried out.

7.3 THE PRINCIPLE OF OPTICAL GENERATION OF LWs

The representation (7.8) can be interpreted as follows. Given an initial plane-wave pulse, a LW generator should transform every monochromatic component of wave number k into a Bessel beam with cone angle $\theta_{\mathrm{LW}}(k)$, determined by Eq. (7.5). If the relative phases between those Bessel beams vanish as well [i.e., the frequency spectrum $\tilde{A}(k)$ is real], the resulting wave field is an LW with a transform-limited central peak. In this section we show that this interpretation can be realized by means of wavelength dispersion of the cone angle of Bessel beam generators.

A monochromatic plane wave can be transformed into a Bessel beam by various optical elements, such as axicons and circular diffraction gratings (see, e.g., [48] and references therein). By illuminating those elements with a plane-wave pulse, we obviously get a superposition of Bessel beams, the support of the angular spectrum of which is determined by the wavelength dispersion of the cone angle of the optical element $\theta(k)$. For example, an axicon is characterized by the complex transmission function $\exp[ik\tan\alpha(1-n(k))]$, where α is the angle formed by the conical surface with a flat surface and $n(k)$ is the refractive index of the axicon (see, e.g., [49]). The stationary phase method [50], used as described in [51], yields $\sin\theta(k)=\tan\alpha(1-n(k))$. Hence, the angular spectrum of plane waves of the corresponding polychromatic wave field can be written as $A(k,\theta)=\tilde{A}(k)\,\delta(\sin\theta-\tan\alpha\,(1-n\,(k)))$, where $\tilde{A}(k)$ is the frequency spectrum of the source pulse. Similarly, a circular grating yields for the cone angle $\sin\theta(k)=2\pi/kd$, where d is the grating constant (first–order diffraction is assumed).

Support of the angular spectrum of an LW (7.5) cannot be approximated by a single diffractive element. However, it can be shown that aside from the conventional

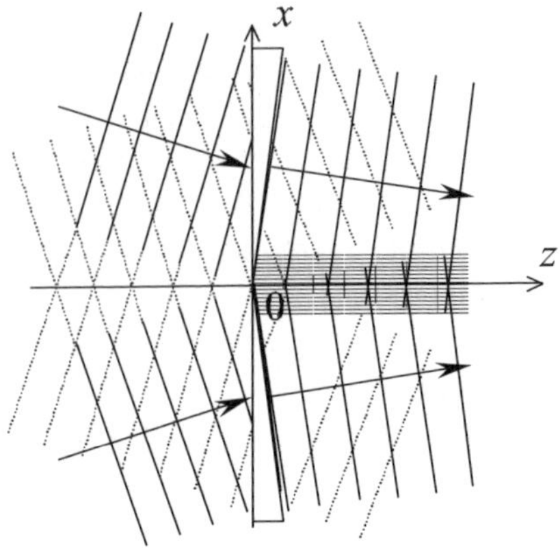

FIGURE 7.4 Qualitative description of a Bessel beam as a superposition of diverging (dotted lines) and converging (solid lines) conical waves in plane $z = 0$ in the process of refraction on an axicon.

configuration, where axicons and circular gratings are used to transform a plane wave into a Bessel beam, those elements can also be used to *change* the cone angle of Bessel beams [27]. Let us place an axicon (or circular grating) onto plane $z = 0$ (see Fig. 7.4). Qualitative analysis by means of ray tracing reveals immediately that the element splits the initial plane wave into converging and diverging waves and changes the cone angle of both conical waves. The converging conical wave forms a Bessel beam in a conical near-axis volume, but the cone angle is now different. The diverging conical component leaves the near-axis region. Hence, axicons and diffraction gratings change the cone angle of a Bessel beam in the sense that the regenerated Bessel beam after the element has a different cone angle. This property allows us to use a set of Bessel beam generators to design more complex supports of the angular spectrum.

Let us find the cone angle of the resulting Bessel beam for a composite optical element, a circular diffraction grating on the surface of an axicon (see Fig. 7.5a). We

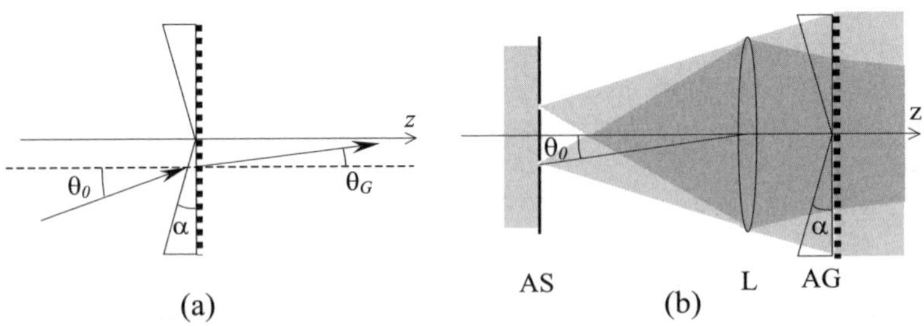

FIGURE 7.5 (a) Circular diffraction grating on the surface of an axicon—a composite optical element, that can be used for generation of LWs; (b) optical setup for generation of LWs. A plane-wave pulse is incident upon an annular slit (AS). A Bessel X-pulse with cone angle θ behind a Fourier lens (L) is incident upon the composite optical element (AG).

assume that a Bessel X-pulse with cone angle θ_0, generated by means of an annular slit and Fourier lens, is incident upon it (see Fig. 7.5b). Snell's law and the grating equation yield the following equation for the cone angle of the resulting Bessel beam:

$$\sin\theta_G() = \frac{2\pi}{kd} + n(k)\sin\left\{-\alpha + \arcsin\left[\frac{1}{n\,(k)}\sin(\theta_0+\alpha)\right]\right\}. \tag{7.12}$$

Here d is the grating constant and $n(k)$ is the refractive index of the axicon material. Sign conventions are chosen so that the angles α, θ_0, and $\theta_G\,(k)$ are positive in Fig. 7.5a. First-order diffraction is assumed. If the angles α and θ_0 are small, so that $\sin x \sim \arcsin x \sim x$, Eq. (7.12) yields

$$\theta_G\,(k) = \frac{2\pi}{kd} + \alpha\,[1 - n\,(k)] + \theta_0. \tag{7.13}$$

The wavelength dispersion of the cone angle of the Bessel beam constituents of an LW $\theta_{\mathrm{LW}}\,(k)$ is determined by Eq. (7.5). Combining this condition with the dispersion of the cone angle of the setup $\theta_G\,(k)$ (7.13), we get

$$\arccos\frac{\gamma\,(k - 2\beta)}{k} = \frac{2\pi}{kd} + \alpha\,(1 - n\,(k)) + \theta_0. \tag{7.14}$$

In the following chapters we will see that the three free parameters (α, θ_0, d) can be adjusted so that the relation (7.14) is satisfied in a very good approximation in a limited spectral range.

7.4 FINITE-ENERGY APPROXIMATIONS OF LWs

As already noted, the total energy content of LWs in Eq. (7.8) is infinite, so they are not realizable in any physical experiment. In trying to find a way to resolve this problem, a natural choice is to consider the approximations that correspond to the specific launching mechanism. In what follows we consider the finite-energy approximations of LWs that are due to the finite aperture of the optical system.

In a Fourier context, the starting point for this discussion is obvious. In this picture any realistic (finite-aperture) optical system generates a superposition of apertured monochromatic Bessel beams, so that the field in the exit plane of the LW generator can be described by

$$\Psi_B\,(\rho, 0, 0; k) = t\,(\rho)\,J_0\,[k\rho\sin\theta_{\mathrm{LW}}\,(k)]\,, \tag{7.15}$$

where $t\,(\rho)$ is the complex-amplitude transmission function of the aperture of the setup and for brevity we restrict ourselves to a cylindrically symmetric case. One just has to show (1) that the resulting superposition of the apertured Bessel beams still represents a wave field that propagates as an LW, and (2) that the superposition has finite energy content.

The behavior of apertured Bessel beams has been discussed in many papers, mostly in terms of Fresnel approximation of scalar diffraction theory (see e.g., [42,43,49,51] and references therein). Here we can just refer to the generally accepted facts that (1) applying a finite aperture to a Bessel beam provides us with a finite-energy-flow wave field that is a very good approximation to the infinite-aperture Bessel beams in Eq. (7.15) in a certain finite-depth near-axis volume (see, e.g., [52–55]), and (2) that the polychromatic superpositions of those apertured Bessel beams approximate very closely the superpositions of nonapertured Bessel beams in this volume [56]. Such behavior can easily be explained in a Fourier context, where a monochromatic Bessel beam is a cylindrically symmetric superposition of plane waves that propagate at an angle θ relative to the z-axis. Indeed, as the apertured plane waves approximate their infinite aperture counterparts very closely in their central parts, one can also observe a very good approximation to the infinite-aperture Bessel beam in this near-axis volume (see, e.g., the near-axis region in Fig. 7.4 and [57] for a more detailed description in terms of diffraction theory). If the cone angle of a Bessel beam is small, as is always the case in paraxial optical systems, the apertured Bessel beam would behave as its infinite-aperture counterpart in Eq. (7.15) for several meters of propagation.

Mathematically the situation can be modeled by applying to the Bessel beams the transmission function $t(\rho)$ of the aperture. In the Weyl context this operation is equivalent to calculating the two-dimensional Fourier transform of the transversal amplitude distribution $t(\rho)\, J_0(\chi_0\rho)$, where χ_0 stands for radial projection of the wave vector of the Bessel beam. Given a Weyl-type angular spectrum of plane waves of the infinite-aperture Bessel beam

$$A(\chi) = K\delta(\chi - \chi_0), \tag{7.16}$$

where $\chi = k\sin\theta$ and K is a constant, the Fourier transform of Eq. (7.15) at point $z = 0, t = 0$ can be found to be

$$\begin{aligned} A_{\mathrm{Ap}B}(\chi) &= \frac{K}{(2\pi)^2} T(\chi) * \delta(\chi - \chi_0) \\ &= \frac{\chi_0 K}{(2\pi)^2} \int_0^{2\pi} d\phi T\left(\sqrt{\chi^2 + \chi_0^2 - 2\chi\chi_0\cos(\phi - \phi_0)}\right), \end{aligned} \tag{7.17}$$

where $T(\chi)$ is the two-dimensional Fourier transform of the transmission function $t(\rho)$ and $*$ denotes convolution (see also [58]). The argument of the function $T(\cdot)$ in Eq. (7.17) is interpreted as being the distance between the points (χ, ϕ) and (χ_0, ϕ_0). As for all convenient apertures, the function $T(\chi)$ is well localized around zero, the major contribution to the integral (7.17) comes from small values of ϕ, and one can write in good approximation

$$A_{\mathrm{Ap}B}(\chi) \approx \frac{\chi_0 K}{2\pi} T(\chi - \chi_0). \tag{7.18}$$

The interpretation of the expression (7.18) is straightforward: The finite aperture gives the support of the angular spectrum of a monochromatic Bessel beam a finite

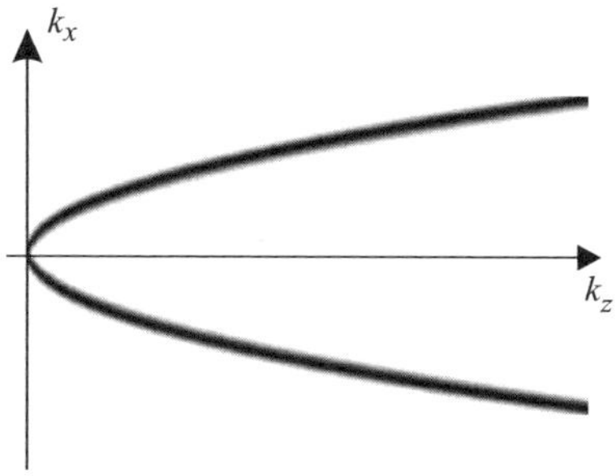

FIGURE 7.6 Qualitative picture of the support of angular spectrums of plane waves of apertured LWs.

"width". The exact form of the support is determined by the complex-amplitude transmission function; however, the well-known set of fundamental Fourier transform pairs provides a good idea of what the support of angular spectrum looks like, without calculations.

Obviously, the finite aperture has a similar effect on the angular spectrum support of an LWs (7.8); the delta function in Eq. (7.3) is substituted by a weighting function and the angular spectrum of plane waves of apertured LWs can be written as

$$A_{ApF}(k,\chi) = \frac{\chi_F(k)\tilde{A}(k)}{(2\pi)^2}\int_0^{2\pi} d\varphi T\left(\sqrt{\chi^2+\chi_F(k)^2-2\chi\,\chi_{LW}(k)\cos\varphi}\right)$$
$$\approx \frac{\chi_F(k)\,\tilde{A}(k)}{2\pi}T\left[\chi-\chi_{LW}(k)\right], \qquad (7.19)$$

where $\chi_{LW}(k) = k\sin\theta_{LW}(k)$. The support of the angular spectrum of plane waves (7.19) of the wave field derived is depicted in Fig. (7.6). In correspondence with Eq. (7.18) it, has a finite thickness.

Obviously, the resulting wave field still has the characteristic narrow central peak. Also, in correspondence with the note at the beginning of the section, the fields of apertured and nonapertured LWs do not differ noticeably in the near-axis volume at $t = 0$. Thus, if the aperture of the system is reasonably large (several millimeters) the only qualitative effect of the finite aperture is the reduced propagation length of the wave field, and in optical setups the finite-energy (physically realizable) approximations to LWs are introduced quite plainly by the finite aperture of the optical system. In the Fourier domain this corresponds to smoothing of the delta function in the angular spectrum of plane waves of the wave fields.

7.5 PHYSICAL NATURE OF PROPAGATION INVARIANCE OF PULSED WAVE FIELDS

To conclude the theoretical part of this review, we interpret some LW properties in the context of classical diffraction theory (see [59] for a relevant discussion).

The first note has to do with the definition of the propagation length of LWs. It has been claimed in several publications (see, e.g., [12]) that LWs propagate over extended distances rather than as defined by the Rayleigh range Z_R—the well-known estimate for the scale length of the falloff in intensity behind a Gaussian aperture in diffraction theory—defined as

$$Z_R = \frac{\pi W_0^2}{\lambda}, \tag{7.20}$$

W_0 being the minimum spot size (radius) of the beam. Indeed, if we replace W_0 in (7.20) with the peak width of the LW, the propagation length of the LW indeed exceeds the propagation length of the corresponding Gaussian pulse by an order of magnitude. However, in our opinion, such estimates are misleading and should not be used without the following additional details.

The effect of diffraction is typically manifested when an obstacle is placed in the path of a light field. For (pulsed) beams, definition (7.20) determines minimal spread, corresponding to the waist radius W_0. In other words, this parameter determines the minimum radius of a circular aperture that can be placed in the path of the beam without significantly distorting its behavior behind the aperture. To compare the Gaussian pulse with an LW, we have to ask: What is the waist radius of an apertured LW?

It appears that the situation is similar to that with monochromatic Bessel beams, for which the energy content in every transverse lobe is approximately constant and equal to the energy content in the central maximum (see, e.g., [59]). In the case of apertured LWs, the amplitude of the branches of the characteristic X-shape fall off approximately as $1/\rho$. Thus, the integrated energy density as the function of radial coordinate ρ is approximately constant, as integration over the polar angle adds the factor ρ to the amplitude of the wave field (see Fig. 7.7). Thus, only a minor part of the energy of the apertured LWs is contained in its central lobe, and if one has to compare the propagation length of the apertured LWs with the Rayleigh range of the Gaussian pulses, one should take $W_0 = D$ (see Fig. 7.8), and in this case the propagation lengths of the pulses are comparable.

As a matter of fact, such a comparison of the Gaussian pulse and apertured LWs is not appropriate, as focusing of the two wave fields is qualitatively different in physical nature. The Gaussian pulses are composed of monochromatic components with curved phase fronts, whereas the phase fronts of the monochromatic components of the LWs are flat. In light of this difference, one can say that the LWs are *never* focused in the conventional sense of the term, and the term *focus wave mode* for the most commonly known LW (see, e.g., [1]) is rather misleading.

In focusing Gaussian pulses, most of its energy content can be concentrated into a single spot for an instant. A cunning mental picture of the energetic properties of focusing in LWs can be acquired if we suppose that we have an ideal (apertured) LW generator and suppose that we illuminate it with a plane-wave pulse. The effect on the LW generator to this monochromatic plane-wave component is that its initial energy in the k-space (a finite-width spot on the k_z-axis) is smeared over the finite-width

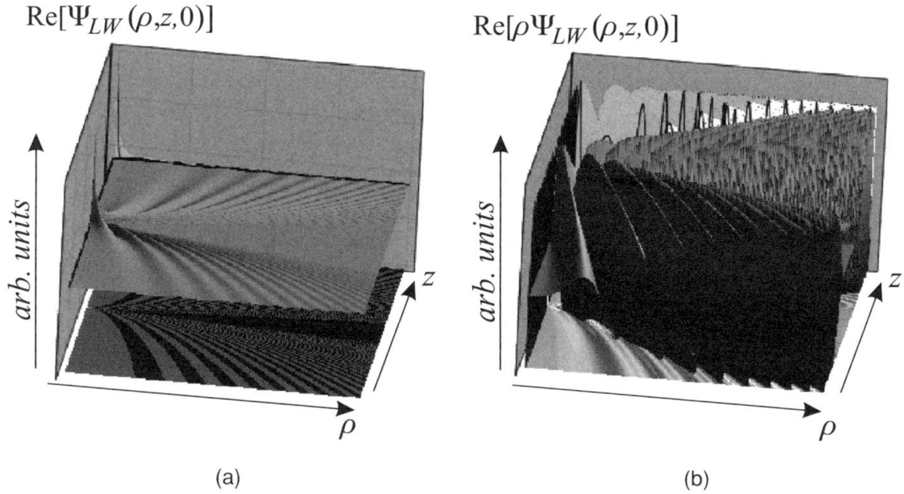

FIGURE 7.7 Comparison of (a) the radial field distribution of a LW with (b) the field that is integrated over the polar angle.

toroid in the k-space. In real position space each point in this toroid will compose an apertured plane wave in the Bessel beam, and the net on-axis amplitude of the Bessel beam is the integrated amplitude over the toroid in k-space (i.e., again the amplitude of the initial monochromatic plane-wave component). Consequently, the amplitude of the central peak of each monochromatic Bessel beam component of the LWs generated is equal to the amplitude of the corresponding monochromatic plane-wave component of the initial plane-wave pulse.

For the superposition of Bessel beams we can say that the only place where the constructive interference occurs is the central spot of the apertured LW. Thus, given an

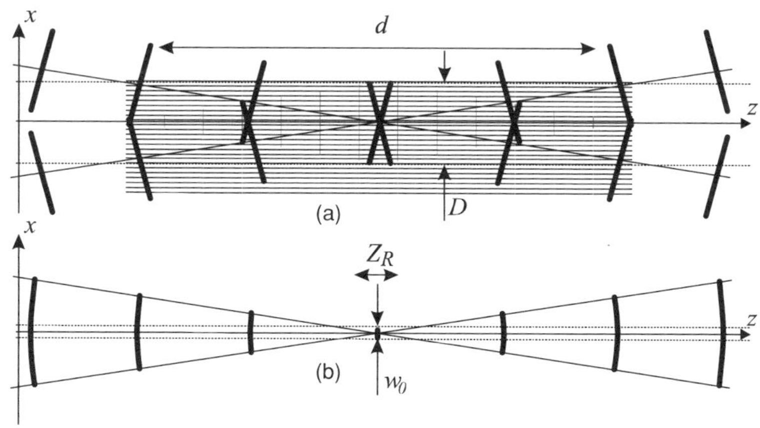

FIGURE 7.8 Comparison of the focusing of (a) LWs and (b) Gaussian pulses (see the text).

initial plane-wave pulse with amplitude A and aperture D, *the amplitude of the central spot of an apertured LW behind an ideal LW generator is that of the initial plane-wave pulse A, the rest of the energy is in the sidelobes of the LWs generated.* This consequence is a good illustration of the qualitative difference between an LW and a Gaussian pulse; due to the curved phase fronts of its monochromatic components, the latter can indeed effectively transfer most of its energy to a single spot for an instant.

7.6 EXPERIMENTS

7.6.1 LWs in Interferometric Experiments

A straightforward method for recording the complicated field shape of a coherent LWs would use a CCD camera with a gate, which should possess temporal resolution and a variable firing delay, both in the subfemtosecond range. As such a gate is not realizable in the optical domain, any workable experimental idea must resort to a field cross-correlation technique [we do not consider here higher (i.e., nonlinear-optical) correlations].

In our experiments [27,31] the wave field under investigation $F(\mathbf{r}, t)$ interferes with a reference wave V_P:

$$V_\Sigma(\mathbf{r}, t) = F(\mathbf{r}, t) + V_P(\mathbf{r}, t). \tag{7.21}$$

For the reference wave we can write

$$V_P(\mathbf{r}, t) = \int_0^\infty dk\, s(k) \upsilon_P(k) \exp[ik(z - c(t + \Delta t))], \tag{7.22}$$

where $s(k)$ is the (generally stochastic) frequency spectrum of the light source, $\upsilon_P(k)$ is the spectral phase shift introduced by the optics in the reference arm of the interferometer, $|\upsilon_P(k)| \equiv 1$, and Δt denotes the variable time delay between the signal and reference wave fields. For the wave field under investigation, we have

$$F(\mathbf{r}, t) = \int_0^\infty dk\, s(k)\, \upsilon_F(k) J_0[kx \sin\theta_{\rm LW}(k)] \exp[ik(z \cos\theta_{\rm LW}(k) - ct)], \tag{7.23}$$

where $\upsilon_F(k)$ is again an undesirable spectral phase shift from the setup, $|\upsilon_F(k)| \equiv 1$ [we have assumed here that the LW generator transforms the input light so that for every spectral component $s(k)$ the amplitude of the central spot of the corresponding Bessel beam is also $s(k)$]. The averaged intensity of the resulting wave field can be expressed as

$$\langle V_\Sigma^* V_\Sigma \rangle = \langle V_P^* V_P \rangle + \langle F^* F \rangle + 2\,\mathrm{Re}\langle F^* V_P \rangle \tag{7.24}$$

(here the exact meaning of the angle brackets depends on the statistical properties of the light source of the experiment).

The quantities $\langle V_P^* V_P \rangle$ and $\langle F^* F \rangle$ denote the time-independent intensity of the wave field. Specifically, the first term in the sum is the uniform intensity of the plane-wave pulse:

$$\langle V_P^* V_P \rangle = \int_0^\infty dk \mathcal{S}(k), \tag{7.25}$$

where again $\mathcal{S}(k) = \langle s^*(k) s(k) \rangle$ is the spectral density. The second term is the time-averaged intensity of F:

$$\langle F^* F \rangle = \int_0^\infty dk \mathcal{S}(k) J_0^2[kx \sin\theta_{\mathrm{LW}}(k)]. \tag{7.26}$$

In principle, the two components can be eliminated from the results by recording them separately and by subtracting them numerically from the interferograms.

From Eqs. (7.22) and (7.23) we can write

$$\begin{aligned} 2\,\mathrm{Re}\langle F^* V_P \rangle = 2\,\mathrm{Re}\Bigg\langle &\int_0^\infty dk_1 s^*(k_1) \upsilon_F^*(k_1) \\ &\times J_0(k_1 \rho \sin\theta_{\mathrm{LW}}(k)) \exp[-ik_1(z\cos\theta_{\mathrm{LW}}(k) - ct)] \\ &\times \int_0^\infty dk_2 s\,(k_2)\,\upsilon_P\,(k_2) \exp[ik_2(z - c(t - \Delta t))] \Bigg\rangle, \end{aligned} \tag{7.27}$$

and the averaging in (7.27) yields

$$\begin{aligned} \langle F^* V_P \rangle = &\int_0^\infty dk\, \mathcal{S}(k) \upsilon_F^*(k) \upsilon_P(k) \\ &\times J_0(kx \sin\theta_{\mathrm{LW}}(k)) \exp[ikz(\cos\theta_{\mathrm{LW}}(k) - 1) + ikc\Delta t] \end{aligned} \tag{7.28}$$

Here we have used the relation

$$\langle s^*(k_1) s(k_2) \rangle = \mathcal{S}(k_1) \delta(k_1 - k_2) \tag{7.29}$$

(For coherent fields the $\delta\,(k_2 - k_1)$ appears as time averaging over the term $\exp\,[ict\,(k_2 - k_1)]$ is carried out; for partially coherent wave fields this condition means frequency noncorrelation [50].) Equation (7.28) can be given the form

$$\begin{aligned} \langle F^* V_P \rangle = \exp(i2\beta\gamma z) &\int_0^\infty dk\, \mathcal{S}(k) \upsilon_F^*(k_1) \upsilon_P(k) \\ &\times J_0(k\rho \sin\theta_{\mathrm{LW}}(k)) \exp\{ik[z(1 - \gamma) - c\Delta t]\}. \end{aligned} \tag{7.30}$$

The mathematical description above yields identical results for coherent and partially coherent fields; that is, the results of such experiments generally do not depend on the correlations between the spectral components of different temporal frequency of the wave fields. It is well known that in any interferometric experiment with averaging in time, the spectral phase information of wave fields does not matter. In other words, the results of the experiments do not depend on whether we use transform-limited femtosecond pulses or a source of stationary white noise.

The latter consequence is of great practical significance. In the numerical example in Fig. 7.2 we used a spectrum that corresponds to a 3-fs laser pulse and showed that the corresponding LWs have good spatial localization. However, computer simulations, or even simple geometrical estimations, show that if the autocorrelation time of the source field τ exceeds $\sim$ 10 fs, the characteristic X-branching occurs too far from the z axis, and in this narrowband limit the resulting wave field would be nothing but trivial interference of quasi-monochromatic plane waves. Thus, the bandwidth of the light source is a very challenging part of the setup. In what follows we add to the reputation of incoherent sources as being the poor man's femtosecond source and confine ourselves to the special case of frequency noncorrelating fields.

In our experiments we realized the optical setups for two LWs: the apertured Bessel X-pulses [27] and apertured LWs [31].

7.6.2 Experiment on Optical Bessel X-Pulses

The class of LWs for which $\beta = 0$ in Eq. (7.5) are called *X-type LWs* (see, e.g., [48,60–64] and references therein). This choice implies that their support of the angular spectrum of plane waves is a cone in k-space (see Fig. 7.3). Consequently, the phase and group velocity of X-type pulses are equal (both necessarily superluminal), and the field propagates without any local changes along the optical axis.

Members of the optically realizable subclass of X-type LWs have been called Bessel X-pulses (see, e.g., [25,26,65–68] and references therein). The mathematical expression for the wave fields reads

$$\Psi_{\mathrm{BX}}(\rho, z, t) = \int_0^\infty dk\ \tilde{A}(k) J_0[k\rho \sin\theta_0] \exp[-ik(z\cos\theta_0 - ct)], \qquad (7.31)$$

where the choice

$$\tilde{A}\,(k) = \frac{\sigma_k}{\sqrt{2\pi}}\sqrt{\frac{k}{k_0}}\exp\left[-\frac{\sigma_k^2\,(k-k_0)^2}{2}\right] \qquad (7.32)$$

leads to a closed-form expression for the wave field. In what follows we give our results of experiments on those Bessel X-wave fields.

Setup Our setup for interferometric experiments on optical Bessel X-pulses is depicted in Fig. 7.9b. Compared to a simple Bessel X-pulse generator, a pinhole is made in the center of the annular ring mask to form the reference field (plane-wave pulse) behind the lens L.

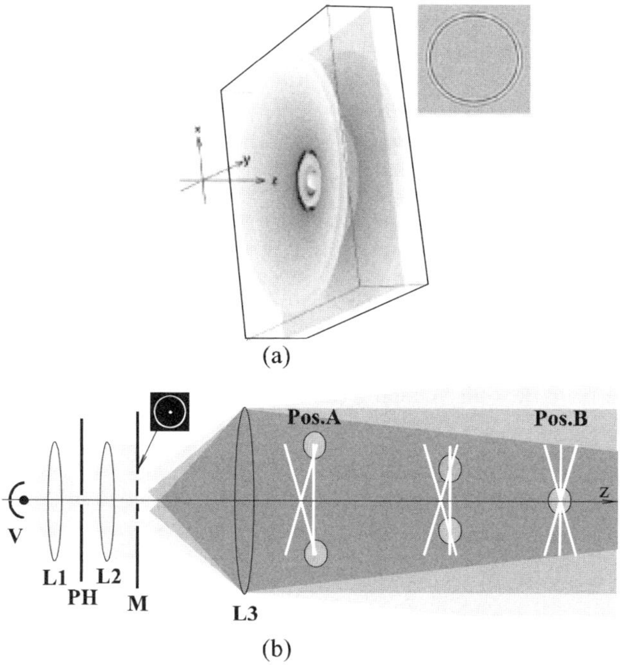

FIGURE 7.9 (a) Intensity profile of a computer-simulated Bessel X-pulse flying in space, shown as surfaces on which the field intensity is equal to a fraction $0.13\left(1/e^2\right)$ of its maximum value in the central point. The field intensity outside the central bright spot has been multiplied by the radial distance to reveal the weak off-axis sidelobes. Inset: Amplitude distribution in the plane, shown as intersecting the pulse. The plots have been computed for a 3-fs near-Gaussian-spectrum source pulse with carrier wavelength $\lambda = 0.6$ μm and the angle $\theta_0 = 0.223°$. For these parameters the dimensions of the plot xyz box are $20 \times 20 \times$ μm. (b) Optical scheme of the experiment. Mutual instantaneous placement of the Bessel X-pulse and the plane-wave pulse is shown for three recording positions (two of which are labeled in accordance with Fig. 7.10). The ovals indicate toroidlike correlation volumes where co-propagating Bessel X- and plane-wave pulses interfere at different propagation distances along the z-axis. L's, lenses; M, mask with Durnin's annular slit and an additional central pinhole for creating the plane wave; PH, cooled pinhole 10 μm in diameter to assure transversal coherence of the light from the source V. In case the source V generates a non-transform-limited pulses or a white continuous-wave noise, the bright shapes depict propagation of the correlation functions instead of the pulses.

For a mathematical description of the situation, we have to choose $\beta = 0$ in Eq. (7.30), that is, $\theta(k) \equiv \theta_0 = \arccos\gamma = \arccos c/v$. Also setting $\Delta t = 0$, we get

$$\langle F^* V_P \rangle = \int_0^\infty dk\mathcal{S}(k)\upsilon_F^*(k_1)\upsilon_P(k)J_0(k\rho\sin\theta_0)\exp[ikz_m(1-\gamma)], \qquad (7.33)$$

where $\mathcal{S}(k)$ is the spectral density of the light source. In Eq. (7.33) z_m denotes the distance along the optical axis of the setup. To understand the significance of this parameter we have to remembe the superluminal group velocity of the Bessel X-pulse; during the flight the latter catch up with the reference plane-wave pulse. In this context the coordinate z_m is the distance of the recording device from this catch-up point (see Fig. 7.9b).

If we now define the general expression for the mutual coherence function [50] of the scalar LWs [37],

$$\Gamma_{\mathrm{LW}}(\mathbf{r}_{1\perp}, \mathbf{r}_{2\perp}, \Delta z, \gamma\Delta z - c\tau) = \exp(-i\beta\gamma\Delta z)\int_0^\infty dk\, |\mathcal{V}_0(k)|^2 \\ \times J_0(k\rho_1 \sin\theta_{\mathrm{LW}}(k))J_0(k\rho_2 \sin\theta_{\mathrm{LW}}(k))\exp[ik(\gamma\Delta z - c\tau)], \quad (7.34)$$

and set the origin of the z-axis as being in the point $z = ct$ so that Δz can be replaced by the distance from the pulse center z_Δ and set $\tau = 0$, $\mathbf{r}_{\perp 1} \equiv \mathbf{r}_\perp, \mathbf{r}_{\perp 2} = 0$, the result reads

$$\Gamma_{\mathrm{LW}}(\mathbf{r}_\perp, 0, z_\Delta) = \exp(i2\beta\gamma z_\Delta)\int_0^\infty dk\, |\mathcal{V}(k)|^2\, J_0(k\rho \sin\theta_{\mathrm{FW}}(k))\exp(ik\gamma z_\Delta). \quad (7.35)$$

If we also set $\beta = 0$ for Bessel X-waves, we get

$$\Gamma_{\mathrm{BX}}(\mathbf{r}_\perp, 0, z_\Delta) = \int_0^\infty dk\, |\mathcal{V}(k)|^2\, J_0(k\rho \sin\theta_0)\exp(ik\gamma z_\Delta). \quad (7.36)$$

Comparing Eqs. (7.33) and (7.36) we see that if $|\mathcal{V}(k)|^2 = \mathcal{S}(k)$ and $\upsilon_F^*(k_1)\upsilon_P(k) \equiv 1$, we have

$$\langle F^* V_P\rangle = \Gamma_{\mathrm{BX}}\left(\mathbf{r}_\perp, 0, z_m\frac{1-\gamma}{\gamma}\right), \quad (7.37)$$

so that

$$z_\Delta = z_m\frac{1-\gamma}{\gamma}. \quad (7.38)$$

The interpretation of the small factor $(1-\gamma)/\gamma$ in (7.38) is that the setup serves as a "z-axis microscope" for recording the mutual coherence function (7.33) along the z-axis, which scales the micrometer-range z-dependence of the field into a centimeter range.

If we compare the expression (7.36) with the hypothetically measurable field distribution given as the real part of the Bessel X-pulse in Eq. (7.31),

$$\Psi_{\mathrm{BX}}(\rho, z, t) = \int_0^\infty dk\, \tilde{A}(k)J_0(k\rho \sin\theta_0)\exp[ik(z\gamma - ct)] \quad (7.39)$$

we conclude that the experiment reveals the entire spatiotemporal structure of the Bessel X field. The natural price we have to pay for resorting to the correlation measurements is replacing the spectrum $\tilde{A}(k)$ with its autocorrelation, which is a minor issue in the case of transform-limited source pulses. Nevertheless, we cannot claim that we actually detect the field under investigation (see also [27] for a relevant discussion). Indeed, as the absolute phases of the plane-wave components are inevitably lost in any linear interferometric experiments, only the spatial amplitude distribution of the wave field can be detected.

Results of the Experiment We took advantage of the insensitivity of Eq. (7.33) to the source-field phase and used a white light noise from a superhigh-pressure Xe arc lamp instead of a laser as the field source to achieve the $\sim$ 3-fs correlation time in our experiment (V in Fig. 7.9b). The recordings at 70 points on the z-axis (from behind the L3 lens up to a point a few centimeters beyond the origin) with a 0.5-cm step were performed with a cooled CCD camera (EDC–1000TE), which has 2.64×2.64 mm working area containing 192×165 pixels and were processed by a PC as follows.

First, the Bessel X field intensity was subtracted (see Fig. 7.10a). Due to its practically even distribution, the same procedure, was found to be unnecessary with the plane-wave field intensity. To reduce noise and the dimensionality of the data array, the polar angle in every recording was averaged by taking advantage of the axial symmetry of the field. Thus, we got a one-dimensional array containing up to 100 significant elements from every 192×165 matrix recorded. Seventy such arrays formed a matrix, which, having the known symmetry of the real part of Eq. (7.33) in mind, was mirrored in the lateral and axial planes. In Fig. 7.10b the result is compared with the Bessel X field distribution in an axial plane, computed from Eq. (7.39) for a model spectrum. The latter was taken as a convex curve covering the entire visible region from blue to near infrared (up to 0.9 μm) in order to simulate the effective light spectrum in the experiment, which is a product of the Xe arc spectrum and the sensitivity curve of the camera. The central (carrier) frequency was chosen corresponding to a wavelength 0.6 μm, which had been determined from the fringe spacing of an autocorrelation pattern recorded for the light source with the same CCD camera. The left- and right-hand tails of the central X-like structure are more conspicuous in the experimental pattern, due to the unevenness of the real Xe arc spectrum. The different scaling of the horizontal axes of the two panels is in accordance with the z-axis magnification factor of the experimental setup.

Observing the obvious agreement between theoretical and experimental patterns, we arrive at the conclusion that we have really recorded the characteristic spatiotemporal profile of an optical realization of a nonspreading axisymmetric Bessel X field.

7.6.3 Experiment on Optical LWs

Three- and Two-Dimensional LWs: The Mathematics of the Experiment To prove the feasibility of the approach for optical generation of apertured LWs described in Eq. (7.8) we implement the setup depicted in Fig. 7.5b and show that the wave field generated indeed behaves as a LW in interferometric experiments. However, the task can be simplified as follows without a loss of generality.

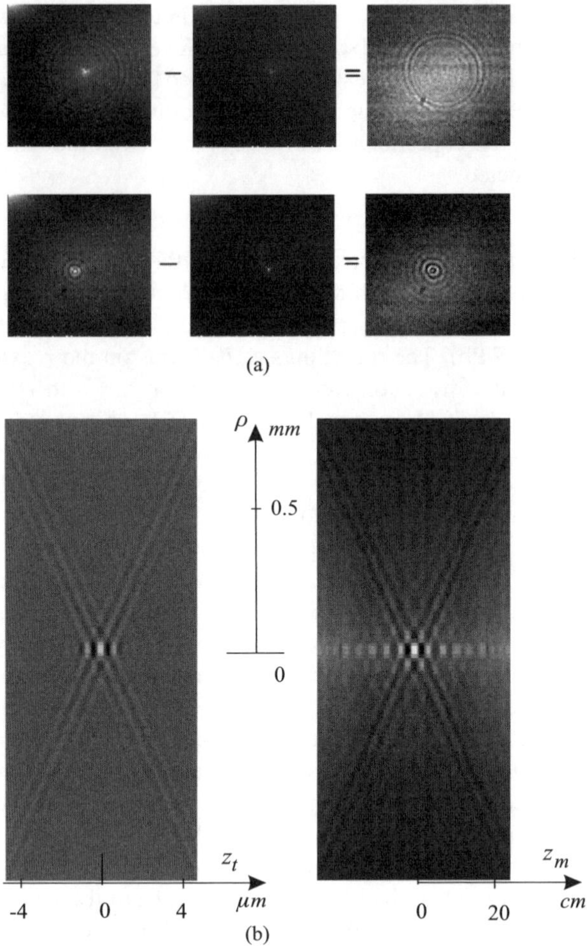

FIGURE 7.10 (a) Samples of the experimental recordings and processing of the intensity distributions measured at positions A and B along the z-axis, as shown in Fig. 7.9b. Left column, total interference pattern of the cross-correlated fields; middle column, lateral intensity distribution of the Bessel X field alone. In the right column gray (about 50%) corresponds to zero level and dark corresponds values. (b) Comparison of the result of the experiment (right panel) with a computer-simulated Bessel X-pulse field.

From the discussions above it is quite clear that all the defining properties of LWs—their propagation invariance, characteristic field distribution, and predetermined group velocity—can be studied in terms of a specific pair of interfering tilted pulses, two-dimensional (2D) LWs [37]:

$$F_{2D}(x, y, z, t) = \int_0^\infty dk\, \tilde{A}(k) \cos[kx \sin\theta_{LW}(k)] \exp[ik(z\cos\theta_{LW}(k) - ct)]. \quad (7.40)$$

In other words, we have shown that the peculiar propagation of LWs is assured exclusively by the specific coupling between the wave number and the direction of propagation of plane-wave components of the LWs as defined by the function $\theta_{\rm LW}(k)$ in Eq. (7.5).

The mutual coherence function for 2D LWs can be deduced easily from the function for three-dimensional (3D) LWs (7.34). In complete analogy, in (7.40) we replace the Bessel function $J_0(\cdot)$ by $\cos(\cdot)$ in Eq. (7.34), choose appropriate coordinates, and obtain [37]

$$\Gamma_{2\rm D}\,(x_1, x_2, \Delta z, \gamma\Delta z - c\tau) = \exp(-i\beta\gamma\Delta z)\int_0^\infty dk\ |\mathcal{V}(k)|^2 \\ \times\ \cos(kx_1\sin\theta_{\rm LW}(k))\cos(kx_2\sin\theta_F(k))\exp[ik(\gamma\Delta z - c\tau)]. \quad (7.41)$$

The mathematical description of experiments with 2D LWs is analogous to that with 3D LWs. In the spatial intensity distribution of the interferometric experiment,

$$\langle V_\Sigma^* V_\Sigma\rangle = \langle V_P^* V_P\rangle + \langle F_{2\rm D}^* F_{2\rm D}\rangle + 2\,{\rm Re}\langle F_{2\rm D}^* V_P\rangle, \quad (7.42)$$

we have for the intensity of the 2D LW,

$$\langle F_{2\rm D}^* F_{2\rm D}\rangle = \int_0^\infty dk\mathcal{S}(k)\cos^2(kx\sin\theta_G(k)), \quad (7.43)$$

where the angular function $\theta_G(k)$ [see Eq. (7.12)] is determined by the specific setup, and $\mathcal{S}(k)$ is the spectral density of the light source. For the third term, instead of Eq. (7.30), we have

$$\langle F_{2\rm D}^* V_P\rangle = \exp(2\beta\gamma z)\int_0^\infty dk\ \mathcal{S}(k)\upsilon_F^*(k_1)\upsilon_P(k) \\ \times\ \cos(k\rho\sin\theta_G(k))\exp\{ik[z(1-\gamma) - c\Delta t]\}. \quad (7.44)$$

In (7.44) for LWs we have to set $\gamma = 1$, so that

$$\langle F_{2\rm D}^* V_P\rangle = \exp(2\beta z)\int_0^\infty dk\mathcal{S}(k)\upsilon_F^*(k_1)\upsilon_P(k)\cos(k\rho\sin\theta_G(k))\exp(-ikc\Delta t). \quad (7.45)$$

In analogy with the case of Bessel X fields, we can define in (7.41), $\Delta z = 0$, $\mathbf{r}_{\perp 2} = 0$, so that the mutual coherence functions of the 2D LW reads

$$\Gamma_{2\rm D}(x, 0, 0, -c\tau) = \int_0^\infty dk\ |\mathcal{V}(k)|^2\cos(kx\sin\theta_G(k))\exp(-ikc\tau). \quad (7.46)$$

Thus, if we record the interference pattern at $z = 0$, a comparison of Eqs. (7.46) and (7.44) yields

$$\langle F_{2\rm D}^* V_P\rangle_{\Delta t} = \Gamma_{2\rm D}(x, 0, 0, -c\Delta t), \quad (7.47)$$

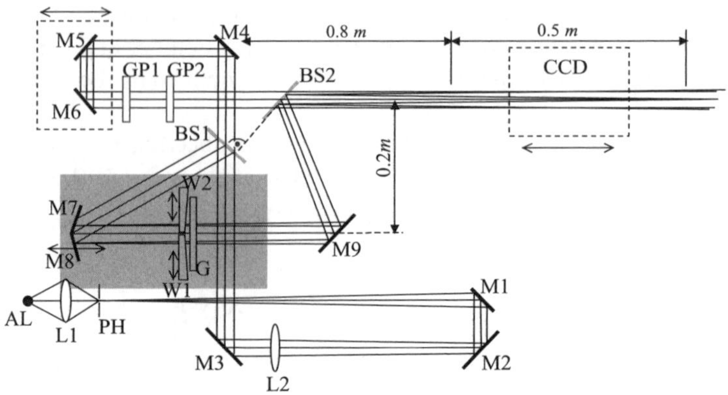

FIGURE 7.11 Experimental setup for generating 2D LWs and recording its interference with plane-wave pulses. The FWM generator can be seen in the gray area; M's, mirrors; L's, lenses; BS's, beamsplitters; W's, wedges; G, diffractional grating; AL, Xe arc lamp; PH, pinhole; GP's, compensating glass plates.

and we can conclude that the mutual coherence function of the 2D LWs can be studied by recording the intensity of the interference picture as the function of the delay Δt between the signal and reference fields.

Note that Eq. (7.45) can be given the form

$$\langle F_{2D}^* V_P \rangle = \int_0^{\infty} dk \mathcal{S}(k) \cos(kx \sin\theta_G(k)) \exp[ikz\cos\theta_G(k) - ikc(t_0 - \Delta t)], \quad (7.48)$$

where $t_0 = z/c$ and the constant has the interpretation of being the time that a wave field propagating at group velocity c travels the distance z to the plane of measurement. The integral expression (7.48) is very similar to (7.40), the one describing the field of the 2D LWs, the only difference being that the theoretical angular function $\theta_{LW}(k)$ is replaced by the $\theta_G(k)$ and the frequency spectrum is replaced by the power spectrum $\mathcal{S}(k)$ in (7.48). Again, as the absolute phases of the plane-wave components are inevitably lost in any linear interferometric experiments, we can detect only the amplitude distribution of the wave field. However, interferograms *do* carry information about the most essential, defining characteristic of the LWs: their *support* of the angular spectrum of plane waves. Indeed, the general structure of the interference patterns in Eq. (7.48) is determined primarily by the angular function $\theta_G(k)$, and it resembles the corresponding transform-limited wave field only if $\theta_G(k) = \theta_{LW}(k)$ in good approximation over the entire bandwidth of the field.

Setup The conversion of the LW generator in Fig. 7.5b to the two-dimensional case is straightforward; we simply replace the axicon and circular diffraction by their one-dimensional counterparts: prisms (wedges) and a diffraction grating. The initial field on the elements is an interfering pair of pulsed plane waves.

The setup of our experiment is depicted in Fig. 7.11. The main part of it is the (2D) LW generator, which consists of the mirrors M7 and M8, of the two wedges W1 and

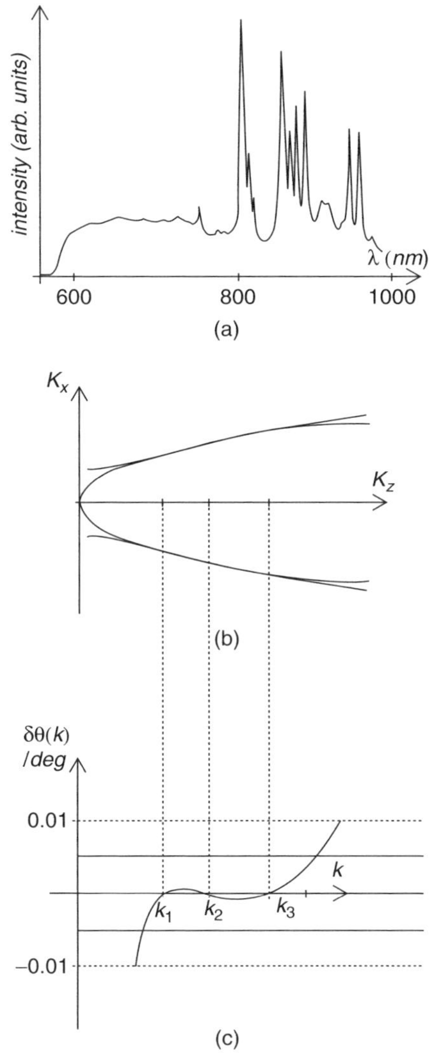

FIGURE 7.12 (a) Power spectrum of the light used in our setup. (b) Angular spectrum of plane waves generated in the setup (solid black line) compared to the theoretical one (dotted gray line). (c) Deviation of support generated, here $k_1 = 7.4 \times 10^6$ rad/m ($\lambda = 849$ nm), $k_2 = 10.6 \times 10^7$rad/m ($\lambda = 593$ nm), $k_3 = 1.6 \times 10^7$rad/m ($\lambda = 381$ nm).

W2, and of a blazed diffraction grating G (see the gray area in Fig. 7.11). The LW generator is placed into an arm of an interferometer, as explained below.

In our experiment we implemented a 2D LW with the following parameters: $\beta = 40\,\text{rad/m}$, $\gamma = 1$ ($v_g = c$) giving $\theta_{\text{LW}}(k_0) \approx 0.23^\circ$ if $k_0 = 7.8 \times 10^6$ rad/m ($\lambda_0 \approx 800$ nm) (see Fig. 7.12a for the spectral density of the light source and Fig. 7.12b for the support of the angular spectrum of plane waves of the specified 2D LW).

The LW generator has three free parameters: α, the apex angle of the wedges; θ_0 the angle that the initial pulsed plane waves subtend with the optical axis of the setup; and d, the groove spacing of the diffraction grating. As to finding the values for the parameters that give the best fit between the support of the angular spectrum of plane waves and the support of the angular spectrum of plane waves of the theoretical LWs in Eq. (7.5),

$$\theta_{\rm LW}(k) = \arccos \frac{k - 2\beta}{k} \tag{7.49}$$

($\gamma = 1$), we combine Eq. (7.49) with Eq. (7.13) and write the following system of equations:

$$\arccos \frac{\gamma(k_m - 2\beta)}{k_m} = \frac{2\pi}{k_m d} + \alpha(1 - n(k_m)) + \theta_0, \quad m = 1, 2, 3. \tag{7.50}$$

We specify the three wave numbers as $k_1 = 7.4 \times 10^6$ rad/m ($\lambda = 849$ nm), $k_2 = 10.6 \times 10^7$rad/m ($\lambda = 593$ nm), $k_3 = 1.6 \times 10^7$rad/m ($\lambda = 381$ nm) and assumed that the wedges are made of the optical glass BK7, for which the refractive index $n(k)$ is known to an accuracy better than 10^{-5}. The system (7.50) then yields

$$\begin{aligned} \alpha &= 1.2044 \times 10^{-2}\,\text{rad} \\ d &= 3.7494 \times 10^{-4}\,\text{m} \\ \theta_0 &= 9.4683 \times 10^{-3}\,\text{rad}\,. \end{aligned} \tag{7.51}$$

As inserted into Eq. (7.12), the maximum deviation $\delta\theta(k) = \theta_{\rm LW}(k) - \theta_G(k)$ for the wavelength dispersion of the cone angle in the spectral range selected is as small as 5×10^{-4} degree, (i.e., $< 0.2\%$). A comparison of the corresponding supports of angular spectrum and the exact form of deviation $\delta\theta(k)$ is depicted in Fig. 7.12b and c. As for rough estimation of the spread of the pulse due to $\delta\theta(k)$, one can estimate the corresponding maximum group velocity dispersion Δv_g and compare this with the mean wavelength of the pulse. The numerical simulations show that the approximation is indeed good enough to propagate the central peak of LWs over several meters.

The LW generator has been placed in what is basically a specially designed and modified Mach–Zehnder interferometer. The interferometer consists of two beamsplitters and of identical broadband nondispersive mirrors. The input field from the light source is split by the beamsplitter BS1 into the fields that travel through the two arms of the interferometer, the one with the LW generator and the arm for the reference beam. Mirrors M5 and M6 form a delay line and they were translated by a *Burleigh Inchworm* linear step motor, the 1-μm translation step of which was reduced to 65 nm by a transmission mechanism. Mirror M7 was translatable continuously to correct for the time shift between the two tilted pulses. Wedges W1 and W2 were transversally translatable to balance the material dispersion they introduce to the plane-wave pulses (see the text below). We used a Kodak Megaplus 1.6i CCD camera with 1534 × 1024 matrix resolution and 10-bit pixel depth. The linear dimensions of the matrix are 13.8 mm (H) × 9.2 mm (V), the pixel size is 9 μm × 9 μm.

Again, in our experiment we used filtered light from a superhigh-pressure Xe arc lamp, giving a ≈ 5 fs correlation time for the input field (see Fig. 7.12a for the power spectrum of the light). To ensure good transversal coherence over the clear aperture of the setup, the required maximum diameter ≈ 15 μm of the pinhole and focal length 2 m of the collimating Fourier lens L1 were estimated from the van Cittert–Zernike theorem [50] for the mean wavelength of the light, $\lambda_0 = 800$ nm. As a result of filtering, the total power of the signal on the ≈ 1.5 cm^2 CCD chip was very low, approximately 0.03 μW.

Due to the short coherence time of the source field, the experiment is highly sensitive to the phase distortions (spectral phase shift) introduced by the dispersive optical elements of the system: the beamsplitters and the LW generator. In the LW generator there are three possible sources of undesirable dispersion: (1) the propagation in the glass substrate of the diffraction grating, (2) the propagation between the grating and the axicon where the support of angular spectrum of the wave field is not appropriate for the free-space propagation [i.e., it does not obey the Eq. (7.5)], and (3) the propagation in the wedges. The beamsplitters in our setup are identical, and if we set them perpendicularly and orient the coated sides so that each beam passes the glass substrate of the beamsplitter twice, the arms of the interferometer remain balanced. The influence of the propagation between the elements can be made neglible by placing them close to each other. The character of the undesirable dispersion in the wedges can be estimated from the following considerations. The entrance wave field on the wedges is the transform-limited Bessel X-pulse, so the on-axis part of the pulse is also transform-limited and should pass through the wedges unchanged (i.e., without any additional spectral phase shift). Consequently, the wedges should be produced and aligned so that their thickness is zero on the optical axis. As the apex angle of the wedges is very small in our setup (≈ 0.7°), this is not a very practical approach and we consider, instead, the finite thickness on the axis as the source of additional spectral phase shift. Thus, the composite spectral phase shift of the LW generator can be described as the phase distortion introduced by the substrate of the diffraction grating and by a glass plate of the material of the wedges, the thickness of which is equal to the thickness of the wedges on the optical axis. We balanced the arms of the interferometer by inserting material dispersion into the reference arm of the setup by means of two appropriate glass plates (GP1 and GP2 in Fig. 7.11a).

Results of the Experiment In the first experiment we recorded the time-averaged interference pattern of the 2D LW and the reference wave field as the function of the time delay between the two. The experiment can be modeled mathematically by varying parameter Δt in Eq. (7.45). We scanned the time delay at three z-axis positions: $z = 0$ cm, $z = 25$ cm, $z = 50$ cm (the origin of the z-axis is about 30 cm

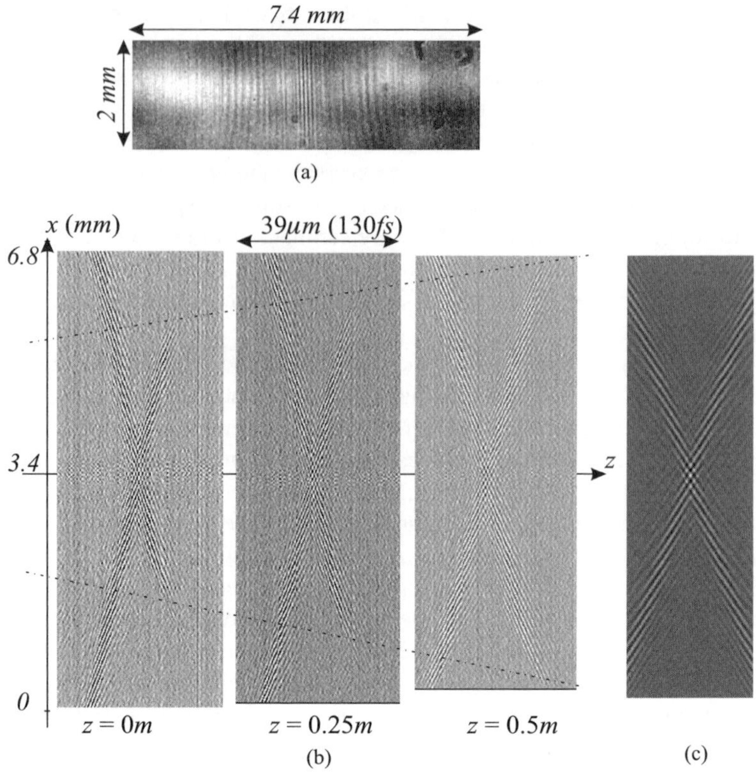

FIGURE 7.13 (a) Typical interferogram in the setup as recorded by the CCD camera. (b) Interference pattern in the setup as a function of the delay between the signal and reference wave fields in three positions of the CCD camera (see the text for more detailed description). (c) Qualitative theoretical simulation of the experiment.

away from the beamsplitter BS2 in Fig. 7.11a). In each experiment we recorded 300 interferograms; the time-delay step was 0.43 fs (0.13 μm).

In a typical interference pattern in our experiment (see Fig. 7.13a) the sharp vertical interference fringes in the center correspond to the second term in the interference sum (7.42); this is the time-averaged propagation-invariant intensity of the 2D LW. The fringes can also be interpreted as the autocorrelation function of the interfering tilted pulses (see Fig. 7.12a for the corresponding power spectrum). In this experiment the intensity of the wave field under study does not carry any important information, so we subtracted it numerically from the results in Fig. 7.13b. The interference fringes that are symmetrical at both sides of the central part correspond to the third, most important term in this sum. It can be seen from Fig. 7.13a that due to the low signal level, the recorded interferograms are quite noisy. To get a better signal-to-noise ratio, we averaged the data in the interferograms over the rows and used the resulting one-dimensional data arrays instead.

The results of the experiment are depicted in Fig. 7.13b. We can see that there is a good qualitative resemblance between the measured $x\Delta t$ plot of the interference pattern and the theoretical field distribution of the 2D LWs in Fig. 7.13c, as predicted by Eq. (7.48). One can clearly recognize the two interfering tilted pulses forming the characteristic X-branching; the phase fronts in the tilted pulses can also be seen. The wave field is definitely transform limited, so we have managed to compensate for the spectral phase shift in the 2D LW generator. We can also see that the interference pattern does not show any spread over the 0.5 m distance; consequently, the wave field does not spread in the course of propagation.

An additional detail can be found in Fig. 7.13b: The tilted pulses do not extend across the entire picture but are cut out (see the dashed lines in Fig. 7.13b). Also, the edges of the tilted pulses move away from the optical axis. This effect can clearly be interpreted as the consequence of the finite extent of the tilted pulses, as illustrated in Fig. 7.4. The dashed lines simply mark the borders of the volume, where the two tilted pulses intersect (i.e., the borders of the volume, where the 2D LWs exist).

In the second experiment we recorded the interference pattern as the function of the propagation distance z. The experiment can be simulated by varying the z coordinate in Eq. (7.45). We recorded 240 interferograms, and the step of the CCD camera position was 3.1 mm. The numerical simulation of the experiment and the results of the experiment are depicted in Fig. 7.14b and c, respectively.

The experiment can be interpreted easily: The position-invariant envelope of the interference pattern is the consequence of the fact that the group velocities of the propagation-invariant 2D LW and the reference field are equal, c, so that the overlapping volume of the two fields do not change in the course of propagation (see Fig. 7.14a). The z-dependent finer structure of the interferograms is the consequence of the fact that the phase velocities of the plane-wave pulse and 2D LW are not equal (i.e., we have also $v^g \neq v^p$ for the phase and group velocities of the 2D LW). The results of the experiment in Fig. 7.14c show good qualitative agreement with the theory.

We can also determine the parameter β from our experiment – the exponent multiplier in Eq. (7.45) reads $\exp(i2\beta z)$; thus, $\beta = \pi/z_0$, where z_0 is the period of the variations along the z-axis. From the result in Fig. 7.14a we estimated $z_0 \approx 7.5$ cm, so that $\beta \approx 42$ rad/m, a result in good agreement with the theory.

Thus, we have shown that the generated wave field has all the characteristic properties described in previous theoretical sections, and the validity of the general idea has been given an experimental proof.

7.7 CONCLUSIONS

The spatial and temporal localization makes the implementation of LW solutions very attractive for applications where the lateral and transversal spread of optical wave fields is a major limitation of the performance (e.g., optical communication,

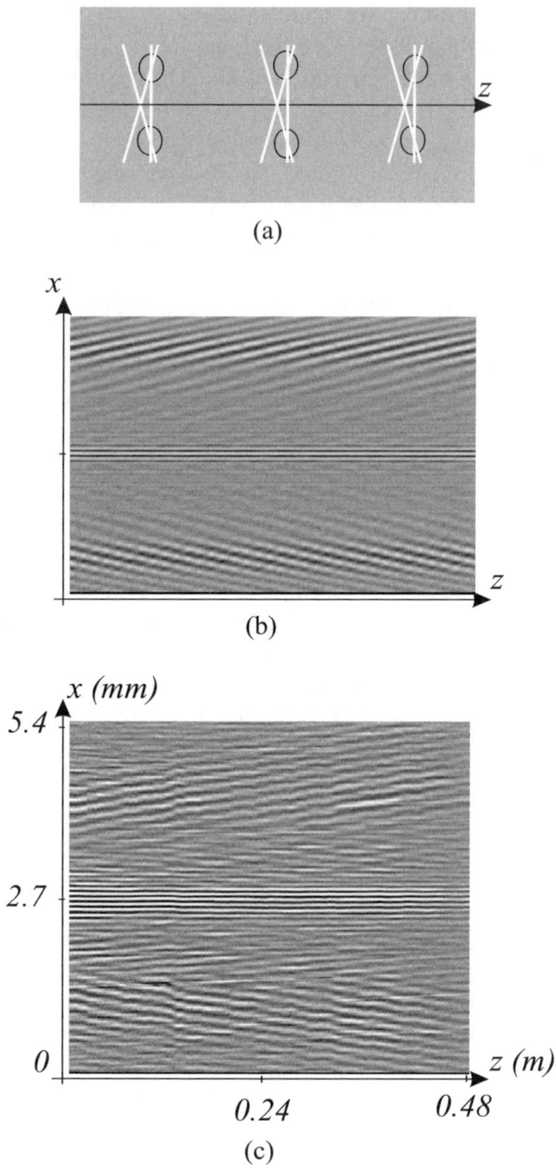

FIGURE 7.14 Interference pattern as a function of the CCD camera position (see the text for a detailed description).

metrology, monitoring, imaging, terahertz and femtosecond spectroscopy, laser particle manipulation and acceleration).

Apart from possible practical applications, the study of LWs has already given some useful lessons concerning fundamental issues of physics. Superluminal movement of X-type LWs caused a lot of discussions, the most important outcome of which was a deeper understanding of what is and what is not in contradiction with the relativistic causality (see [68–71] and references therein). Superficially understood taboo against superluminal movements may even hinder progress in the physics. For example, if Sommerfeld's work of 1904 on conical radiation of a superluminal charge hadn't been forgotten after 1905, the Cherenkov effect would have been discovered and explained much earlier (see [67,68,71] and references therein).Tachyon studies around the 1980s would have benefited considerably from research in LWs if the latter had been developed earlier (see [61] and references therein). Finally, something concerning future developments in quantum optics: A fundamental notion of photon localizability is revised thanks to LW studies [72].

We sincerely hope that our results reviewed here have been helpful for better understanding of the subject and may serve as a sound ground for future experiments and also for practical applications of those fascinating wave fields.

Acknowledgments

This research was supported by the Estonian Science Foundation. We are grateful to all those who initiated and carried out the idea for this book and prepared it for publication.

REFERENCES

1. J. N. Brittingham, *J. Appl. Phys.* **54**, 1179 (1983).
2. R. W. Ziolkowski, *J. Math. Phys.* **26**, 861 (1985).
3. E. Heyman, *IEEE Trans. Antennas Propag.* **AP-37**,1604 (1989).
4. I. M. Besieris, A. M. Shaarawi, and R. W. Ziolkowski, *J. Math. Phys.* **30**, 1254 (1989).
5. R. Donnelly and R. Ziolkowski, *Proc. R. Soc. London A* **440**, 541 (1993).
6. R. Donnelly and R. Ziolkowski, *Proc. R. Soc. London A* **437**, 673 (1992).
7. A. M. Shaarawi, R. W. Ziolkowski, and I. M. Besieris, *J. Math. Phys.* **36**, 5565 (1995).
8. R. W. Ziolkowski, *Phys. Rev. A* **39**, 2005 (1989).
9. P. L. Overfelt, *Phys. Rev. A* **44**, 3941 (1991).
10. P. L. Overfelt, *J. Opt. Soc. Am. A* **18**,1905 (2001).
11. A. M. Shaarawi, I. M. Besieris, and R. W. Ziolkowski, *J. Appl. Phys.* **65**, 805 (1989).
12. R. W. Ziolkowski and D. K. Lewis, *J. Appl. Phys.* **68**, 6083 (1990).
13. R. W. Ziolkowski, D. K. Lewis, and B. D. Cook, *Phys. Rev. Lett.* **62**, 147 (1989).
14. R. W. Ziolkowski, *Phys. Rev. A* **44**, 960 (1991).
15. R. W. Ziolkowski, *IEEE Trans. Antennas. Propag.* **40**, 888 (1992).
16. R. W. Ziolkowski, I. M. Besieris, and A. M. Shaarawi, *J. Opt. Soc. Am. A* **10**, 75 (1993).
17. A. M. Shaarawi, I. M. Besieris, R. W. Ziolkowski, and S. M. Sedky, *J. Opt. Soc. Am. A* **12**, 1954 (1995).
18. A. M. Shaarawi, S. M. Sedky, R. W. Ziolkowski, and F. M. M. Taiel, *J. Opt. Soc. Am. A* **13**, 1712 (1996).

19. S. M. Sedky, A. M. Shaarawi, I. M. Besieris, and F. M. M. Taiel, *J. Opt. Soc. Am. A* **13**, 1719 (1996).
20. A. M. Shaarawi, S. M. Sedky, F. M. M. Taiel, R. W. Ziolkowski, and I. M. Besieris, *J. Opt. Soc. Am. A* **13**, 1827 (1996).
21. A. M. Shaarawi, *J. Opt. Soc. Am. A* **14**, 1804 (1997).
22. P. L. Overfelt, *J. Opt. Soc. Am. A* **14**, 1087 (1997).
23. P. L. Overfelt, *J. Opt. Soc. Am. A* **16**, 2239 (1999).
24. V. V. Borisov and A. P. Kiselev, *Appl. Math. Lett.* **13**, 83 (2000).
25. H. Sõnajalg and P. Saari, *Opt. Lett.* **21**, 1162 (1996).
26. H. Sõnajalg, M. Rätsep, and P. Saari, *Opt. Lett.* **22**, 310 (1997).
27. P. Saari and K. Reivelt, *Phys. Rev. Lett.* **79**, 4135 (1997).
28. K. Reivelt and P. Saari, *J. Opt. Soc. Am. A* **17**, 1785 (2000).
29. K. Reivelt and P. Saari, *Phys. Rev. E* **65**, 046622 (2002).
30. K. Reivelt, *Opt. Express* **10**, 360 (2002).
31. K. Reivelt and P. Saari, *Phys. Rev. E* **66**, 056611 (2002).
32. K. Reivelt and P. Saari, *Opt. Lett.* **29**, 1176 (2004).
33. I. Alexeev, K. Y. Kim, and H. M. Milchberg, *Phys. Rev. Lett.* **88**, 073901 (2002).
34. R. Grunwald, V. Kebbel, U. Griebner, U. Neumann, A. Kummrow, M. Rini, E. T. J. Nibbering, M. Piché, G. Rousseau, and M. Fortin, *Phys. Rev. A* **67**, 063820 (2003).
35. P. Di Trapani, G. Valiulis, A. Piskarskas, O. Jedrkiewicz, J. Trull, C. Conti, and S. Trillo, *Phys. Rev. Lett.* **91**, 093904 (2003).
36. R. Butkus, S. Orlov, A. Piskarskas, V. Smilgevicius, and A. Stabinis, *Opt. Commun.* **244**, 411–421 (2005).
37. K. Reivelt, and P. Saari, arXiv.org e-Print archive, physics/0309079, 2003; http://arxiv.org/abs/physics/0309079.
38. I. Besieris, M. Abdel-Rahman, A. Shaarawi, and A. Chatzipetros, *Prog. Electromagn. Res.* **19**, 1 (1998).
39. P. Saari and K. Reivelt, *Phys. Rev. E* **69**, 036612 (2004).
40. M. Zamboni-Rached, A. M. Shaarawi, and E. Recami, *J. Opt. Soc. Am. A* **21**, 1564 (2004).
41. J. Salo and M. M Salomaa, *J. Opt. A Pure. Appl. Opt.* **3**, 366 (2001).
42. A. T. Friberg, *J. Opt. Soc. Am. A* **12**, 743 (1996).
43. R. Piestun, Y. Y. Schechner, and J. Shamir, *J. Opt. Soc. Am. A* **17**, 294 (2000).
44. M. A. Porras, R. Borghi, and M. Santarsiero, *Phys. Rev. E* **62**, 5729 (2000).
45. C. J. R. Sheppard, *J. Opt. Soc. Am. A* **18**, 1 (2001).
46. P. Saari, *Opt. Express* **8**, 590 (2001).
47. P. Saari, *Laser Phys.* **12**, 219 (2002).
48. J. Fagerholm, A. T. Friberg, J. Huttunen, D. Morgan, and M. M. Salomaa, *Phys. Rev. E* **54**, 4347 (1996).
49. R. M. Hermann and T. A Wiggins, *J. Opt. Soc. Am. A* **8**, 932 (1991).
50. L. Mandel and E. Wolf, *Optical Coherence and Quantum Optics*, Cambridge University Press, Cambridge, 1995.
51. A. Vasara, J. Turunen, and A. T. Friberg, *J. Opt. Soc. Am. A* **6**, 1748 (1989).
52. J. Durnin, J. J. Miceli, Jr., and J. H. Eberly, *Phys. Rev. Lett.* **58**, 1499 (1987).

53. J. H. McLeod, *J. Opt. Soc. Am.* **44**, 592 (1954).
54. H. S. Lee, B. W. Stewart, D. Will, and H. Fenichel, *Appl. Phys. Lett.* **59**, 3096 (1991).
55. V. P. Koronkevich, I. A. Mikhaltsova, E. G. Churin, and Y. I. Yurlov, *Appl. Opt.* **34**, 5761 (1995).
56. S. Chávez-Cerda, G. S. McDonald, and G. H. C. New, *Opt. Commun.* **123** , 225 (1996).
57. R. Borghi, M. Santarsiero, and F. Gori, *J. Opt. Soc. Am. A* **14**, 23 (1997).
58. Z. Bouchal and J. Wagner, *Opt. Commun.* **176**, 299 (2000).
59. B. Hafizi and P. Sprangle, *J. Opt. Soc. Am. A* **8**, 705 (1991).
60. J.-Y. Lu and J. G. Greenleaf, *IEEE Trans. Ultrason. Ferroelectr. Freq. Control* **39**, 19 (1992).
61. E. Recami, *Physica A* **252**, 586 (1998).
62. J.-Y. Lu and S. He, *Opt. Commun.* **161**, 187 (1999).
63. J. Salo, J. Fagerholm, A. T. Friberg, and M. M. Salomaa, *Phys. Rev. E* **62**, 4261 (2000).
64. A. M. Shaarawi, B. H. Tawfik, and I. M. Besieris, *Phys. Rev. E* **66**, 046626 (2002).
65. P. Saari, in O. Svelto, S. De Silvestri, and G. Denardo, Eds., *Ultrafast Processes in Spectroscopy*, Plenum Press, New York, 1996 pp. 151–156.
66. P. Saari and H. Sõnajalg, *Laser Phys.* **7**, 32 (1997).
67. P. Saari, in T. Elsaesser, J. G. Fujimoto, D. A. Wiersma, and W. Zinth, Eds., *Ultrafast Phenomena XI*, Springer-Verlag, New York, 1998, pp. 121–123.
68. P. Saari, in *Time's Arrows, Quantum Measurement and Superluminal Behavior*, Scientific Monographs: Physics Sciences Series, Italian CNR Publishing, Rome, 2001, p. 37; see also http://xxx.lanl.gov/abs/physics/ 0103054.
69. H. Ringermacher and L. R. Mead, *Phys. Rev. Lett.* **87**, 059402 (2001).
70. W. A. Rodrigues, Jr., D. S. Thober, and A. L. Xavier, Jr., *Phys. Lett.* **A 284**, 217 (2001).
71. P. Saari, in *Ultrafast Photonics*, Institute of Physics Publishing, Philadelphia, PA, 2004, pp. 317–340.
72. P. Saari, M. Menert, and H. Valtna, *Opt. Commun.* **246**, 445 (2005).

CHAPTER EIGHT

Optical Wave Modes: Localized and Propagation-Invariant Wave Packets in Optically Transparent Dispersive Media

MIGUEL A. PORRAS
Universidad Politécnica de Madrid, Madrid, Spain

PAOLO DI TRAPANI
Università degli studi dell' Insubria, Como, Italy

WEI HU
South China Normal University, Guangzhou, P. R. China

8.1 INTRODUCTION

Many applications of beams and pulses of light in modern optical technology, such as in long-distance communications,and laser micromachining, deep-field microscopy, plasma-channel generation, and laser writing of waveguides, gratings, and photonic crystals (to cite only a few), will benefit heavily from the use of *particle-like waves*. This term refers to waves that are capable of defeating diffraction spreading and dispersion broadening induced by the host material medium, while maintaining their spatiotemporal localization along sufficiently large, ideally infinite propagation distances.

Localized Waves, Edited by Hugo E. Hernández-Figueroa, Michel Zamboni-Rached, and Erasmo Recami

In the field of nonlinear optics, the achievement of localized and stationary propagation relies on the development of spatiotemporal *solitary waves* or *solitons* (also called "light bullets"), in which linear spreading (diffraction and material dispersion) is balanced by the compression supplied by the nonlinear response of the material. Although one-dimensional (temporal) solitary propagation has been achieved and has found broad technological applications in fiber communications, extension to the three-dimensional case has proved elusive to date, due to intrinsic soliton instabilities.

In linear optics, instead, the road toward the light bullet has followed the course of source design and spatiotemporal wave conformation by optical elements. Interest in the linear approach grew considerably from the introduction, or revival [1], of the Bessel or axicon beam [2,3], a (weakly) localized Bessel-shaped interference pattern that propagates without any diffraction change and from its generation in practical settings (see, e.g., [2,4–6]). Ideally, the Bessel beam is generated by the superposition of monochromatic plane waves with wave vectors distributed over the surface of a cone. Many generalizations and modifications of Bessel beams have been introduced, as Bessel beams of higher order, vectorial Bessel beams, apodized (finite power) Bessel–Gauss beams [7], partially coherent conical beams [8], Hankel beams [9], and the intriguing property of self-reconstruction of Bessel beams beyond obstacles [10] have found unexpected application as optical tweezers for microscopic particle trapping and manipulation [11].

X-waves, introduced first in acoustics [12,13] and later demonstrated in optics [14] with the name Bessel X-waves, represent the generalization of Bessel beams to the nonmonochromatic case (see [15] for a review), and hence the first type of truly spatiotemporal localized and stationary wave packet. At the same time, focus wave modes, introduced by different means [16,17] as a new type of spatiotemporal wave localization, were later demonstrated to belong also to the polychromatic Bessel beam family [18] and were subsequently demonstrated in real setups [19].

Whereas for X-waves all monochromatic Bessel beam constituents have the same cone angle, focus wave modes present inherently cone-angle dispersion, that is, dependence of cone angle with frequency. This difference is essential for further generalization of polychromatic Bessel beams for the achievement of localization and stationarity in a dispersive material. In free-space or dispersion-less materials, cone-angle dispersion is not needed for stationarity but is essential in dispersive materials. Stationarity requires the introduction of a suitable amount of cone-angle dispersion that balances material dispersion. Localized and stationary propagation in dispersive media was first observed in [20,21] with Bessel X-pulses endowed by a small amount of cone-angle dispersion. The same effect was described further to also be attainable with other cone-angle dispersion configurations, as that in pulsed Bessel beams [22,23], envelope X-waves [24], spatiotemporal Gauss–Laguerre waves in media with anomalous dispersion [25], and to be possible also for broadband subcycle pulses [26].

The scope of this chapter is to collect, characterize, and classify all types of localized optical wave packets that present stationary propagation in a dispersive transparent medium. Throughout this chapter these waves are referred to as *wave modes* [27]. They key tool for the classification is the angular or transversal dispersion curve.

First, we consider narrowband wave modes. These are shown (Section 8.2) to belong to two broad categories: hyperbolic modes, with X-shaped dispersion curve and spatiotemporal structure, if material dispersion is normal, or elliptic modes, with an O-shaped dispersion curve and spatiotemporal form, if material dispersion is anomalous [28]. In Section 8.3 we show that each wave mode can adopt the approximate form of (1) a pulsed Bessel beam (PBB), (2) an envelope focus wave mode (eFWM), or (3) an envelope X (eX) wave in normally dispersive media [envelope O (eO) wave in anomalously dispersive media], according to whether the mode bandwidth makes phase mismatch (PM), group-velocity mismatch (GVM) (with respect to a monochromatic plane wave), or defeated group velocity dispersion (GVD), respectively, to be the dominant mode characteristic on propagation. This classification allows us to understand the spatiotemporal features of wave modes in terms of a few parameters (the characteristic PM, GVM, and GVD lengths), including modes with mixed pulsed Bessel, focus wave mode, and X-like (O-like) structure.

In Section 8.4 we also describe broadband wave modes, whose dispersion curves and spatiotemporal forms are much more complex. The simple, symmetric, X-like, and O-like dispersion curves are actually approximations in narrow spectral bands to strongly asymmetric dispersion curves. Dispersion curves in broad spectral regions present asymmetric X or O-like forms, "fishlike" structure (formed by a closed loop plus a tail), or a single tail.

Our description is always paraxial. Although conical waves are usually described as nonparaxial waves, the paraxial approach turns out to lead to simpler expressions in terms of parameters linked directly to the physically relevant properties of the mode and the medium. We also use this approach because of its wider use in nonlinear optics, where unexpectedly, wave modes are finding wide application for understanding nonlinear phenomena, such as spatiotemporal effects in second harmonic generation [29–31] and collapse and filamentation of light pulses in Kerr media [32–34]. Wave modes appear to act as a kind of attractor of the wave dynamics described by the nonlinear Schrödinger equation, and have in fact been seen to be generated spontaneously in a number of nonlinear optics experiments [35–37], as also seen from numerical simulations [34,38]. For completeness, in Section 8.6 we compare the various paraxial and nonparaxial approaches.

8.2 LOCALIZED AND STATIONARITY WAVE MODES WITHIN THE SVEA

We start by considering the propagation of a three-dimensional wave packet $E(x, y, z, t) = A(x, y, z, t)\exp(-i\omega_0 t + ik_0 z)$ of a certain optical carrier frequency ω_0, subject to the effects of diffraction and dispersion of the material medium. Within the SVEA, and up to second order in dispersion, propagation of the wave packet is ruled by the equation

$$\partial_z A = \frac{i}{2k_0}\Delta_\perp A - i\frac{k_0''}{2}\partial_\tau^2 A, \qquad (8.1)$$

where z is the propagation direction, $\tau = t - k_0' z$ is the local time, $\Delta_\perp \equiv \partial_x^2 + \partial_y^2$, and $k_0^{(i)} \equiv \partial_\omega^{(i)} k(\omega)|_{\omega_0}$, with $k(\omega)$ the propagation constant in the medium. Equation (8.1) is valid for a narrow envelope spectrum $\hat{A}(\boldsymbol{x}_\perp, z, \Omega)$ around $\Omega \equiv \omega - \omega_0 = 0$, that is, for bandwidths

$$\Delta\Omega \ll \omega_0. \tag{8.2}$$

Wave modes are stationary and localized solutions of Eq. (8.1) in the wide sense that the *intensity* does not depend on z in a reference frame moving at *some velocity*. These solutions must then be of the form

$$A(x, y, \tau, z) = \Phi(x, y, \tau + \alpha z) \exp(-i\beta z). \tag{8.3}$$

The SVEA condition that $|\partial_z A| \ll k_0 |A|$ requires the free parameters α and β to be small, in the sense that

$$|\alpha| \ll k_0', \tag{8.4}$$
$$|\beta| \ll k_0, \tag{8.5}$$

so that the group velocity $1/(k_0' - \alpha)$ of the wave mode differs slightly from that of a plane pulse of the same carrier frequency in the same material, $1/k_0'$, and so does the phase velocity.

Wave modes may also be introduced in the spatiotemporal frequency domain as localized waves for which the axial component k_z of the wave vector of the various monochromatic and plane-wave constituents varies linearly with frequency. This definition in Fourier space is equivalent to the ansatz (8.3) in space and time: Writing Eq. (8.3) as

$$A(x, y, \tau, z) = \frac{1}{(2\pi)^3} \int_{-\infty}^{\infty} d\Omega\, dk_x\, dk_y \hat{\Phi}(k_x, k_y, \Omega) \exp[i(k_x x + k_y y)] \\ \times \exp\{-i\Omega[t - (k_0' - \alpha)z]\} \exp(-i\beta z)\,, \tag{8.6}$$

where $\hat{\Phi}$ is the spatiotemporal Fourier transform of Φ, and because $E = A \exp(-i\omega_0 t + k_0 z)$, it follows that the axial component of the wave vector of any monochromatic and plane-wave constituent is given by the linear relation

$$k_z(\Omega) = (k_0 - \beta) + (k_0' - \alpha)\Omega\,. \tag{8.7}$$

Equation (8.1) with the wave-mode ansatz (8.3) yields

$$\Delta_\perp \Phi - k_0 k_0''\, \partial_\tau^2 \Phi + 2ik_0\alpha\, \partial_\tau \Phi + 2k_0 \beta \Phi = 0 \tag{8.8}$$

for the reduced envelope Φ, or for its spatiotemporal spectrum $\hat{\Phi}$,

$$\left[-k_\perp^2 + K^2(\Omega)\right] \hat{\Phi} = 0, \tag{8.9}$$

where $k_\perp = (k_x^2 + k_y^2)^{1/2}$ and where

$$K(\Omega) = \sqrt{2k_0\left(\beta + \alpha\Omega + \frac{1}{2}k_0''\Omega^2\right)}. \tag{8.10}$$

This relation, of utmost importance in this chapter, and will be referred to as the (transversal) dispersion relation or curve of the wave mode since it relates the modulus K of the transversal component of the wave vector with the detuning Ω of each monochromatic wave component from the carrier frequency ω_0. The general solution of Eq. (8.9) is given by

$$\hat{\Phi}(k_x, k_y, \Omega) = (2\pi)^3 \hat{g}(k_x, k_y, \Omega)\delta[k_\perp^2 - K^2(\Omega)], \tag{8.11}$$

where $\hat{g}(k_x, k_y, \Omega)$ is a spectral amplitude and $\delta(\cdot)$ is the Diracs delta function.

A single localized wave mode is determined by choices of α, β and of a spectral amplitude $\hat{g}(k_x, k_y, \Omega)$ yielding a localized wave. Choices of the form $\hat{g}(k_x, k_y, \Omega) = \hat{g}(k_\perp, \Omega)$ lead to cylindrical symmetric or conical wave modes, which are of particular relevance in practice. Conical wave modes in space and time can be calculated by inverse spatial Fourier transform of Eq. (8.11) from the expression

$$\begin{aligned} A(r, \tau, z) &= \Phi_{\alpha,\beta}(r, \tau + \alpha z)\exp(-i\beta z) \\ &= \frac{1}{2\pi}\int_{K(\Omega)\,\mathrm{real}} d\Omega\; \hat{f}(\Omega) J_0[K(\Omega)r]\exp[-i\Omega(\tau + \alpha z)]\exp(-i\beta z), \end{aligned} \tag{8.12}$$

where the integral is limited to values of Ω such that $K^2(\Omega)$ is positive [i.e., $K(\Omega)$ is real, as indicated], $J_0(\cdot)$ is the Bessel function of the first kind and zero order, $r = (x^2 + y^2)^{1/2}$, and $\hat{f}(\Omega) = \hat{g}(K(\Omega), \Omega)/2$. In Eq. (8.12), the cylindrical symmetric wave modes exhibit their conical nature. They appear as superpositions of monochromatic Bessel beams whose cone-angle frequency dependence $\theta \simeq K(\Omega)/k(\omega_0 + \Omega)$ is determined by the dispersion relation (8.10). As $\hat{g}$, the function $\hat{f}$ is in practice an arbitrary function of detuning Ω related to the on-axis temporal form of the wave mode.

8.2.1 Dispersion Curves Within the SVEA

Most of the relevant properties of a wave mode are determined by its dispersion relation $K(\Omega)$. As shown in Fig. 8.1, the form of the dispersion curve $K(\Omega)$ reflects the underlying hyperbolic or elliptic geometries of the wave equation (8.1) in the respective cases of propagation in media with normal or anomalous dispersion.

For normal dispersion ($k_0'' > 0$), the dispersion curve is hyperbolic, often referred to as X-like. $K(\Omega)$ is a single-branch vertical hyperbola if $\beta > \alpha^2/2k_0''$ (Fig. 8.1a), and a two-branch horizontal hyperbola if $\beta < \alpha^2/2k_0''$ (Fig. 8.1b).

For anomalous dispersion ($k_0'' < 0$), $K(\Omega)$ takes real values only if $\beta > \alpha^2/2k_0''$, in which case the dispersion curve is an ellipse (Fig. 8.1c), or O-like, for short. It is

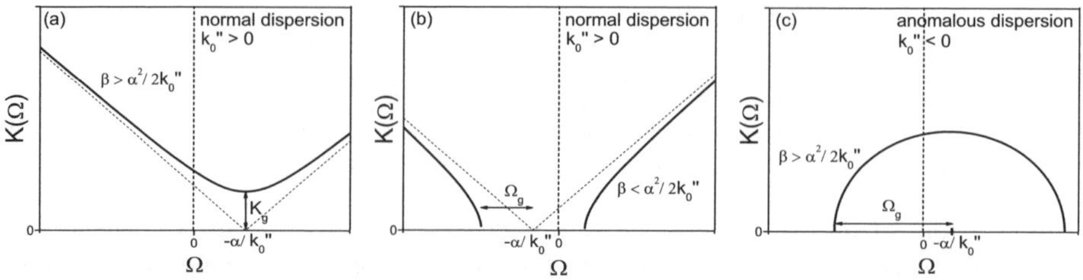

FIGURE 8.1 X- and O-like dispersion curves of wave modes in a medium with (a) normal dispersion and $\beta > \alpha^2/2k_0''$; (b) normal dispersion and $\beta < \alpha^2/2k_0''$; (c) anomalous dispersion and $\beta > \alpha^2/2k_0''$ (exists only in this case).

also convenient to introduce the (real or imaginary) frequency and radial wave vector gaps

$$\Omega_g \equiv \sqrt{\frac{\alpha^2}{k_0''^2} - \frac{2\beta}{k_0''}}, \qquad K_g \equiv \sqrt{-k_0 k_0'' \Omega_g^2}. \tag{8.13}$$

When Ω_g and K_g are real, they represent actual frequency and radial wave vector gaps in the dispersion curve $K(\Omega)$, as illustrated in Fig. 8.1. In any case, their moduli characterize the scales of variation of the frequency and radial wave vector in the dispersion curves.

8.2.2 Impulse-Response Wave Modes

Closely connected with the dispersion curves are the *impulse response* wave modes $\Phi^{(i)}_{\alpha,\beta}(r, \tau + \alpha z)$, or modes with $\hat{f}(\Omega) = 1$. As seen in Fig. 8.2, the structure of $\Phi^{(i)}_{\alpha,\beta}$ in space and time closely resembles that of the dispersion curve in the K–Ω plane, but at radial and temporal scales of variation determined by the reciprocal quantities $|K_g|^{-1}$ and $|\Omega_g|^{-1}$, respectively. Equation (8.12) with $\hat{f}(\Omega) = 1$ and the change $\Omega' = \Omega + \alpha/k_0''$ yields

$$\Phi^{(i)}_{\alpha,\beta}(r, \tau + \alpha z) = \frac{1}{2\pi} \int_{K(\Omega')\,\mathrm{real}} d\Omega'\, J_0\left[K(\Omega')r\right] \exp[-i\Omega'(\tau + \alpha z)] \times \exp\left[i\frac{\alpha}{k_0''}(\tau + \alpha z)\right], \tag{8.14}$$

where

$$K(\Omega') = \sqrt{k k_0'' \left(\Omega'^2 + \frac{2\beta}{k_0''} - \frac{\alpha^2}{k_0''^2}\right)}. \tag{8.15}$$

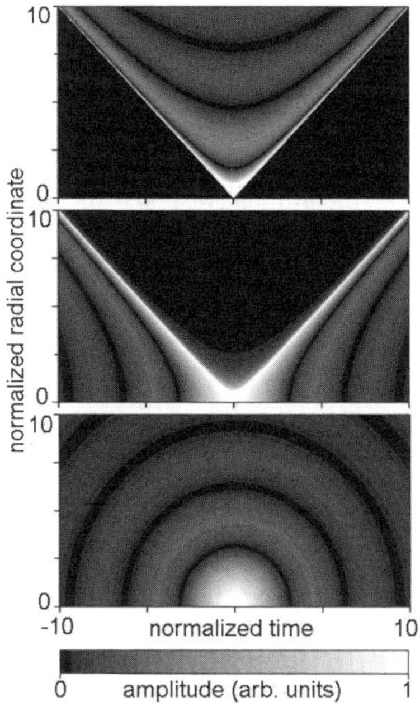

FIGURE 8.2 Contour plot for the amplitude $|\Phi^{(i)}_{\alpha,\beta}|$ of the impulse response wave modes. From top to bottom: normal dispersion ($k_0'' > 0$) with $\beta > \alpha^2/2k_0''$ (Ω_g imaginary, transversal wave vector gap K_g real); normal dispersion ($k_0'' > 0$) with $\beta < \alpha^2/2k_0''$ (frequency gap Ω_g real, K_g imaginary); anomalous dispersion ($k_0'' < 0$) with $\beta > \alpha^2/2k_0''$ (Ω_g and K_g real). Normalized local time and radial coordinate are defined as $|\Omega_g|(\tau + \alpha z)$ and $|K_g|r$, respectively.

The integral in Eq. (8.14) can be performed in all possible cases from formulas 6.677.3 (for $k_0'' > 0, \beta > \alpha^2/2k_0''$), 6.677.2 (for $k_0'' > 0, \beta < \alpha^2/2k_0''$), and 6.677.6 (for $k_0'' < 0$, $\beta > \alpha^2/2k_0''$) of [39], to yield the closed-form expression for impulse response modes:

$$\Phi^{(i)}_{\alpha,\beta}(r, \tau + \alpha z) = \frac{1}{2\pi}\left[\frac{\exp\left\{i\left[\sqrt{2\beta/k_0'' - \alpha^2/k_0''^2}\sqrt{k_0 k_0'' r^2 - (\tau+\alpha z)^2}\right]\right\}}{\sqrt{k_0 k_0'' r^2 - (\tau+\alpha z)^2}} + \text{c.c.}\right] \times \exp\left[\frac{i\alpha}{k_0''}(\tau+\alpha z)\right], \tag{8.16}$$

or in terms of the frequency and radial wave vector gap,

$$\Phi^{(i)}_{\alpha,\beta} = \frac{1}{2\pi}\left[\Omega_g \frac{\exp(iR)}{iR} + \text{c.c.}\right]\exp\left[\frac{i\alpha}{k_0''}(\tau+\alpha z)\right], \tag{8.17}$$

where $R = [K_g^2 r^2 + \Omega_g^2(\tau+\alpha z)^2]^{1/2}$.

As shown in Fig. 8.2 (top), for $k_0'' > 0$ and $\beta > \alpha^2/2k_0''$ (Ω_g imaginary and K_g real), the impulse response, K-gap, X-like wave mode is singular in the cone $r = |(\tau + \alpha z)|/\sqrt{k_0 k_0''}$, is zero for $r < |(\tau + \alpha z)|/\sqrt{k_0 k_0''}$ (within the cone), and decays as $1/r$ for $r > |(\tau + \alpha z)|/\sqrt{k_0 k_0''}$ (out of the cone). The radial beatings in this region, of period $2\pi/K_g$, are a consequence of the radial wave vector gap K_g.

Figure 8.2 (middle) shows the impulse response mode for $k_0'' > 0$ and $\beta < \alpha^2/2k_0''$ (Ω_g real and K_g imaginary). This Ω-gap, X-like wave mode is, as in the previous case, singular at the cone $r = |(\tau + \alpha z)|/\sqrt{k_0 k_0''}$, but damped oscillations are now temporal, of period $2\pi/\Omega_g$, as corresponds to the frequency gap Ω_g in the dispersion curve. Out of the cone $[r > |(\tau + \alpha z)|/\sqrt{k_0 k_0''}]$ the mode is exponentially localized.

Modes in media with anomalous dispersion [i.e., with $k_0'' < 0$ and $\beta > \alpha^2/2k_0''$ (real Ω_g and K_g)] exhibit rather different characteristics [Fig. 8.2 (bottom)]. These modes are no longer singular and of X-type, but regular and, say, O-like. The damped oscillations decay temporally and radially as $1/t$ and $1/r$, respectively, with periods $2\pi/\Omega_g$ and $2\pi/K_g$. The absence of singularities is a consequence of the actual limitation that the elliptic dispersion curve imposes on the uniform spectrum $\hat{f}(\Omega) = 1$.

8.3 CLASSIFICATION OF WAVE MODES OF FINITE BANDWIDTH

Numerical integration of Eq. (8.12) with a given dispersion curve (specified by the values of α, β, and k_0'') but different (bell-shaped) spectral amplitude functions $\hat{f}(\Omega)$ also having different (but finite) bandwidths $\Delta\Omega$ [alternatively, numerical integration of

$$\Phi_{\alpha,\beta}(r, \tau + \alpha z) = \int_{-\infty}^{\infty} d\sigma\ f(\sigma)\Phi^{(i)}_{\alpha,\beta}(r, \tau + \alpha z - \sigma), \tag{8.18}$$

where $f(\tau)$ is the inverse Fourier transform of $\hat{f}(\Omega)$], shows much richer and complex spatiotemporal features than is the case with infinite bandwidth. These features depend strongly on the choice of the spectral bandwidth $\Delta\Omega$, while no essentially new properties arise from the specific choice of $\hat{f}(\Omega)$ (Gaussian, Lorentzial, two-side exponential, etc.). Wave modes with finite bandwidth may exhibit mixed, more-or-less pronounced radial and temporal oscillations, along with incipient or strong X-wave (O-wave), focus wave mode or Bessel structure, as explained throughout this section (see also the figures that follow). The purpose of this section is to perform a simple comprehensive classification of wave modes in dispersive media. In the remainder of this chapter, $\Delta\Omega$ will refer to any suitable definition of the half-width of the bell-shaped spectral amplitude function $\hat{f}(\Omega)$.

Given a wave mode of parameters α and β satisfying conditions (8.4) and (8.5), propagating in a dispersive material with GVD k_0'' and some spectral bandwidth $\Delta\Omega$ satisfying (8.2), we have found it convenient to define the following three characteristic lengths:

1. The mode PM length

$$L_p \equiv \frac{1}{\beta}. \tag{8.19}$$

2. The mode walk-off, or GVM length,

$$L_w \equiv \frac{1}{\alpha \Delta\Omega}, \tag{8.20}$$

measuring, respectively, the axial distances at which the mode becomes phase mismatched and walks off with respect to a plane pulse of the same spectrum in the same medium.

3. The GVD length

$$L_d \equiv \frac{1}{k_0''(\Delta\Omega)^2}, \tag{8.21}$$

or distance at which the mode (invariable) duration differs significantly from that of the (broadening) plane pulse. Note that, as defined, L_p, L_w, and L_d can be positive or negative.

In terms of the mode lengths, the transversal dispersion relation (8.10) takes the form

$$K(\Omega) = \sqrt{2k_0\left(L_p^{-1} + L_w^{-1}\Omega_n + \frac{1}{2}L_d^{-1}\Omega_n^2\right)}, \tag{8.22}$$

where $\Omega_n = \Omega/\Delta\Omega$ is the normalized detuning, which ranges in $[-1, +1]$ for Ω within the bandwidth $\Delta\Omega$. Then they are the values of the mode lengths L_p, L_w, and L_d that determine the form of the dispersion curve within the spectral bandwidth, and hence the parameters that determine the spatiotemporal structure of the mode, as shown throughout this section. We analyze here three extreme cases:

$\lvert L_p\rvert \ll \lvert L_w\rvert, \lvert L_d\rvert$	PM-dominated case
$\lvert L_w\rvert \ll \lvert L_p\rvert, \lvert L_d\rvert$	GVM-dominated case
$\lvert L_d\rvert \ll \lvert L_p\rvert, \lvert L_w\rvert$	GVD-dominated case

that represent three well-defined, opposite experimental situations, and that also allow us to understand, at least qualitatively, the features of general, intermediate cases. The limiting cases in which $|L_p|$, $|L_w|$, and $|L_d|$ are infinitely small compared with the other two lengths correspond, respectively, to constant, parabolic, and linear (with normal dispersion) or elliptic (with anomalous dispersion) dispersion curves [see Fig. 8.3 (second row)], and lead to the prototype PBB, eFWM, and eX or eO wave modes, respectively [see Fig. 8.3 (first row)].

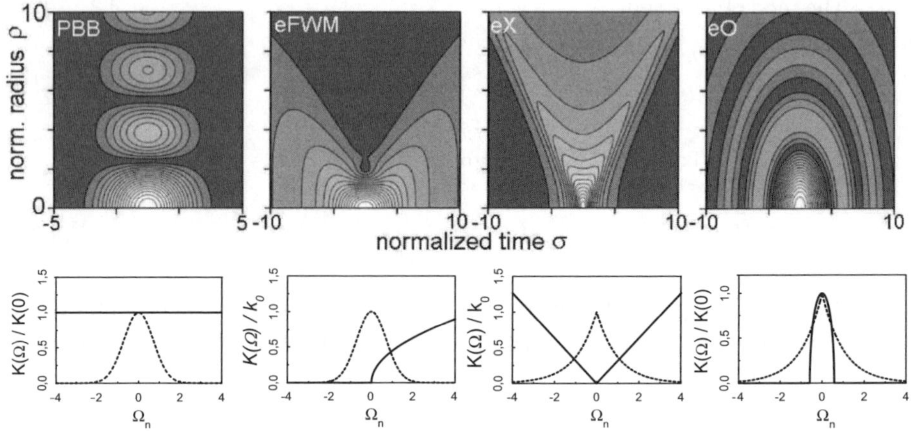

FIGURE 8.3 *Top*: Contour plots of the amplitude of the prototype PBB, eFWM, eX, and eO wave modes. *Bottom*: Corresponding constant, parabolic, linear, and elliptical dispersion curves (solid curves) within the spectra (dashed curves). Normalized coordinates are $\sigma = (\tau + \alpha z)\,\Delta\Omega$, $\rho = r/r_0$, with $r_0 = (2k_0\beta)^{-1/2}$.

Generally speaking, modes of sufficiently long duration belong to, or participate mostly of, the PM-dominated (PBB-like) case, modes of some (still unspecified) intermediate duration belong to the GVM-dominated (eFWM-like) case, and extremely short modes to the GVD-dominated [eX-like (eO-like)] case, since L_p is independent on bandwidth but L_p and L_d are inversely proportional to $\Delta\Omega$ and $\Delta\Omega^2$, respectively. Depending, however, on the relative values of α, β, and k_0'' (particularly when one or two of them are very small), the GVM-dominated case, even the PM-dominated case, can extend down to the single-cycle regime, or on the contrary, the GVM-dominated case, even the GVD-dominated case, can apply to considerably long modes.

8.3.1 Phase-Mismatch-Dominated Case: Pulsed Bessel Beam Modes

Consider first modes with $|L_p| \ll |L_w|, |L_d|$. When $L_p > 0$, the dispersion curve within the spectral bandwidth can be approached by the real constant value $K(\Omega) \simeq (2k_0 L_p^{-1})^{1/2}$, or

$$K(\Omega) \simeq \sqrt{2k_0\beta} \qquad \text{if } \beta > 0, \tag{8.23}$$

regardless of whether the exact dispersion curve is an actual hyperbola or ellipse (as in Fig. 8.4), that is, independent of the sign of material group velocity dispersion. Wave modes under these conditions can have only superluminal phase velocity ($\beta > 0$) but can have super- or subluminal group velocity ($\alpha > 0$ or $\alpha < 0$, respectively), and will adopt, from Eqs. (8.12) and (8.23), the approximate factorized form

$$\Phi_{\alpha,\beta}(r, \tau + \alpha z) \simeq f(\tau + \alpha z) J_0\left(\sqrt{2k_0\beta}\, r\right) \tag{8.24}$$

of a PBB of transversal size on the order of $(2k_0\beta)^{-1/2}$.

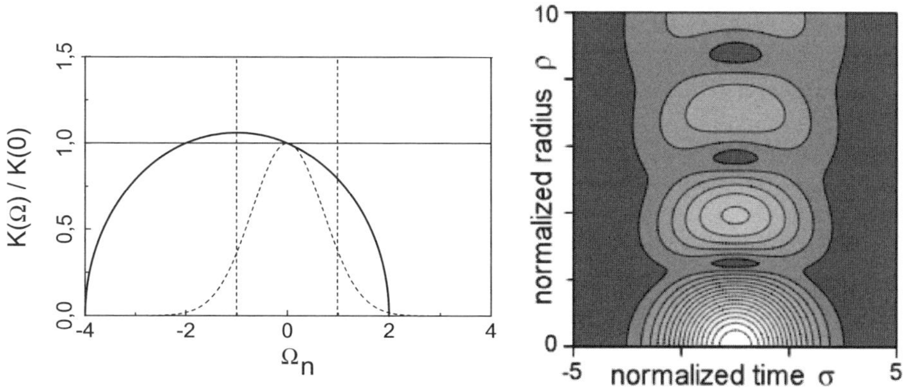

FIGURE 8.4 *Left*: Limiting dispersion curve for $|L_p|/|L_w| \to 0$, $|L_p|/|L_d| \to 0$ (thin line), for $L_p/L_w = -0.25$, $L_p/L_d = -0.25$ (thick curve), and Gaussian spectrum $\hat{f}(\Omega) = \exp[-(\Omega/\Delta\Omega)^2]$ (i.e., $f(\tau) \propto \exp[-(2\Delta\Omega\tau)^2]$) (dashed curve) for calculation of the wave modes. *Right*: Contour plot of the amplitude $|\Phi_{\alpha,\beta}|$ of the wave mode with the Gaussian spectrum and with $L_p/L_w = -0.25$, $L_p/L_d = -0.25$, calculated numerically from Eq. (8.12). Normalized coordinates are $\sigma = (\tau + \alpha z)\,\Delta\Omega$, $\rho = r/r_0$, with $r_0 = (2k_0\beta)^{-1/2}$.

Figure 8.3 shows the prototype PBB of this kind of wave mode [Eq. (8.24)] with a Gaussian spectrum $\hat{f}(\Omega)$, that is, the limiting case $|L_p/L_w| = 0$, $|L_p/L_d| = 0$. In Fig. 8.4 we also show, for comparison, the wave mode with $|L_p/L_w| = 0.25$, $|L_p/L_d| = 0.25$, and with the same Gaussian spectrum, obtained numerically from Eq. (8.12). We see that the wave mode preserves a spatiotemporal structure similar to that of the prototype PBB of Fig. 8.3, even if $|L_p|$ is not much smaller, but simply smaller than $|L_w|$ and $|L_d|$. Small differences can be understood as incipient focus wave mode and O-wave-type behavior, as described in the following sections.

8.3.2 Group-Velocity-Mismatch-Dominated Case: Envelope Focus Wave Modes

When $|L_w| \ll |L_p|, |L_d|$, the dispersion curve within the bandwidth can be approached by the horizontal parabola $K(\Omega) \simeq (2k_0 L_w^{-1}\Omega_n)^{1/2}$ with vertex at $\Omega = 0$, or

$$K(\Omega) \simeq \sqrt{2k_0\alpha\Omega}, \tag{8.25}$$

regardless of whether material dispersion is normal (as in Fig. 8.5) or anomalous. For modes with superluminal group velocity ($\alpha > 0$), the horizontal parabola is right-handed (as in Fig. 8.5) and is left-handed for subluminal modes ($\alpha < 0$). In any case, their spatiotemporal form can be approached by Eq. (8.12), with $K(\Omega)$ given by Eq. (8.25). Moreover, with the two-sided exponential spectrum

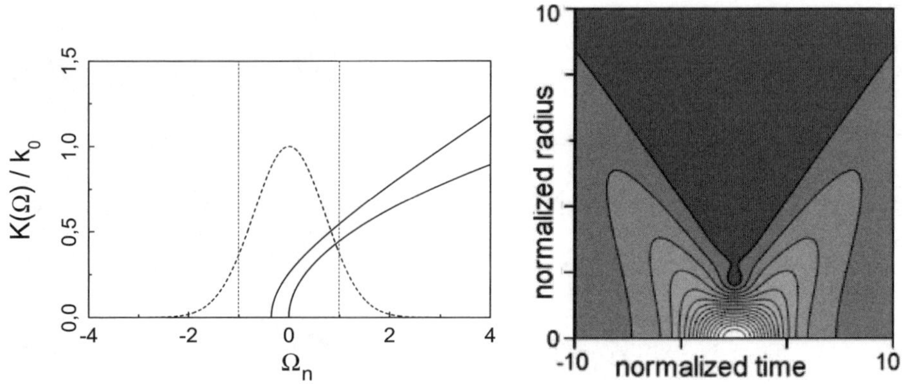

FIGURE 8.5 *Left*: Limiting dispersion curve for $L_w = 10/k_0$, $L_w/L_p \to 0$, $L_w/L_d \to 0$ (thin curve), for $L_w = 10k_0$, $L_w/L_p = 1/3$, $L_w/L_d = 1/3$ (thick curve), and Gaussian spectrum $\hat{f}(\Omega) = \exp[-(\Omega/\Delta\Omega)^2]$ (dashed curve) for calculation of the wave modes. *Right*: Contour plots of the amplitude $|\Phi_{\alpha,\beta}|$ of the wave mode with the Gaussian spectrum and with $L_w/L_p = 1/3$, $L_w/L_d = 1/3$, calculated numerically from Eq. (8.12). Normalized coordinates are $\sigma = (\tau + \alpha z)\Delta\Omega$, $\rho = r/r_0$, with $r_0 = (2/k_0\Delta\Omega|\alpha|)^{1/2}$.

$\hat{f}(\Omega) = (2\pi/\Delta\Omega)\exp(-|\Omega|/\Delta\Omega)$, Eq. (8.12) yields

$$\Phi_{\alpha,\beta}(r, \tau + \alpha z) \simeq \frac{-i\tau_0}{\tau + \alpha z - i\tau_0} \exp\left[\frac{ik_0|\alpha|r^2}{2(\tau + \alpha z - i\tau_0)}\right] \tag{8.26}$$

for superluminal modes ($\alpha > 0$), and the complex conjugate of the right-hand side of Eq. (8.26) for subluminal modes ($\alpha < 0$). In Eq. (8.26), $\tau_0 \equiv (\Delta\Omega)^{-1}$ characterizes the mode duration. The mode spot size at pulse center ($\tau + \alpha z = 0$) can be characterized by $r_0 = (2/k_0\Delta\Omega|\alpha|)^{1/2}$.

The functional form of the reduced envelope in Eq. (8.26) is similar to the fundamental Brittigham–Ziolkowski focus wave mode (FWM) [16,17], and as such will be called the envelope focus wave mode (eFWM). There are, however, important physical differences between them, which can be understood for the respective expressions of the complete fields E of both types of waves:

$$E_{\alpha,\beta}(r, z, t) \simeq \frac{-i\tau_0}{\tau + \alpha z - i\tau_0} \exp\left[\frac{ik_0|\alpha|r^2}{2(\tau + \alpha z - i\tau_0)}\right] \exp(-i\beta z) \times \exp(-i\omega_0 t + ik_0 z), \tag{8.27}$$

for the envelope focus wave mode, and

$$E(r, z, t) = \frac{-i\tau_0}{\tau - i\tau_0} \exp\left[\frac{ik_0 r^2}{2c(\tau - i\tau_0)}\right] \exp(-i\omega_0 t - ik_0 z), \tag{8.28}$$

with $k_0 = \omega_0/c$, for the fundamental FWM [17]. The fundamental FWM is a localized, stationary free-space wave whose envelope propagates at luminal group velocity c, whereas the carrier oscillations back-propagate at the same velocity c. The eFWM is also a stationary localized wave with the same intensity distribution as that of the fundamental FWM, but propagates in a dispersive medium with super- or subluminal group velocity $1/(k_0' - \alpha)$, and carrier oscillations propagate in the same direction.

Figure 8.3 shows the prototype eFWM of this type of wave modes, obtained from numerical integration of Eq. (8.12) with the approximate dispersion curve $K(\Omega) = \sqrt{2k_0\alpha\Omega}$ (i.e., in the limiting case $|L_w/L_p| = 0$, $|L_w/L_d| = 0$) and a Gaussian spectrum. To pursue the validity of the model eFWM to describe this type of wave mode, we have also evaluated the wave mode field in some nonlimiting cases with the same Gaussian spectrum. Even for the relatively large ratios $|L_w/L_p| = 1/3$, $|L_w/L_d| = 1/3$, in which case the dispersion curve departs significantly from the limiting one (the thick and thin curves in Fig. 8.5), the wave mode calculated (Fig. 8.5) is nearly indistinguishable from the prototype eFWM (Fig. 8.3), exhibiting a similar eFWM structure along with some incipient eX-wave behavior because of the actual hyperbolic form (not parabolic) of the dispersion curve, as explained in the next section.

8.3.3 Group-Velocity-Dispersion-Dominated Case: Envelope X- and Envelope O-Modes

Normal Group-Velocity Dispersion: Envelope X-Waves We consider finally modes with $|L_d| \ll |L_p|, |L_w|$, or modes of sufficiently short duration, or propagating in a medium with large enough GVD. When material dispersion is normal ($k_0'' > 0$), the dispersion curve within the bandwidth approaches the X-shaped curve

$$K(\Omega) \simeq \sqrt{k_0 k_0''}|\Omega| \tag{8.29}$$

of the limiting case $|L_d/L_p|, |L_d/L_w| = 0$ [Fig. 8.6 (thin line)]. The actual dispersion curve of a mode may be shifted slightly toward negative frequencies (as in Fig. 8.6) or positive frequencies for modes with superluminal ($\alpha > 0$) or subluminal ($\alpha < 0$) group velocity, respectively. For modes with $\beta > 0$, $K(\Omega)$ is real everywhere (Fig. 8.6), but for modes with $\beta < 0$ there is a narrow frequency gap about $\Omega = 0$. A prototype wave mode for this case can be obtained by introducing the approximate dispersion curve of Eq. (8.29) into Eq. (8.12). With the two-sided exponential spectrum $\hat{f}(\Omega) = (2\pi/\Delta\Omega)\exp(-|\Omega|/\Delta\Omega)$, we obtain

$$\Phi_{\alpha,\beta}(r, \tau + \alpha z) \simeq \mathrm{Re}\left\{\frac{\tau_0}{\sqrt{k_0 k_0'' r^2 + [\tau_0 + i(\tau + \alpha z)]^2}}\right\}, \tag{8.30}$$

where $\tau_0 \equiv (\Delta\Omega)^{-1}$ measures the pulse duration. Equation (8.30) is the eX-wave described in [24] as an exact, stationary, and localized solution of the paraxial wave equation with luminal phase and group velocities ($\alpha = \beta = 0$) in media with normal GVD. The eX-wave (8.30) is understood here as an approximate expression for modes with α, β such that $|L_d/L_p| \ll 1$, $|L_d/L_w| \ll 1$. The spatiotemporal form of the eX-wave is shown in Fig. 8.3. For $L_d/L_p = 1/6$ ($\beta > 0$), $L_d/L_w = 1/6$ (thick curve

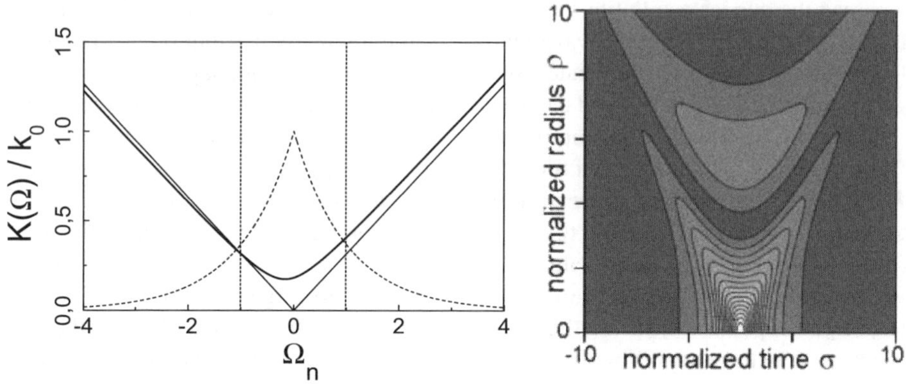

FIGURE 8.6 *Left*: Limiting dispersion curve for $L_d = 10/k_0$, $L_d/L_p \to 0$, $L_d/L_w \to 0$, with $L_d > 0$ (thin line), for $L_d = 10/k_0$, $L_d/L_p = 1/6$, $L_d/L_w = 1/6$ (thick curve), and two-sided exponential spectrum $\hat{f}(\Omega) = \exp(-|\Omega|/\Delta\Omega)$ (dashed curve). *Right*: Contour plot of the amplitude $|\Phi_{\alpha,\beta}|$ of the wave mode with the two-sided exponential spectrum and with $L_d/L_p = 1/6$, $L_d/L_w = 1/6$. Normalized coordinates are $\sigma = (\tau + \alpha z)\Delta\Omega$, $\rho = r/r_0$, with $r_0 = (k_0 k_0'' \Delta\Omega^2)^{-1/2}$.

in Fig. 8.6), the mode retains an X-shaped structure (Fig. 8.6) despite the fact that the dispersion curve differs significantly from the limiting curve (the thin curve in Fig. 8.6). Incipient PBB behavior, or radial oscillations, originates from the nearly horizontal dispersion curve in the central part of the spectrum.

Anomalous Group-Velocity Dispersion: Envelope O-Waves When $|L_d| \ll |L_p|, |L_w|$ but GVD is anomalous, the dispersion curve within the bandwidth can be approached by the ellipse centered approximately on $\Omega = 0$ (Fig. 8.7, thick and thin curve), given by the expression

$$K(\Omega) \simeq \sqrt{2k_0(\beta - |k_0''|\Omega^2/2)}. \tag{8.31}$$

Note that the term with β, no matter how small, must be retained to reproduce the real-valued part of the dispersion curve. The group velocity of the mode can be slightly subluminal ($\alpha < 0$) or superluminal ($\alpha > 0$), as in Fig. 8.7, but the phase velocity of these modes is always superluminal ($\beta > 0$). An approximate analytical expression for this type of mode can be obtained by introducing the approximate dispersion curve of Eq. (8.31) into Eq. (8.12). Under the condition $|L_d| \ll |L_p|$, the frequency gap $\Omega_g \simeq \sqrt{2\beta/|k_0''|}$ is much smaller than $\Delta\Omega$, so that the amplitude spectrum $\hat{f}(\Omega)$ can be assumed to take a constant value in the integration domain of integral in Eq. (8.12), which then yields the expression

$$\Phi_{\alpha,\beta} \simeq \frac{1}{\sqrt{k_0|k_0''|r^2 + (\tau+\alpha z)^2}} \sin\left[\sqrt{2\beta/|k_0''|}\sqrt{k_0|k_0''|r^2+(\tau+\alpha z)^2}\right], \tag{8.32}$$

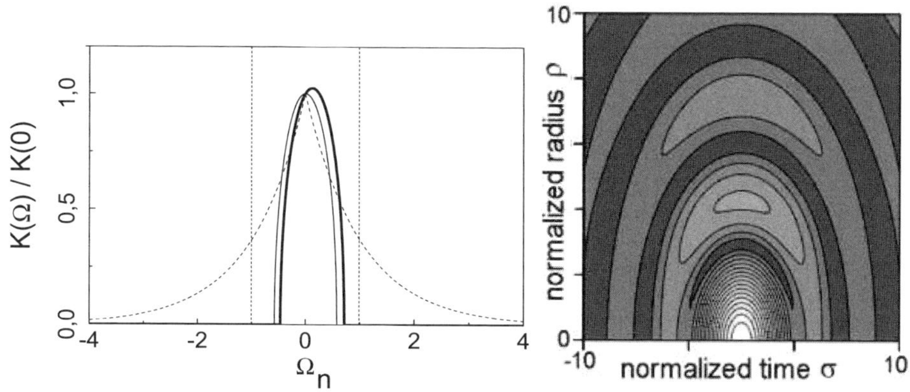

FIGURE 8.7 *Left*: Limiting dispersion curve for $|L_d|/|L_p| \to 0$, $|L_d|/|L_w| \to 0$, with $L_d < 0$ (thin curve), for $L_d/L_p = -1/6$, $L_d/L_w = -1/8$ (thick curve), and the two-sided spectrum $\hat{f}(\Omega) = \exp(-|\Omega|/\Delta\Omega)$ (dashed curve). *Right*: Contour plot of the wave mode with $L_d/L_p = -1/6$, $L_d/L_w = -1/8$ and the exponential spectrum, calculated numerically from Eq. (8.12). Normalized coordinates are $\sigma = (\tau + \alpha z)\sqrt{2\beta/|k_0''|}$, $\rho = \sqrt{2k_0\beta}r$.

which has the same form as the O-type impulse response mode in media with anomalous dispersion. Figure 8.3 shows its spatiotemporal form. For comparison, the wave mode with $L_d/L_p = -1/6$, $L_d/L_w = -1/8$ [Fig. 8.7 (thick curve)] and the two-sided exponential spectrum [Fig. 8.7 (dashed curve)] was calculated from Eq. (8.12), and its O-shaped spatiotemporal form is also depicted in Fig. 8.7.

8.4 WAVE MODES WITH ULTRABROAD BANDWIDTH

Wave modes have been described above within the SVEA. In some circumstances (see below) wave modes with ultrabroad bandwidth, comparable to that of single-cycle pulses, have been observed, in which case the SVEA is no longer accurate. In this section we extend the description to a frame compatible with these broad spectra. Broadband wave modes show much richer spatiotemporal structures and dispersion curves than the simple X- and O-like narrowband wave modes, as shown in this section.

Our basic propagation equation is now

$$\partial_z A = \frac{i}{2k_0}\left(1 + i\frac{k_0'}{k_0}\partial_\tau\right)^{-1} \Delta_\perp A + iDA, \tag{8.33}$$

derived in [40] within the slowly evolving wave approximation (SEWA) (but without performing the final approximation $k_0'/k_0 \simeq 1/\omega_0$ [41]), which describes the propagation of pulses with duration Δt (FWHM of $|A|^2$) as short as one carrier period $2\pi/\omega_0$ under the effects of material dispersion up to the desired order, diffraction in

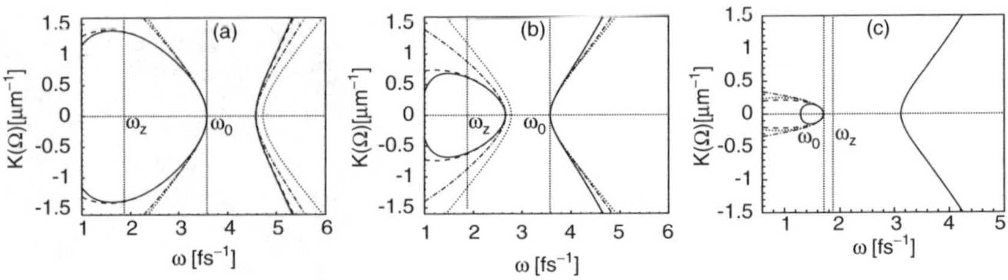

FIGURE 8.8 Dispersion curves of the wave modes in the normal [(a) and (b) $\lambda = 0.527$ μm] and anomalous [(c) $\lambda = 1.7$ μm] dispersive region for water. (a) $\alpha = -0.032$, $\beta = 0$; (b) $\alpha = 0.022$, $\beta = 0$; (c) $\alpha = -0.001$, $\beta = 0$.

the paraxial approximation, and spatiotemporal focusing (dependence of diffraction with frequency) up to first order. In Eq. (8.33),

$$D = \sum_{m=2}^{\infty} \frac{k_0^{(m)}}{m!} (i\partial_\tau)^m \tag{8.34}$$

is the dispersion operator.

Following a procedure similar to that described in Section 8.2 and using the wave mode ansatz (8.3) in Eq. (8.33), we obtain

$$\Delta_\perp \Phi + 2k_0 \left(1 + i\frac{k_0'}{k_0}\partial_\tau\right) [\beta + i\alpha\partial_\tau + iD]\Phi = 0\,, \tag{8.35}$$

from which the dispersion relation can be found to be

$$K(\Omega) = \sqrt{2(k_0 + k_0'\Omega)\left[\beta + \alpha\Omega + \hat{D}(\Omega)\right]}\,, \tag{8.36}$$

where

$$\hat{D}(\Omega) = \sum_{m=2}^{\infty} \frac{k_0^{(m)}}{m!} \Omega^m = k(\omega_0 + \Omega) - k_0 - k_0'\Omega\,. \tag{8.37}$$

Wave modes within the SEWA have the same expressions (8.11) and (8.12) as SVEA modes, but the dispersion relation is more accurate. Compared with the SVEA dispersion relation in Eq. (8.10), there is a new term in Eq. (8.36) standing for first-order spatiotemporal focusing for broadband pulses. From the conditions of validity of the SEWA, the parameters α and β should satisfy the same requirements as in Eqs. (8.4) and (8.5), whereas the condition (8.2) of narrow temporal spectrum is no longer required.

As an example, Fig. 8.8 shows wave-mode dispersion curves up to different degrees of approximation for water in normal dispersion [527-nm carrier wavelength, (a) and (b)] and anomalous dispersion [1100 nm, (c)]. One of the interesting features of

SEWA wave modes is that their dispersion curves (solid curves) can reproduce the structure of the broadband spatiotemporal spectra observed for light filaments [34,38], which suggests spontaneous formation of wave modes in the filamentation process. While dispersion curves within the SVEA (dotted curves) are always symmetric (X- or O-like) in the respective cases of normal and anomalous dispersion, SEWA dispersion curves are able to describe the strongly asymmetric structures observed. It is worth noting that the inclusion of third- or higher-order dispersion terms in SVEA wave modes (dash–dotted curves) renders the dispersion curves asymmetric but does not significantly improve the fitting to the observations. The inclusion of spatiotemporal focusing up to the first order is essential to obtain good agreement when dealing with such broad spectra. Generally, first-order spatiotemporal focusing and third-order dispersion suffice to reproduce the spectra observed when most of the spectrum lies on either the normal or anomalous spectral regions (dashed curves in Fig. 8.8a and b), but not when the spectrum covers both normal and anomalous regions (dashed curve in Fig. 8.8c, where the tail is missing), in which case all orders of dispersion are generally needed. In conclusion, the description of wave modes within the SEWA approximation appears to be the simplest, minimal extension of the SVEA description that is able to reproduce the observations in nonlinear experiments involving ultrabroad spectra.

8.4.1 Classification of SEWA Dispersion Curves

In this section we classify SEWA wave modes according to the possible forms of their dispersion curves. Whereas SVEA dispersion curves are always X- or O-shaped, SEWA dispersion curves are much more complex, including distorted X and O shapes, "fishlike" and single-branch forms. For our classification, the only assumption on the medium is that second-order dispersion vanishes, as much, at only one frequency in the transparency spectral region of interest, and the only simplifying assumption is that $\beta \simeq 0$ [i.e., the dispersion curve passes through $(K = 0, \Omega = 0)$], which appears to correspond (as in Fig. 8.8) with the physically relevant situations.

The shape of the dispersion curve is seen to be determined primarily by its zeros. From Eq. (8.36) with $\beta = 0$, the zeros of $K(\Omega)$ are those of the function

$$\hat{h}(\Omega) = \alpha\Omega + \hat{D}(\Omega) = k(\omega_0 + \Omega) - k_0 - (k_0' - \alpha)\Omega \tag{8.38}$$

(we do not consider the zero at $\Omega = -k_0/k_0' \simeq -\omega_0$, i.e., at the dc component $\omega = 0$).

Distorted X- and O-Like Wave Modes Suppose first that GVD is normal (anomalous) for all frequencies of interest. This means that the concavity $\hat{h}''(\Omega) = k''(\omega_0 + \Omega)$) of $\hat{h}(\Omega)$ is always positive (negative), that the slope $\hat{h}'(\Omega) = k'(\omega_0 + \Omega) - (k_0' - \alpha)$ of $\hat{(\Omega)}$ is a monotonous function, crossing consequently the Ω-axis only once at a frequency close to the carrier (since $|\alpha|$ is small), and that $\hat{h}(\Omega)$ presents a single minimum (maximum) for normal (anomalous) dispersion at that frequency. Since Eq. (8.38) also requires that $\hat{h}(\Omega)$ cross the Ω-axis at the carrier frequency $\Omega = 0$, it

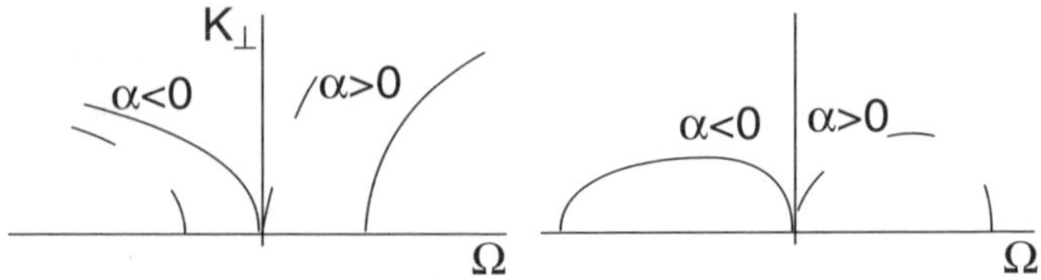

FIGURE 8.9 Typical dispersion curves of wave modes in a medium with only (a) normal or (b) anomalous dispersion.

follows that $\hat{h}(\Omega)$ must cross it once more, being positive outside (inside) the interval joining these two zeros for normal (anomalous) dispersion.

In conclusion, the dispersion curve K for normal dispersion is real outside this interval, resembling a more or less distorted letter X with a frequency gap between the carrier and another frequency (Fig. 8.9a). For anomalous dispersion, K is real between the carrier frequency and another frequency, displaying a distorted O-like form (Fig. 8.9b). The smaller the $|\alpha|$, the smaller the X gap and O width, reaching zero for $\alpha = 0$. Distortion from X- and O-like SVEA wave modes is due to the effects of higher-order dispersion and the spatiotemporal focusing term, which always flattens the low-frequency part of the dispersion curve.

Fishlike and Single-Branch Wave Modes Suppose now the more general case that GVD changes from anomalous to normal at only one frequency ω_z, and assume that $k_0'''(\omega_z) > 0$, as in fused silica, water, and almost all optical materials of interests. This means that the $k'(\omega)$ presents only one minimum at ω_z: in particular, for any carrier frequency $k_0' - k'(\omega_z) > 0$.

In this case, the concavity of $\hat{h}(\Omega)$ ($\hat{h}''(\Omega) = k''(\omega_0 + \Omega)$) changes from negative to positive only once at the flex point $\Omega_z = \omega_z - \omega_0$, where the slope takes its minimum (positive or negative) value $\hat{h}'(\Omega_z) = k'(\omega_z) - (k_0' - \alpha)$. At the same time, from Eq. (8.38), $\hat{h}(\Omega)$ crosses the Ω-axis at the carrier frequency with slope $\hat{h}'(0) = \alpha$ (positive or negative). Thus, depending on whether α is smaller than zero, lies between 0 and $k_0' - k'(\omega_z)$, or is larger than $k_0' - k'(\omega_z)$, the dispersion curve may adopt the following forms.

Fishlike Wave Modes If $\alpha < 0$ (subluminal wave modes), the slopes of $\hat{h}(\Omega)$ at the carrier ω_0 and at the flex point ω_z are both negative. Thus, $\hat{h}(\Omega)$ must cross the Ω-axis three times, and the dispersion curve $K(\Omega)$ adopts the fishlike structure illustrated in Fig. 8.10a and b (top), consisting of a closed loop and an isolated tail. This structure is the same regardless of whether the carrier frequency lies on the normal or anomalous spectral regions. However, the fishlike wave mode with carrier in anomalous dispersion behaves as a distorted O-like wave mode plus an added branch in the sense that the O narrows as $\alpha \to 0$ and the branch remains far from the O. In normal dispersion, the fishlike wave mode behaves as a distorted X-like wave

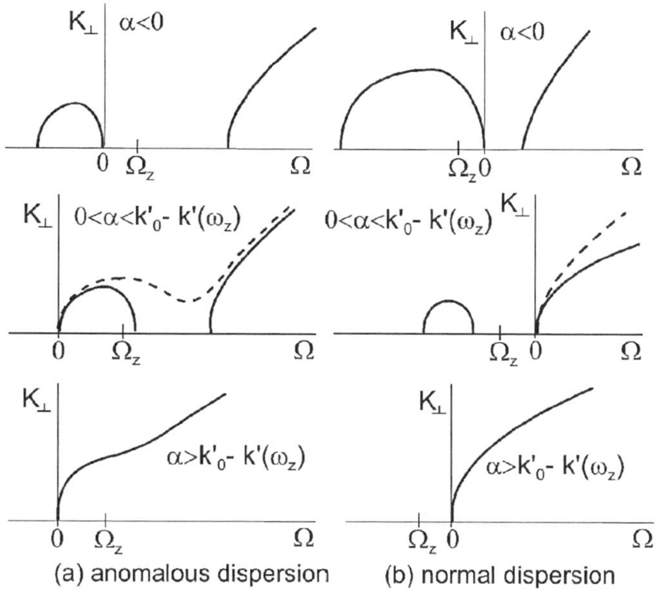

FIGURE 8.10 Typical dispersion curves of wave modes in a medium with one zero second-order dispersion frequency.

mode featuring a closed branch, since the gap shortens for $\alpha \to 0$ and the tail closes far from the gap.

Transition Wave Modes If $0 < \alpha < k_0' - k'(\omega_z)$ (slightly superluminal wave modes), the slope of $\hat{h}(\Omega)$ at the carrier ω_0 is positive, but is negative at the flex point ω_z. In this case, $\hat{h}(\Omega)$ will cross the Ω-axis only once for the smaller values of α, and three times for the larger values, the transition taking place at some α that can only be determined from knowledge of the material dispersion curve $k(\omega)$ at all frequencies. Accordingly, the dispersion curve for the smaller values of α will present a fishlike structure similar to that in the previous case, as shown in Fig. 8.10a and b (middle, solid curves). However, these dispersion curves differ substantially from the preceding case in that the O (in the anomalous region) and the gap (in the normal region) are placed in opposite sides of the carrier, and that the tail approaches the O (anomalous) and the tail closing moves toward the gap (normal). For the larger values of α, the O and tail join together (anomalous), and the tail closing reached the gap and disappeared (normal), the dispersion curve featuring only one branch, as shown in Fig. 8.10a and b (middle, dashed curves).

Tail-like Wave Modes If $\alpha > k_0' - k'(\omega_z) > 0$ (superluminal wave modes), the slopes of $\hat{h}(\Omega)$ at the carrier and at the flex point are both positive, which means that $\hat{h}(\Omega)$ is a growing function everywhere that crosses the Ω-axis only once at the carrier. The same applies to the dispersion curve K, which exhibits in anomalous

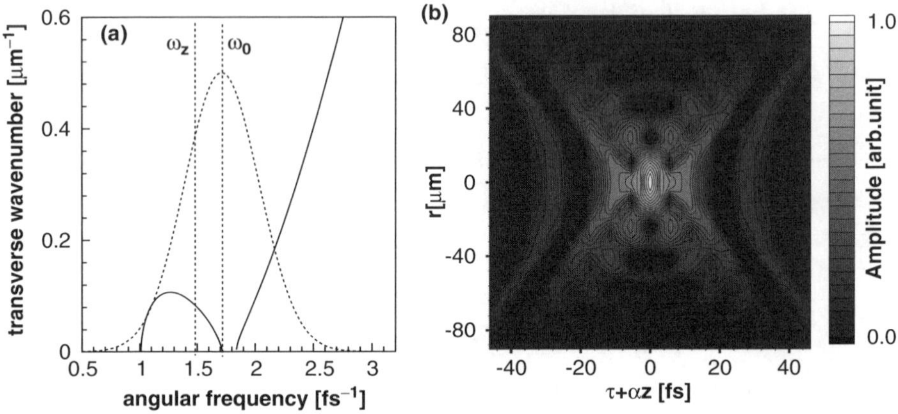

FIGURE 8.11 Dispersion curve (a) and contour plot of the amplitude $|A|$ (b) of a single-cycle fishlike wave mode in the normal dispersion region of fused silica ($\omega_0 = 1.71$ fs^{-1}, $\omega_z = 1.48$ fs^{-1}, $\alpha = -0.001$ fs/μm, $\beta = 0$). The thin-dotted curve is the spectrum of a single-cycle pulse ($\Delta t = 3.67$ fs).

and normal cases the form of a single monotonously increasing tail, as illustrated in Fig. 8.10a and b (bottom).

Further details can be known only from the specific form of the material dispersion. In the infrequent situation that $k'''(\omega_z) < 0$ at the zero GVD frequency ω_z [$k'(\omega)$ is maximum at ω_z, and hence $k'_0 - k'(\omega_z) < 0$], all possible forms of the dispersion curves are identical but reflected about zero dispersion frequency Ω_z, and fishlike wave modes are superluminal ($\alpha > 0$), transition wave modes are slightly subluminal [$k'_0 - k'(\omega_z) < \alpha < 0$], and tail-like wave modes are subluminal [$\alpha < k'_0 - k'(\omega_z)$].

Finite bandwidth SEWA wave modes can be classified in a way similar to that in Section 8.3. The only essentially different situation takes place when the bandwidth covers all three zeros of the dispersion curve. This may occur when the carrier frequency ω_0 is close to ω_z and α is small. As an example, Fig. 8.11 shows the dispersion curve and spatiotemporal distribution of a single-cycle fishlike wave mode in a normal dispersion region of fused silica at carrier wavelength 1100 nm nearby the zero-GVD wavelength 1273 nm. As seen, the spatiotemporal structure of the fishlike wave mode is a combination of an X-wave and an O-wave.

8.5 ABOUT THE EFFECTIVE FREQUENCY, WAVE NUMBER, AND PHASE VELOCITY OF WAVE MODES

One of the most striking features of wave modes is that their group velocity $1/(k'_0 - \alpha)$ is independent of the spectrum $\hat{f}(\Omega)$. This property also holds for beating-like wave modes when discrete spectra are taken, even if composed of only two frequencies. However, for the phase velocity, this is no longer true. The frequency $\omega_0 + \Omega$ of the actual carrier oscillations within the envelope of a wave mode is seen to depend on the spectrum $\hat{f}(\Omega)$, and so does the axial wave number $k_z(\Omega) = k_0 - \beta + (k_0' - \alpha)\Omega$ and the phase velocity $v_p = (\omega_0 + \Omega)/k_z(\Omega)$ along the propagation direction z. In this section we provide physically meaningful definitions of effective frequency, wave number, and phase velocity of a wave mode of spectrum $\hat{f}(\Omega)$.

We assume that $\hat{f}(\Omega)$ represents the spectrum of a bell-shaped pulse envelope $f(\tau)$. *In the core of the wave mode* (around the peak of the pulse), the effective frequency (shift from ω_0) is often characterized by the "gravity center" of the wave-mode spectrum: namely,

$$\Omega_{\rm core} = \frac{\int_{K(\Omega)\,{\rm real}} d\Omega |\hat{f}(\Omega)|\Omega}{\int_{K(\Omega)\,{\rm real}} d\Omega |\hat{f}(\Omega)|}, \tag{8.39}$$

which may differ from the central frequency of $\hat{f}(\Omega)$ because of the frequency gaps in the dispersion curve $K(\Omega)$. This yields an effective wave number

$$k_{\rm core} = k_0 - \beta + \Omega_{\rm core}(k_0' - \alpha) \tag{8.40}$$

and a phase velocity $v_p = (\omega_0 + \Omega_{\rm core})/k_{\rm core}$. This phase velocity reduces to $\omega_0/(k_0 - \beta)$ if the effective core frequency is $\Omega_{\rm core} = 0$ [e.g., if $\hat{f}(\Omega)$ is centered on $\Omega = 0$ and $K(\Omega)$ does not have gaps].

If the dispersion curve has at least one zero with nonnegligible spectral amplitude $\hat{f}$, the frequency of the oscillations in the trailing and leading parts of the wave mode may differ significantly from that in the peak. In fact, the temporal behavior of the wave mode at large $|\tau|$ and on-axis ($r = 0$) is found from Eq. (8.12) to be determined by the asymptotic expression

$$\Phi(0, \tau + \alpha z) \sim -\frac{i}{\tau + \alpha z} \sum_{j=1}^{m} \hat{f}(\Omega_j) \exp[-i\Omega_j(\tau + \alpha z)], \tag{8.41}$$

where m is the number of zeros Ω_j of the dispersion curve with significant amplitude $\hat{f}(\Omega_j)$. The field amplitude decays as $1/(\tau + \alpha z)$, while it oscillates at an effective frequency that can be determined from

$$\Omega_{\rm clad} = \frac{\sum_{j=1}^{m} |\hat{f}(\Omega_j)|\Omega_j}{\sum_{j=1}^{m} |\hat{f}(\Omega_j)|}. \tag{8.42}$$

Then the effective wave number is given by

$$k_{\rm clad} = k_0 - \beta + \Omega_{\rm clad}(k_0' - \alpha), \tag{8.43}$$

and the phase velocity in the trailing and leading parts of the wave mode by $v_p = (\omega_0 + \Omega_{\rm clad})/k_{\rm clad}$.

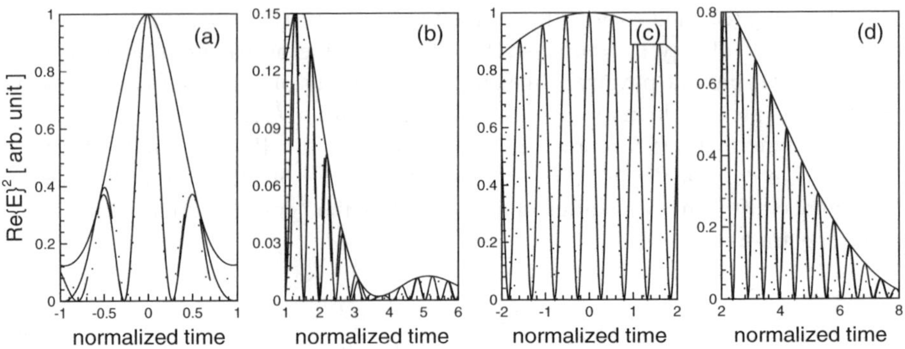

FIGURE 8.12 (a) and (b): On-axis real axial field Re E (solid curves) of the wave mode in Fig. 8.8a. (c) and (d): The same for the wave mode in Fig. 8.8c. The normalized time is $(\tau + \alpha z)/\Delta t$.

If $\Omega_{\text{core}} = \Omega_{\text{clad}}$, the entire wave mode is characterized by a single effective frequency. This type of "good" wave mode can be constructed by an adequate choice of the spectral amplitude $\hat{f}(\Omega)$. For instance, good X or O wave modes require $\hat{f}$ to be centered on the X-gap or O-band.

In Fig. 8.12 we verify that these definitions provide physically meaningful values of wave-mode frequencies. The solid curves represent the time evolution of the on-axis field Re E of the wave modes in Fig. 8.8a (X-like wave mode) and Fig. 8.12c (O-like wave mode). The dotted curves represent oscillations at ω_0, and the dashed curves represent oscillations at Ω_{core} in parts (a) and (c) for the inner parts of these wave modes, and oscillations at Ω_{clad} in parts (b) and (d) for their outer parts. For the X-like wave mode [(a) and (b)], the inner and outer carrier oscillations of the wave mode differ significantly, and are seen to be suitably characterized by Ω_{core} and Ω_{clad}. The same can be said for the O-like wave mode, although in this case both frequencies are very close.

8.6 COMPARISON BETWEEN EXACT, SEWA, AND SVEA WAVE MODES

A wave mode, independent of the approach used, is a localized wave in space and time in which the axial projection of its monochromatic plane-wave constituents presents a linear dependence with frequency:

$$k_z(\Omega) = (k_0 - \beta) + (k_0' - \alpha)\Omega \,. \tag{8.44}$$

In the cylindrically symmetric case, wave modes are often expressed as the superposition of Bessel beams:

$$E(r, z, t) = \frac{1}{2\pi} \int_{K\ \text{real}} d\omega\, \hat{f}(\omega - \omega_0) J_0[K(\omega)r] \exp(ik_z(\omega)z) \exp(-i\omega t). \tag{8.45}$$

Introducing the linear relation (8.44) in Eq. (8.45), we obtain

$$A(r, \tau, z) = \frac{1}{2\pi} \int_{K(\Omega)\,\mathrm{real}} d\Omega\, \hat{f}(\Omega) J_0[K(\Omega)r] \exp[-i\Omega(\tau + \alpha z)] \exp(-i\beta z), \tag{8.46}$$

which is indeed identical for all approaches to wave modes [see Eq. (8.12)]. What differs in the various approaches to wave modes is the accuracy of the dispersion relation $K(\Omega)$, as required by the physical conditions.

In exact, nonparaxial wave modes, the transversal wave vector is determined simply by the constraint $k_z^2(\Omega) + K^2(\Omega) = k^2(\omega_0 + \Omega)$, or

$$(k_0 - \beta) + (k_0' - \alpha)\Omega = \sqrt{k^2(\omega_0 + \Omega) - K^2}\,, \tag{8.47}$$

leading directly to

$$K(\Omega) = \sqrt{k^2(\omega_0 + \Omega) - [(k_0 - \beta) + (k_0' - \alpha)\Omega]^2}\,, \tag{8.48}$$

which is the exact nonparaxial dispersion relation.

In SEWA wave modes, instead, the paraxial approximation for each frequency is performed in Eq. (8.47):

$$(k_0 - \beta) + (k_0' - \alpha)\Omega \simeq k(\omega_0 + \Omega) - \frac{K^2}{2k(\omega_0 + \Omega)}, \tag{8.49}$$

and spatiotemporal coupling, or dependence of diffraction with frequency, is retained up to its first order by approaching $k(\omega_0 + \Omega) \simeq k_0 + k_0'\Omega$ in the denominator of Eq. (8.49), that is,

$$(k_0 - \beta) + (k_0' - \alpha)\Omega \simeq k(\omega_0 + \Omega) - \frac{K^2}{2(k_0 + k_0'\Omega)}, \tag{8.50}$$

obtaining

$$K(\Omega) = \sqrt{2(k_0 + k_0'\Omega)\left[k(\omega_0 + \Omega) - (k_0 - \beta) - (k_0' - \alpha)\Omega\right]}\,, \tag{8.51}$$

which is the same as Eq. (8.36). The dispersion relation within the SVEA approximation results from further neglecting the spatiotemporal focusing term $k_0'\Omega \simeq 0$ in Eq. (8.50). Up to second order in dispersion, we obtain the relation

$$K(\Omega) = \sqrt{2k_0 \left(\beta + \alpha\Omega + \frac{1}{2}k_0''\Omega^2\right)}\,, \tag{8.52}$$

which is the same as Eq. (8.10).

8.7 CONCLUSIONS

In this chapter we have reviewed and classified wave modes, or the ultrashort polychromatic Bessel beams featuring localization and stationarity (diffraction- and dispersion-free) in a dispersive, transparent optical material, propagating with slightly super- or subluminal velocities. The key tool for the analysis is the transversal dispersion curve $K(\Omega)$ [or angular dispersion curve $\theta(\Omega) \simeq K(\Omega)/k_0$], which allows us also to understand the spatiotemporal wave-mode structure. The dispersion curves of SVEA wave modes are X- or O-like. Their spatiotemporal structure has been classified into three broad categories: PBB-like, eFWM-like, and eX-like (eO-like) modes, depending on the relative strength of their phase and group velocity mismatches with respect to a plane pulse, and their defeated GVD, as measured by the mode phase-mismatch length L_p, group-mismatch length L_w, and the dispersion length L_d. The dispersion curves of broadband wave modes, described from the SEWA approximation, are much more complex. In a medium with a zero-dispersion wavelength, they are fishlike or tail-like, in most cases reduced to distorted X and O wave modes if the bandwidth covers only the normal or anomalous dispersion regions. We believe that understanding the counterintuitive, particlelike properties of these fascinating waves is essential for the success of their technologic applications.

REFERENCES

1. J. A. Stratton, *Electromagnetic Theory*, McGraw-Hill, New York, 1941.
2. J. Durnin, J. J. Miceli, and J. H. Eberly, *Phys. Rev. Lett.* **58**, 1499 (1987).
3. J. Durnin, *J. Opt. Soc. Am. A* **4**, 651 (1987).
4. J. Turunen, A. Vasara, and A. T. Friberg, *Appl. Opt.* **27**, 3959 (1988).
5. A. Vasara, J. Turunen, and A. T. Friberg, *J. Opt. Soc. Am. A* **6**, 1748 (1989).
6. L. Niggl, T. Lanzl, and M. Maier, *J. Opt. Soc. Am. A* **14**, 27 (1997).
7. F. Gori, G. Guattari, and C. Padovani, *Opt. Commun.* **64**, 491 (1987).
8. A. V. Shchegrov and E. Wolf, *Opt. Commun.* **25**, 141 (2000).
9. S. Chávez-Cerda, G. S. McDonald, and G. H. C. New, *Opt. Commun.* **123**, 225 (1996).
10. Z. Bouchal, J. Wagner, and M. Chlup, *Opt. Commun.* **151**, 207 (1998).
11. V. Garcés-Chávez, D. McGloin, H. Melville, W. Sibbett, and K. Dholakia, *Nature* **419**, 145 (2002).
12. J.-Y. Lu and J. F. Greenleaf, *IEEE Trans. Ultrason. Ferroelectr. Freq. Control* **37**, 438 (1990).
13. J.-Y. Lu and J. F. Greenleaf, *IEEE Trans. Ultrason. Ferroelectr. Freq. Control* **39**, 19 (1992).
14. P. Saari and K. Reivelt, *Phys. Rev. Lett.* **79**, 4135 (1997).
15. J. Salo, J. Fagerholm, A. T. Fribeg, and M. M. Solamaa, *Phys. Rev. E* **62**, 4261 (2000).
16. J. N. Brittingham, *J. Appl. Phys.* **54**, 1179 (1983).
17. R. W. Ziolkowski, *Phys. Rev. A* **39**, 2005 (1989).
18. K. Reivelt and P. Saari, *J. Opt. Soc. Am. A* **17**, 1785 (2000).

19. K. Reivelt and P. Saari, *Phys. Rev. E* **66**, 05661 (2002).
20. H. Sonajalg and P. Saari, *Opt. Lett.* **21**, 1162 (1996).
21. H. Sonajalg, M. Ratsep, and P. Saari, *Opt. Lett.* **22**, 310 (1997).
22. M. A. Porras, *Opt. Lett.* **26**, 1364 (2001).
23. M. A. Porras, R. Borghi, and M. Santarsiero, *Opt. Commun.* **206**, 235 (2002).
24. M. A. Porras, S. Trillo, and C. Conti, *Opt. Lett.* **28**, 1090 (2003).
25. S. Longhi, *Phys. Rev. E* **68**, 066612 (2003).
26. S. Orlov, A. Piskarskas, and A. Stabinis, *Opt. Lett.* **27**, 2167, 2103 (2002).
27. M. A. Porras and P. Di Trapani, *Phys. Rev. E* **69**, 066606 (2004).
28. G. Valiulis, Optical Nonlinear Process Research Unit, internal report (unpublished).
29. P. Di Trapani, G. Valiulis, A. Piskarskas, O. Jedrkiewicz, J. Trull, C. Conti, and S. Trillo, *Phys. Rev. Lett.* **91**, 093904 (2003); see also *Phys. Rev. Focus*, (Sept. 4, 2003), http://focus.aps.org/story/v12/st7.
30. O. Jedrkiewicz, J. Trull, G. Valiulis, A. Piskarskas, C. Conti, S. Trillo, and P. Di Trapani, *Phys. Rev. E* **68**, 026610 (2003).
31. G. Valiulis, J. Kilius, O. Jedrkiewicz, A. Bramati, S. Minardi, C. Conti, S. Trillo, A. Piskarskas, and P. Di Trapani, in *OSA Trends in Optics and Photonics* (*TOPS*), QELS 2001 Technical Digest, Vol. 57, Optical Society of America, Washington, DC, 2001.
32. M. A. Porras, A. Dubietis, E. Kucinskas, F. Bragheri, V. Degiorgio, A. Couairon, D. Faccio, and P. Di Trapani, *Opt. Lett.* **30**, 3398 (2005).
33. D. Faccio, F. Bragheri, and M. Cherchi, *J. Opt. Soc. Am. B* **21**, 296 (2004).
34. M. Kolesik, E. M. Wright, and J. V. Moloney, *Phys. Rev. Lett.* **92**, 253901 (2004).
35. D. Faccioaaa, P. Di Trapani, S. Minardi, A. Bramati, F. Bragheri, C. Liberale, V. Degiorgio, A. Dubietis, and A. Matijosius, *J. Opt. Soc. Am. B* **22**, 862 (2005).
36. D. Faccio, M. A. Porras, A. Dubietis, F. Bragheri, A. Couairon, and P. Di Trapani, *Phys. Rev. Lett.* **96**, 193901 (2006).
37. M. A. Porras et al., Characterization of O-shaped conical emission of light filaments in fused silica, *J. Opt. Soc. Am. B* **24**, 581 (2007).
38. M. Kolesik, E. M. Wright, and J. V. Moloney, *Opt. Express* **13**, 10729–10741 (2005).
39. I. S. Gradshteyn and I. M. Ryzhik, *Table of Integrals, Series and Products*, 4th ed., Academic Press, New York, 1965.
40. T. Brabec and F. Krausz, *Phys. Rev. Lett.* **78**, 3282 (1997).
41. M. A. Porras, *Phys. Rev. E* **60**, 5069 (1999).

CHAPTER NINE

Nonlinear X-Waves

CLAUDIO CONTI
Research Center Enrico Fermi, Rome, Italy, and Research Center SOFT
INFM-CNR, University La Sapienza, Rome, Italy

STEFANO TRILLO
University of Ferrara, Ferrara, Italy

9.1 INTRODUCTION

A nonmonochromatic superposition of propagation-invariant beams is obviously a propagation-invariant beam itself. However, this is no longer the case when the superposition principle ceases to be valid, as in the case of wave propagation in nonlinear media. On the other hand, it is well known that nonlinearity can counteract beam spreading, being at the very origin of those strongly confined propagation-invariant wave packets known as *solitons* or *solitary waves* (SWs) [1,2]. Unfortunately, SWs typically involve a reduced number of dimensions, whereas the observation of a three-dimensional (3D) SW has been elusive to date [3,4]. Among various reasons behind this, the most relevant are (1) the fact that 3D SWs are affected by intrinsic instabilities, and (2) (more specific to optics) 3D SWs require not only a nonlinear medium (and hence very high laser intensities) but also anomalous material dispersion (i.e., $k'' \equiv d^2k/d\omega^2 < 0$ where k is the wave number).

In contrast with SW, localized waves (LWs) are 3D propagation-invariant wave packets that do not rely on any nonlinearity. They have been observed in several contexts (see other chapters of this book) and in principle they should not exist in a nonlinear medium since they rely on the superposition principle. In this respect the experimental observation of optical 3D propagation-invariant pulses, called *light*

Localized Waves, Edited by Hugo E. Hernández-Figueroa, Michel Zamboni-Rached, and Erasmo Recami

bullets, in a nonlinear medium with normal dispersion ($k'' > 0$) (i.e., in a regime where both LW and SW were not expected to exist) really appeared to be an astonishing result [5,6]. The spatiotemporal far-field spectrum of the light bullets observed clearly appeared as an X [7], and an accurate and sophisticated experimental re-construction (by tomography) of the spatiotemporal profile unveiled the double conical structure of the wave packet [8,9], in agreement with theoretical and numerical analysis [10–16]. Thus, naturally, the light bullet was dubbed a *nonlinear X-wave*. This result opened up many different roads of investigation, aimed primarily at providing a physical and mathematical background to a special class of nonlinear 3D beams whose features recall both SW (they form spontaneously at high intensity and propagate without sensible distortion) and linear LW (the structure is seemingly that of conical waves), thus constituting a sort of *chimera*.

Thereafter the field of nonlinear X-waves (NLXs) has grown rapidly, on one hand, linking linear LW solutions in dispersive media to the existing literature and the plethora of electromagnetic LWs investigated mainly as solutions to Maxwell's equations in vacuum (see other chapters in this book) [17–29] and on the other hand, attempting to extend the idea of "non-bell-shaped" localized solutions to the very active field of nonlinear optics, which deals with solitary waves, including periodic media and different types of nonlinearities [30–47]. Importantly, numerical simulations, experiments, and theory concurred to establish that NLXs can also be an effective paradigm for interpreting ultrafast laser propagation in nonlinear media with an intensity-dependent refractive index (i.e., Kerr media). Given the facts that any material displays such nonlinearity at sufficiently high intensities and that many experiments have been done in water or air (due to the virtual absence of a damage threshold), these results have many implications in biophysical or remote environmental sensing applications [15,16,48–56].

Moreover, the formal analogy between nonlinear optics in Kerr media and the time evolution of the semiclassical wave function of ultracold bosons led to the prediction of the existence of *Matter X-waves* [57]. Matter waves are a natural manifestation of the large-scale coherence of an ensemble of atoms populating a fundamental quantum state. The observation of Bose–Einstein condensates (BECs) in dilute ultracold alkalis [58,59] has initiated the exploration of many intriguing properties of matter waves, whose macroscopic behavior can be described successfully via a mean-field approach in terms of a single complex wave function [60]. Large-scale coherence effects are usually observed by means of 3D magnetic or optical confining potentials in which BECs are described by their ground-state wave function. Trapping can also occur in free space (i.e., without a trap) through the mutual compensation of the leading-order (i.e., two-body) interaction potential and kinetic energy, leading to matter SWs. This phenomenon, however, has been observed only in one dimension [61]. In two and three dimensions free-space localization cannot occur, due to the instability of SWs, and even in a trap collapse usually prevents stable formation of BECs [62,63], necessitating stabilizing mechanisms [64]. For these reasons, the use of periodic potentials induced by optical lattices [65], where the behavior of atoms mimics that of electrons in crystals or photons in periodic media [66,67] is attracting a great deal of interest: for instance, in one-dimensional (1D) (elongated) lattices, 1D SWs can form and are referred to as *gap solitons* [68]. Notably enough, in the presence of a 1D lattice potential, the 3D dynamics of the fundamental state wave function formally obeys the same NLX model in optics. Specifically, under conditions for which the Bloch state associated with the lattice has a negative effective mass, the natural state of BECs is a localized *matter X-wave* characterized by a peculiar biconical shape. The atoms are organized in this way in the absence of any trap, solely as the result of the strong anisotropy between the 1D periodic modulation and the free motion in the 2D transverse plane. In this respect, NLXs appear as novel nontrivial localized states of ultracold atoms.

This chapter is aimed at reviewing many of the aforementioned aspects of NLXs. A strong emphasis is given to the theory, especially to "X-wave-oriented" analytical techniques that directly involve LWs in nonlinear dynamics. This approach has to be compared with plane-wave-oriented approaches. Given the fact that many of the well-known classifications of LW (detailed in this book) lose their meaning when dealing with nonlinear effects, we will roughly indicate with "X-waves" those solutions whose intensity profile travels undistorted, thus using LWs and X-waves as being synonymous.

Outline In Section 9.2 we review the basic models for nonlinear optical propagation and BECs; in Section 9.3 we introduce *envelope X-waves* [i.e., linear X-waves in the paraxial and slowly varying envelope approximation (SVEA)] and the corresponding X-wave expansion that leads to finite-energy solutions; Section 9.4 is devoted to a discussion of how nonlinear processes foster the exponential amplification of LWs at the expense of a pump beam; in Section 9.5 we apply X-wave expansion to derive general results for the nonlinear propagation and generation of LW; Section 9.6 deals with numerical results concerning the exact profile of X-waves in nonlinear media; in Section 9.7 we discuss an approach to the highly nonlinear dynamical processes observed in experiments: *coupled X-wave theory*; in Section 9.8 we briefly review experimental results; and conclusions and perspectives are drawn in Section 9.9.

9.2 NLX MODEL

The reference model for NLX pertains to a medium where the nonlinearity is responsible for direct self-action on a wave-packet envelope. In nonlinear optics, self-action occurs typically through a third-order effect (the dielectric polarization depends on the electric field cube) that involves degenerate four-photon mixing interactions $\omega = \omega - \omega + \omega$ among photons at the generic frequency ω within the bandwidth of the field. This is well known to be equivalent to a refractive index varying according to the Kerr law $n(\omega) = n_0(\omega) + n_{2I} I$, where $n_0(\omega)$ is the linear or low-intensity index, I is the optical intensity, and n_{2I} is known as the Kerr nonlinear index coefficient. In such a medium, under standard approximations (paraxial and SVEA), an electric field with linearly polarized unit vector $\mathbf{e}$ and a spatiotemporal spectrum localized around the propagation direction $(k_x, k_y, k_z) = (0, 0, k) \equiv \mathbf{k}$ and vacuum carrier wavelength

$\lambda = 2\pi c/\omega_0$,

$$\mathbf{E} = \Re\left[\mathbf{e}\sqrt{\frac{2Z_0}{n(\omega)}}A(\mathbf{r}T)\exp(ikz - i\omega_0 T)\right], \tag{9.1}$$

has its complex electric field envelope A obeying the equation (see, e.g., [69])

$$i\frac{\partial A}{\partial z} + ik'\frac{\partial A}{\partial T} - \frac{k''}{2}\frac{\partial^2 A}{\partial T^2} + \frac{1}{2k}\nabla_\perp^2 A = -k_0 n_{2I}|A|^2 A, \tag{9.2}$$

where $\nabla_\perp^2 \equiv \partial_x^2 + \partial_x^2$, $k = k_0 n_0 = \omega_0 n_0/c$ is the wave number at carrier angular frequency ω_0, and the prime denotes the derivative with respect to ω calculated at ω_0. Note that the square root containing the vacuum impedance Z_0 in Eq. (9.1) is such that the square modulus of the envelope gives the intensity (i.e., $I = |A|^2$) directly.

Although in this chapter we restrict primarily to the model (9.2), we observe that its validity extends well beyond Kerr media and even nonlinear optics. (As discussed below, it is also approximately valid for a quadratically nonlinear medium and within a perturbative formulation when dealing with ultracold gases.) We point out that even in the case of Kerr media (and within paraxial SVEA), it does not take into account many phenomena that are expected and observed when dealing with the propagation of femtosecond pulses in water or air, such as those related to plasma formation or higher-order dispersion. Although many generalizations of the models and the exploration of related issues are indeed possible within the paradigms of LWs, Eq. (9.2) is very rich and describes the essential physics of LWs, thus appearing as the most relevant basic model for NLXs.

Quadratic Media and Second Harmonic Generation As mentioned in Section 9.1, media for the second harmonic generation (SHG) are relevant for NLXs, because they were employed in the first experimental demonstration [12]. In the most general case the nonlinear interaction in a quadratically nonlinear medium (i.e., a medium in which any component of the nonlinear polarization is expressed in terms of the products of two electric field components) commits together three optical carriers ω_i $(i = 1, 2, 3)$ such that $\omega_3 = \omega_1 + \omega_2$. At variance with Kerr media, where only one angular frequency is involved, the model is given by three coupled envelope equations of the Eq. (9.2) type. For the partially degenerate case of SHG, when $\omega_1 = \omega_2 = \omega$ and $\omega_3 = 2\omega$, only two equations have to be considered, as reported in the literature (see [11,13,14,30,70]). However, in this chapter, we restrict ourselves to Kerr media while recalling that in suitable regimes, Eq. (9.2) is also a good approximation of the nonlinear dynamics of the fundamental beam at ω in quadratic media (see, e.g., [1]).

Bose–Einstein Condensation In the presence of a lattice, the Gross–Pitaevskii equation models the mean-field dynamics of ultracold bosons (see, e.g., [60]). As discussed in more detail in [57], in the presence of an external lattice (which can be induced on the atom cloud by the interference of counter-propagating laser beams), the Gross–Pitaevskii equation is reduced to an equation similar to Eq. (9.2):

$$i\hbar\partial_t\phi + \frac{\hbar^2}{2m}\left(\nabla_\perp^2 - \frac{m}{m_e}\partial_z^2\right)\phi - \frac{3a}{2}|\phi|^2\phi = 0, \tag{9.3}$$

where m and a are the boson mass and the "scattering length," respectively, and m_e is the negative effective mass, which depends on the lattice period and the strength of the corresponding potential, as investigated experimentally in [71]. A *matter X-wave*, a solution of Eq. (9.3) as detailed below, entails localization in both momentum and configuration space, thus being a clear signature of a Bose condensed gas as well as anisotropy in the distribution function detected in early experiments [60]. Unlike any other form of BEC, including solitons, matter X-waves can be observed in free space and in a noninteracting regime, where they are the natural basis to describe the coherent properties of atom wave packets. To fix the notation, we make explicit reference to optics in the following sections, where experimental evidence has already been reported. Results for BECs are obtained via the formal substitution $A \to \phi$, $z \to t$, $T \to z$, $k' \to 0$, $k'' \to \hbar/m_e$, $k \to m/\hbar$, $k_0 n_{2I} \to 3a/2\hbar$.

9.3 ENVELOPE LINEAR X-WAVES

Envelope X-waves are linear X-waves in the paraxial slowly varying envelope approximation, or in other words, solutions of our model (9.2) in the limit of zero intensity ($n_{2I} = 0$), which reads explicitly as

$$i\partial_z A + ik'\partial_T A - \frac{k''}{2}\partial_T^2 A + \frac{1}{2k}\nabla_\perp^2 A = 0. \tag{9.4}$$

The case of interest here is that of normal dispersion $k'' > 0$, yielding a hyperbolic operator. Equation (9.4) can be simplified further by introducing the retarded time $t \equiv T - k'z$, so that Eq. (9.4) can be rewritten as

$$i\partial_z A - \frac{k''}{2}\partial_t^2 A + \frac{1}{2k}\nabla_\perp^2 A = 0. \tag{9.5}$$

Looking for "stationary solutions" [i.e., solutions traveling with natural group velocity $1/k'$, such that $\partial_z = 0$ in Eq. (9.6)] yields an equation that is formally identical to the one obtained from the scalar wave equation when considering standard LWs:

$$\nabla_\perp^2 A - kk''\partial_t^2 A = 0. \tag{9.6}$$

Thus, all the solutions that are known from the theory of LWs in vacuum gives corresponding envelope X-wave solutions. More general axisymmetric envelope

X-waves can be sought in the form

$$A = \psi(t - \beta z, r)\exp(ik_z z), \tag{9.7}$$

which corresponds to an intensity profile $I = |A|^2$ traveling as an invariant object in the z-direction with group velocity $(1/k' + 1/\beta)^{-1}$ measured in the laboratory frame. In the case $k' = 0$ (e.g., for BEC waves) these solutions still represent 3D bullets. Introducing a retarded time $t_\beta \equiv t - \beta z$, the equation for ψ reads as

$$-k_z\psi - i\beta\partial_{t_\beta}\psi - \frac{k''}{2}\partial_{t_\beta}^2\psi + \frac{1}{2k}\nabla_\perp^2\psi = 0. \tag{9.8}$$

Let ψ represent a superposition of monochromatic Bessel beams, $J_0(\sqrt{kk''}\alpha r)$ $\exp(-i\omega t_\beta)$; with α in frequency units, the corresponding spatiotemporal dispersion relation is

$$-k_z - \beta\omega + \frac{k''}{2}\omega^2 = \frac{k''\alpha^2}{2}. \tag{9.9}$$

In order to have a continuous spectrum along ω, the left-hand side must be positive, thus ensuring the absence of evanescent waves. This can be achieved in the simplest way by letting $k_z = -\beta^2/2k''$, which gives

$$\left(\omega - \frac{\beta}{k''}\right)^2 = \alpha^2. \tag{9.10}$$

Equation (9.10) implies the existence of two types of X-waves that depend parametrically on the inverse velocity β:

$$\begin{aligned}\psi_/(t_\beta, r; \beta) &= \textstyle\int_0^\infty e^{-i[(\beta/k'')+\alpha]t_\beta} J_0(\sqrt{k''k}\,\alpha r) f_/(\alpha)\, d\alpha \\ \psi_\backslash(t_\beta, r; \beta) &= \textstyle\int_0^\infty e^{-i[(\beta/k'')-\alpha]t_\beta} J_0(\sqrt{k''k}\,\alpha r) f_\backslash(\alpha)\, d\alpha,\end{aligned} \tag{9.11}$$

with the corresponding "spectra" $f_/(\alpha)$ and $f_\backslash(\alpha)$. They will be denoted as *slash* and *backslash X-waves*, because of the shape of their spatiotemporal frequency content, as discussed below. A general linear X-wave solution characterized by the inverse differential velocity β is given by

$$A_X = e^{-i(\beta^2/2k'')z}[\psi_/(t - \beta z, r; \beta) + \psi_\backslash(t - \beta z, r; \beta)], \tag{9.12}$$

which can be rewritten as

$$A_X = e^{-i(\beta/k'')t + i(\beta^2/2k'')z}[\varphi_/(t - \beta z, r) + \varphi_\backslash(t - \beta z, r)], \tag{9.13}$$

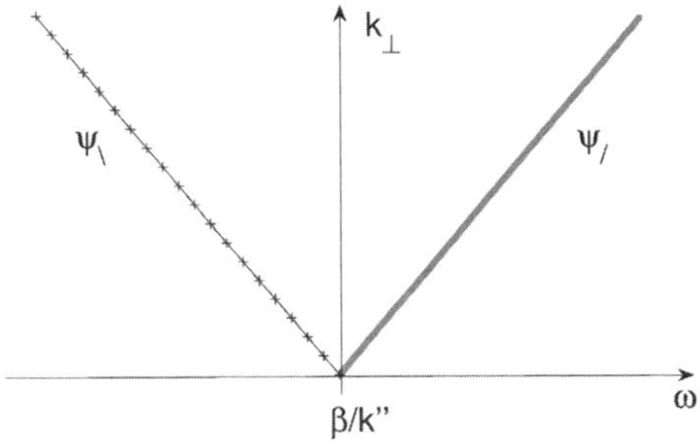

FIGURE 9.1 Spatiotemporal spectra of slash and backslash X-waves.

with

$$\varphi_X(t_\beta, r) = \int_0^\infty e^{\mp i\alpha t_\beta} J_0(\sqrt{k''k}\,\alpha r) f_X(\alpha)\, d\alpha. \tag{9.14}$$

Here, X stands for either / or \, and φ_X corresponds to X-waves of the scalar wave equation (see, e.g., [72] or other chapters of this book). Note that f_X is complex valued. The spatiotemporal spectrum of $A_X(r, t, z)$ is centered at the shifted central frequency β/k'', determined by the X-wave velocity, and it looks like an X in the angle-frequency plane, as shown in Fig. 9.1 for $k_\perp > 0$ (positive angles). The following relation is useful for any $\psi_X(t - \beta z, r; \beta)$:

$$\mathcal{L}\psi_X \equiv \left(i\partial_z + \frac{1}{2k}\nabla_\perp^2 - \frac{k''}{2}\partial_t^2 \right) \psi_X = -\frac{\beta^2}{2k''}\psi_X. \tag{9.15}$$

Therefore, an X-wave can also be defined as a solution of Eq. (9.6) of the type $A_X = C(z, \beta)\psi_X(t - \beta z, r)$, with C obeying the equation

$$i\frac{\partial C}{\partial z} - \frac{\beta^2}{2k''}C = 0, \tag{9.16}$$

which turns out to be a useful information when one formulates a nonlinear approach.

The invariant envelope X-waves contain, in general, infinite energy. This is due to the idealized situation (never achieved in experiments) of a precisely defined velocity (or inverse differential velocity β). Any finiteness introduced by the experimental setup, as, for example, the spatial extension of the sample, will in general cause the spectrum line shape around the X to fade, determining uncertainty in β. In the following, we show how to introduce a packet of X-waves, with velocities around a given value and finite energy. Considering an X-wave with a specific velocity is as idealized as considering an elementary particle of given momentum.

9.3.1 X-Wave Expansion and Finite-Energy Solutions

X-Wave Transform The general solution of Eq. (9.4) can be expressed by the Fourier–Bessel spectrum of the field at $z = 0$, denoted by $S(\omega, k_\perp) = \mathcal{B}[A](\omega, k_\perp, 0)$:

$$A(r, z, t) = \int_0^\infty \int_{-\infty}^\infty S(\omega, k_\perp) J_0(k_\perp r) e^{i(k_z z - \omega t)} k_\perp \, dk_\perp \, d\omega, \qquad (9.17)$$

with $k_z = -k_\perp^2/2k + k''\omega^2/2$. By a simple variable change, the field can be written as a superposition of slash X-waves traveling with different velocities. If we set

$$\omega = \alpha + \frac{\beta}{k''}, \qquad k_\perp = \sqrt{k\,k''}\,\alpha, \qquad (9.18)$$

we obtain

$$A(r, z, t) = \int_{-\infty}^\infty e^{-i(\beta^2/2k'')z} \Psi_/(t - \beta z, r; \beta) \, d\beta, \qquad (9.19)$$

while

$$\Psi_/(t - \beta z, r; \beta) = \int_0^\infty X_/(\alpha, \beta) J_0(\sqrt{k\,k''}\,\alpha r) e^{-i(\alpha + \beta/k'')(t - \beta z)} \, d\alpha \qquad (9.20)$$

and

$$X_/(\alpha, \beta) \equiv \mathcal{X}_/[A(r, t, z = 0)](\alpha, \beta) \equiv k\alpha S\left(\alpha + \frac{\beta}{k''}, \sqrt{k\,k''}\alpha\right). \qquad (9.21)$$

An equivalent representation is obtained by backslash X-waves. The variable change (9.18) corresponds to spanning the $(\omega, k_\perp)$ space by oblique (slash) parallel lines in (9.17). Equation (9.20) is a formulation of the *X-wave transform,* introduced in [73] and indicated as $\mathcal{X}[A](\alpha, \beta)$. Hence, the spatiotemporal evolution due to the interplay of diffraction and dispersion can be represented by one-dimensional propagation of packets with various velocities.

Orthogonal X-Waves and Finite-Energy Solutions An arbitrary pulsed beam can be expressed as a superposition of X-waves traveling with various velocities. Conversely, such a superposition can be used to construct new classes of physically realizable finite-energy X-waves. To this extent, orthogonal X-waves, introduced in [74] for the wave equation, are a fruitful approach (see also [16]). With reference to two (either slash or backslash) X-wave solutions of Eq. (9.8), denoted by A_X and B_X, with inverse differential velocities β and β' and spectra f and g, respectively, the inner product can be defined as the integral of $B_X^* A_X$ with respect to x, y, t, extended to the entire space. A specific class of spectra, denoted as f_p with $p = 0, 1, 2, \ldots,$

satisfies the orthogonality condition ($A_q = B_X$ and $A_p = A_X$):

$$< A_q(r,t,z,\beta)|A_p(r,t,z,\beta') >= \delta_{pq}\delta(\beta-\beta'). \tag{9.22}$$

When $X(\alpha,\beta) = C(\beta)f_p(\alpha)$, $\mathcal{E}_\beta = |C(\beta)|^2$ and the resulting beam is a finite-energy X-wave that spreads according to a prescribed velocity distribution function $C(\beta)$; this corresponds to the existence of solutions with an arbitrary depth of focus. This type of wave packet can be written, with reference to slash X-waves, as

$$A = \int_{-\infty}^{\infty} C(\beta,z)\psi_/{}^{(q)}(r, t-\beta z;\beta)\, d\beta, \tag{9.23}$$

with C obeying Eq. (9.16). Introducing a suitable Fourier transform with respect to the variable β,

$$c(s,z) = \frac{1}{2\pi\sqrt{k''}}\int_{-\infty}^{\infty} C(\beta,z)e^{i(\beta/k'')s}\, d\beta, \tag{9.24}$$

we find, from Eq. (9.16), that c obeys the equation

$$i\frac{\partial c}{\partial z} + \frac{k''}{2}\frac{\partial^2 c}{\partial s^2} = 0. \tag{9.25}$$

Therefore, the 3D linear propagation of an X-wave packet in a normally dispersive medium can be reduced to that of a 1D pulse with anomalous dispersion. The energy is

$$\mathcal{E} = 2\pi\int_0^\infty\int_{-\infty}^{\infty}|A(r,t,z)|^2 r\, dr\, dt = \int_{-\infty}^{\infty}|C(\beta)|^2\, d\beta = 2\pi\int_{-\infty}^{\infty}|c(s)|^2\, ds. \tag{9.26}$$

In the following, the (slash) X-wave, with spectrum

$$f_0(\alpha) = \frac{\sqrt{k}}{\pi}\Delta\alpha\, \exp(-\Delta\alpha) \tag{9.27}$$

and spatiotemporal profile

$$\varphi_/^{(0)} = -\frac{\sqrt{k}}{\pi}\frac{\Delta}{[1-k\,k''r^2/(s-i\Delta)^2]^{3/2}(s-i\Delta)^2}, \tag{9.28}$$

is the simplest X-wave with finite power (i.e., the integral of $|\varphi_/^{(0)}|^2$ over transverse variable converges; see also [21]). This X-wave is shown in Fig. 9.2.

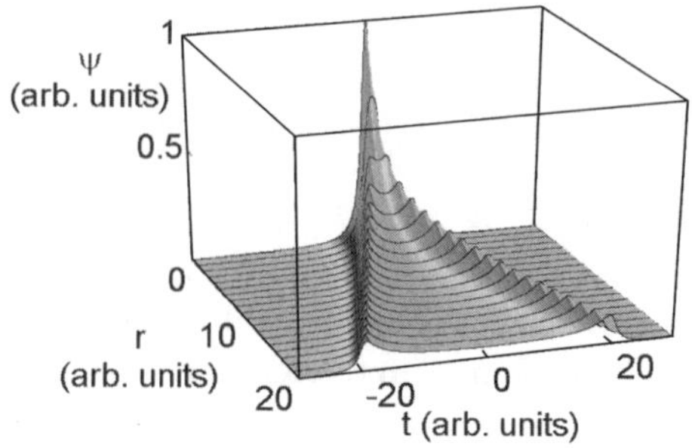

FIGURE 9.2 Spatiotemporal profile of the fundamental X-wave (9.2).

9.4 CONICAL EMISSION AND X-WAVE INSTABILITY

So far we have dealt with linear X-waves of a simplified scalar model, *prolegomena* to the nonlinear regime. The fundamental physical result is that in nonlinear media, X-waves are generated spontaneously during propagation of a bell-shaped (in space and time) pulse. The universal mechanism of spectral wave reshaping in nonlinear media is the *modulational instability* (MI), which describes the amplification of plane waves at the expense of a pump beam. The features of MI depend on the number of effective dimensions and the sign of dispersion; in the most general case (i.e., with a propagation coordinate z), two transverse coordinates x, y and time t are taken into account and MI in a normally dispersive medium takes the name *conical emission* (CE). CE describes a well-known and experimentally investigated effect: the generation of far-field rings with different colors when an intense ultrashort pulse propagates in a nonlinear medium [55,75–78]. The analysis of CE is technically simple and relies on a linearized perturbative approach [11]. In the following, we reformulate this treatment to see how X-waves can be directly involved [15]. First, we rewrite Eq. (9.2) in terms of dimensionless variables by setting $u = A/\sqrt{I_0}$, $(\xi, \eta) = (x, y)/W_0$, $\zeta = z/Z_{\mathrm{df}}$, $\tau = (t - k'z)/T_0$, $\Gamma = Z_{\mathrm{df}}/Z_{\mathrm{nl}}$, where I_0 is a reference intensity (e.g., the input beam peak intensity), W_0 is a reference waist (e.g., the input beam waist), $Z_{\mathrm{df}} = 2kW_0^2$, and $Z_{\mathrm{nl}} = (k_0 n_{2I} I)^{-1}$ are diffraction and nonlinear length scales, respectively, and $T_0 = (k'' Z_{\mathrm{df}}/2)^{1/2}$. The normalized equation reads as

$$i\partial_\zeta u + \left(\partial_\xi^2 + \partial_\eta^2\right) u - \partial_\tau^2 u + \Gamma |u|^2 u = 0. \tag{9.29}$$

Restricting ourselves to solutions traveling at the group velocity of the medium ($\beta = 0$), we consider X-waves $u = \psi(\xi, \eta, \zeta)\exp(i\kappa\zeta)$ in the linear case ($\Gamma = 0$), where ψ obeys the equation

$$\left(\partial_\xi^2 + \partial_\eta^2 - \partial_\tau^2\right)\psi = \kappa\psi. \tag{9.30}$$

Explicit solutions can be found, for example, by introducing the complex variable $v = (\Delta - i\tau)^2 + \xi^2 + \eta^2$, and looking for a solution $\psi = \psi(v)$ of the equation

$$6\partial_v \psi + 4v\partial_{vv}\psi = \kappa\psi, \tag{9.31}$$

for example, the real-valued solution (with parameters κ and Δ)

$$\psi_\kappa = \Re\left[\frac{\exp(-\sqrt{\kappa v})}{\sqrt{v}}\right]. \tag{9.32}$$

It is not difficult to recognize in Eq. (9.32) some well-studied X-wave solutions for the scalar wave equation (see the other chapters in this book). Next, starting from the exact plane-wave solution $u = u_0 \exp(iu_0^2 z)$ of Eq. (9.29), with u_0 a constant, the solution of the nonlinear wave equation is written as

$$u = [u_0 + \epsilon(\xi, \eta, \tau, \zeta)] \exp(iu_0^2 \zeta),$$

and at first order in the perturbation ϵ, the governing equation reads as

$$i\partial_\zeta \epsilon + \partial_\xi^2 \epsilon + \partial_\eta^2 \epsilon - \partial_\tau^2 \epsilon + \Gamma u_0^2(\epsilon + \epsilon^*) = 0. \tag{9.33}$$

If we set $\epsilon = \mu(z)\psi(\xi, \eta, \tau) + \nu^*(z)\psi(\xi, \eta, \tau)^*$ with ψ solution of Eq. (9.30), and separate terms weighting ψ and ψ^* in Eq. (9.33), we end up with the coupled equations

$$\begin{aligned} +i\partial_\zeta \mu + \kappa\mu + a_0^2\Gamma(\mu + \nu) &= 0 \\ -i\partial_\zeta \nu + \kappa\nu + a_0^2\Gamma(\mu + \nu) &= 0. \end{aligned} \tag{9.34}$$

Simple arguments lead to the conclusion that an exponentially growing solution such that $u, v, \epsilon \propto \exp(\gamma z)$ with $\gamma = \sqrt{-\kappa(\kappa + 2\chi a_0^2)}$ does exist: Given a pump beam in a Kerr medium (either focusing or defocusing) there exist X-waves that get amplified along propagation. Their spectrum is given by Eq. (9.30) and plotted in Fig. 9.3. In a focusing (defocusing) medium, the amplification range involves $-2a_0^2 < \kappa < 0$ $(0 < \kappa < 2aa_0^2)$. In Fig. 9.4, two examples of X-waves with different κ are given.

This simple perturbative analysis is confirmed by the numerical simulation of Eq. (9.2), which is reported in Fig. 9.5 and involves the early stage of propagation (i.e., it is limited to samples with dimension on the order of the diffraction length) of a Gaussian beam with

$$A(x, y, z = 0) = \sqrt{I_0} \exp\left(-\frac{x^2 + y^2}{2W_0^2} - \frac{t^2}{2T_p^2}\right),$$

whose FWHM spot size and duration (i.e., measured at "full width at half-maximum" of the intensity profile) are 70 μm and 100 fs, respectively. The other parameters are as follows: $n_0 = 1.5$, $n_{2I} = 2.5 \times 10^{-20}$; $k'' = 360 \times 10^{-28}$ in MKSA units; and

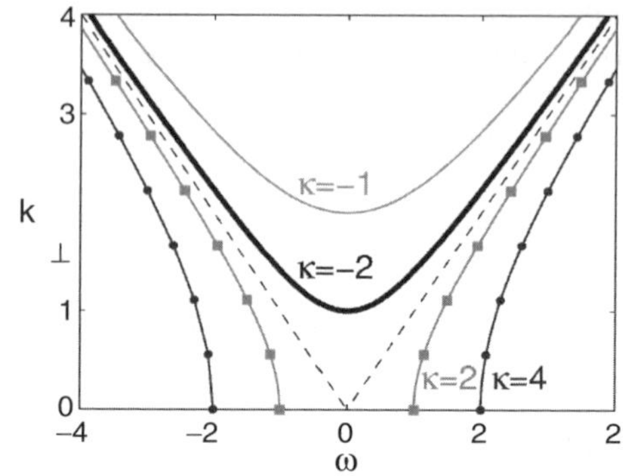

FIGURE 9.3 Spatiotemporal spectrum for various X-waves ψ_κ. The dashed line is the spectrum of the simplest X-wave ($\kappa = 0$).

the peak power is $P = 1.5P_c$, with $P_c = (0.61\lambda)^2\pi/(8n_{2I}n_0) \cong 2.6\,\text{MW}$ the critical power for self-focusing [15].

The arguments reported are limited to perturbative analysis and deal with a beam that is essentially a pump wave, with a small superimposed *X-wave halo* (see [15], and [79] for quadratic media). The halo may be clearly evident in the experiments and in the simulations, because it decays slowly far from the beam center. However,

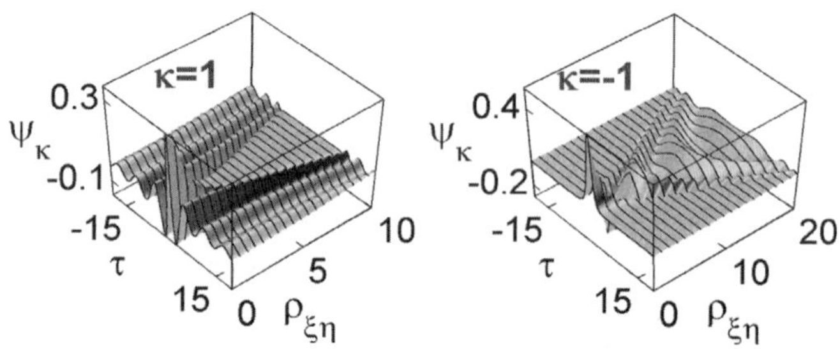

FIGURE 9.4 Profiles of the X-waves ψ_κ for $\Delta = 1$ and $\kappa = \pm 1$ ($\rho_{\xi\eta}^2 = \xi^2 + \eta^2$).

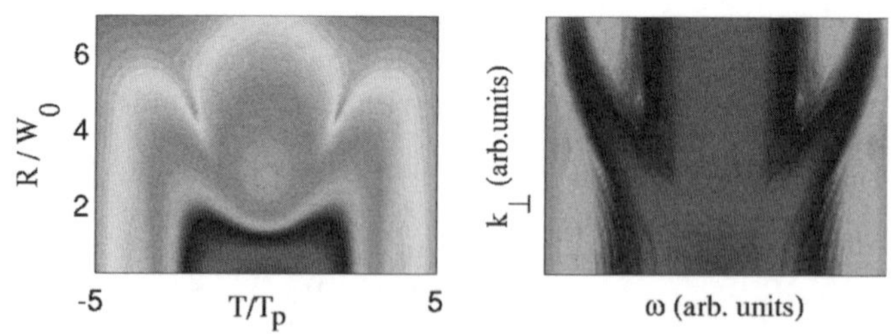

FIGURE 9.5 Spatiotemporal intensity profile (left panel) and spectral content (right panel) of a Gaussian pulsed beam after propagation in a nonlinear Kerr medium (parameters in the text).

regimes investigated experimentally involve dramatic reshaping processes whose analysis must be tackled with a different approach, as discussed below.

9.5 NONLINEAR X-WAVE EXPANSION

We observe that X-waves are localized objects that exist even for a vanishing nonlinear response. Conversely, in the field of nonlinear wave propagation, SWs are known as nonperturbative solutions that do not exist in the absence of nonlinearity and for which a straightforward perturbative expansion (i.e., in a power series of the relevant nonlinear coefficient, as, e.g., n_{2I}) does not apply (see, e.g., [80]). In this respect, NLXs, for which perturbative expansions are meaningful, are strikingly different from SWs. In particular, with a simple argument, we can arrive at the following interesting conclusion [16]: *If the pulsed-beam incident on a nonlinear sample is an X-wave, the nonlinearity just "dresses" the linear X-waves*; that is, the nonlinearity does not destroy the propagation-invariant nature of X-waves. Additionally, even in the case for which the input beam is not an X-wave but rather, undergoes a negligible diffraction and dispersion (to some extent the incident beam approximates a propagation-invariant pulsed beam), an X-wave is generated. Notably enough, this result is not limited to a perturbative regime.

Indeed, by rewriting our starting model (9.2) as

$$i\partial_z A + \frac{1}{2k}\nabla_\perp^2 A - \frac{k''}{2}\partial_t^2 A = \chi \mathcal{P}_{\rm NL}(z, t, r), \tag{9.35}$$

where $\mathcal{P}_{\rm NL}(z, t, r)$ is a nonlinear source weighed by the coefficient χ, one finds after straightforward expansion of A in powers of χ that the relevant equation still has the form of Eq. (9.35) at any order in χ, where the right-hand side stems from solutions at a lower order. If $\chi \mathcal{P}_{\rm NL}(z, t, r) = \mathcal{P}(t - \bar{\beta} z, r)$, the evolution according to Eq. (9.35) always provides a spatiotemporal spectrum corresponding to an X-wave [Eq. (9.9)]. Thus, if an X-wave is taken as a solution of the linear model ($\chi = 0$), the nonlinearity plays the role of "dressing" that solution, assuring that it continues to exist. Since this result is valid at any order in χ, it turns out that linearly self-invariant beams are very robust against the nonlinearity, even beyond first-order perturbation.

9.5.1 Some Examples

1. *X-waves propagating in a nonlinear medium.* $\mathcal{P}$ can be interpreted as a function of the X-wave field and its complex conjugate and travels unchanged since it depends on $t - \beta z$.

2. *Harmonic generation.* This entails a polarization $\mathcal{P}$, which is some power of the pump beam, traveling at the group velocity of the fundamental frequency; if diffraction and dispersion of the pump (in the absence of nonlinearity) is negligible, $\mathcal{P} = \mathcal{P}(T - k'_p z, r)$, where k'_p is the inverse group velocity of the pump.

3. *Kerr media.* Equation (9.2) holds; for the solution at the lowest order ($n_{2I} = 0$), it is possible to take either an X wave packet A_X centered around $\bar{\beta}$ (so that we fall in the case 1) or a wide pulsed beam with negligible diffraction and dispersion (with $\bar{\beta} = 0$) (as in case 2). Higher orders are obtained in the form (9.35): At first order, the right-hand side is proportional to $A_X^2 A_X^*$. As a result of the following analysis, the correction to A_X is still a progressive undistorted wave traveling at the same velocity. At the next order in χ, the situation is the same and the same argument applies, leading us to conclude that an X-wave is obtained again.

We can summarize by saying that whenever the polarization $\mathcal{P}$ can be approximated as a function of a retarded time $t - \bar{\beta}z$ only, an X-wave emerges after propagation in a nonlinear sample at any order of nonlinear effects.

9.5.2 Proof

By writing A as a superposition of slash X-waves,

$$A = \int_0^\infty \int_{-\infty}^\infty C(\alpha_1, \beta_1, z) J_0(\sqrt{kk''}\alpha_1 r) e^{-i(\alpha_1+\beta_1/k'')(t-\beta_1 z)}\, d\alpha_1\, d\beta_1, \qquad (9.36)$$

and inserting Eq. (9.36) into Eq. (9.35), we find that

$$i\frac{\partial C}{\partial z} - \frac{\beta^2}{2k''}C = \mathcal{X}_/[P](\alpha, \beta) e^{i(\alpha+\beta/k'')(\bar{\beta}-\beta)z}, \qquad (9.37)$$

with (α, β) related to $(\omega, k_\perp)$ through Eqs. (9.18) and

$$\mathcal{X}_/[P](\alpha, \beta) = k\alpha\mathcal{B}[P]\left(\alpha + \frac{\beta}{k''}\sqrt{k\,k''}\alpha\right). \qquad (9.38)$$

Equation (9.37) can be integrated readily by taking $C = 0$ at $z = 0$, obtaining

$$C(\alpha, \beta, z) = -i\mathcal{X}_/[P]\frac{\sin gz}{g} \exp\left\{i\left[\left(\alpha + \frac{\beta}{k''}\right)(\bar{\beta} - \beta) - \frac{\beta^2}{4k''}\right]z\right\}, \qquad (9.39)$$

with

$$g = \frac{1}{2}\left[\left(\alpha + \frac{\beta}{k''}\right)(\bar{\beta} - \beta) + \frac{\beta^2}{2k''}\right]. \qquad (9.40)$$

Equation (9.39) can be interpreted in the $(\omega, k_\perp)$ or (α, β) planes and shows that for large propagation distances, C tends to a Dirac δ centered at $g = 0$. Therefore, the propagation acts as a spatiotemporal filter, selecting specific combinations of frequencies and wave vectors that correspond to the condition $g = 0$, or explicitly in

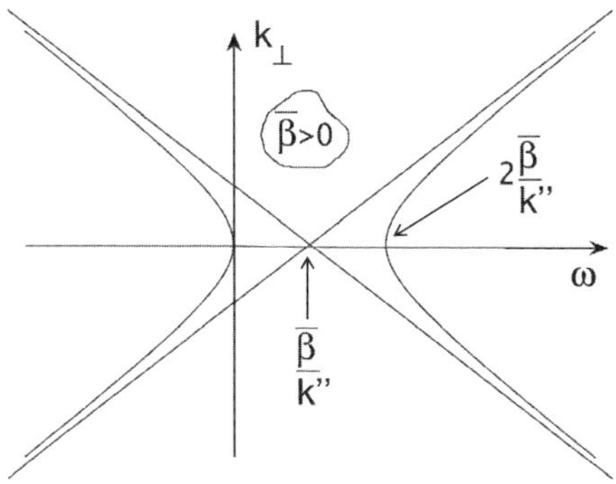

FIGURE 9.6 Sketch of the spatiotemporal spectrum (symmetrized for $k_\perp < 0$) generated when $\bar{\beta} > 0$. The straight lines are slash and backslash spectra.

the $\omega - k_\perp$ plane,

$$\frac{k''}{2}\left(\omega - \frac{\bar{\beta}}{k''}\right)^2 - \frac{k_\perp^2}{2k} = \frac{\bar{\beta}^2}{2k''}. \tag{9.41}$$

This curve represent a hyperbola, as sketched in Fig. 9.6. It is apparent that Eq. (9.41) is the dispersion relation corresponding to Eq. (9.8), so the resulting pulsed beam is a progressive undistorted wave traveling at inverse differential velocity $\bar{\beta}$.

9.5.3 Evidence

In the experimental results reported so far, there is a clear evidence of this hyperbola in the spectral domain (see, e.g., [7,55]), as well as of a structured spectrum [78,81] that corresponds to the splitting phenomenon described below. The asymptotes over which energy is concentrated correspond to slash and backslash X-waves. Thus, in any nonlinear process that can be reduced to Eq. (9.35), X-wave packets are generated spontaneously. When $\bar{\beta} = 0$, Eq. (9.41) yields an exact X-shaped spectrum. The specific features will, in general, depend on the source spectrum [16].

The previous argument supports the existence and spontaneous formation of nonlinear X-waves. However, an element is still missing: exact solutions. So far no analytic solutions are known for NLXs, but numerical analysis can be used as described below.

9.6 NUMERICAL SOLUTIONS FOR NONLINEAR X-WAVES

Nonlinear X-waves are found as solutions to nonlinear wave equations. Let us consider again the normalized equation (9.29):

$$i\partial_\zeta u + \partial_\xi^2 u + \partial_\eta^2 u - \partial_\tau^2 u + \Gamma |u|^2 u = 0. \tag{9.42}$$

NLXs with velocity $1/k'$ ($\beta = 0$) have the form $u = w(\xi, \eta, \tau)\exp(-i\beta_\zeta \zeta)$ with the real-valued profile w satisfying the equation

$$\beta_\zeta w + \partial_\xi^2 w + \partial_\eta^2 w - \partial_\tau^2 w + \Gamma w^3 = 0. \tag{9.43}$$

The number of parameters in Eq. (9.43) can be reduced further by introducing an additional rescaling that discriminates between the cases $\beta_\zeta = 0$ and $\beta_\zeta \neq 0$, respectively. In the latter case, we set $\upsilon \equiv |\beta_\zeta|^{1/2}\tau$, $\gamma \equiv \Gamma/|\beta_\zeta|$, $b \equiv \beta_\zeta/|\beta_\zeta|$, and limiting ourselves to circularly symmetric solutions $w = f(\rho, \upsilon)$ that depend on $\rho = |\beta_\zeta|^{1/2}\rho_{\xi\eta} = [|\beta_\zeta|(\xi^2 + \eta^2)]^{1/2}$, we find that

$$bf + \frac{1}{\rho}\frac{\partial f}{\partial \rho} + \frac{\partial^2 f}{\partial \rho^2} - \frac{\partial^2 f}{\partial \upsilon^2} + \gamma f^3 = 0, \tag{9.44}$$

where $b = \text{sign}(\beta_\zeta)$. Equation (9.44) also holds true in the case $\beta_\zeta = 0$, with $b = 0$, $\gamma = \Gamma$, $\upsilon \equiv \tau$, and $\rho = \rho_{\xi\eta}$. Both families of NLXs depend on the single parameter γ.

The degree of arbitrariness hidden by boundary conditions on f is removed by imposing $f(\rho, \pm\infty) = f(\infty, \upsilon) = 0$ and $df(0, \upsilon)/d\rho = 0$, which give a localized solution in 3D. We also fix the additional constraint $\max(f) = 1$, which does not limit the generality since it can always be satisfied by the free parameter I_0 in the normalization leading to Eq. (9.29).

Finding numerically X-wave solutions of Eq. (9.44) is a nontrivial task because of the slowly decaying tails in the radial (ρ) direction. A trick that can be used consists in interpretating Eq. (9.44) as an evolution problem in the ρ variable. In fact, by making a suitable discretization in the υ coordinate, one reduces the original problem to a set of coupled equations in the variable ρ that can be solved by standard ODE algorithms (care is needed at the first step at $\rho = 0$, where boundary conditions must be used to avoid singularities). The υ discretization can be done using pseudospectral techniques such as Chebychev polynomials or fast Fourier transform (we employed both techniques and found comparable performances).

It turns out that if an appropriate guess is given for the field at $\rho = 0$, the numerical code provides a slowly decaying field with respect to υ at any ρ. Conversely, a wrong guess furnishes an exploding solution as ρ increases. By such an approach, the richness of the nonlinear X-wave solutions can simply be unveiled, without resorting to shooting or iterative techniques, which result in computationally expensive and ill-conditioned problems. We found that an effective guess is given by the real or imaginary part of a linear exact X-wave solution (which is a solution as $\gamma = 0$):

$$u = \frac{1}{\sqrt{(\Delta - i\upsilon)^2 + \rho^2}}, \tag{9.45}$$

where Δ is a length scale with respect to υ for the profile at $\rho = 0$. The real (imaginary) part provides an even (odd) profile at $\rho = 0$, which is inherited by nonlinear solutions, as discussed below (we take $\Delta = 1$ in the following). Once a solution for $\gamma \neq 0$

is found for a given set of parameters, we enlarged the integration domain to the maximum dimensions made available by our computational resources to verify that the function is indeed localized. In practice, this amounts to increasing the maximum value of ρ in our ODE solver and to enlarge the domain in υ to avoid boundary effects in the ρ evolution.

9.6.1 Bestiary of Solutions

Trial Classification Once the guess at $\rho = 0$ is fixed, the solution depends on the two parameters b and γ. An open issue is the classification of nonlinear X-waves. For the moment they can be divided roughly into two classes, distinguished by the two values $b = 0$ and $b = 1$. The former correspond to a spatiotemporal spectrum in a low-intensity regime (i.e., $\gamma \to 0$) lying around the spectral lines $k_\rho^2 = k_\upsilon^2$ (with obvious notation). Conversely, the latter is characterized by $k_\rho^2 - k_\upsilon^2 = b$, and hence the spectrum is not contiguous to the origin in the low-intensity limit. As a result, the beam profile exhibits oscillations regardless of the value of γ, at variance with the case $b = 0$, where such oscillations appear at large values of γ (see the figures that follow).

Infinite (Nonnumerable) Number of Solutions Notably, NLXs exist for both focusing and defocusing nonlinearities (the latter case being relevant for BECs), due to the fact that linear solutions exist; additionally, it is expected that the infinity of solutions in the linear case reflects an infinity of solutions in the nonlinear case. In other words, differently from SWs, which exist only in the presence of a nonlinear response and such that only one solution exists for a fixed set of parameters (e.g., waist), X-waves are innumerable. A rigorous proof of this fact is still to come.

Infinite Energy Even if, numerically, energy is necessarily finite (as the spatial grid adopted in the calculation), it is well established that Eq. (9.44) does not admit finite-energy solutions [82]. Additionally, given the fact that the tails in any direction must decay according to the linear equation, because of the corresponding small values of $|u|^2$, so that the nonlinear terms are negligible, the numerical solution decays slowly with respect to ρ, and hence the NLXs found numerically have infinite energy [for $\rho \in [0, \infty)$ and $\upsilon \in (-\infty, \infty)$].

Focusing Nonlinearity In Fig. 9.7 we show the effect of the nonlinear response on the profile of the fundamental X-wave (9.45). We consider the case $b = 0$ such that Eq. (9.45) is an exact solution for $\gamma = 0$ (in the absence of nonlinearity). Then we report the X-wave profile obtained numerically for increasing positive γ (focusing medium). It is clearly evident that the nonlinearity concentrates the energy around the central spot up to a region at high γ, where oscillations start to take place. Figure 9.8 displays the same result for odd X-waves. Figures 9.9 and 9.10 show solutions with $b = 1$, which also display oscillations in the linear regime ($\gamma = 0$), with frequency

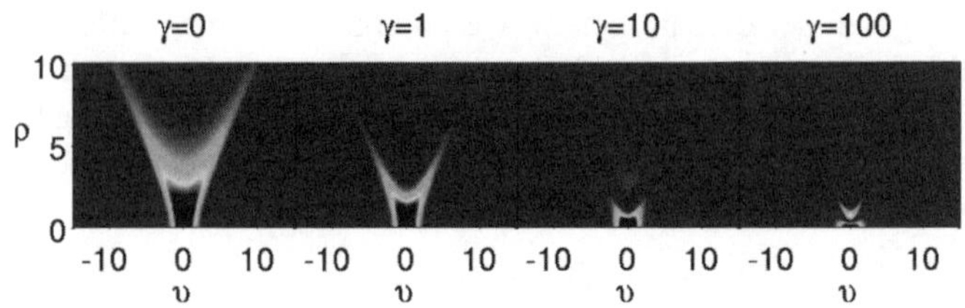

FIGURE 9.7 "Dressing" of a fundamental even X-wave for increasing nonlinearity; the intensity f^2 is shown versus ρ and υ ($\Delta = 1, b = 0$).

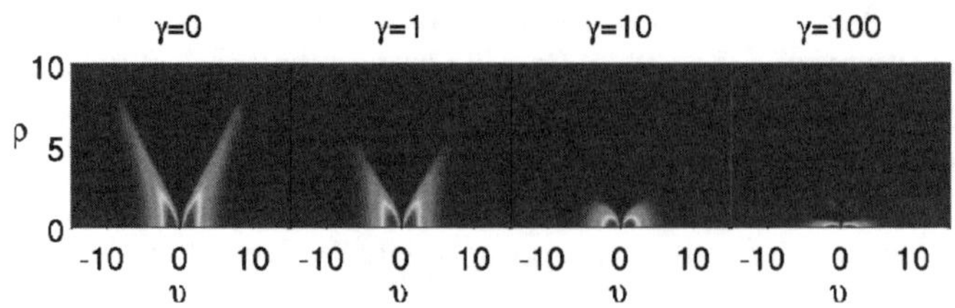

FIGURE 9.8 "Dressing" of a fundamental odd X-wave for increasing nonlinearity; the intensity f^2 is shown versus ρ and υ ($\Delta = 1, b = 0$).

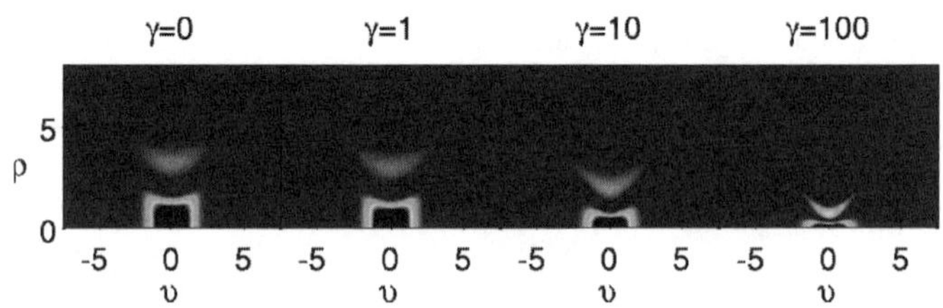

FIGURE 9.9 "Dressing" of a fundamental even X-wave with $b = 1$ for increasing nonlinearity; the intensity f^2 is shown versus ρ and υ ($\Delta = 1, b = 1$).

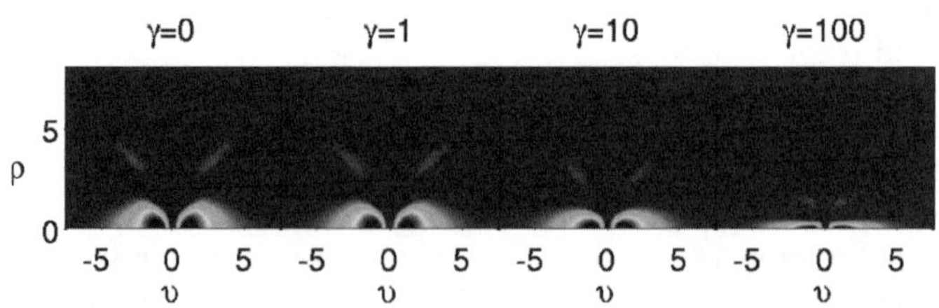

FIGURE 9.10 "Dressing" of a fundamental even X-wave with $b = 1$ for increasing nonlinearity; the intensity f^2 is shown versus ρ and υ ($\Delta = 1, b = 1$).

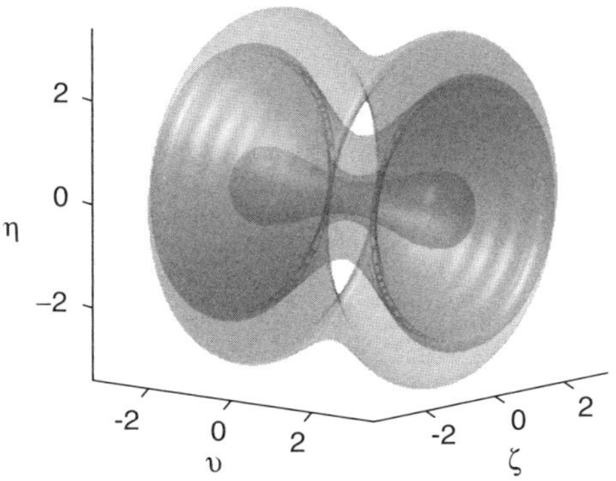

FIGURE 9.11 Isosurface (at level 0.01 of u^2) of an even NLX solution with parameters $\gamma = 100$, $b = 0$, $\Delta = 1$, $\kappa = 0$. Note that this type of solution can be viewed as multiple fundamental X-waves nested one into the other.

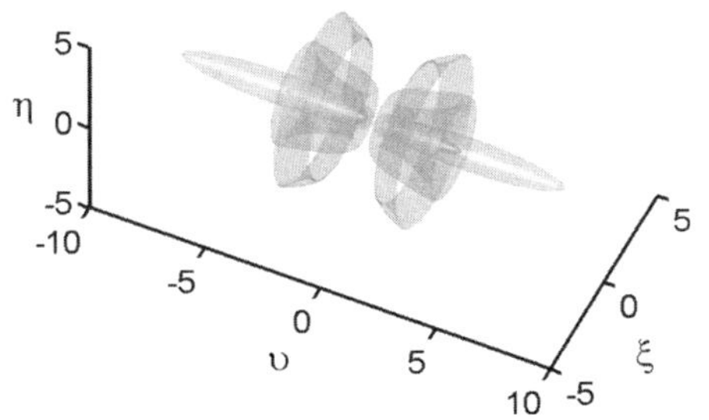

FIGURE 9.12 Isosurface of an odd NLX solution with parameters $\gamma = 100$, $b = 1$, $\Delta = 1$, $\kappa = 0$; the inner surface corresponds to level 0.06 of u^2 and the outer surface to level 0.014 of u^2. This type of X-wave is formed by symmetric bullets that travel locked together (high-intensity area in the middle surface, in red) in the presence of an X-wave halo.

increasing with γ. The 3D representations of these numerical solutions are reported in Figs. 9.11 and 9.12.

Defocusing Nonlinearity The situation for the defocusing case ($\gamma = -1$) is more tricky. This case is of particular interest for matter X-waves, as discussed above. We consider only solutions with an even profile at $\rho = 0$. When $b = 0$ we were not able to find localized solutions starting from the guess given by Eq. (9.45). Conversely, results with this guess when $b = 1$ are obtained, but at variance with the focusing case, we were able to find solutions only up to some critical value of $|\gamma|$ ($|\gamma| \cong 2$). Nevertheless, using a different guess, the "κ–X wave" from Eq. (9.32) with $\kappa = 1$, solutions are also found for $b = 0$ and for higher values of $|\gamma|$, as shown in Fig. 9.13. We observe that in the defocusing cases, solutions get less localized as the amount of nonlinearity is increased (as physically reasonable). In Fig. 9.14 we show a 3D representation of an NLX in the defocusing case.

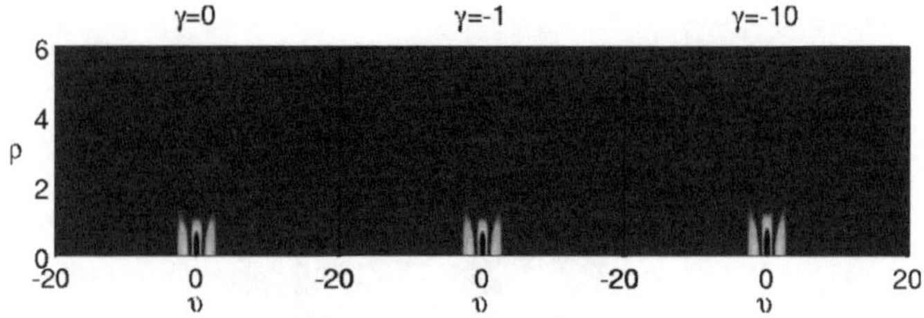

FIGURE 9.13 "Dressing" of even X-waves in Eq. (9.32) with $\kappa = \Delta = 1$ and $b = 0$ for increasing defocusing nonlinearity.

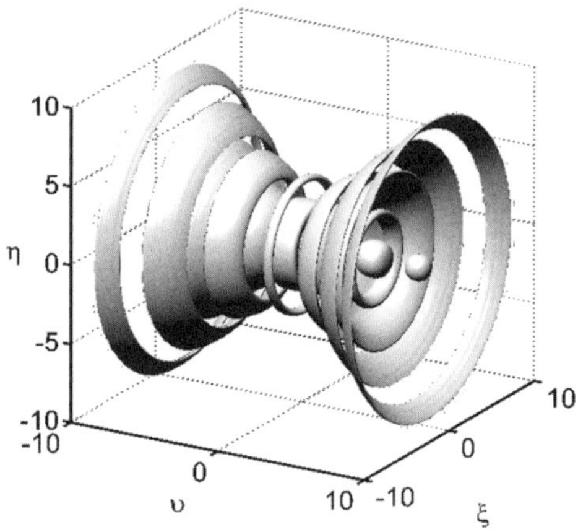

FIGURE 9.14 Isosurface (u^2, level 0.001) for an X-wave in a defocusing medium with parameter $b = 1, \kappa = 1, \Delta = 1, \gamma = -1$. Note both the radial and the longitudinal modulation of the energy distribution, which are due to nonvanishing κ and b parameters. This type of solution can be regarded as a distribution of bullets (clearly evident at $\rho = 0$) with a modulated X-wave halo.

Remark The examples reported are clearly not conclusive. Although there is numerical evidence of the existence of nonlinear X-wave solutions, their multitude and complexity call for an analytical approach. So far, the existence of NLX solutions can only be inferred from the regularity of the reported solutions; they cannot be envisaged for sure.

9.7 COUPLED X-WAVE THEORY

Exact NLX solutions describe a stationary regime: a 3D bullet of energy (or density for BECs) traveling without distortion. However, NLXs may generate spontaneously during nonlinear propagation, and this is a highly dynamic process that in some experimentally and numerically investigated cases does not result in a propagation-invariant regime. Indeed, it has been observed that even if an X-shaped spectrum is obtained, this does not necessarily correspond to a stable X-shaped packet but to a periodic redistribution of energy in and out of a conical region [49,51,55]. This type of behavior leads to consider nonlinear dynamics of X-waves.

To fix the ideas, one can take into account one or more X-waves generated spontaneously and then exhibiting some dynamics, or some X-wave generator device (i.e., an axicon) shooting X-waves into a nonlinear medium, inside which they interact (an experiment that so far has not been performed). In all these cases the natural approach is to develop a coupled X-wave theory which is reviewed briefly below (for details, see [16]).

We again consider Eq. (9.2), with $X(\alpha, \beta, z) = f_p(\alpha)C(\beta, z)$ being the X-wave transform of solution A, which represents an X-wave packet centered around medium group velocity (i.e., $\bar{\beta} = 0$). If C peaks around $\bar{\beta} = 0$, all components travel at nearly the same velocity and one obtains *coupled X-wave equations* in the approximate form

$$\begin{aligned} &i\partial_z C(\beta, z) - \frac{\beta^2}{2k''} C(\beta, z) \\ &\quad = -\frac{k\, n_{2I}}{n_0} \int \chi(\beta + \beta_1 - \beta_2 - \beta_3) C(\beta_1)\, C(\beta_2)\, C(\beta_3)^*\, d\vec{\beta}, \end{aligned} \tag{9.46}$$

where the interaction kernel $\chi(\gamma)$ is the Fourier transform of the quantity $\sigma(s)$ defined below. Taking the Fourier transform of Eq. (9.46) [see Eq. (9.24)], we find that

$$i\frac{\partial c}{\partial z} + \frac{k''}{2}\frac{\partial^2 c}{\partial s^2} + k_0\, n_{2I}\sigma(s)|c|^2\, c = 0. \tag{9.47}$$

Hence, *the evolution of an X-wave packet in a nonlinear Kerr medium can be approximated by an effective (1 + 1D) nonlinear Schrödinger equation with a nonhomogeneous nonlinear coefficient.* The latter, given by $\sigma(s)$, has the dimensions of an inverse area and is expressed by the Fourier transform of the kernel χ:

$$\sigma(s) = 4\pi^2 k'' \int_{-\infty}^{\infty} \chi(\gamma) e^{i\gamma\, s/k''}\, d\gamma = \int_0^{\infty} |2\pi\sqrt{k''}\, \varphi_l{}^{(p)}|^4 r\, dr. \tag{9.48}$$

$\sigma(s)$ is the spatial self-overlap of the component X-wave profile at $\beta = 0$. Any solution c of the nonlinear equation (9.47), with C given by Eq. (9.24), generates a solution A of the envelope equation (9.2):

$$A = \int C(\beta, z)\psi_l{}^{(p)}(r, t - \beta z, \beta)\, d\beta = \int c(s, z)\xi_l{}^{(p)}(r, t, s, z)\, ds, \tag{9.49}$$

with

$$\xi_l{}^{(p)}(r, t, s, z) = \frac{1}{\sqrt{k''}} \int e^{-i(s+t)(\beta/k'') + i(\beta^2/k'')z} \varphi_l{}^{(p)}(r, t - \beta z)\, d\beta. \tag{9.50}$$

Thus, the (3D + 1) nonlinear evolution problem (9.2) is reduced (under suitable approximations) to a (1 + 1D) model (9.47). Two applications are discussed below.

9.7.1 Fundamental X-Wave and Fundamental Soliton

Let us consider the previous results for the fundamental X-shaped profile [see Eq. (9.27)]. It can be found that $\sigma(s)$ is a bell-shaped function [16]; when the on-axis temporal dynamics is dominated by the c envelope (i.e., the temporal width of c is smaller that Δ), it can be approximated by $\sigma(0) \cong W_0^{-2}$, where W_0 is the spatial beam waist. The effective nonlinear Schrödinger equation can be rewritten as

$$i\frac{\partial c}{\partial z} + \frac{k''}{2}\frac{\partial^2 c}{\partial s^2} + \frac{k_0 n_{2I}}{W_0^2}|c|^2 c = 0, \qquad (9.51)$$

so that, compared with the plane-wave coefficient, the effective nonlinear coefficient has the additional factor W_0^{-2}. This reflects the compensation of diffraction due to X-waves, which behave as "modes of free space." $\sigma(s)$ resembles the mode overlap in a waveguide from coupled mode theory. Equation (9.51) is well known from soliton theory, and a number of nontrivial exact solutions can be built (see, e.g., [80]). The fundamental soliton [when $\sigma(s) \cong \sigma(0)$] can be expressed in terms of the peak c_0^2 of the energy distribution function; it reads

$$A = \int c_0 \left(\sqrt{\frac{c_0^2 k_0 n_{2I}\sigma(0)}{k''}} s\right) \exp\left[i\frac{c_0^2 k_0 n_{2I}\sigma(0)}{2} z\right] \xi_l^{(0)}(r, t, z, s)\, ds. \qquad (9.52)$$

Equation (9.52) expresses the "dressing" mechanism associated with nonlinearity: the latter acts only on the shape of the envelope, which even in the linear limit would travel almost undistorted. Despite this remarkable difference with solitary waves, the (approximate) validity of an integrable model such as Eq. (9.51) seems to establish a strong link with solitons.

9.7.2 Splitting and Replenishment in Kerr Media as a Higher-Order Soliton

The self-trapped behavior of the 3D beam given by Eq. (9.52) is ultimately related to the X-shape. More interesting dynamics can be described referring to multisoliton solutions of the integrable equation (9.51). The $N > 1$ solitons (see, e.g., [80,83]) are natural concepts in explaining splitting and replenishment, investigated numerically in [49] and experimentally in [52]. It is noteworthy that breathing linear X-waves have been reported [25,28]. Using higher-order soliton solutions for (9.51) (see, e.g., [83]), one can build an approximate *breathing nonlinear X-wave* [16]. Figure 9.15 shows an example of the corresponding spatiotemporal profile. The periodic depletion and replenishment of the X-shaped distribution are apparent. During propagation, the breather solution pulsates and the beam evolves, retaining most of its energy localized,

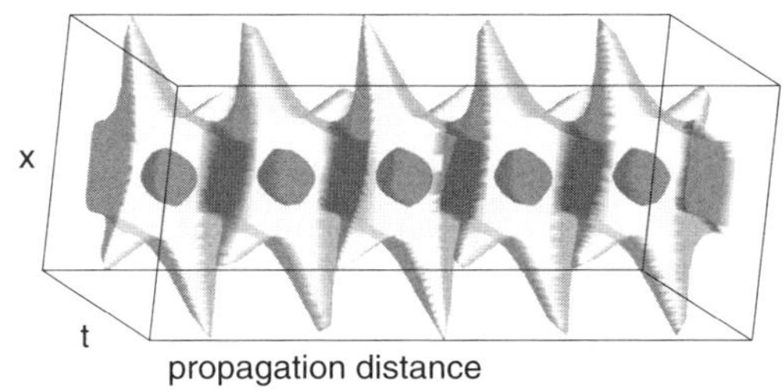

FIGURE 9.15 Typical spatiotemporal profile (at $y = 0$) of a breathing X-wave. Two isosurfaces are displayed: the darkest corresponds to higher intensity.

but exhibiting the nontrivial nonlinear dynamics of the X-wave. It is also noticeable that higher-order solitons exhibit the spectral splitting typically described in numerical simulations.

Before concluding, it is fruitful to summarize the splitting/replenishment phenomenon commonly observed in experiments in Kerr media in terms of NLXs. At the beginning, a wide bell-shaped pulsed beam evolves into an X-wave, owing to the spatiotemporal pattern formation of X-wave instability [15]. Once the envelope width is reduced sufficiently, the increased intensity through compression feeds the generation of a higher-order soliton, or breather. After some spatiotemporal oscillations, several mechanisms may intervene to stop the periodic behavior of, for example, losses (eventually, of nonlinear origin, such as two-photon absorption) or simply that for large propagation distances, the nonlinear response average out, due to the sliding between components of the finite-energy X-wave packet.

9.8 BRIEF REVIEW OF EXPERIMENTS

9.8.1 Angular Dispersion

The key feature that identifies X-waves is the *angular dispersion* (AD), the dependence of the temporal frequency on angle. Using AD it is possible to moderate, and in principle make vanishing any tendency to delocalize energy due to material dispersion [19,84–87]. Before invoking X-waves explicitly, the role of AD was considered with the aim of obtaining broadband phase matching in SHG experiments [88]. AD has been adopted in all modern commercial optical-parametric amplifiers that exploit second-order nonlinearity to achieve tunable light sources: a very active field of research, linking together localized waves and laser physics [10,31,89–92]. AD was also exploited for the observation of temporal solitons in quadratic media [93].

9.8.2 Nonlinear X-Waves in Quadratic Media

Experiments leading to the first observation of NLXs were aimed at the generation of light bullets starting from a Gaussian pulse displaying AD, in order to compensate the group-velocity difference between two harmonics (a first harmonic, FH, and a second harmonic, SH) that get involved in the nonlinear process. During these experiments, it was found that AD-free input beams lead to unexpected spatiotemporal localization. Numerical simulations showed the emergence of an X-type structure, and the results were explained by the existence of a stationary nonlinear X-wave acting as an attractor for nonlinear dynamics [5,6,12]. The observations reported were then confirmed by more accurate experiments, which lead ultimately to the spatiotemporal tomography of the nonlinear X-wave generated [8,9]. Related experimental results, with emphasis on the conical emission processes, were reported in [94–96].

9.8.3 X-Waves in Self-Focusing of Ultrashort Pulses in Kerr Media

Recent investigations of ultrashort pulse propagation in Kerr media with normal ($k'' > 0$) dispersion [33,49,51–55,97] have shown that light filaments formed in condensed matter have a conical oscillating tail [51] and a complicated spatiotemporal structure [52]. Numerical simulations [49,50] show that the dynamics is governed by cycles of pulse splitting and subsequent temporal replenishment of the on-axis pulse. Past the first splitting point the field pattern shows evidence of X-waves. Since the latter is highly dynamical, the X features are best seen in the spectral domain (the spectrally resolved far field), where the wave exhibits characteristic tails that follow the X-shaped dispersion curve (see Fig. 9.6). In typical experimentally retrieved spectra, X-shaped tails are clearly evident (as, e.g., [55,98]). The spectral arms observed are due to conical emission [15,75,99]; the corresponding arms follow the asymptotic lines $k_\perp = \sqrt{k_0 k''}(\omega - \omega_0)$ characteristic of the linear dispersion relationship (see Fig. 9.6), while at low spatial frequencies ($k_\perp \cong 0$) there is evidence of the spectral gap predicted [55]. Indeed, the locus of points θ and Ω corresponding to maximum intensity in the arms fits the hyperbolic curve of Eq. (9.41) [16]. The splitting and replenishment scenario is well settled (qualitatively, to say the least) in the higher-order soliton proposed above. These results provide a clear indication that NLXs can be taken as a "paradigm" for interpreting ultrafast dynamics of laser pulses in a condensed matter, however many issues are still to be deepened.

9.9 CONCLUSIONS

In this chapter we provided an introduction to the rapidly growing field of NLXs, from which many other lines of investigation recently originated, as being due to the intense research activity in modern photonics. Related most closely to this chapter are the O-waves (analogs of envelope X-waves but with anomalous dispersion), including nonlinear effects [34,56], as well as the deepening of (2 + 1D) models [29,70].

Fields that are still at an embryonic stage but that are potentially very productive are NLXs of discrete models [32,35] and in dissipative systems (e.g., optical resonators) [36–38].

Many results are currently being published that concern LWs and NLXs in photonic crystals (i.e., media displaying a periodic refractive index; see other chapters in this book and [39–44,46,47]).

In the authors' opinion the most relevant challenges, from a theoretical perspective, are certainly those related in finding exact solutions for NLXs. For what concerns experiments, perhaps the most fascinating issue is the observation of matter X-waves, while a lot of interesting investigations are still to be performed in nonlinear optics: first, those related to many theoretical predictions, and then topics such as harmonic generation by an X-wave and the nonlinear interaction between LWs and related issues. A particular mention goes to quantum effects (as in [100–102]), which are relevant for both matter and optical X-waves; the interplay between squeezing and entanglement due to nonlinear processes and (3D + 1) propagation-invariant properties is still largely unexplored and may have relevant implications in fundamental physics and quantum information.

REFERENCES

1. S. Trillo and W. Torruealls, Eds., *Spatial Solitons*, Springer-Verlag, Berlin, 2001.
2. Y. Kivshar and G. P. Agrawal, *Optical Solitons*, Academic Press, San Diego, CA, 2003.
3. B. A. Malomed, D. Mihalache, F. Wise, and L. Torner, Spatiotemporal optical solitons, *J. Opt. B Quantum Semiclass. Opt.* **7**, R53–R72 (2005).
4. F. Wise and P. Di Trapani, Spatiotemporal solitons: the hunt for light bullets, *Opt. Photon. News*, pp. 28–32 (2002).
5. G. Valiulis, J. Kilius, O. Jedrkiewicz, A. Bramati, S. Minardi, C. Conti, S. Trillo, A. Piskarskas, and P. Di Trapani, Space-time nonlinear compression and three dimensional complex trapping in normal dispersion, in *Quantum Electronics and Laser Science Conference* Vol. 57 of *Trends in Optics and Photonics*, Optical Society of America, Washington, DC, 2001, Postdeadline Papers QPD10-1, arXiv:physics/311081.
6. C. Conti, P. Di Trapani, S. Trillo, G. Valiulis, S. Minardi, and O. Jedrkiewicz, Nonlinear X-waves: light bullets generation in normally dispersive media? In *Quantum Electronics and Laser Science Conference Trends in Optics and Photonics*, Optical Society of America, Washington, DC, 2002, p. QTuJ6.
7. K. Kriger, Bullets of light, *Phys. Rev. Focus* **12**, story 7 (2003); http://focus.aps.org/story/v12/st7.
8. J. Trull, O. Jedrkiewicz, P. D. Trapani, A. Matijosius, A. Varanavicius, G. Valiulis, R. Danielius, E. Kucinskas, A. Piskarskas, and S. Trillo, Spatiotemporal three-dimensional mapping of nonlinear X waves, *Phys. Rev. E* **69**, 026607 (2004).
9. O. Jedrkiewicz, J. Trull, G. Valiulis, A. Piskarskas, C. Conti, S. Trillo, and P. D. Trapani, Nonlinear X waves in second-harmonic generation: experimental results, *Phys. Rev. E* **68**, 026610 (2003).
10. S. Orlov, A. Piskarskas, and A. Stabinis, Focus wave modes in optical parametric generators, *Opt. Lett.* **27**, 2103–2105 (2002).
11. S. Trillo, C. Conti, P. D. Trapani, O. Jedrkiewicz, J. Trull, G. Valiulis, and G. Bellanca, Colored conical emission by means of second-harmonic generation, *Opt. Lett.* **27**, 1451–1453 (2002).

12. P. Di Trapani, G. Valiulis, A. Piskarskas, O. Jedrkiewicz, J. Trull, C. Conti, and S. Trillo, Spontaneously generated X-shaped light bullets, *Phys. Rev. Lett.* **91**, 093904 (2003).
13. C. Conti, S. Trillo, G. Valiulis, A. Piskarskas, O. Jedrkiewicz, J. Trull, and P. Di Trapani, Nonlinear electromagnetic X-waves, *Phys. Rev. Lett.* **90**, 170406 (2003), arXiv:physics/0204066.
14. C. Conti and S. Trillo, X-waves generated at the second harmonic, *Opt. Lett.* **28**, 1251–1253 (2003), arXiv:physics/0208097.
15. C. Conti, X-wave-mediated instability of plane waves in Kerr Media, *Phys. Rev. E* **68**, 016606 (2003).
16. C. Conti, Generation and nonlinear dynamics of X waves of the Schrödinger equation, *Phys. Rev. E* **70**, 046613 (2004).
17. M. A. Porras, S. Trillo, C. Conti, and P. Di Trapani, Paraxial envelope X-waves, *Opt. Lett.* **28**, 1090–1092 (2003).
18. M. A. Porras and P. D. Trapani, Localized and stationary light wave modes in dispersive media, *Phys. Rev. E* **69**, 066606 (2004), arXiv:physics/0309084v1.
19. M. A. Porras, G. Valiulis, and P. Di Trapani, Unified description of Bessel X waves with cone dispersion and tilted pulses, *Phys. Rev. E* **68**, 016613 (2003).
20. S. Longhi, Spatio-temporal Gauss–Laguerre in dispersive media, *Phys. Rev. E* **68**, 066612 (2004).
21. A. Ciattoni, C. Conti, and P. Di Porto, Universal space time properties of X-waves, *J. Opt. Soc. Am. A* **21**, 451–455 (2004).
22. A. Ciattoni, C. Conti, and P. Di Porto, Vector electromagnetic X waves, *Phys. Rev. E* **69**, 036608 (2004).
23. A. Ciattoni, and P. Di Porto, One-dimensional nondiffractive pulses, *Phys. Rev. E* **69**, 056611 (2004).
24. P. Saari and K. Reivelt, Generation and classification of localized waves by Lorentz transformations in Fourier space, *Phys. Rev. E* **69**, 036612 (2004).
25. M. Zamboni-Rached, A. M. Shaarawi, and E. Recami, Focused X-shaped pulses, *J. Opt. Soc. Am. A* **21**, 1564–1574 (2004), arXiv:physics/0309098v1.
26. M. Zamboni-Rached, H. E. Hernández-Figueroa, and E. Recami, Chirped optical X-shaped pulses in material media, *J. Opt. Soc. Am. A* **21**, 2455–2463 (2004). arXiv:physics/0405059v2.
27. R. Grunwald, V. Kebbel, U. Griebner, U. Neumann, A. Kummrow, M. Rini, E. T. J. Nibbering, M. Piché, G. Rousseau, and M. Fortin, Generation and characterization of spatially and temporally localized few-cycle optical wave packets, *Phys. Rev. A* **67**, 063820 (2003).
28. A. M. Shaarawi, I. M. Besieris, and T. M. Said, Temporal focusing by use of composite X-waves, *J. Opt. Soc. Am. A* **20**, 1658–1665 (2003).
29. Y. Kominis, N. Moshonas, P. Papagiannis, K. Hizanidis, and D. Christodoulides, Continuous-wave-controlled nonlinear X-wave generation, *Opt. Lett.* **30**, 2924–2926 (2005).
30. S. Longhi, Parametric amplification of spatiotemporal localized envelope waves, *Phys. Rev. E* **69**, 016606 (2004).
31. R. Butkus, S. Orlov, A. Piskarskas, V. Smilgevicius, and A. Stabinis, Phase matching of optical X-waves in nonlinear crystals, *Opt. Commun.* **244**, 411–421 (2005).

32. S. Droulias, K. Hizanidis, J. Meier, and D. Christodoulides, X-waves in nonlinear normally dispersive waveguide arrays, *Opt. Express* **13**, 1827–1832 (2005).
33. M. A. Porras, A. Parola, D. Faccio, A. Dubietis, and P. D. Trapani, Nonlinear unbalanced bessel beams: stationary conical waves supported by nonlinear losses, *Phys. Rev. Lett.* **93**, 153902 (2004).
34. M. A. Porras, A. Parola, and P. Di Trapani, Nonlinear unbalanced O waves: nonsolitary conical light bullets in nonlinear dissipative media, *J. Opt. Soc. Am. B* **22**, 1406–1416 (2005).
35. P. G. Kevrekidis, J. Gagnon, D. J. Frantzeskakis, and B. A. Malomed, X,y,z-waves: extended structures in nonlinear lattices, arXiv:cond-mat/0603397 (2006).
36. K. Staliunas, and M. Tlidi, Hyperbolic transverse patterns in nonlinear optical resonators, *Phys. Rev. Lett.* **94**, 133902 (2005).
37. J. M. Soto-Crespo, P. Grelu, and N. Akhmediev, Optical bullets and "rockets" in nonlinear dissipative systems and their transformations and interactions, *Opt. Express* **14**, 4013–4025 (2006).
38. C. T. Zhou, M. Y. Yu, and X. T. He, X-wave solutions of complex Ginzburg–Landau equations, *Phys. Rev. E* **73**, 026209 (2006).
39. D. N. Christodoulides, N. K. Efremidis, P. Di Trapani, and B. A. Malomed, Bessel X-waves in 2D and 3D bidispersive optical systems, *Opt. Lett.* **29**, 1446–1448 (2004).
40. S. Longhi and D. Janner, X-shaped waves in photonic crystals, *Phys. Rev. B* **70**, 235123 (2004).
41. S. Longhi and D. Janner, Diffraction and localization in low-dimensional photonic bandgaps, *Opt. Lett.* **29**, 2653–2655 (2004).
42. S. Longhi, D. Janner, and P. Laporta, Propagating pulsed Bessel beams in periodic media, *J. Opt. B Quantum Semiclass. Opt.* **6**, 477–481 (2004).
43. S. Longhi, Localized and nonspreading spatiotemporal Wannier wave packets in photonic crystals, *Phys. Rev. E* **71**, 016603 (2005).
44. O. Manela, M. Segev, and D. N. Christodoulides, Nondiffracting beams in periodic media, *Opt. Lett.* **30**, 2611–2613 (2005).
45. P. Polesana, D. Faccio, P. Di Trapani, A. Dubietis, R. Piskarskas, A. Varanavicius, A. Gaizauskas, and A. Piskarskas, High localization, focal depth and contrast by means of nonlinear Bessel beams, *Opt. Express* **13**, 6110 (2005).
46. K. Staliunas and R. Herroro, Nondiffractive propagation of light in photonic crystals, *Phys. Rev. E* **73**, 016601 (2006).
47. K. Staliunas, C. Serrat, R. Herrero, C. Cojocaru, and J. Trull, Subdiffractive light pulses in photonic crystals, *Phys. Rev. E* **74**, 016605 (2006).
48. L. Bergé, K. Germaschewski, R. Grauer, and J. J. Rasmussen, Hyperbolic shock waves of the optical self-focusing with normal gvd, *Phys. Rev. Lett.* **89**, 153902 (2002).
49. M. Kolesik, E. M. Wright, and J. V. Moloney, Dynamic nonlinear X-waves for femtosecond pulse propagation in water, *Phys. Rev. Lett.* **92**, 253901 (2004), ArXiv:physics/0311021.
50. M. Kolesik, E. M. Wright, and J. V. Moloney, Interpretation of the spectrally resolved far field of femtosecond pulses propagating in bulk nonlinear dispersive media, *Opt. Express* **13**, 10729–10741 (2005).

51. A. Dubietis, E. Gaizauskas, G. Tamosauskas, and P. D. Trapani, Light filaments without self-channeling, *Phys. Rev. Lett.* **92**, 253903 (2004).

52. A. Matijosius, J. Trull, P. Di Trapani, A. Dubietis, R. Piskarskas, A. Varanavicius, and A. Piskarskas, Nonlinear space-time dynamics of ultrashort wave packets in water, *Opt. Lett.* **29**, 1123–1125 (2004).

53. D. Faccio, A. Matijosius, A. Dubietis, R. Piskarskas, A. Varanavicius, E. Gaizauskas, A. Piskarskas, A. Couairon, and P. D. Trapani, Near- and far-field evolution of laser pulse filaments in Kerr media, *Phys. Rev. E* **72**, 037601 (2005).

54. A. Couairon, E. Gaizauskas, D. Faccio, A. Dubietis, and P. D. Trapani, Nonlinear X-wave formation by femtosecond filamentation in Kerr media, *Phys. Rev. E* **73**, 016608 (2006).

55. D. Faccio, M. A. Porras, A. Dubietis, F. Bragheri, A. Couairon, and P. Di Trapani, Conical emission, pulse splitting, and X-wave parametric amplification in nonlinear dynamics of ultrashort light pulses, *Phys. Rev. Lett.* **96**, 193901 (2006).

56. M. Porras, A. Dubietis, E. Kucinskas, F. Bragheri, A. Couairon, D. Faccio, and P. Di Trapani, From X- to O-shaped spatiotemporal spectra of light filaments in water, *Opt. Lett.* **30**, 3398–3400 (2005).

57. C. Conti and S. Trillo, Nonspreading wave packets in three dimensions formed by an ultracold Bose gas in an optical lattice, *Phys. Rev. Lett.* **92**, 12040 (2004), arXiv:physics/0308051.

58. M. H. Anderson, J. R. Ensher, M. R. Matthews, C. E. Wieman, and E. A. Cornell, Observation of Bose–Einstein condensation in a dilute atomic vapor, *Science* **269**, 198–201 (1995).

59. K. B. Davis, M. O. Mewes, M. R. Andrews, N. J. van Druten, D. S. Durfee, D. M. Kurn, and W. Ketterle, Bose–Einstein condensation in a gas of sodium atoms, *Phys. Rev. Lett.* **75**, 3969–3973 (1995).

60. F. Dalfovo, S. Giorgini, L. Pitaevskii, and S. Stribngari, Theory of Bose–Einstein condensation in trapped gases, *Rev. Mod. Phys.* **71**, 463–512 (1999).

61. L. Khaykovich, F. Schreck, G. Ferrari, T. Bourdel, J. Cubizolles, L. Carr, Y. Castin, and C. Salomon, Formation of matter-wave bright solitons, *Science* **296**, 1290–1293 (2002).

62. L. Pitaevskii, Dynamics of collapse of a confined Bose gas, *Phys. Lett. A* **221**, 14–18 (1996).

63. L. Bergé, Stability criterion for attractive Bose–Einstein condensates, *Phys. Rev. A* **62**, 23607 (2000).

64. H. Saito and M. Ueda, Dynamically stabilized bright solitons in a 2D Bose–Einstein condensate, *Phys. Rev. Lett.* **90**, 40403 (2003).

65. B. P. Anderson and M. Kasevich, Macroscopic quantum interference from atomic tunnel arrays, *Science* **282**, 1686–1689 (1998).

66. E. A. Ostrovskaya and Y. Kivshar, Matter-wave gap solitons in atomic band-gap structures, *Phys. Rev. Lett.* **90**, 160407 (2003).

67. S. Burger, F. S. Cataliotti, C. Fort, F. Minardi, M. Inguscio, M. L. Chiofalo, and M. P. Tosi, Superfluid and dissipative dynamics of a Bose–Einstein condensate in a periodic optical potential, *Phys. Rev. Lett.* **86**, 4447–4450 (2001).

68. S. Pötting, P. Meystre, and E. M. Wright, Atomic solitons in optical lattices, in R. Slusher and B. Eggleton, Eds., *Nonlinear Photonic Crystals* Vol. 10 of *Photonics*, Springer-Verlag, New York, 2003, p. 301.

69. J. Moloney and A. Newell, *Nonlinear Optics*, Westview Press, New York, 2003.
70. P. V. Larsen, M. P. Sorensen, O. Bang, W. Z. Krolikowski, and S. Trillo, Nonlocal description of X waves in quadratic nonlinear materials, *Phys. Rev. E* **73**, 036614 (2006).
71. I. Carusotto, D. Embriaco, and G. C. LaRocca, Nonlinear atom optics and bright-gap-soliton generation in finite optical lattices, *Phys. Rev. A* **65**, R053611 (2002).
72. P. Saari, Superluminal localized waves of electromagnetic field in vacuo, in *Time's Arrows: Quantum Measurement and Superluminal Behavior*, Scientific Monographs, Physics Sciences Series, Italian CNR, Rome, 2001, pp. 37–48, arXiv:physics/0103054.
73. J. Lu and A. Liu, An X wave transform, *IEEE Trans. Ultrason. Ferroelectr. Freq. Control* **47**, 1472–1480 (2000).
74. J. Salo and M. M. Salomaa, Orthogonal X waves, *J. Phys. A Math. Gen.* **34**, 9319–9327 (2001).
75. L. W. Liou, X. D. Cao, C. J. McKinstrie, and G. P. Agrawal, Spatiotemporal instabilities in dispersive nonlinear media, *Phys. Rev. A* **46**, 4202–4208 (1992).
76. G. G. Luther, A. C. Newell, J. V. Moloney, and E. M. Wright, Short-pulse conical emission and spectral broadening in normally dispersive media, *Opt. Lett.* **19**, 789–791 (1994).
77. O. G. Kosareva, V. P. Kandidov, C. Y. Brodeur, A. Chien, and S. L. Chin, Conical emission from laser–plasma interactions in the filamentation of powerful ultrashort laser pulses in air, *Opt. Lett.* **22**, 1332–1334 (1997).
78. E. T. J. Nibbering, P. F. Curley, G. Grillon, B. S. Prade, M. A. France, F. Salin, and A. Mysyrowicz, Conical emission from self-guided femtosecond pulses in air, *Opt. Lett.* **21**, 62–64 (1996).
79. S. Longhi, Localized subluminal envelope pulses in dispersive media, *Opt. Lett.* **29**, 147–149 (2004).
80. P. G. Drazin, and R. S. Johnson, *Solitons: An Introduction*, Cambridge University Press, Cambridge, 1989.
81. D. Faccio, P. Di Trapani, S. Minardi, A. Bramati, F. Bragheri, C. Liberale, V. Degiorgio, A. Dubietis, and A. Matijosius, Far-field spectral characterization of conical emission and filamentation in Kerr media, arXiv:physics/0405086v1 (2004).
82. L. Berge, Wave collapse in physics: principles and applications to light and plasma waves, *Phys. Rep.* **286**, 259–370 (1998).
83. G. Agrawal, *Nonlinear Fiber Optics*, 3rd ed., Academic Press, San Diego, CA, 2001.
84. O. E. Martinez, Pulse distortions in tilted pulse schemes for ultrashort pulses, *Opt. Commum.* **59**, 229–232 (1986).
85. S. Orlov, A. Piskarskas, and A. Stabinis, Localized optical subcycle pulses in dispersive media, *Opt. Lett.* **27**, 2167–2169 (2002).
86. S. Szatmari, P. Simon, and M. Feuerhake, Group-velocity-dispersion-compensated propagation of short pulses in dispersive media, *Opt. Lett.* **21**, 1156–1158 (1996).
87. H. Sonajalg, M. Ratsep, and P. Saari, Demonstration of the Bessel-X pulse propagating with strong lateral and longitudinal localization in a dispersive media, *Opt. Lett.* **22**, 310–312 (1997).
88. G. Szabo and Z. Bor, Broadband frequency doubler for femtosecond pulses, *Appl. Phys. B* **50**, 51–54 (1990).

89. R. Danielius, A. Piskarskas, P. Di Trapani, A. Andreoni, C. Solcia, and P. Foggi, Matching of group velocities by spatial walk-off in collinear three-wave interaction with tilted pulses, *Opt. Lett.* **21**, 973–975 (1996).

90. J. Trull, O. Jedrkiewicz, P. Di Trapani, A. Matijosius, A. Varanavicius, G. Valiulis, R. Danielius, E. Kucinskas, A. Piskarskas, and S. Trillo, Spatiotemporal three-dimensional mapping of nonlinear X waves, *Phys. Rev. E* **69**, 026607 (2004).

91. P. Di Trapani, A. Andreoni, G. P. Banfi, C. Solcia, R. Danielius, A. Piskarskas, P. Foggi, M. Monguzzi, and C. Sozzi, Group-velocity self-matching of femtosecond pulses in noncollinear parametric generation, *Phys. Rev. A* **51**, 3164–3168 (1995).

92. P. Di Trapani, A. Andreoni, P. Foggi, C. Solcia, R. Danielius, and A. Piskarskas, Efficient conversion of femtosecond blues pulses by travelling-wave parametric generation in non-collinear phase matching, *Opt. Commun.* **119**, 327–332 (1995).

93. P. Di Trapani, D. Caironi, G. Valiulis, A. Dubietis, R. Danielius, and A. Piskarskas, Observation of temporal solitons in second-harmonic generation with tilted pulses, *Phys. Rev. Lett.* **81**, 570–573 (1998).

94. H. Zeng, K. Wu, H. Xu, and J. Wu, Seeded amplification of colored conical emission via spatiotemporal modulational instability, *Appl. Phys. Lett.* **87**, 061102 (2005).

95. H. Zeng, J. Wu, H. Xu, and K. Wu, Generation and weak beam control of two-dimensional multicolored arrays in a quadratic nonlinear medium, *Phys. Rev. Lett.* **96**, 083902 (2006).

96. W. Su, L. Qian, H. Luo, X. Fu, H. Zhu, T. Wang, K. Beckwitt, Y. Chen, and F. Wise, Induced group-velocity dispersion in phase-mismatched second-harmonic generation, *J. Opt. Soc. Am. B* **23**, 51–55 (2006).

97. D. Faccio, P. Di Trapani, S. Minardi, A. Bramati, A. Bragheri, F. Bragheri, C. Liberale, V. Degiorgio, A. Dubietis, and A. Matijosius, Far-field spectral characterization of conical emission and filamentation in Kerr media, *J. Opt. Soc. Am. B* **22**, 862–869 (2005).

98. J. Liu, H. Schroeder, S. L. Chin, R. Li, W. Yu, and Z. Xu, Space-frequency coupling, conical waves, and small-scale filamentation in water, *Phys. Rev. A* **72**, 053817 (2005).

99. G. G. Luther, A. C. Newell, and J. V. Moloney, The effects of normal dispersion on collapse events, *Physica D* **74**, 59–73 (1994).

100. P. Saari, M. Menert, and H. Valtna, Photon localization barrier can be overcome, *Opt. Commun.* **246**, 445–450 (2005).

101. C. Conti, Quantum X-waves in Kerr media and the progressive undistorted squeezed vacuum, arXiv:quant-ph/0409130 (2004).

102. C. Conti, Generation of entangled 3D localized quantum wave-packets via optical parametric amplification, arXiv:quant-ph/0309069 (2003).

CHAPTER TEN

Diffraction-Free Subwavelength-Beam Optics on a Nanometer Scale

SERGEI V. KUKHLEVSKY
University of Pécs, Pécs, Hungary

10.1 INTRODUCTION

Diffraction is one of the fundamental laws of physics. It affects, without exception, all classical and quantum mechanical fields. Owing to the law, it is impossible in quantum mechanics to define at a given time, to an arbitrary degree of accuracy both the position of a matter wave packet and its direction. Because of the diffraction phenomenon, light spreads out in all directions after passing a slit smaller than its wavelength. The harder one tries to decrease the beam transverse dimension by narrowing the slit, the more it broadens out. Similarly, the beam width increases dramatically with increasing distance from the slit. Thus, the diffraction imposes a fundamental limit on the transverse dimension of a beam at a given distance from a subwavelength aperture and consequently limits the resolution capabilities and makes more difficult the position requirements of subwavelength-beam optical devices, such as near-field scanning optical microscopes and spectroscopes (see, e.g., [1–6]).

To avoid the diffractive broadening of a beam in free space, one can, generally speaking, use two main approaches. A light beam can be confined to subwavelength transverse dimensions by multiple internal reflections at boundaries of a high-refractive-index waveguide [7–16]. Unfortunately, the guided beam diffracts out in all directions after passing the waveguide output aperture. Another approach uses

Localized Waves, Edited by Hugo E. Hernández-Figueroa, Michel Zamboni-Rached, and Erasmo Recami

diffraction-free beams [17–41]. A nondiffractive beam (e.g., a Bessel-type beam) propagates in frcc space without diffractive broadening. The diffraction-free beams (continuous waves and pulses), whose widths greatly exceed the wavelength, have been well understood and realized experimentally (see e.g., [17–41] and references therein). Recently, the concept of diffraction-free beams has been extended to a subwavelength regime [42]. It was recognized that the basic transverse shape of a Bessel-type beam $U_m(r, \phi, z) = \exp(i\phi m) J_n(r/\Lambda) \exp(-i\beta z)$ having subwavelength ($\Lambda < \lambda$) characteristic transverse dimension and purely imaginary propagation constant $\beta = \sqrt{(n2\pi/\lambda)^2 - (1/\Lambda)^2}$ is preserved, while the intensity decays exponentially with increasing longitudinal distance. Here, (r, ϕ, z) are cylindrical coordinates defined in the usual way, and n is the refraction index of the medium. Ruschin and Leizer [42] also indicated that at the condition $\lambda/n2\pi < \Lambda < \lambda$, a subwavelength ($\Lambda < \lambda$) Bessel-type beam propagates in a high-refraction-index medium as an ordinary (nonevanescent) Bessel-type beam. Although a pulsed Bessel-type beam has not been studied in [42], the condition of its exponential decay or nonevanescent propagation can be understood easily, considering the relevant properties of a subwavelength cylindrical waveguide [43,44] and taking into consideration the concept of the Fresnel waveguide light source [45–48]. More recently, several groups reported very important results related to theoretical principles and practical realization of diffraction-free subwavelength pulses [49–51]. Orlov et al. [49] reported on the generation of a pulse with normalized duration $\Delta\omega/\omega_0 \sim 2$. Orlov et al. [50] also suggested practical setups for the generation of a pulse with an 8.8-fs FWHM duration and a beam width of ~4 μm. Grunwald et al. [51] described the experimental detection of subcycle and subwavelength waves in a sub-10-fs region.

In the present chapter we consider the problem of generation of a subwavelength nanometer-sized diffraction-free beem using a recently established relation between the waveguide and free-space optics [45–48]. The waveguide [7–16] and diffraction-free beam [17–41] optics are usually considered to be independent of each other. It was shown, however, that an arbitrary scalar field (continuous wave or pulse) confined by an arbitrary-shaped steplike or gradient high-refraction-index material waveguide, whose width Λ exceeds the wavelength λ, could be reproduced in free space by a Fresnel light source of this waveguide [45–48]. For instance, the eigenmodes of slab, cylindrical, elliptic, and coaxial multimode waveguides are reproduced in free space by a Fresnel-waveguide source [45–48] like the plane cosine [18], Bessel [23], Mathieu [34], and (TE_{mn}, TM_{mn}, TEM)-type [9,45–48] diffraction-free beams, respectively. Other examples of the Fresnel-waveguide fields reproducing in free space are the coherent, partially coherent, on-axis and off-axis, self-imaging, fractal, continuous, and pulsed waves confined by straight, bent, and tapered waveguides with a steplike or gradient refraction index [45–48,52].

The chapter is organized as follows. In Section 10.2 we discuss the natural spatial and temporal broadening of light beams and the physical mechanisms that contribute to the diffraction- and distortion-free propagation of light pulses in free space. The physics of diffraction-free beam optics in the overwavelength domain ($\Lambda > \lambda$) is described in Section 10.3. In Section 10.4, the Fresnel-waveguide concept is extended to the subwavelength nanometer-scale domain ($\Lambda < \lambda$). A summary and conclusions are given in Section 10.5. The brief description of the model describing the electric and magnetic fields of a continuous wave and an ultrashort pulse used in construction of the Fresnel light source of a subwavelength waveguide is presented in the Appendix.

10.2 NATURAL SPATIAL AND TEMPORAL BROADENING OF LIGHT WAVES†

In this section we consider several general properties of the optical beams which are important for understanding the concept of subwavelength nanometer-sized diffraction-free beams. The physical mechanisms that contribute to the natural spatial and temporal broadening of continuous waves and light pulses in free space are discussed.

The natural spatial and temporal broadening of light beams in free space are closely connected with the spatial and temporal uncertainties of a wave packet. The spatial ($\Delta\mathbf{x} \sim 1/\Delta\mathbf{k}$) and temporal ($\Delta t \sim 1/\Delta\omega$) uncertainties of a wave packet are universal laws of physics and some of the best understood. A detailed description of the phenomena can be found in optical textbooks. The uncertainty relations affect all classical and quantum mechanical fields without exception. For instance, given the de Broglie postulate, $\mathbf{p} = \hbar\mathbf{k}$, the spatial uncertainty in quantum mechanics is determined by the Heisenberg relation, $\Delta\mathbf{x} \sim \hbar/\Delta\mathbf{p}$. Because of the uncertainty, it is impossible to define at a given time to an arbitrary degree of accuracy both the position of a wave packet, $\Delta\mathbf{x}$, and its direction, $\Delta\mathbf{k}/k$. The temporal uncertainty ($\Delta t \sim 1/\Delta\omega$), which in quantum mechanics is often called the *fourth Heisenberg uncertainty relation* ($\Delta t \sim \hbar/\Delta E$), shows that for a wave packet, the smaller the frequency uncertainty $\Delta\omega$, the longer the packet duration Δt. The mathematical principle behind the spatial and temporal uncertainties of wave packets involves a well-known relation between the widths of two functions which are Fourier transforms of each other. Owing to the relation, the greater the energy uncertainty ΔE of a quantum system, the more rapid is the time evolution Δt. Another example is diffraction phenomenon: The limitation of the phase space ($\Delta\mathbf{k}$) of a wave by the x_{ap} and y_{ap} boundaries of an aperture causes the spatial broadening of the wave in the $\{x\}\{y\}$ domain. According to the space-frequency uncertainty, the beam angle divergence $\Delta\mathbf{k}/k$ increases with decreasing region of spatial localization $\Delta\mathbf{x}$. Similarly, the decrease in the spectral region $\Delta\mathbf{k}$ causes an increased value $\Delta\mathbf{x}$. According to the quantum-mechanics operator formalism, the spatial and temporal uncertainties arise because the conjugate variables must satisfy the Heisenberg postulate.

The spatial $\Delta\mathbf{x}$ and temporal Δt uncertainties of a wave packet are usually considered to be independent of each other. In frames of such consideration, the unlimited small temporal and spatial uncertainties ($\Delta t \to 0$, $\Delta\mathbf{x} \to 0$) of a wave packet can be achieved simultaneously. This can be realized, for instance, by using

†This section is reprinted from *Optics Communications*, Vol. 209, S. V. Kukhlevsky and G. Nyitray, Correlation between spatial and temporal uncertainties of a wave packet, pp. 377–382. Copyright © 2002, with permission from Elsevier.

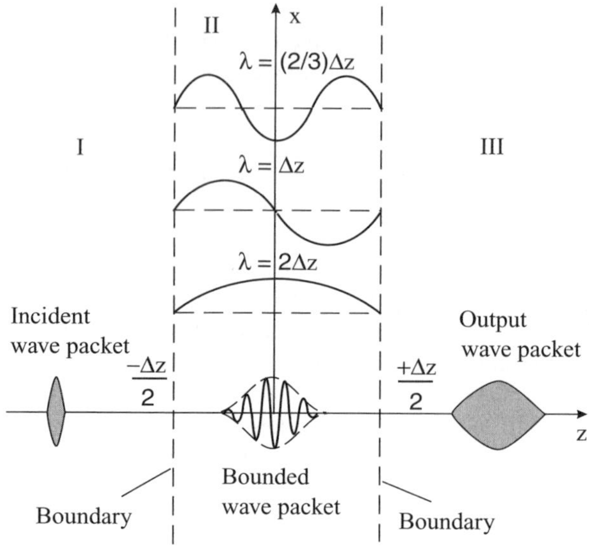

FIGURE 10.1 Propagation of a wave packet through bounded space.

an ultrashort wave packet ($\Delta t \to 0$, $\Delta\omega \to \infty$) passing through a small aperture ($\Delta\mathbf{x} \to 0$, $\Delta\mathbf{k} \to \infty$). The value of the central wavelength λ_0 of the wave packet can be arbitrary. The following simple consideration shows that the spatial ($\Delta\mathbf{x} \sim 1/\Delta\mathbf{k}$) and temporal ($\Delta t \sim 1/\Delta\omega$) uncertainties of a wave packet correlate with each other: The smaller the uncertainty $\Delta\mathbf{x}$, the bigger the uncertainty Δt. The value of the uncertainty Δt is determined by the width $\Delta\mathbf{x}$ and the value of the central wavelength λ_0 of the wave packet.

Let us investigate the propagation of a wave packet of light through the bounded space. For the sake of simplicity, we consider a one-dimensional scalar packet passing through the space confined by two boundaries, as shown in Fig. 10.1. At a point $P(x, z)$ of region I, an incident wave packet $\psi(P, t)$ of duration Δt_I can be presented in the form of the Fourier time expansion:

$$\psi(P, t) = \int_0^\infty \psi(P, \omega) \exp(-i\omega t)\, d\omega, \tag{10.1}$$

where $\omega = ck_z$ is the light frequency. The propagation of the Fourier components $\psi(P, \omega(k_z))$ in regions I, II, and III is governed by the Helmholtz equation

$$[\nabla_z^2 + k_z^2]\psi(P, k_z) = 0, \tag{10.2}$$

with boundary conditions imposed. In nonbounded regions I and III, the Fourier spectra is continuous with the value $\omega(k_z) \in [0, \infty]$. In confined space II, the values of momenta $k_z = \pi n/\Delta z$ are discrete ($n = 1, 2, \ldots$) and a nonzero lower limit

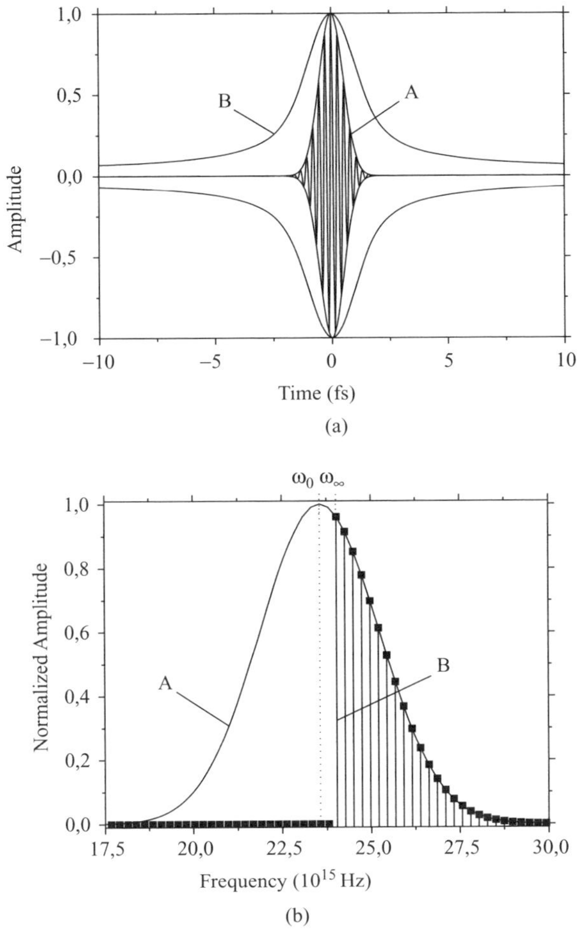

FIGURE 10.2 (a) Amplitudes of the original (A) and modified (B) wave packets. (b) Fourier spectra of the original (A) and modified (B) wave packets. Here $\psi(P,t) = \exp[-2\ln(2)(t/\tau)2]\exp(i\omega_0 t)$; $\tau = 1$ fs is the wave-packet duration and ω_0 is the central frequency of the wave packet. The figure demonstrates the broadening of the femtosecond pulse by cutting off the wave packet spectra at the frequency ω_{co}.

$k_z^{\min} = \pi/\Delta z$ exists due to the boundaries (see Fig. 10.1). Compared with regions I and III, the spectral width in bounded region II is smaller, $\omega \in [\omega(k_z^{\min)}, \infty]$. According to the Fourier analysis, the decrease in the spectral width $\Delta\omega$ of the wave packet in region II (the spectral interval changes from $[0, \infty]$ to $[\omega(k_z^{\min}), \infty]$ causes increasing packet duration $\Delta t_{\text{II,III}}$ in regions II and III. As an example, Fig. 10.2 demonstrates the increasing width of a femtosecond wave packet by cutting off the wave-packet spectra at the frequency $\omega_{co} = \omega(k_z^{\min})$. The cutoff frequency ω_{co} depends on the boundary conditions. The approximate value is given by $\omega(k_z^{\min}) = c\pi/\Delta z$. The cutoff frequency increases with decreasing width Δz of the packet. Thus, the temporal uncertainty $\Delta t_{\mathrm{II,III}}$ is limited by the spatial uncertainty Δz. Such behavior indicates the correlation between the spatial $\Delta z \sim 1/\Delta k_z$ and the temporal $\Delta t \sim 1/\Delta\omega$ uncertainties; the smaller uncertainty, Δz, and the bigger uncertainty, Δt. In the case of a wave packet confined in three directions, the nonzero lower limits of the momenta $\mathbf{k}$ at the x, y, and z directions are given, respectively, by $k_x^{\min} = \pi/\Delta x$, $k_y^{\min} = \pi/\Delta y$ and $k_z^{\min} = \pi/\Delta z$, and the cutoff frequency is given by

$$\omega(\mathbf{k}^{\min}) = c\left(\left(\frac{\pi}{\Delta x}\right)^2 + \left(\frac{\pi}{\Delta y}\right)^2 + \left(\frac{\pi}{\Delta z}\right)^2\right)^{1/2}. \tag{10.3}$$

It can easily be demonstrated, using Eq. (10.3) and the relation $\Delta t \sim 1/\Delta\omega$, that the space–time uncertainty of a wave packet is given by

$$\Delta t \sim \frac{1}{\omega_0 + \Delta\omega/2 - c((\pi/\Delta x)^2 + (\pi/\Delta y)^2 + (\pi/\Delta z)^2)^{1/2}}, \tag{10.4}$$

where Δt is the duration of the spatially localized wave packet, and ω_0 and $\Delta\omega$ are the central frequency and spectral width of the free-space (unbounded) packet [53]. The temporal uncertainty Δt of a wave packet increases with decreasing spatial uncertainties Δx, Δy, and Δz. The increase in the temporal uncertainty Δt of the wave packet caused by decreasing its spatial uncertainty $\Delta\mathbf{x}$ can be reduced by increasing the central frequency ω_0 of the packet. In the limit $\omega_0 \to \infty$, the light wave packets are described by geometrical optics. We believe that the space–time uncertainty relation (10.4) can be considered as a general property of a wave packet, because it is based on the basic properties of the Fourier transformation.

The spatial, temporal, and space–time uncertainty relations are in agreement with results of the exact analytical and numerical solutions of Maxwell's equations for both the overwavelength ($\Lambda > \lambda$) and subwavelength ($\Lambda < \lambda$) domains. As an example, in the overwavelength domain, the angle divergence $\Delta\mathbf{k}/k$ determined by the spatial uncertainty $\Delta\mathbf{x} \sim 1/\Delta\mathbf{k}$ is in agreement with the diffractive broadening of a beam predicted by Fresnel–Kirchhoff diffraction theory [54]. In the case of a plane wave $E(z,t) \sim E(t)\exp(ik_z z)$, the pulse $E(t)$ of duration Δt transforms into the field $E(\omega)$ localized in the Fourier spectral region $\Delta\omega$. The regions of localization Δt and $\Delta\omega$ are given by the temporal uncertainty relation ($\Delta t \sim 1/\Delta\omega$). One of the most important consequences of the space–time uncertainty (10.4) is that the duration Δt of a wave (pulse) localized in the space increases with decreased region of spatial localization $\Delta\mathbf{x}$. In the overwavelength domain, such behavior is known for the Gaussian beams [54,55]. The numerical models of subwavelength diffraction (e.g., [2,44,56–61]) also demonstrate the spatial and time–space uncertainty correlations affecting the spatial and temporal broadening of a subwavelength nanometer-sized light beam. As an example of computer simulation, Figs. 10.4 and 10.5 show the intensity distributions of a continuous wave and light pulse diffracted by a subwavelength nanometer-sized

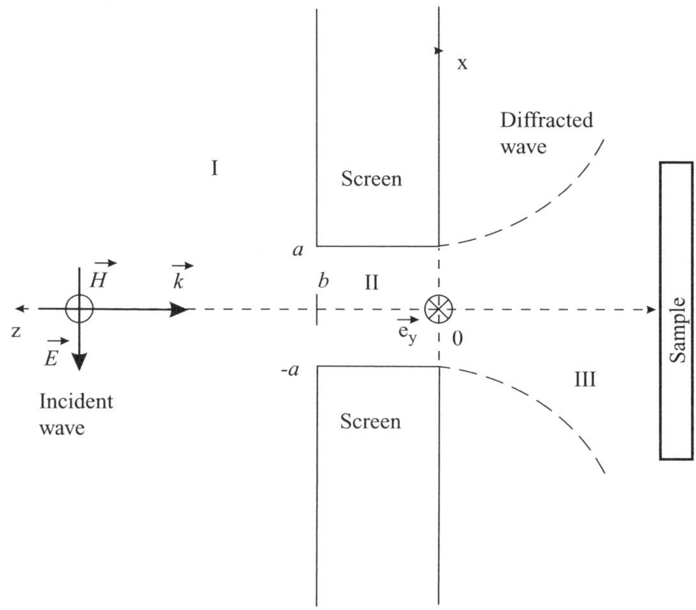

FIGURE 10.3 Diffraction of a continuous wave by a subwavelength nanometer-sized slit (waveguide) in a perfectly conducting screen and its implications to near-field optical microscopy. The slit width and length are $2a$ and b, respectively.

slit (waveguide) in a perfectly conducting thick screen (Fig. 10.3). The details of computations and the model are presented in the Appendix. Figure 10.4 shows the energy flux distribution S_z of a continuous wave diffracted by a subwavelength slit. The energy flux distributions S_z and S_x of a femtosecond pulse passed through the slit

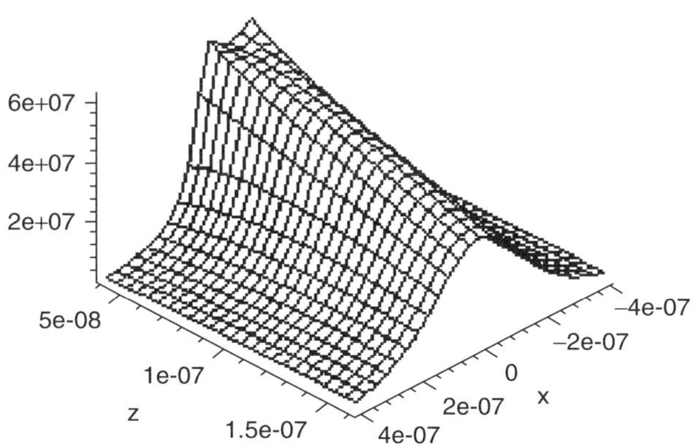

FIGURE 10.4 Energy flux distributions S_z of a continuous wave diffracted by a slit of width $2a = 50$ nm in a perfectly conducting screen of thickness $b = 25$ nm. The wavelength is $\lambda = 500$ nm and the flux is in arbitrary units. The values x and z are in meters.

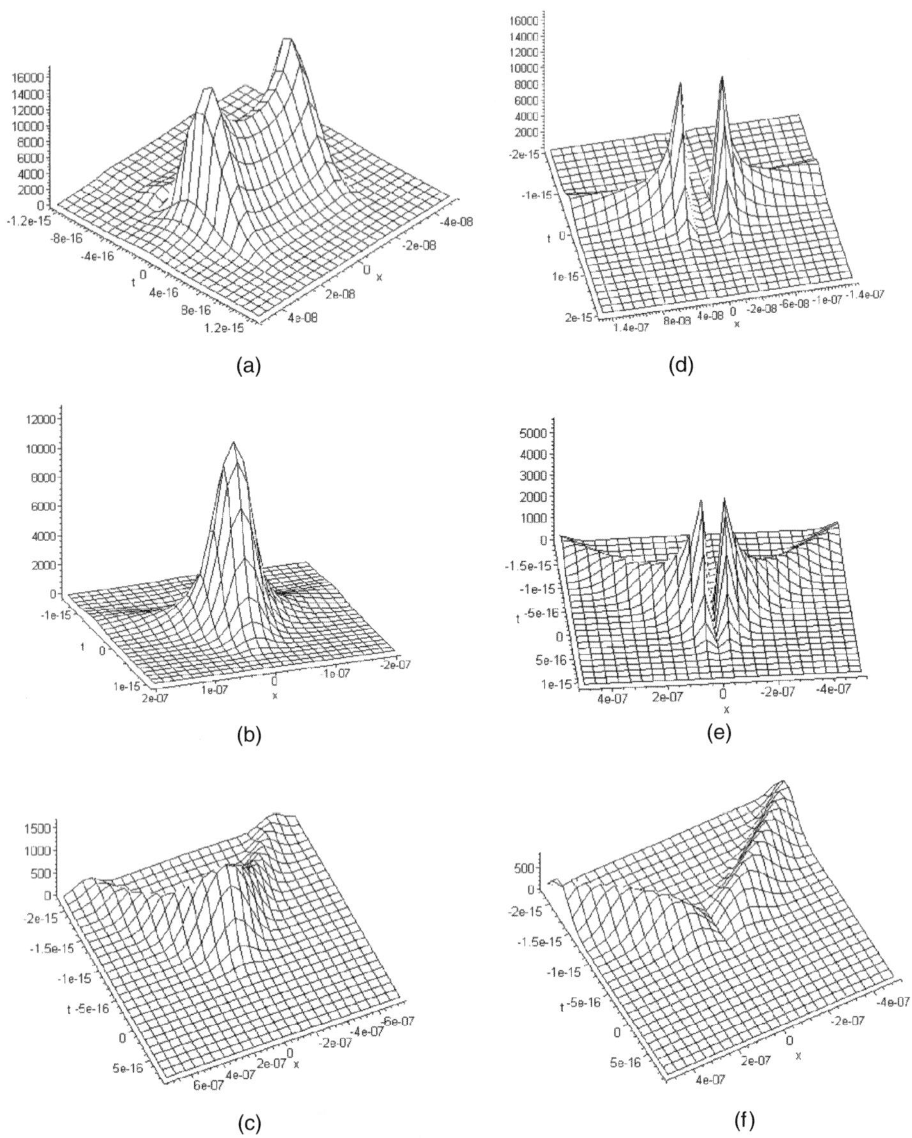

FIGURE 10.5 Energy flux distributions S_z (a, b, c) and S_x (d, e, f) of a 0.5-fs pulse diffracted by a slit of width $2a = 50$ nm in a perfectly conducting screen of thickness $b = 25$ nm at three distances z from the screen: (a,d) $z = -a/10$, (b,e) $z = -a$, and (c,d) $z = -10a$ [59]. Here, the central wavelength of the wave packet is $\lambda_0 = 500$ nm and the flux is in arbitrary units. The values t and x are in femtoseconds and meters, respectively.

are demonstrated in Fig. 10.5. We notice the spatial and temporal broadenings of the light beams, which are in agreement with the spatial and space–time uncertainties for the continuous wave and wavepacket. The figures also demonstrate the feasibility of the femtosecond temporal resolution together with the confinement of light to the nanometer scale.

10.3 DIFFRACTION-FREE OPTICS IN THE OVERWAVELENGTH DOMAIN†

The physics of diffraction-free beam optics in the overwavelength domain ($\Lambda > \lambda$) is described in this section. Bounded optical fields confined by multiple internal reflections at the waveguide boundaries and optical beams propagating in free space have been known for decades and descriptions of many ways to treat the problem appear in the literature. The behavior of such waves is governed by Maxwell's wave equations with boundary conditions imposed. According to the conventional theory of optical waveguides (solving a boundary condition problem), a wave arriving at the guide entrance and satisfying the wave equation and the boundary conditions of the guide propagates without diffractive broadening through the optical conduit as a superposition of the waveguide eigenmodes. In free space the wave equation and the boundary conditions are different from those of the waveguide. A wave propagates between different locations in free space as the optical beam diffracts and broadens. It is usually thought that fields confined by optical waveguides are different in principle from optical beams propagating in free space. We show using the scalar diffraction theory and the method of images that the arbitrary scalar field confined by the optical waveguide can be generated in free space by the appropriate light source.

Let us first consider, for the sake of simplicity, the propagation of waves through a plane waveguide (Fig. 10.6). The guide, which consists of a core and a cladding, is tapered with taper angle γ. The complex index of refraction $n = n_r - in_{im}$ changes abruptly from n_1 to n_2 at the guide boundary. Here n_{im} is the absorption index of the medium. We consider the guide of length z_L with the dimensions $2a$ and $2b$ of the entrance and exit, respectively. In the case of harmonic fields, a wave $E_0(P,t)$ at the point $P(x,z)$ of the guide entrance ($x \in [-a, a]$ and $z = 0$) is given by $E_0(P,t) = E_0(x)\exp[i(-\omega t + \phi_0(x))]$, where x and z are the coordinates of point P in the coordinate system (X, Z); ω and $\phi_0(x)$ are the wave frequency and the phase, respectively. In accordance with the Huygens–Fresnel principle, which is usually regarded as a form of the Helmholtz–Kirchhoff integral theorem, every point $P(x,z)$ of the wave $E_0(P,t)$ can be considered as the center of a secondary spherical wave. When the secondary wave reaches the core–cladding boundary, it is split into two waves: a transmitted (leaky) wave proceeding into the second medium, and a reflected wave propagating back into the first medium. According to the method of images, the secondary wave reflected can be represented as a wave emerging from the respective

† This section is reprinted with permission from S. V. Kukhlevsky, G. Nyitray, and V. L. Kantsyrev, *Physical Review E*, Vol. 64, 026603 (2001). Copyright © 2001 by the American Physical Society.

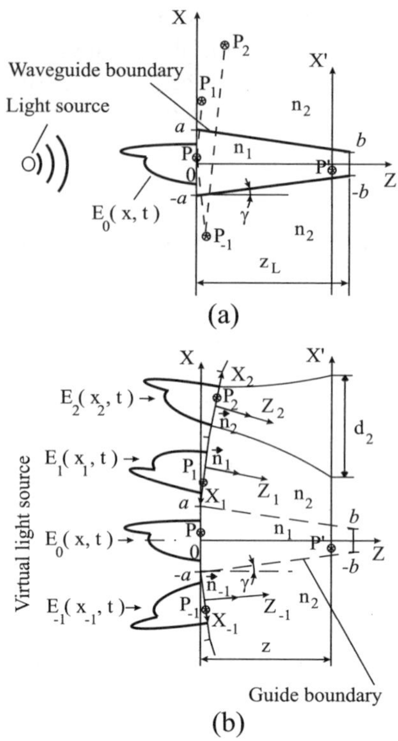

FIGURE 10.6 Construction of a Fresnel waveguide light source for a plane-parallel waveguide that has total-reflection walls.

point $P_1(x_1, z_1)$ of the free-space virtual source (Fig. 10.6). The amplitude and phase of the wave are determined by the Fresnel field reflectivity and the phase change associated with the reflection [54]. After m reflections, the field $E_0(P, t)$ can be represented as the field (beam) $E'_m(P', t)$ emerging from the mth zone of the virtual source having the field distribution $E_m(P_m, t)$. The field distribution in this zone is given by $E_m(P_m, t) = R_m E_0(x_m) \exp[i(\phi_0(x_m) + \Delta\phi_m - \omega t)]$, where the amplitude and phase are determined by the reflectivity $R_m = \prod_{j=1}^{m} R_j(\Phi_j, n_1, n_2)$ and the phase change $\Delta\phi_m = \sum_{j=1}^{m} \Delta\phi_j(\Phi_j, n_1, n_2)$ for the m reflections. Here $R_j(\Phi_j, n_1, n_2)$ and $\Delta\phi_j(\Phi_j, n_1, n_2)$ are, respectively, the Fresnel field reflectivity and the phase change for the beam $E'_j(P', t)$ emerging from the jth zone of the virtual source and reflected at the glancing angle Φ_j [54]. The field $E'_m(P', t)$ at a general point P' of the guide core is given by the Fresnel–Kirchhoff integral [54,62]:

$$E'_m(P', t) = \frac{1}{\sqrt{2i\lambda}} \int_{x_m^{\min}}^{x_m^{\max}} \frac{\exp[ikr(P'_m, P_m)]}{r(P'_m, P_m)} (1 + \cos\Theta_m) E_m(P_m, t)\, dx_m, \quad (10.5)$$

with $P_m = P(x_m, z_m)$ and $P'_m = P'(x'_m, z'_m)$, where (x_m, z_m) and (x'_m, z'_m) are the coordinates of points P and P' in the coordinate system (X_m, Z_m), respectively.

Points $P_m = P(x_m, z_m)$ are images of points $P = P(x, z)$. The transformation can be presented as the rotation, translation, and inversion of the coordinate system (X, Z):

$$\begin{pmatrix} x_m \\ z_m \end{pmatrix} = \begin{pmatrix} \cos 2m\gamma, & (-1)^m \sin 2m\gamma \\ (-1)^m \sin 2m\gamma, & \cos 2m\gamma \end{pmatrix} \begin{pmatrix} x_m^{t,i} \\ z_m^t \end{pmatrix} \tag{10.6}$$

$$x_m^{t,i} = \left\{ x - a \left[1 + \cos 2m\gamma - 2(\delta_{1m} - 1) \sum_{j=1}^{m-1} \cos 2j\gamma \right] \right\} (-1)^m \tag{10.7}$$

$$z_m^t = z - a \left[\sin 2m\gamma - 2(\delta_{1m} - 1) \sum_{j=1}^{m-1} \sin 2j\gamma \right]. \tag{10.8}$$

Here $r(P'_m, P_m)$ is the distance between points P'_m and P_m; Θ_m is the angle that the line $(P'_m P_m)$ makes with the unit normal $\vec{e}_m$ to the line $(x_m^{\min} x_m^{\max})$; λ is the wavelength; $[x_m^{\min}, x_m^{\max}]$ is the mth zone; δ_{1m} is the Kronecker symbol; the top and bottom signs are used for $x > 0$ and $x < 0$, respectively. The total field $E'(P', t)$ at point P' is found by summing the contributions from the M zones of the virtual source:

$$E'(P', t) = \sum_{m=-M}^{M} E'_m(P', t), \tag{10.9}$$

where M is the number of zones (beams) that contribute energy into the field $E'(P', t)$. The number M depends on the direction of propagation and the divergence angle of the beams $E'_m(P', t)$. These two parameters are determined by the value γ and the transverse dimensions $d_m(\lambda, a, z)$ of the beams $E'_m(P', t)$. Thus, an arbitrary wave $E_0(P, t)$ propagates down the waveguide as the superposition of the "transient" modes $E'_m(P', t)$, which diffract in the off-axis direction and interfere with each other. It should be mentioned that the correspondence between free-space modes and waveguide modes is also the basis of the nonorthogonal (transient) mode formalism derived in [63] to describe excess quantum noise in unstable resonators.

The analysis above shows that guided waves that have an arbitrary duration, field distribution, degree of coherence, and direction of propagation at the guide entrance can be generated in free space by the appropriate equivalent source. The waves guided by material waveguides can be produced in free space like a diffraction-free beam provided that an appropriate launch pattern containing multiple virtual sources (Fresnel waveguide light source) can be constructed. As an example, Fig. 10.7 shows the intensity distributions calculated for the waveguide eigenmode TE_0, the transient modes $E'_m(x', t)$, and the superposition of transient modes $E'(P', t) = \sum_{m=-M}^{M} E'_m(P', t)$ of a planar hollow waveguide that has total-reflection walls. We notice that the distributions of the waveguide eigenmode TE_0 and the superposition in the core region of transient modes (diffraction-free beam) $E'(P', t)$ are indistinguishable. Another example (Fig. 10.8) shows the propagation of a 10-fs pulse, which is guided by

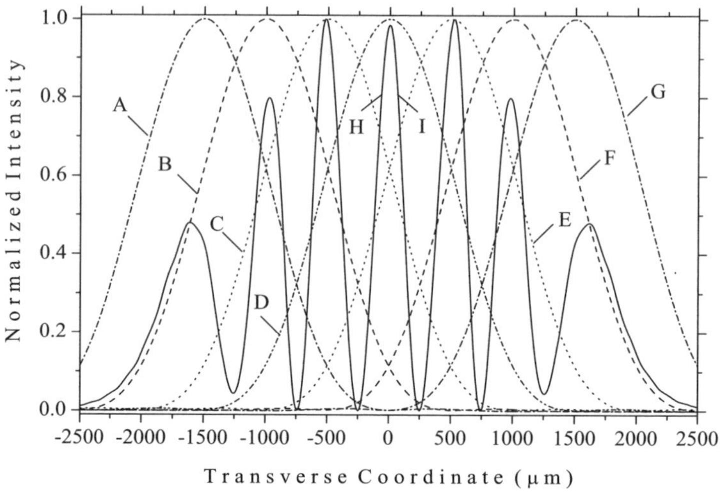

FIGURE 10.7 Normalized intensity distributions of the eigenmode TE_0, the transient modes $E'_m(x', t)$, and superposition of transient modes $E'(P', t) = \sum_{m=-3}^{3} E'_m(P', t)$, corresponding to a material planar waveguide. Curves A, B, C, D, E, F, and G are, respectively, the normalized intensities of the transient modes having $m = -3, -2, -1, 0, 1, 2,$ and 3. In the core region $x \in [-a, a]$), curves H and I are, respectively, the normalized intensities of the TE_0 eigenmode and the superposition of transient modes (diffraction-free beam) $E'(P', t)$. The intensity distributions of the transient modes and their superposition were calculated for the distance $z = 1$ m from the virtual source using the parameters $\lambda = 500$ nm and $2a = 500$ μm.

a planar material waveguide or the respective Fresnel waveguide. The input pulse $E_0(x, z = 0, t)$ at the guide entrance is presented in the form of the Fourier integral:

$$E_0(P, t) = \int_{-\infty}^{\infty} E_0(P, \omega) \exp(-i\omega t)\, d\omega. \tag{10.10}$$

Using Eqs. (10.5–10.9) for the input harmonic field $E_0(P, \omega) \exp(-i\omega t)$ and substituting the result into Eq. (10.10), we get the field distribution of the pulse inside the material and virtual waveguides. The computations were performed for a planar hollow waveguide with total-reflection walls, guide thickness $2a = 2b = 500$ μm and length $z_L = 10$ cm. At the guide entrance the Gaussian-shaped pulse matches the profile of the TE_0 and has the plane-wave front

$$E_0(x, z = 0, t) = E_0(x) \exp[-2\ \ln(2)(t/\tau_0)^2] \exp(i[kz - \omega_0 t + \phi_o(x)]), \tag{10.11}$$

where $\tau_0 = 10$ fs, $\lambda_0 = 2\pi c/\omega_0 = 500$ nm, and $\phi_0(x) = \text{const}$. Figure 10.8 shows the intensity distributions calculated for the input pulse, the transient modes $E'_m(x', t)$, and the superposition of transient modes $E'(P', t) = \sum_{m=-M}^{M} E'_m(P', t)$. The input pulse

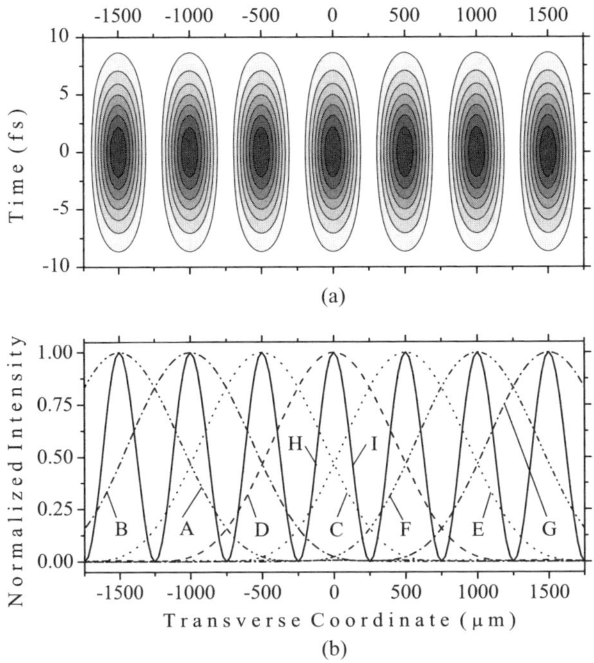

FIGURE 10.8 (a) Normalized intensity distributions calculated for the input pulse, the transient modes $E'_m(x', t)$, and superposition of the transient modes $E'(P', t) = \sum_{m=-12}^{12} E'_m(P', t)$. (b) Curves A, B, C, D, E, F, and G are, respectively, the normalized intensities of the transient modes having $m = -3, -2, -1, 0, 1, 2,$ and 3. Curves H and I in the core region $x \in [-a, a]$ are, respectively, normalized intensities of the TE_0 eigenmode and the superposition (diffraction-free light pulse) $E'(P', t)$ of the transient modes. The intensity distributions of the transient modes and their superposition were calculated for the distance $z = 0.8$ m from the virtual source. The curves A, B, C, D, E, F, G, H, and I show the intensities at the time $t = 0$ fs.

matches the eigenmode TE_0, $E_0(x) = TE_0(x)$. We notice that the pulse guided by the material and the diffraction-free pulsed beam produced by the Fresnel-waveguide light source are indistinguishable.

The correspondence between the guided and free-space waves can easily be demonstrated for many other particular fields, such as the Bessel (diffraction-free), self-imaging, ultrashort, solitonlike, partially coherent waves and laser fractals [45,46,48]. Such fields are attracting continuous interest because of their importance for basic physics and applications in technology [23,64–76]. The method can be extended to guides of other shapes: for instance, polygonal guides [48]. In the case of curved shapes (circular, elliptical, coaxial, or arbitrary-shaped guides), finding the equivalent source is an interesting mathematical problem. In principle, it can be solved as the problem of polygonal guides with the number of sides $N \to \infty$. The correspondence between the waveguide eigenmodes and diffraction-free fields is not unexpected. Indeed, it is a well-known fact that an eigenmode is a propagation-invariant solution inside a waveguide, while a nondiffracting wave is a similar solution in free space. For instance, the Bessel-type diffraction-free beam is the free-space equivalent of the eigenmode of the cylindrical waveguide.

10.4 DIFFRACTION-FREE SUBWAVELENGTH-BEAM OPTICS ON A NANOMETER SCALE†

In this section, the Fresnel waveguide concept is extended to the subwavelength nanometer-scale domain ($\Lambda < \lambda$). We consider the problem of generation of a subwavelength nanometer-sized diffraction-free beam by using the relation between the waveguide and free-space optics [45–48,52], which demonstrated that an arbitrary scalar wave confined by an arbitrary material waveguide, whose width exceeds the wavelength λ, could be reproduced in free space by a Fresnel light source of this waveguide.

The basic concept of the Fresnel waveguide light source considered in Section 10.3 is quite simple. The concept is demonstrated in Fig. 10.9 for a plane-parallel waveguide than has total-reflection walls. In the Fresnel waveguide approach, the boundaries of the waveguide are replaced by virtual sources. A diffraction-free time-harmonic

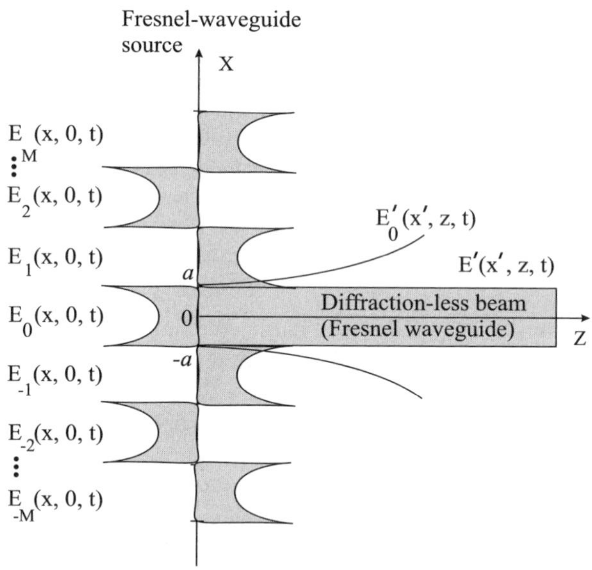

FIGURE 10.9 Construction of a Fresnel waveguide light source for a plane-parallel waveguide that has total-reflection walls.

†This section is reprinted from *Optics Communications*, Vol. 231, S. V. Kukhlevsky and M. Mechler, Diffraction-free subwavelength-beam optics at nanometer scale, pp. 35–43. Copyright © 2004, with permission from Elsevier.

beam $E'(x', z, \omega)$ confined by the waveguide is supported in free space at points (x', z) by the constructive interference and diffraction of multiple beams $E'_n(x', z, \omega)$ of the Fresnel source of the waveguide:

$$E'(x', z, \omega) = \sum_{n=-M}^{M} E'_n(x', z, \omega), \tag{10.12}$$

where the number $2M + 1$ of the beams $E'_n(x', z, \omega)$ depends on their widths at a distance z from the source; $n = 0, \pm 1, \pm 2, \ldots, \pm M$; $z > 0$ and $|x'| < a$. The beam $E'_n(x', z, \omega)$ emerges from the nth zone of the Fresnel source having the field distribution

$$E_n(x, 0, \omega) = E_0(x_n, 0, \omega) \exp(i\pi n), \tag{10.13}$$

which is obtained by the periodic ($x_n = x \pm 2na$) translation of the field $E_0(x, 0, \omega)$ and the πn-change of its phase; $E_0(x, 0, \omega)$ is the field at the input aperture ($z = 0$, $|x| < a$). Thus, the Fresnel waveguide field $\sum_{n=-M}^{M} E'_n(x', z, \omega)$ is constructed by the periodic translation and phase change of the beam $E'_0(x', z, \omega)$ emerging from the waveguide aperture. In the case of a diffraction-free pulse confined by the waveguide, the input field $E_0(x, 0, t)$ is composed in the form of a Fourier time expansion (for details, see the Appendix):

$$E_0(x, 0, t) = \int_{-\infty}^{\infty} E_0(x, 0, \omega) \exp(-i\omega t)\, d\omega. \tag{10.14}$$

The field $E'(x', z, t)$ of the diffraction-free pulse is found by using Eqs. (10.12) and (10.13) for each ω-Fourier component of the wave packet (10.14). According to Eqs. (10.12)–(10.14), a diffraction-free pulse confined by the waveguide is supported in free space by the constructive interference of $(2M + 1)$ beams (pulses) of a Fresnel source of the waveguide. Notice that in contrast to the case of the infinite-width Bessel-type beams [26], the finite-width Fresnel waveguide source of the respective finite-length waveguide generates a field with finite energy.

The Fresnel waveguide approach was originally developed using the Helmholtz–Kirchhoff integral theorem, which fails when the waveguide width $2a$ is close to the wavelength λ. It is surprising that the Fresnel waveguide approach in a general form [the periodic translation and phase change of the output beam, Eqs. (10.12)–(10.14)] also provides a solution to the problem in the case of subwavelength nanometer-sized waveguides as will be shown below. From a theoretical point of view, the most important questions are the degree of spatial collimation and temporal broadening of the wave at different distances z from the Fresnel source of a subwavelength nanometer-sized waveguide. To address these questions, the Fresnel waveguide field $E'(x', z, t)$ was constructed by the periodic translation and phase change of the single-output beam (pulse) $E'_0(x', z, t)$ launched from a subwavelength nanometer-sized hollow waveguide with perfectly conducting walls [77]. The beam $E'_0(x', z, t)$ is formatted

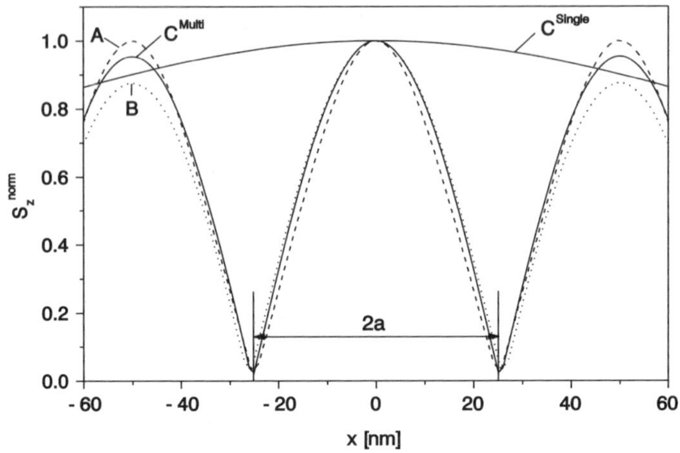

FIGURE 10.10 Energy flux distribution $S_z(x', z)$ of a diffraction-free subwavelength beam $E'(x', z, t)$ at distances z from the Fresnel source of the waveguide: A, a; B, $3a$; and C^{multi}, $6a$. The flux $S_z(x', z)$ at the distances z is normalized to the one at the coordinate $(x = 0, z = 0)$. A single beam $E'_0(x', z, t)$ used for construction of the source is produced by a 50-nm slit (waveguide) in a perfectly conducting screen of thickness $b = 50$ nm. The normalized energy flux distribution C^{single} of this beam is shown at $z = 6a$ for comparison. The number $2M + 1$ of used beams for A and B is 51, for C^{multi} is 501; $a = 0.05\lambda$ and $\lambda = 500$ nm. The flux C^{single} at the distances z is normalized to the one at the coordinate $x = 0$, $z = 0$.

by transmission of a continuous or pulsed wave through the waveguide. The transmitted field $E'_0(x', z, t)$ was determined by solving Maxwell's equations numerically (one cannot expect analytical solutions on the subwavelength scale). A Fresnel waveguide beam $E'(x', z, t)$ in the "not-too-distant" field regime ($z = 0.1 - 0.5\lambda$) was computed for different values of continuous-wave wavelength λ, incident-pulse duration τ, central wavelength λ_0, and waveguide width $2a$ and length b. The details of the computations of the electric $\vec{E} = (E_x, 0, E_z)$ and magnetic $\vec{H} = (0, H_y, 0)$ field distributions of the transmitted beam are presented in the Appendix. As an example, Fig. 10.10 presents the energy flux distributions $S_z = (c/8\pi)\,\mathrm{Re}(\vec{E} \times \vec{H}^*)_z$ of a diffraction-free nanometer-sized beam produced by a Fresnel source of a subwavelength ($a = 0.05\lambda$) waveguide, at three distances z from the Fresnel waveguide source. The parameters of the waveguide and the incident wave are given in the figure caption. In the figure, the flux of a single beam used in construction of the Fresnel source is shown at $|z| = 6a$ for comparison. The flux distributions of the diffraction-free beam in the Fresnel waveguide region $[-a, a]$ are practically undistinguished at the various distances z. The width of the central beam (Fresnel waveguide field) is a few times smaller than the single beam. We also notice that the sidelobe intensity is approximately identical to that of the central maximum. The full width at half-maximum (FWHM) and the energy flux of the diffraction-less subwavelength beam versus the distance z from the Fresnel waveguide source are shown in Fig. 10.11 in comparison with that of the single beam produced by the material

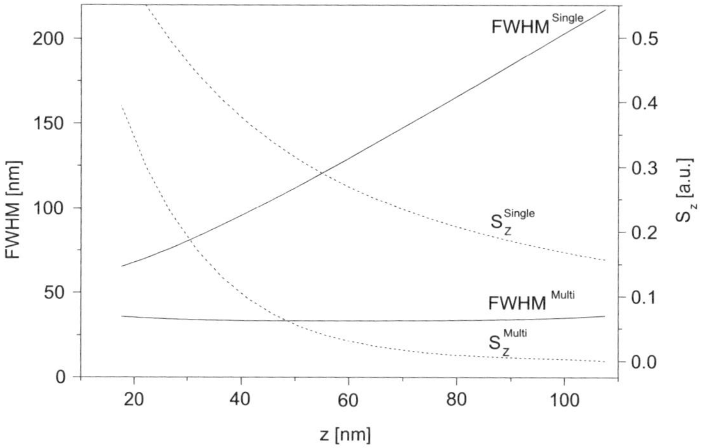

FIGURE 10.11 Full width at half-maximum $(\mathrm{FWHM})^{\mathrm{multi}}$ and energy flux $S_z^{\mathrm{multi}}(x' = 0, z)$ of a diffraction-less subwavelength beam $E'(x', z, t)$ versus distance z from the Fresnel source of the waveguide. The width and flux of a single beam $E_0'(x', z, t)$ used for construction of the source are shown for comparison. Here, the waveguide parameters are the same as indicated in Fig. 10.2. The number $2M + 1$ of used beams is 501; $a = 0.05\lambda$ and $\lambda = 500$ nm.

waveguide (slit). The width of the diffraction-free central beam remains practically constant, while the width of the single beam increases drastically with increasing distance z. The results presented in Figs. 10.10 and 10.11 demonstrate theoretically the feasibility of diffraction-free subwavelength-beam optics on a nanometer scale.

Although the Fresnel waveguide source producing a subwavelength nanometer-sized diffraction-free continuous wave is already an unexpected finding, our analysis showed that such a source can also be constructed for ultrashort (near-single-cycle) pulses. The Fresnel waveguide $E'(x', z, t)$ was constructed by the periodic translation and phase change of the single output pulse $E_0'(x', z, t)$ launched from the subwavelength nanometer-sized waveguide. The Fresnel waveguide field $E'(x', z, t)$ was computed for different values of the distance $z(z = 0.1 - 0.5\lambda)$, incident-pulse duration τ, central wavelength λ_0, and waveguide width $2a$ and length b. As an example, Fig. 10.12 shows the electric field $|E_x|$ computed for the diffraction-free subwavelength nanometer-sized pulse formatted by the Fresnel-waveguide source at the distance $z = 3a$. The pulse used for construction of the Fresnel waveguide is formatted by transmission of the femtosecond (near-single-cycle) pulse through the waveguide. Parameters of the incident pulse and waveguide are described in the figure caption and the Appendix. We notice that the diffraction-free pulse produced by the Fresnel waveguide source propagates in free space practically without spatial and temporal broadenings. The following brief analysis helps us to understand the conditions of the diffraction- and distortion-free propagation. In the general case, the pulse profile and duration are modified by the waveguide transmission function $T = T(\lambda)$. According to [44], the coefficient $T = T(\lambda)$ of the relatively short ($b < 50$ nm) waveguide is

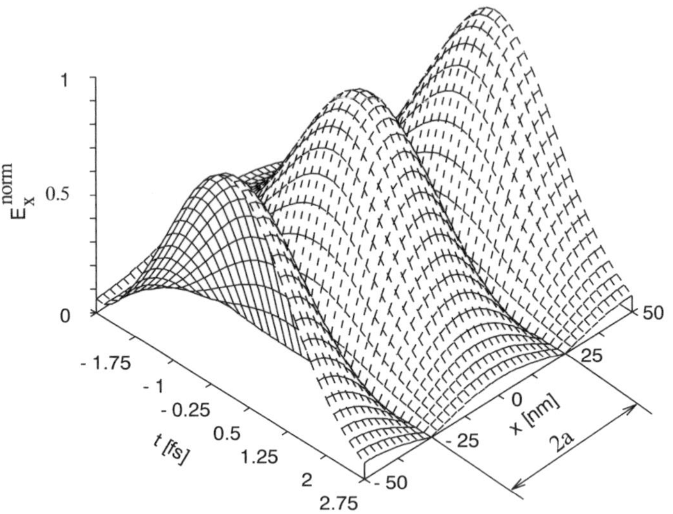

FIGURE 10.12 Electric field $|E_x|$ computed for a diffraction-less femtosecond (near-single-cycle) pulse $E'(x', z, t)$ produced by the Fresnel source of a subwavelength waveguide at the distance $z = 3a$. The field $|E'(x', z = 3a, t)|$ is normalized to the field at the coordinate $x = 0, z = 0$. The parameters of the waveguide are as indicated in Fig. 10.2. A single pulse $E'_0(x', z, t)$ used for construction of the source has the pulse length $\tau = 2$ fs and central wavelength of the wave packet $\lambda_0 = 500$ nm. The number of beams (pulses) is $2M + 1 = 18$; $a = 0.05\lambda_0$.

approximately constant in the spectral band $\Delta\lambda \approx [300$ nm, 1000 nm$]$. The width of the Fourier spectra of a wave packet having femtosecond-scale ($\tau > 2$ fs) duration and central wavelength $\lambda_0 \sim 500$ nm does not exceed the dispersion-free band $\Delta\lambda$. Therefore, the Fresnel waveguide source provides the distortion- and dispersion-free properties of a diffraction-free pulse. Decreased input-pulse duration and/or increased the waveguide length leads to spatial and temporal distortions of the pulse. In the case of a waveguide filled by the medium whose refraction index depends on the wavelength, the dispersion should also lead to pulse distortion. The distortion of single-cycle pulses propagating in dispersive mediam was studied in [38,39,49]. It should also be noted that diffraction-free subwavelength nanometer-sized pulses (see, e.g., Fig. 10.12) are not in conflict with the natural space–time uncertainty of the wave packet considered in Section 10.2. For a single pulse, the correlation between spatial and temporal uncertainties of a wave packet is caused by a natural limitation of the spectral width of a diffracted pulsed beam. Owing to the multiple-beam character of the dispersion-free Fresnel waveguide source, the pulse is localized in space and time at the expense of increasing the number of beams (pulses) supporting a diffraction-free wave packet in free space. It should be also noted that the Fresnel-waveguide beam could not be shrunk down to infinitesimal width. When the beam width approaches the distance between atoms in the waveguide wall of the Fresnel waveguide source, the influence of the discrete structure of the crystal grating on the beam formation should limit the beam width achievable.

The analysis above is applicable directly to two-dimensional near-field scanning optical microscopy. The potential applications, however, are not limited to the near-field microscopy. Broadly speaking, the near-field phenomenon concerns all physical phenomena involving evanescent electromagnetic waves [78]. The approach could be used for many subwavelength-photonic purposes, such as sensors, communications, and optical switching devices. Some of the three-dimensional potential applications can be hampered by the fact that localization to a subwavelength scale is in one dimension, only, and that as a consequence, the sidelobe intensity is approximately identical to that of the central maximum (see Figs. 10.10 and 10.12). To solve this problem one should modify the approach to three-dimensional subwavelength Fresnel waveguides. Fresnel sources of three-dimensional waveguides, whose width exceeds the wavelength λ, were recently described in [48]. The accuracy and stability of the model presented above must be considered before the results are used for a particular experimental situation. The advantages and limitations of the model have been discussed in detail in [44,56,58]. We consider here only the main limitations of the model. The model provides stable solutions with accuracy that is limited only by the speed and memory of the computer used [44,56,58]. The principle of a subwavelength nanometer-sized diffraction-free beam is based on construction of the Fresnel waveguide $E'(x', z, t)$ by the periodic translation and phase change of a single-output continuous wave or pulse $E'_0(x', z, t)$ launched from a subwavelength nanometer-sized waveguide. The change of phase from 0 to π on the nanometer scale, which is equivalent to a field sign change from + to −, can be obtained simply by variation of the waveguide length. The model uses waveguides having perfectly conductive walls of infinitesimal thickness $d_w \rightarrow 0$. A Fresnel waveguide source can be constructed experimentally in the form of a multiple-waveguide structure using the periodic translation of a single waveguide. Perfect conductivity of the waveguide walls prevents beam coupling and leaking. The required waveguides can be made of conventional perfectly conductive (at low temperatures) materials. Perfectly reflecting photonic crystals, air- and dielectric-guide bends or hybrid heterostructures (see e.g., [79–81] and references therein) also have potential for construction of Fresnel waveguide sources at a nanometer scale. However, in the context of current technology, the use of more conventional materials such as metal films [82–85] for the construction of waveguides is more practical. In such a case, one should take into account the finite conductivity of the metal, which can lead to beam coupling and leaking through the spacing between adjacent waveguides. The perfect-conductivity assumption should remain valid as far as thickness of the metallic waveguide wall exceeds the extinction length for a wave within the metal. For instance, perfect conductivity is a very good approximation in a situation involving an aluminum waveguide that has walls with thickness $d_w > 25$ nm and wavelengths in the visible spectral region [44]. These requirements should be taken into account when a Fresnel waveguide light source is being constructed.

10.5 CONCLUSIONS

We have shown that a subwavelength nanometer-sized beam propagating without diffractive broadening can be produced in free space by the interference of multiple beams of a Fresnel light source of the respective material. The source can be constructed not only for continuous waves but also for ultrashort pulses. The results demonstrate theoretically the feasibility of diffraction-free subwavelength-beam optics on a nanometer scale for both continuous waves and near-single-cycle pulses. The approach extends the operational principle of near-field subwavelength-beam optics such as near-field scanning optical microscopy and spectroscopy to the "not-too-distant" field regime (0.1 to about 0.5 wavelength).

APPENDIX

The integral approach of Neerhoff and Mur [56,58], which was used in this chapter to find the amplitude distributions of a continuous wave and an ultrashort pulse diffracted by a subwavelength slit in a perfectly conducting thick screen can be summarized as follows. In region I, a continuous time-harmonic plane wave falls normally onto the slit (waveguide) in the $x-z$ plane, as shown in Fig. 10.3. The slit width and screen thickness are $2a$ and b, respectively. The magnetic field of the incident wave is assumed to be time harmonic and both polarized and constant in the y-direction:

$$\vec{H}(x, y, z, t) = U(x, z)\exp(-i\omega t)\vec{e}_y. \tag{10A.1}$$

The electric field $\vec{E}$ of the incident wave is found by using the Maxwell equations for the magnetic field $\vec{H}$. The Green's function approach [56,58] uses multipole expansion of the field with the Hankel functions in regions I and III and with waveguide eigenmodes in region II. The expansion coefficients are determined using the standard boundary conditions for a perfectly conducting screen. In region III, the amplitude distributions of the magnetic $\vec{H}(x, z, t)$ and electric $\vec{E} = E_x\vec{e}_x + E_y\vec{e}_y$ fields of the diffracted wave are given by

$$\vec{H}(x, z, t) = i\frac{a}{N}\frac{\epsilon_3}{\epsilon_2}\sum_{j=1}^{N} H_0^{(1)}[k_3((x-x_j)^2+z^2)^{1/2}](D\vec{U}_0)_j\exp(-i\omega t)\vec{e}_y, \tag{10A.2}$$

$$E_x(x, z, t) = -\frac{a}{N}\frac{\sqrt{\epsilon_3}}{\epsilon_2}\sum_{j=1}^{N}\frac{z}{((x-x_j)^2+z^2)^{1/2}}H_1^{(1)}[k_3((x-x_j)^2+z^2)^{1/2}] \times (D\vec{U}_0)_j\exp(-i\omega t), \tag{10A.3}$$

$$E_z(x, z, t) = \frac{a}{N}\frac{\sqrt{\epsilon_3}}{\epsilon_2}\sum_{j=1}^{N}\frac{x-x_j}{((x-x_j)^2+z^2)^{1/2}}H_1^{(1)}[k_3((x-x_j)^2+z^2)^{1/2}] \times (D\vec{U}_0)_j\exp(-i\omega t), \tag{10A.4}$$

where ϵ_i and k_i are, respectively, the permittivity and wave number in the regions $i =$ I, II, and III; $x_j = 2a(j - 1/2)/N - a$, with $j = 1, 2, \ldots, N$ and $N > 2a/z$; $H_0^{(1)}$ and $H_1^{(1)}$ are the Hankel functions. The coefficients $(D\vec{U}_0)_j$ are computed by solving a set of the Neerhoff and Mur coupled integral equations [56,58]. For more details on the model and the solution of integral equations, see [44,56,58,59].

In the case of light pulses, the magnetic field of the incident pulse is assumed to be Gaussian-shaped in time and both polarized and constant in the y-direction:

$$\vec{H}(x, y, z, t) = U(x, z) \exp[-2 \ln(2)(t/\tau)^2] \exp(-i\omega_0 t)\vec{e}_y, \qquad (10A.5)$$

where τ is the pulse duration and $\omega_0 = 2\pi c/\lambda_0$ is the central frequency. The pulse is composed in the wave-packet form of a Fourier time expansion [45]:

$$\vec{H}(x, y, z, t) = \int_{-\infty}^{\infty} \vec{H}(x, y, z, \omega) \exp(-i\omega t)\, d\omega. \qquad (10A.6)$$

The field distribution of the diffracted pulse is found by using the expressions (10A.2–10A.4) for each Fourier ω-component of the wave packet (10A.5) (see [44,59]). In the computations, we used a discrete fast Fourier transform (FFT) of the function $\vec{H}(x, y, z, t)$ instead of the integral composition (10A.6). The spectral interval $[\omega_{\min}, \omega_{\max}]$ and the sampling points ω_i were optimized by matching the FFT composition to the original function (10A.5).

Acknowledgments

I wish to thank all my students who have been associated with the research described in this chapter. They are named as coauthors in the references below. The author is especially grateful to his colleagues K. Janssens, V. L. Kantsyrev, and O. Samek for their substantial contributions to the present study. The support of the research by the Fifth Framework of the European Commission (Contract NG6RD-CT-2001-00602) is gratefully acknowledged. The study was supported in part by the Hungarian Scientific Research Foundation (OTKA, Contracts T046811 and M045644), and the Hungarian R&D Office (KPI, Contract GVOP-3.2.1.-2004-04-0166/3.0).

REFERENCES

1. J. W. Strutt (L. Rayleigh), *Philos. Mag.* **44**, 28 (1897).
2. H. A. Bethe, Theory of diffraction by small holes, *Phys. Rev.* **66**(7–8), 163–182 (Oct. 1944).
3. E. A. Ash and G. Nicholls, Super-resolution aperture scanning microscope, *Nature* **237**, 510–512 (June 1972).
4. A. Lewis and K. Lieberman, Near-field optical imaging with a non-evanescently excited high-brightness light source of sub-wavelength dimensions, *Nature* (*London*) **354**, 214–216 (Nov. 1991).

5. E. Betzig, J. K. Trautman, T. D. Harris, J. S. Weiner, and R. L. Kostelak, Breaking the diffraction barrier: optical microscopy on a nanometric scale, *Science* **251**, 1468 (Mar. 1991).
6. D. W. Pohl and D. Courjon, Ed., *Near Field Optics*, Vol. 242 of NATO ASI Series E, Kluwer, Dordrecht, The Netherlands, 1993.
7. M. J. Adams, *An Introduction to Optical Waveguides*, Wiley, New York, 1981.
8. A. W. Snyder and J. D. Love, *Optical Waveguides Theory*, Chapman & Hall, London, 1983.
9. D. S. Jones, *Acoustic and Electromagnetic Waveguides*, Clarendon Press, Oxford, 1986.
10. A. Yariv, *Quantum Electronics*, 3rd ed., Wiley, New York, 1989.
11. W. C. Chew, *Waves in Inhomogeneous Media*, Van Nostrand Reinhold, New York, 1990.
12. C. Vassallo, *Optical Waveguides Concepts*, Elsevier Science, Amsterdam, The Netherlands, 1991.
13. D. Marcuse, *Theory of Dielectric Optical Waveguides*, 2nd ed., Academic Press, San Diego, CA, 1991.
14. B. E. A. Saleh and M. C. Teich, *Fundamentals of Photonics*, Wiley, New York, 1991.
15. C. Tsao, *Optical Fibre Waveguide Analysis*, Oxford University Press, New York, 1992.
16. K. Okamoto, *Fundamentals of Optical Waveguides: Optics and Photonics*, Academic Press, San Diego, CA, 2000.
17. H. Bateman, *Electrical and Optical Wave Motion*, Cambridge University Press, Cambridge, 1915.
18. J. A. Stratton, *Electromagnetic Theory*, McGraw-Hill, New York, 1941.
19. R. Courant and D. Hilbert, *Methods of Mathematical Physics*, Vol. 2, Wiley, New York, 1966.
20. J. N. Brittingham, Focus waves modes in homogeneous Maxwell's equations: transverse electric mode, *J. Appl. Phys.* **54**, 1179–1189 (Mar. 1983).
21. R. W. Ziolkowski, Exact solutions of the wave equation with complex source locations, *J. Math. Phys.* **26**, 861–863 (1985).
22. I. M. Besieris, A. M. Shaarawi, and R. W. Ziolkowski, A bidirectional traveling plane wave representation of exact solutions of the scalar wave equation, *J. Math. Phys.* **30**, 1254–1269 (June 1989).
23. J. Durnin, J. J. Miceli, and J. H. Eberly, Diffraction-free beams, *Phys. Rev. Lett.* **58**, 1499–1501 (Apr. 1987).
24. F. Gori, G. Guattari, and C. Padovani, Bessel–Gauss beams, *Opt. Commun.* **64**, 491–495 (1987).
25. R. M. Herman and T. A. Wiggins, Production and uses of diffractionless beams, *J. Opt. Soc. Am. A* **8**, 932 (June 1991).
26. B. Hafizi and P. Sprangle, Diffraction effects in directed radiation beams, *J. Opt. Soc. Am. A* **8**, 705 (May 1991).
27. Z. Bouchal, Dependence of Bessel beam characteristics on angular spectrum phase variations, *J. Mod. Opt.* **40**, 1325–1329 (1993).
28. J. Ruschin, Modified Bessel nondiffracting beams, *J. Opt. Soc. Am. A* **11**, 3224 (Dec. 1994).
29. S. Chávez-Cerda, G. S. McDonald, and G. H. C. New, Nondiffracting beams: travelling, standing, rotating and spiral waves, *Opt. Commun.* **123**, 225–233 (Feb. 1996).

30. P. Saari and K. Reivelt, Evidence of X-shaped propagation-invariant localized light waves, *Phys. Rev. Lett.* **79**, 4135–4138 (Nov. 1997).
31. Z. Bouchal, J. Wagner, and M. Chlup, Self-reconstruction of a distorted nondiffracting beam, *Opt. Commun.* **151**, 207–211 (June 1998).
32. J. Arlt and K. Dholakia, Generation of high-order Bessel beams by use of an axicon, *Opt. Commun.* **177**, 297–301 (2000).
33. V. Jarutis, R. Paskauskas, and A. Stabinis, Focusing of Laguerre–Gaussian beams by axicon, *Opt. Commun.* **184**(1–4), 105–112 (2000).
34. J. C. Gutiérrez-Vega, M. D. Iturbe-Castillo, G. A. Ramírez, E. Tepichín, R. M. Rodríguez-Dagnino, S. Chávez-Cerda, and G. H. C. New, Experimental demonstration of optical Mathieu beams, *Opt. Commun.* **195**, 35–40 (2001).
35. M. A. Porras, Diffraction-free and dispersion-free pulsed beam propagation in dispersive media, *Opt. Lett.* **26**, 1364–1366 (Sept. 2001).
36. M. Zamboni-Rached, E. Recami, and H. E. Hernández-Figueroa, Superluminal localized solutions to Maxwell equations propagating along a normal-sized waveguide, *Phys. Rev. E* **64**, 066603 (Dec. 2001).
37. M. Zamboni-Rached, K. Z. Nóbrega, E. Recami, and H. E. Hernández-Figueroa, Superluminal X-shaped beams propagating without distortion along a coaxial guide, *Phys. Rev. E* **66**, 046617 (Oct. 2002).
38. M. A. Porras, Diffraction effects in few-cycle optical pulses, *Phys. Rev. E* **65**, 026606 (Feb. 2002).
39. M. A. Porras, R. Borghi, and M. Santarsiero, Suppression of dispersive broadening of pulses with Bessel–Gauss beams, *Opt. Commun.* **206**, 235–241 (2002).
40. J. C. Gutiérrez-Vega, M. D. Iturbe-Castillo, and S. Chávez-Cerda, Alternative formulation for invariant optical fields: Mathieu beams, *Opt. Lett.* **25**, 1493–1495 (Oct. 2003).
41. M. A. Porras and I. Gonzalo, Control of temporal characteristics of Bessel-X pulses, *Opt. Commun.* **217**, 257–264 (Mar. 2003).
42. S. Ruschin and A. Leizer, Evanescent Bessel beams, *J. Opt. Soc. Am. A* **15**, 1139–1143 (May 1998).
43. N. A. McDonald, Electric and magnetic coupling through small apertures in shield walls of any thickness. *IEEE Trans. Microwave Theory Tech.* **20**(10), 689–695 (Oct. 1972).
44. S. V. Kukhlevsky, M. Mechler, L. Csapo, and K. Janssens, Near-field diffraction of fs and sub-fs pulses: super-resolution of nsom in space and time, *Phys. Lett. A* **319**(5–6), 439–447 (Dec. 2003).
45. S. V. Kukhlevsky, Optical waveguide fields as free-space waves, *Europhys. Lett.* **54**, 461–467 (2001).
46. S. V. Kukhlevsky, G. Nyitray, and V. L. Kantsyrev, Fields of optical waveguides as waves in free space, *Phys. Rev. E* **64**, 026603 (Aug. 2001).
47. J. Canning, Diffraction-free mode generation and propagation in optical waveguides, *Opt. Commun.* **207**, 35–39 (June 2002).
48. S. V. Kukhlevsky and G. Nyitray, Reproduction in free space of fields confined by 3-dimensional optical waveguides, *J. Mod. Opt.* **50**(13), 2043–2056 (Sept. 2003).
49. S. Orlov, A. Piskarskas, and A. Stabinis, Localized optical subcycle pulses in dispersive media, *Opt. Lett.* **27**, 2167–2169 (Dec. 2002).

50. S. Orlov, A. Piskarskas, and A. Stabinis, Focus wave modes in optical parametric generators, *Opt. Lett.* **27**, 2103–2105 (Dec. 2002).
51. R. Grunwald, V. Kebbel, U. Griebner, U. Neumann, A. Kummrow, M. Rini, E. T. J. Nibbering, M. Piche, G. Rousseau, and M. Fortin, Generation and characterization of spatially and temporally localized few-cycle optical wave packets, *Phys. Rev. A* **67**, 063820 (June 2003).
52. S. V. Kukhlevsky and G. Nyitray, Stopping of diffraction-free beams by interference collapse, *Opt. Commun.* **218**(4), 213–219 (Apr. 2003).
53. S. V. Kukhlevsky and G. Nyitray, Correlation between spatial and temporal uncertainties of a wave packet, *Opt. Commun.* **209**, 377–382 (Aug. 2002).
54. M. Born and E. Wolf, *Principle of Optics*, Pergamon Press, Oxford, 1980.
55. S. Akturk, X. Gu, P. Gabolde, and R. Trebino, The general theory of first-order spatio-temporal distortions of Gaussian pulses and beams, *Opt. Express* **13**, 8642–8661 (Oct. 2005).
56. F. L. Neerhoff and G. Mur, Diffraction of plane electromagnetic waves by a slit in a thick screen placed between two different media, *Appl. Sci. Res.* **28**(73), 78–88 (1973).
57. R. F. Harrington and D. T. Auckland, Electromagnetic transmission through narrow slots in thick conducting screens, *IEEE Trans. Antennas Propag.* **28**, 616–622 (Sept. 1980).
58. E. Betzig, A. Harootunian, A. Lewis, and M. Isaacson, Near-field diffraction by a slit: implications for superresolution microscopy, *Appl. Opt.* **25**(12), 1890–1900 (June 1986).
59. S. V. Kukhlevsky and M. Mechler, Detailed structure of femtosecond and sub-femtosecond pulses diffracted by a nanometre-sized aperture, *J. Opt. A Pure Appl. Opt.* **5**(3), 256–262 (May 2003).
60. M. Labardi, M. Zavelani-Rossi, D. Polli, G. Cerullo, M. Allegrini, S. De Silvestri, and O. Svelto, Characterization of femtosecond light pulses coupled to hollow-pyramid near-field probes: localization in space and time, *Appl. Phys. Lett.* **86**, 031105 (Jan. 2005).
61. Y. Ben-Aryeh, Evanescent waves and super-resolution effects, *Int. J. Quantum Inf.* **3**(1), 111–115 (June 2005).
62. J. W. Goodman, *Statistical Optics*, Wiley-Interscience, New York, 1985.
63. A. E. Siegman, Excess spontaneous emission in non-Hermitian optical systems: I. Laser amplifiers, *Phys. Rev. A* **39**, 1253–1263 (Feb. 1989).
64. T. Wulle and S. Herminghaus, Nonlinear optics of Bessel beams, *Phys. Rev. Lett.* **70**, 1401–1404 (Mar. 1993).
65. S. Klewitz, P. Leiderer, S. Herminghaus, and S. Sogomonian, Tunable stimulated Raman scattering by pumping with Bessel beams, *Opt. Lett.* **21**, 248 (Feb. 1996).
66. Z. Jaroszewicz, A. Kolodziejczyk, A. Kujawski, and C. Gomez-Reino, Diffractive patterns of small cores generated by interference of Bessel beams, *Opt. Lett.* **21**, 839 (June 1996).
67. R. Borghi and M. Santarsiero, M 2 factor of Bessel–Gauss beams, *Opt. Lett.* **22**, 262–264 (Mar. 1997).
68. V. E. Peet and R. V. Tsubin, Third-harmonic generation and multiphoton ionization in Bessel beams, *Phys. Rev. A* **56**, 1613–1620 (Aug. 1997).
69. D. Ding and Z. Lu, Response to “Comment on ‘The second harmonic component in the Bessel beam,’” *Appl. Phys. Lett.* **71**, 723–724 (Aug. 1997).
70. G. P. Karman and J. P. Woerdman, Fractal structure of eigenmodes of unstable-cavity lasers, *Opt. Lett.* **23**, 1909–1911 (Dec. 1998).

71. G. P. Karman, G. S. McDonald, G. H. C. New, and J. P. Woerdman, Laser optics: fractal modes in unstable resonator, *Nature* **402**, 138–138 (Nov. 1999).
72. W. D. Montgomery, Self-imaging objects of infinite aperture, *J. Opt. Soc. Am.* **57**, 772–778 (June 1967).
73. R. Sudol and B. J. Thompson, Explanation of the lau effect based on coherence theory, *Opt. Commun.* **31**, 105–110 (1979).
74. A. W. Lohmann and J. O. Castaneda, *Opt. Acta* **30**(4), 475–479 (Apr. 1983).
75. G. Indebetouw, Polychromatic self-imaging, *J. Mod. Opt.* **35**(2), 243–252 (Feb. 1998).
76. G. A. Edgar, *Measure, Topology and Fractal Geometry*, Springer-Verlag, New York, 1994.
77. S. V. Kukhlevsky and M. Mechler, Diffraction-free subwavelength-beam optics at nanometer scale, *Opt. Commun.* **231**, 35–43 (Dec. 2004).
78. C. Girard and A. Dereux, Near-field optics theories, *Rep. Prog. Phys.* **59**, 657–699 (May 1996).
79. D. V. Skryabin, F. Luan, J. C. Knight, and P. St. J. Russel, Soliton self-frequency shift cancellation in photonic crystal fibers, *Science* **301**, 1705–1708 (Sept. 2003).
80. A. Chutinan, S. John, and O. Toader, Diffractionless flow of light in all-optical microchips, *Phys. Rev. Lett.* **90**, 123901 (Mar. 2003).
81. D. G. Ouzounov, F. R. Ahmad, D. Muller, N. Venkataraman, M. T. Gallagher, M. G. Thomas, J. Silicox, K.W. Koch, and A. L. Gaeta, Generation of megawatt optical solitons in hollow-core photonic band-gap fibers, *Science* **301**(5640), 1702–1704 (Sept. 2003).
82. H. J. Lezec, A. Degiron, E. Devaux, R. A. Linke, L. Martín-Moreno, F. J. García-Vidal, and T. W. Ebbesen, Beaming light from a subwavelength aperture, *Science* **297**(5582), 820–822 (Aug. 2002).
83. L. Martín-Moreno, F. J. García-Vidal, H. J. Lezec, A. Degiron, and T. W. Ebbesen, Theory of highly directional emission from a single subwavelength aperture surrounded by surface corrugations, *Phys. Rev. Lett.* **90**, 167401 (Apr. 2003).
84. A. Dogariu, M. Hsu, and L. J. Wang, Reducing far-field diffraction by structured apertures, *Opt. Commun.* **220**, 223–228 (May 2003).
85. T. W. Ebbesen, H. J. Lezec, H. F. Ghaemi, T. Thio, and P. A. Wolff, Extraordinary optical transmission through sub-wavelength hole arrays. *Nature (London)* **391**, 667–669 (Feb. 1998).

CHAPTER ELEVEN

Self-Reconstruction of Pulsed Optical X-Waves

RUEDIGER GRUNWALD, UWE NEUMANN, UWE GRIEBNER, GÜNTER STEINMEYER, GERO STIBENZ, AND MARTIN BOCK
Max-Born-Institute for Nonlinear Optics and Short-Pulse Spectroscopy, Berlin, Germany

VOLKER KEBBEL
Automation and Assembly Technologies, Bremen, Germany

11.1 INTRODUCTION

The phenomenon of a partial or fairly complete reconstruction of propagating complex wave fields is well known from progressive nondistorted beams [1], focus wave modes [2–4], diffraction-free beams [5], nonspreading beams [6], and packet- or particlelike [7] waves. For reasons of simplicity, members of the entire class of approximately propagation-invariant solutions of Maxwell's equations or the Helmholtz wave equation, including spatiotemporally localized X-waves [6,8–10], are referred to here as *nondiffracting waves*. Such beams show unique properties such as stable intensity profiles over relatively large propagation distances, axial field components, or superluminal group velocities [11–14]. They are regarded as having a great potential for harnessing light applications in nonlinear spectroscopy, microstructuring, measuring techniques, information technologies, and biophotonics. For a high-resolution confinement of photons in space and time or a reliable transfer of information, the influence of distortions has to be minimized. Therefore, the study of spatiotemporal

Localized Waves, Edited by Hugo E. Hernández-Figueroa, Michel Zamboni-Rached, and Erasmo Recami

and spatiospectral transfer functions and self-reconstructing properties of wave fields is of considerable practical interest.

In the particular case of Bessel-like beams [15], theoretical and experimental evidence of self-healing or self-reconstruction behavior after significant distortions has been found [16]. It was shown that an initial intensity pattern following a zero-order Bessel distribution is clearly recovered in shape and contrast after propagating a certain path length. Self-reconstruction was also verified for Bessel beams in nonlinear media [17] and nondiffracting vortex beams [18].

Furthermore, self-reconstruction of higher-order Bessel beams with central intensity minima ("bottle beams") is of utmost importance for guiding and manipulating microscopically small objects in scattering or turbulent environments as in biochemistry and medicine [19,20]. Recently, the manipulation of particles and cells was demonstrated with polychromatic Bessel-like beams [21]. In the field of ultrafast optical communication and processing, the robust transfer of image data ("flying images") by encoding patterns into nondiffracting waves was proposed [22,23]. Furthermore, concepts for using pulsed X-waves [24] and arrays of nonlinear localized waves [25,26] as carriers of information were developed.

For special types of waves and wave packets, a theoretical description was given based on sampling of the spatial spectrum [27] and the coupling of spatial and temporal frequency spectra [28]. However, the self-reconstruction behavior of ultrashort-pulsed nondiffracting beams has not been well investigated up to now.

For complex spatiotemporal structures as they are generated by ultrashort-pulse lasers at pulse durations of about 10 fs or less, no reliable experimental data concerning the self-reconstruction were available. Recently, the self-reconstruction of arrays of nondiffracting optical few-cycle wave packets with center wavelengths in the near infrared was analyzed [29].

Here we give an overview on the state of the art of shaping and characterizing self-reconstructing ultrashort-pulse localized wave packets. Particular attention is directed to the concept of small-angle beam shaping with thin-film components, self-reconstruction of pulsed Bessel-like X-waves in space and time, and truncated Bessel–Gauss beams. The importance of advanced detection methods such as spatiotemporal autocorrelation and highly resolved spectral mapping is explained. The spatiospectral transfer of Bessel-like beams and creation of self-reconstructing nondiffracting images are analyzed for the first time.

11.2 SMALL-ANGLE BESSEL-LIKE WAVES AND X-PULSES

For shaping a Bessel-like wave, a conically converging angular distribution is necessary that can be realized in the Fourier plane of a circular slit, using either holographic elements or cone-shaped refractive or reflective components (see, e.g., [5]). Ideal (theoretical) Bessel beams are formed by illuminating infinitely extended beam shapers with monochromatic plane waves of homogeneous intensity distribution. Without any diffraction, an infinitely narrow angular spectrum is generated from a perfect cone. Under linear-optical conditions (vacuum, low intensity), Bessel beams represent propagation-invariant, radially symmetric solutions of the Helmholtz wave equation with field distributions characterized by Bessel functions of infinite transversal extension. Because of the inevitably finite transversal extension of experimentally achievable Bessel-like beams from refractive or diffractive axicons or as generated by Fourier imaging of circular slits, their axial depth remains limited under real-world conditions [30]. X-shaped wave packets in space and time (pulsed X-waves or X-pulses) result from the spectral interference of broadband polychromatic Bessel-like beams [31]. This effect was originally found in acoustics [8]. The X-shape is asymmetric in time if the initial phase distribution is nonplanar. At center wavelengths of about 800 nm (Ti : sapphire laser), the X-shape can be observed with sufficiently high contrast at pulse durations < 20 fs, corresponding to a spectral FWHM of > 50 nm. Recently, we reported on the first direct detection of free-space, linear-optical X-pulses using a second-order spatiotemporal autocorrelation [32]. In contrast to correlation experiments reported earlier that employed lamp light of short coherence time [6], and also in contrast to the indirect indication of X-pulses at pulse durations in the 200-fs range [33], highly intense few-cycle wave packets in sub-10-fs range are now available. At the same time, X-pulses were also obtained by spontaneous beam shaping at high field intensities in nonlinear media [34].

The axial characteristics of Bessel-like beams can be modified significantly by deviations from a pure conical phase distribution (e.g., by applying axicons of Gaussian phase shape [35]) by radially changing parameters of the incident beam (phase, spectrum, intensity) or by breaking the radial symmetry as in the case of Mathieu beams [36]. Near-paraxial nondiffracting beams with extremely extended interference zones result from shaping the beam with flat thin-film axicon structures of extremely small conical angles [37–39]. Here superluminal properties are less relevant. One of the advantages of small-angle structures for ultrashort pulses is the very slight difference in the propagation time of different radial parts of the initial wave packets (low geometrical dispersion).

Bessel-like beams that are generated by *Gaussian phase profiles* are substantially different from focused Gaussian beams or those types of Bessel-like beams that are obtained by illuminating conical axicons with plane waves or Gaussian intensity distributions [40]. Because of the vanishing edges of ultraflat Gaussian-shaped thin-film structures, diffraction is reduced (self-apodization). The zone of maximum stability (minimum change of the pattern with the propagation distance) is more extended than in the case of conical axicons. In our experiments, Bessel-like beams with confocal parameters ranging from several millimeters to the 1-m range were generated. Figure 11.1 shows a typical intensity distribution of a part of an individual beamlet from an array of Bessel-like beams generated from a 10-fs-Ti : sapphire laser beam (center wavelength 790 nm) with Gaussian-shaped microaxicons. The period of these fused silica structures was 405 μm and the maximum thickness was 5.7 μm. The spatiotemporal structure of such Bessel-like beams was analyzed by a spatially resolved second-order collinear autocorrelation setup [29]. A two-dimensional projection of the resulting three-dimensional autocorrelation function is plotted in Fig. 11.2. Second harmonic generation in a BBO crystal was used to obtain the second-order autocorrelation function.

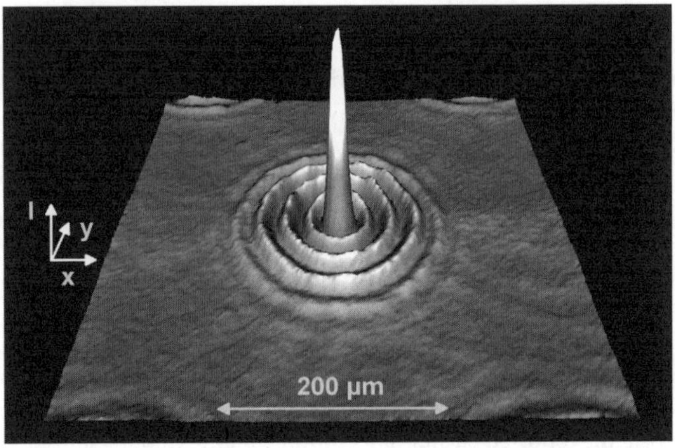

FIGURE 11.1 Intensity profile of a microscopic Bessel-like beam. To generate the pattern, the beam of a 10-fs Ti : sapphire laser was spatially multiplexed with an array of Gaussian-shaped thin-film microaxicons (fused silica layer on a subtrate of identical material, structure height 5.7 μm). The field of view was magnified with a microcope objective (20×) and imaged onto a high-resolution CCD camera.

By reducing the conical angle below a critical value, even fringe-less beams can be generated [38]. Another method to generate fringe-less beams from Bessel beams is self-apodized truncation, where the nondiffracting and self-reconstructing propagation characteristics are maintained at a maximum spectral bandwidth for the central part of the beam (see Section 11.4).

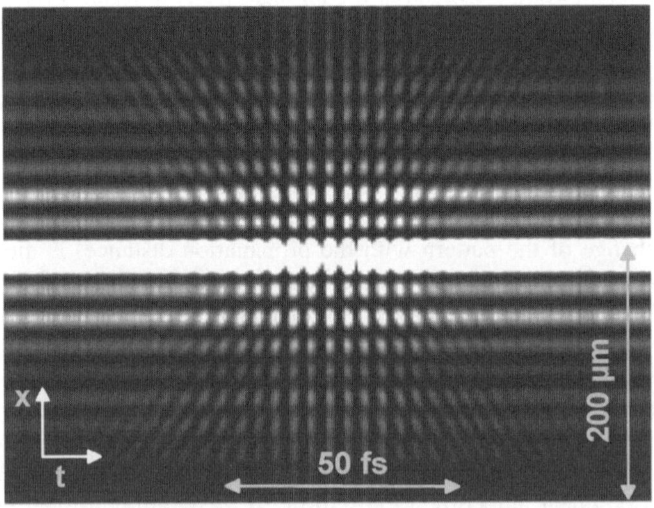

FIGURE 11.2 Spatiotemporal autocorrelation of an undistorted pulsed small-angle Bessel-like X-wave (pulse duration < 10 fs) which was linear optically generated with refractive thin-film axicons and propagated through air. The relevant X-shape is clearly visible.

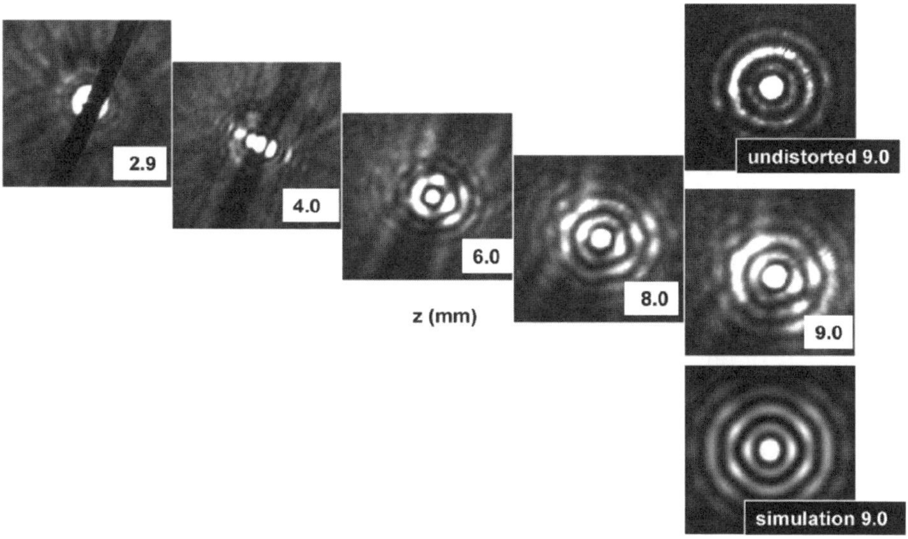

FIGURE 11.3 Self-reconstruction of a 10-fs Bessel beam after significant distortion during propagation in air. The transversal intensity pattern was detected at different distances z between 2.9 and 9 mm and compared to the undistorted case and a numerical simulation assuming a realistic spectrum [40]. The beam was shaped by a single microaxicon from a femtosecond Ti : sapphire laser beam after inserting a 30-μm-thick gold wire perpendicular to the optical axis (laser: center wavelength 790 nm, pulse duration 12 fs, axicon: fused silica on quartz, sagittal height 5.7 μm, vertex directed to the detector, detection: time integrated, CCD camera, 20× magnification, distance of wire axis from axicon vertex: 2.9 mm, field of view: 320 × 320 μm^2).

11.3 SELF-RECONSTRUCTION OF PULSED BESSEL-LIKE X-WAVES

The combination of spatial localization (propagation invariance), stability (self-reconstruction), and advantageous features of ultrashort laser pulses in the few-cycle range (temporal localization, spectral bandwidth, high intensity) is an extraordinarily attractive goal [29]. As shown theoretically as well in first experiments, such pulsed high-power X-waves can propagate like solitons [41–43]. Applications such as X-pulse communication [26] and sensing through dispersive media [29] were proposed. The particular task for all of these concepts is to detect the X-structure of single or arrayed self-reconstructing wave packets after passing a zone of distortion.

As a first step toward robust ultrafast multichannel processors, we investigated the response of Bessel-like beams to significant disturbances. In the vicinity of a microaxicon, we placed a shading gold wire of 30 μm diameter in the center of the beam (Fig. 11.3), obscuring about 40% of the beam energy.

The time-integrated intensity pattern was monitored on a high-resolution CCD camera (1 million pixels). In pictures captured at different distances up to 9 mm, recovery of the spatial beam structure is evident. This process starts at about 5 mm and improves with increasing distance. For the "optimum distance" of 9 mm for the undistorted propagation (highest intensity in the center, minimum sensitivity against axial displacement, best fit of a Bessel distribution), the distorted beam was compared to the undistorted beam, both experimentally and in a numerical simulation based on Rayleigh–Sommerfeld diffraction theory. As can be seen in Fig. 11.3, the qualitative and quantitative agreement of theory and experiment is excellent.

In contrast to spatial and temporal *self-imaging* (Talbot effect [44]), where the constructive interference of diffracted waves leads to a replication of periodic phase or amplitude patterns at discrete axial distances, the *self-reconstruction* addressed here refers to the self-completion process of significantly disturbed Bessel-like waves within a finite zone of pseudo-nondiffracting propagation. In a simplified geometrical picture of self-reconstruction *in space*, the self-reconstruction situation can be explained simply by means of the *shadow* of an opaque object [16]. In a conical wave, any radial information is transformed into an axial one. If the shadowing object is small against the radial extension of the wave, only part of the corresponding axial extension (Bessel zone) is obscured. It has to be noted that for extremely small or distributed objects, however, the model may fail completely. In general, diffraction, scattering, absorption, and dispersion have to be carefully taken into account. All these influences modify the structure of the wave packet in space and time. For sufficiently long pulse durations, spatial and temporal parameters can be separated, whereas they are coupled in the case of few-cycle pulses. This coupling manifests itself in the spatiotemporal modifications caused by the obscuring object (i.e., the gold wire in our experiment). Numerical simulation of the self-reconstruction of pseudo-X-waves was performed on the basis of the well-known angular-spectrum representation with plane waves [40,45]. First- and second-order spatiotemporal autocorrelation distributions computed at realistic spectral and angular parameters are plotted in Fig. 11.4a and b.

These theoretical profiles correspond to experiments with and without nonlinear frequency conversion (detection directly at the fundamental laser wavelength and after passing a second harmonic crystal). For second harmonic generation (SHG), thin BBO crystals (10 to 100 μm thickness) and ZnO nanolayers (<1 μm thickness) were used. In comparison to the theory, experimental data for the first-order autocorrelation (Fig. 11.5a and b) show more complicated features in the space–time plot after passing a certain propagation length, which probably result from a nonplanarity of the real incident wavefront, aberrations of the optical system, diffraction effects, and other deviations from the idealized propagation model.

Despite the high degree of distortion (reduction of contrast, change in spatial frequencies, asymmetry), the specific information on an X-shape in space and time is not totally lost. It can be recovered by adapted image-processing procedures with a high selectivity to fringes, lines, tilts, and spatial frequencies (e.g., Hough transform, Radon transform, Fourier transform). As a first step, improvement of the line contrast due to simple edge sharping and contrast enhancement is demonstrated in Fig. 11.5c and d. Here residual inclined lines of the X-structure within the distorted interference pattern are clearly indicated. In further steps, these lines have to be recognized by advanced data cluster analysis methods.

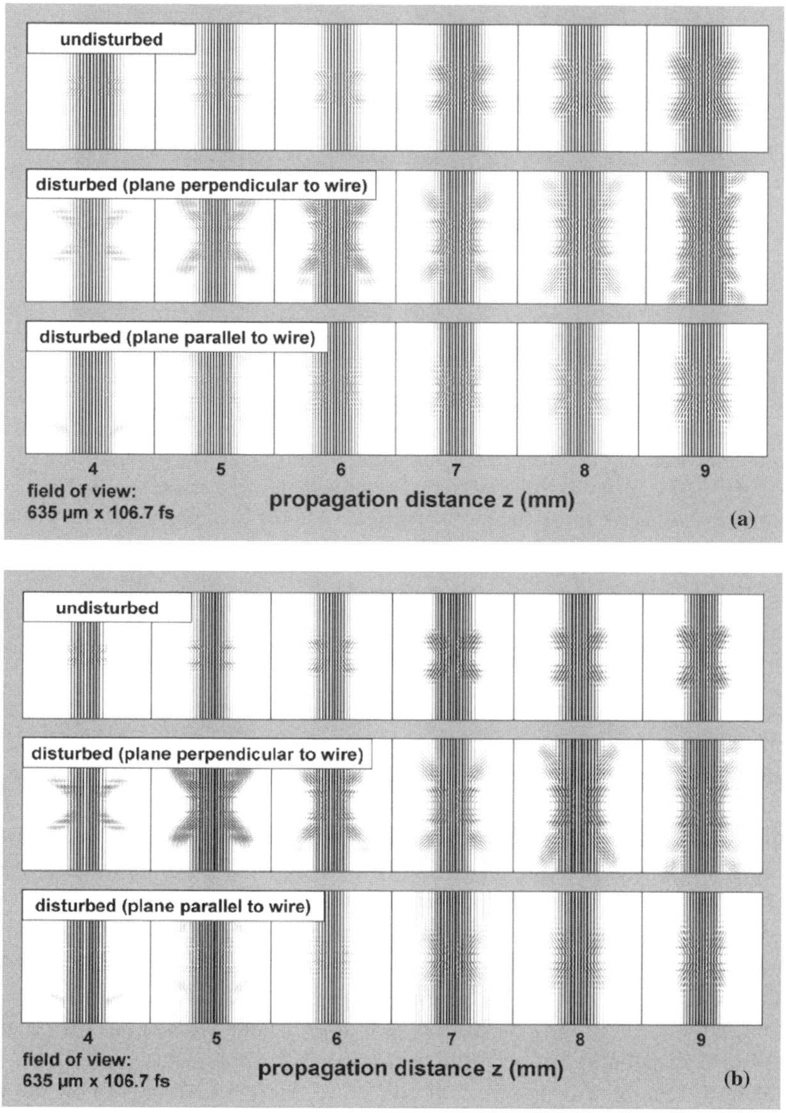

FIGURE 11.4 Numerical simulation of the spatiotemporal self-reconstruction of Bessel-like few-cycle X-waves represented by the (a) first-order and (b) second-order spatiotemporal autocorrelation functions (gray-scale inverted). The parameters were chosen to fit the experimental conditions for shaping X-waves with Gaussian-shaped microaxicons (fused silica, structure depth 5.7 μm, spectral data from measurements of the Ti : sapphire laser oscillator at a pulse duration of 12 fs). The field distortion was simulated by an absorbing shading element of 30×635 μm (corresponding to the wire in the experiments; see Fig. 11.3).

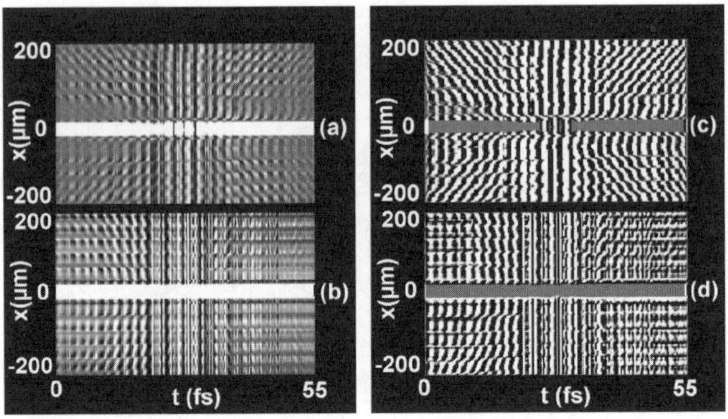

FIGURE 11.5 Spatiotemporal first-order autocorrelation of an undistorted (a) and distorted (b) Bessel-like few-cycle X-pulse with partial self-reconstruction (corresponding to the obscuration geometry used for the experiment demonstrated in Fig. 11.3, Ti : sapphire laser oscillator, pulse duration 12 fs, axial distance $z = 9$ mm, cut through the 3D data set in a plane perpendicular to the wire). Image improvement of patterns (a) and (b) by simple edge sharpening, and contrast enhancement is shown in (c) and (d), respectively.

11.4 NONDIFFRACTING IMAGES

Two-dimensional image information can be transferred simultaneously in a pixellated geometry (i.e., by an array configuration) in either a binary mode (the separate channels carry the information "X" and "no-X") or in a more complex fashion (spectral, gray scale, temporal or polarization encoding). In any case, the necessary condition to be fulfilled is that the X-structure or a combination of X-structure and other parameters be reliably detectable with a sufficiently high signal-to-noise ratio. The principle is shown schematically in Fig. 11.6 for one possible scheme [29]. Other approaches can work with single Bessel beams using information encoding in the spectrum or in the spectral phase [29].

In the experiments, the setup shown in Fig. 11.6 was used. A pseudoreflective liquid-crystal-on-silicon (LCoS) display was used to encode spatial information in the amplitude distribution of an ultrashort-pulse Ti : sapphire laser oscillator (pulse duration 10 fs, center wavelength 790 nm, pulse energy 4 nJ, repetition frequency 87 MHz). The display was incorporated into a spatial light modulator system LC-R 3000 (HoloEye Photonics) with 1920 × 1200 pixels, a fill factor of 92%, and a pixel size of $9.5 \times 9.5\ \mu m^2$. Eight-bit gray-scale patterns at a 60-Hz refresh rate were encoded from bitmap data and transferred to the SLM head. Thus, the 256 gray values are converted into voltage signals, modulating phase and amplitude simultaneously (maximum phase shift about 2π). The laser beam was attenuated by the Fresnel reflection at two glass prisms and illuminated the SLM at a relatively small angle (about 10°). At a distance of 15 mm, a hexagonal thin-film microaxicon array (fused silica on 1-mm-thick silica substrate, period 405 μm, structure depth about 3 μm) was inserted, transforming the beam into a matrix of subbeams of Bessel X type. Because

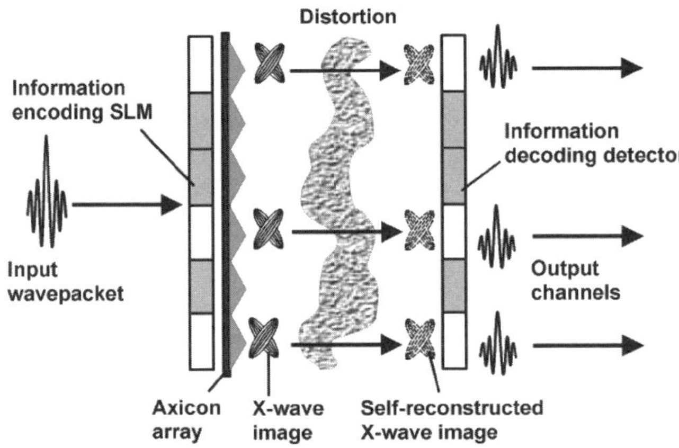

FIGURE 11.6 Scheme proposed for a robust high-speed optical information transfer based on the self-reconstruction of ultrashort-pulsed Bessel X-waves in space and time. The approach is based on splitting the beam into an array of Bessel beams and addressing these subbeams by an adapted spatial light modulator (SLM) of broad-bandwidth spectral transfer. X-shape detection can be realized by a second-order autocorrelation of the beam array with a replica generated in a reference arm of a collinear autocorrelator using the nonlinear absorption in a semiconductor detector array or second harmonic generation in a thin nonlinear crystal (see also [29]).

of the beam diameter of 2 mm, a maximum number of about 25 subbeams could be addressed without an additional telescope.

After the reflection, the gray-scale-encoded beam array propagated as a discrete nondiffracting image. The intensity patterns were magnified by an achromatic 10x microscope objective and imaged with a zoom lens onto a high-resolution CCD-camera (Voßkühler CCD-4000, 2048 $\times$ 2048 pixels, pixel size 7.4 $\times$ 7.4 μm^2, 12 bit, $\geq$ 60 dB). A NIR polarizer (Codixx, maximum contrast ratio 1 : 1000) was placed in front of the camera. The best contrast of the images was found to be 0.74, and the transfer ratio for the FWHM bandwidth of the oscillator spectrum in the beam center amounts to about 75%. The time-integrated patterns in Fig. 11.7 show the propagation of discrete nondiffracting subbeams forming the letter E with (below) and without (above) significant distortion by an amplitude grating. After a propagation depth of a few millimeters, the fringelike structure of the distorted beam can be recognized again. The spatiotemporal autocorrelation of nondiffracting images and the detailed study of spatiospectral transfer functions will be a subject of further investigation in the near future.

11.5 SELF-RECONSTRUCTION OF TRUNCATED ULTRABROADBAND BESSEL–GAUSS BEAMS

As recently demonstrated, Bessel-like nondiffracting beams can be readily transformed into single-maximum beams with ultrabroadband spectral transfer functions

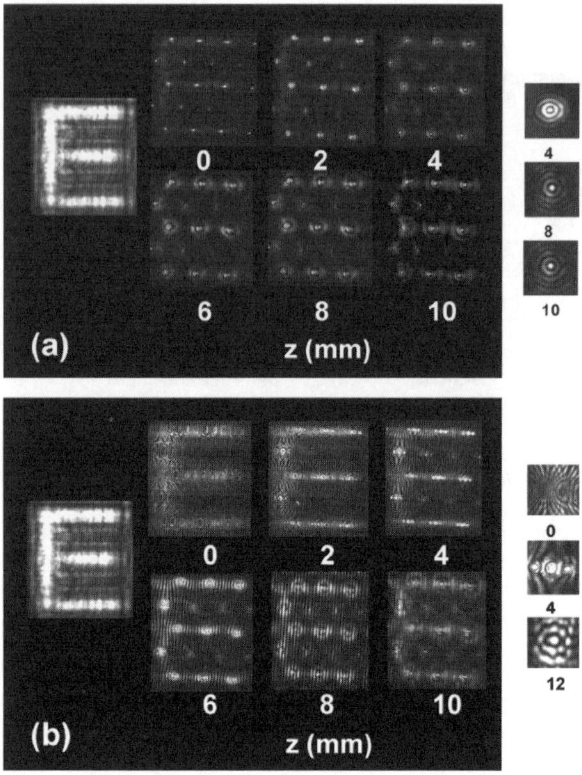

FIGURE 11.7 Generation of ultrashort-pulse nondiffracting images from Ti : sapphire laser wave packets. (a) Patterns of discrete nondiffracting subbeams (spatially multiplexed spatiotemporal localized X-waves) were shaped by combining a liquid-crystal-on-silicon spatial light modulator with thin-film axicon arrays. One can recognize the Bessel fringes (insets on the right side, above). (b) Self-reconstruction of nondiffracting images after a significant distortion by an amplitude grating structure. After a certain propagation depth, the pattern returns to a fringelike structure.

by self-apodized truncation. The condition for self-apodized truncation is fulfilled if the radius of a truncating diaphragm is matched to the radius of the first minimum of the Bessel fringe pattern [46]. Thus, diffraction is minimized at the edges of the diaphragm. In contrast to focused Gaussian beams, the spectral characteristics are less sensitive against the propagation distance. It has to be noted that this is valid only under the assumption of a homogeneous initial spatiospectral profile, whereas inhomogeneities of the initial radial distribution are transformed into spectral modulations along the axis.

The self-reconstruction behavior of truncated Bessel-like beams was not published up to now. Here we report on first experiments with few-cycle small-angle Bessel–Gauss beams of relatively large focal zones. Bessel–Gauss beams were generated by concave gold-coated metal axicons [39,46]. The X-pulse structure of such beams

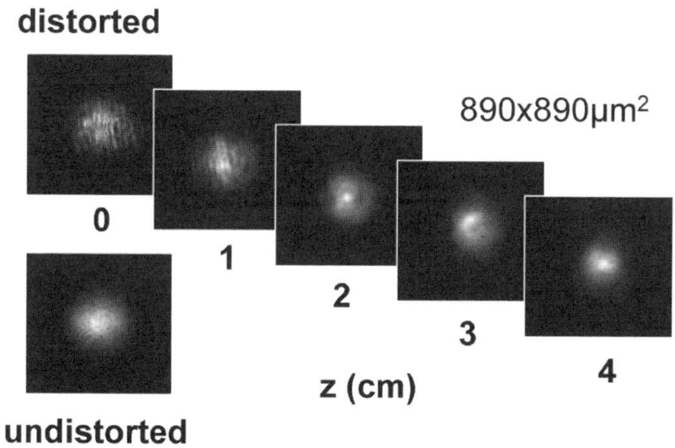

FIGURE 11.8 Self-reconstruction of a truncated few-cycle Bessel–Gauss beam generated from a Ti : sapphire laser oscillator (pulse duration 10 fs, time-integrated intensity patterns detected at different axial distances from a truncating diaphragm). The field of view in the parts of the camera frames presented is $890 \times 890\ \mu m^2$. The ratio of the Rayleigh length to the initial waist radius of the central lobe (in the plane of the diaphragm) was determined to be 440 : 1. The distortion was produced by a statistical phase object (scratched polymer foil). The return from the distorted to the smoothed beam structure can be recognized clearly.

was detected by second-order spatiotemporal autocorrelation experiments [32,47]. The axicon used for beam shaping had a diameter of 1 cm and a conical angle of 0.027°. The typical fringe structure of a Bessel–Gauss beam [48] was generated from a distance of $z = 40$ cm up to a distance of 150 cm. A circular aluminum diaphragm with a conical bore (minimum inner diameter 500 μm) was shifted along the axis to fit the first minimum of the intensity pattern optimally. Thus, the slight change of the spatial frequency of the radial intensity distribution could be exploited for the adjustment. The pattern was detected with a CCD camera to ensure an optimum adjustment (centered beam in the hole). The laser oscillator was operated at a pulse duration of 10 fs. Time-integrated intensity patterns were detected at different axial distances from a truncating diaphragm (Fig. 11.8). The field of view in the parts of the camera frames presented is $890 \times 890\ \mu m^2$.

The distortion was produced by an object that generates a statistical phase (scratched polymer foil). The return from the distorted to the smoothed beam structure can be recognized clearly. A Rayleigh length of 13 cm was measured. The ratio of the Rayleigh length to the initial waist radius of the central lobe (in the plane of the diaphragm) was determined to be 440 : 1. By spatiospectral mapping with a fiber-based grating spectrometer, it was found that the spectral bandwidth of a 10-fs pulse was maintained in the truncated beam even after passing the Rayleigh length. Therefore, the investigated truncated Bessel–Gauss beams can be regarded to be progressive undistorted beams, also with respect to their spectral characteristics i.e., they are *spectrally nondiffracting* beams). This property of localized X-waves is of particular interest for applications in spectroscopy and ultrafast communication. Currently, the *spectral self-reconstruction* of distorted Bessel–Gauss beams [49] and the transfer of information encoded in the spectral map is investigated.

11.6 CONCLUSIONS

The propagation properties investigated in both and significantly distorted ultrashort-pulse localized wave packets indicate a partial self-reconstruction of the spatial and temporal characteristics for all experiments performed. Even in the case of strong distortions, specific information such as the fringe structure in space, the X-shape in space–time coordinates, or spectral fingerprints could be recognized. In combination with successive image processing, the self-reconstruction effect allows us to overcome particular limits of classical imaging systems. With the help of established procedures for contrast enhancement or pattern filtering, the signal-to-noise ratio can be improved drastically. Self-reconstruction was also obtained for single-maximum nondiffracting beams generated by the self-apodized truncation of extended Bessel–Gauss beams. Nondiffracting images (proposed by Saari as "flying images" [22]) were generated by a pseudoreflective spatial light modulator and fractionized by an array of small-angle microaxicons. The robustness of such programmed beam patterns was tested for the first time. The essential image information could be reconstructed as well.

Future applications in such fields as microbiology, optical computing, data storage, measuring techniques, or microstructuring will take advantage of the new prospects to tailor self-reconstructing wave fields. The miniaturization in integrated optical chips and implementation of real-time all-optical processors are challenges for further developments. By enabling highly robust, reliable, ultrafast transfer of information in the spatial, temporal, and/or spectral domain, self-reconstruction may contribute in a specific way to next-generation optical technologies.

Acknowledgments

The authors thank all colleagues who contributed to the success of the work by support, exciting discussions, or helpful criticism: in particular, T. Elsässer, E. T. J. Nibbering, M. Wörner, K. Reimann, S. Langer, M. Tischer, R. Müller (Max-Born-Institute, Berlin), W. Jüptner (BIAS, Bremen), M. Piché, M. Fortin, G. Rousseau, J.-L. Néron (Laval University, Quebec), H.-J. Kühn (Berliner Glas), G. Wernicke (Humboldt University, Berlin), P. Herman and J. Li (University of Toronto), C. Conti (University of Rome), E. Recami (University of Bergamo), K. Mann, B. Schäfer (Laser Laboratory, Göttingen), W. Seeber (Friedrich-Schiller-University, Jena), A. Lohmann (University of Erlangen), J. Jahns, H. Knuppertz (FernUniversität, Hagen), A. Richter (Technical High School, Wildau), S. Osten, S. Krüger (HoloEye Photnics AG, Berlin), and H. Weber (Technical University, Berlin). We gratefully acknowledge financial support from BMBF in projects 01M3025C and 13N7474/7, in the German–Canadian Collaboration Program in project CAN 00/016, and from DFG in the projects Gr1782/2-1 and Gr1782/7-1.

REFERENCES

1. J. A. Stratton, *Electromagnetic Theory*, McGraw-Hill, New York, 1941, p. 356.
2. J. N. Brittingham, Focus wave modes in homogeneous Maxwell equations: transverse electric mode, *J. Appl. Phys.* **54**, 1179–1181 (1983).
3. R. W. Ziolkowski, Exact solutions of the wave equation with complex source locations, *J. Math. Phys.* **26**, 861–863 (1985).
4. K. Reivelt and P. Saari, Experimental demonstration of realizability of optical focus wave modes, *Phys. Rev. E* **66**, 056611 (2002).
5. J. Durnin, J. J. Miceli, and J. H. Eberly, Diffraction-free beams, *Phys. Rev. Lett.* **58**, 1499–1501 (1987).
6. P. Saari and K. Reivelt, Evidence of X-shaped propagation-invariant localized light waves, *Phys. Rev. Lett.* **79**, 4135–4138 (1997).
7. P. A. Belanger, Packetlike solutions of the homogeneous wave equation, *J. Opt. Soc. Am. A* **1**, 723–724 (1984).
8. J.-Y. Lu and J. F. Greenleaf, Nondiffracting X-waves: exact solutions to free-space scalar wave equation and their finite aperture realizations, *IEEE Trans. Ultrason. Ferroelectr. Freq. Contrd* **39**, 19–31 (1992).
9. P. Saari, Relativistic Doppler effect, aberration and Gouy effect on localized waves, *Atti Fond. Giorgio Ronchi*, **58**(6), 729–754 (2003).
10. P. Saari and K. Reivelt, Generation and classification of localized waves by Lorentz transformations in Fourier space, *Phy. Rev. E* **69**, 036612 (2004).
11. W. A. Rodrigues and J.-Y. Lu, On the existence of undistorted progressive waves (UPWs) of arbitrary speeds in nature, *Found. Phys.* **27**, 435–508 (1997).
12. M. Zamboni-Rached, E. Recami, and F. Fontana, Superluminal localized solutions to the Maxwell equations propagating through normal (non-evanescent) regions, *Ann. Fond. Louis de Broglie* **26** (spec. publ.), 541–554 (2001).
13. E. Recami, M. Zamboni-Rached, K. Z. Nóbrega, C. A. Dartora, and H. E. Hernández-Figueroa, On the localized superluminal solutions to the Maxwell equations, *IEEE J. Sel. Top. Quantum Electron.* **9**, 59–73 (2003).
14. I. Alexeev, K. Y. Kim, and H. M. Milchberg, Measurement of the superluminal group velocity of an ultrashort Bessel beam pulse, *Phys. Rev. Lett.* **88**, 073901 (2002).
15. R. M. Herman and T. A. Wiggins, Production and uses of diffractionless beams, *J. Opt. Soc. Am. A* **8**, 982–942 (1991).
16. Z. Bouchal, J. Wagner, and M. Chlup, Self-reconstruction of a distorted nondiffracting beam, *Opt. Commun.* **151**, 207–211 (1998).
17. S. Sogomonian, S. Klewitz, and S. Herminghaus, Self-reconstruction of a Bessel beam in a nonlinear medium, *Opt. Commun.* **139**, 313–319 (1997).
18. Z. Bouchal, Resistance of nondiffracting vortex beam against amplitude and phase perturbations, *Opt. Commun.* **210**, 155–164 (2002).
19. V. Garcés-Chávez, D. McGloin, H. Melville, W. Sibbett, and K. Dholakia, Simultaneous micromanipulation in multiple planes using a self-reconstructing light beam, *Nature* **419**, 145 (2002).
20. D. McGloin and K. Dholakia, Bessel beams: diffraction in a new light, *Contemp. Phys.* **46**, 15–18 (2005).

21. P. Fischer, C. T. A. Brown, J. E. Morris, C. López-Mariscal, E. M. Wright, W. Sibbett, and K. Dholakia, White light propagation invariant beams, *Opt. Express* **13**, 6657–6666 (2005).
22. P. Saari, Spatially and temporally nondiffracting ultrashort pulses, in O. Svelto, S. De Silvestri, and G. Denardo, Eds., *Ultrafast Processes in Spectroscopy*, Plenum Press, New York, 1996, pp. 151–156.
23. Z. Bouchal, Controlled spatial shaping of nondiffracting patterns and arrays, *Opt. Lett.* **27**, 1376–1387 (2002).
24. J. Lu and S. He, Optical X wave communications, *Opt. Commun.* **161**, 187–192 (1999).
25. S. Minardi, A. Varanavicus, A. Piskarskas, and P. di Trapani, A compact, multi-pixel parametric light source, *Opt. Commun.* **224**, 301–307 (2003).
26. Z. Xu, Y. V. Kartashov, and L. Torner, Reconfigurable soliton networks optically-induced by arrays of nondiffracting Bessel beams, *Opt. Express* **13**, 1774–1779 (2005).
27. Z. Bouchal, Self-reconstruction ability of wave field, *Proc. SPIE* **4356** (2001), 217–224.
28. Z. Bouchal and M. Bertolotti, Self-reconstruction of wave packets due to spatio-temporal couplings, *J. Mod. Opt.* **47**, 1455–1467 (2000).
29. R. Grunwald, U. Griebner, U. Neumann, and V. Kebbel, Self-reconstruction of ultrashort-pulse Bessel-like X-waves, CLEO/QELS 2004, San Francisco, CA, May 16–21, 2004, *Conference Digest* (CD-ROM), paper CMQ7.
30. K. Reivelt and P. Saari, Bessel–Gauss pulse as an appropriate mathematical model for optically realizable localized waves, *Opt. Lett.* **29**, 1176–1178 (2004).
31. M. Piché, G. Rousseau, C. Varin, and N. McCarthy, Conical wave packets: their propagation speed and longitudinal fields, *Proc. SPIE* **3611**, 332–343 (1999).
32. R. Grunwald, V. Kebbel, U. Griebner, U. Neumann, A. Kummrow, M. Rini, E. T. J. Nibbering, M. Piché, G. Rousseau, and M. Fortin, Generation and characterization of spatially and temporally localized few-cycle optical wavepackets, *Phys. Rev. A* **67**, 063820 (2003).
33. H. Sõnajalg, M. Rätsep, and P. Saari, Demonstration of the Bessel–X pulse propagating with strong lateral and longitudinal localization in a dispersive medium, *Opt. Lett.* **22**, 310–312 (1997).
34. P. di Trapani, G. Valiulis, A. Piskarskas, O. Jedrkiewicz, J. Trull, C. Conti, and S. Trillo, Spontaneously generated X-shaped light bullets, *Phys. Rev. Lett.* **91**, 093904-1 (2003).
35. R. Grunwald, U. Griebner, F. Tschirschwitz, E. T. J. Nibbering, T. Elsaesser, V. Kebbel, H.-J. Hartmann, and W. Jüptner, Generation of femtosecond Bessel beams with micro-axicon arrays, *Opt. Lett.* **25**, 981–983 (2000).
36. J. C. Gutiérrez-Vega, M. D. Iturbe-Castillo, and S. Chávez-Cerda, Alternative formulation for invariant optical fields: Mathieu beams, *Opt. Lett.* **25**, 1493–1495 (2000).
37. R. Grunwald, U. Griebner, E. T. J. Nibbering, A. Kummrow, M. Rini, T. Elsaesser, V. Kebbel, H.-J. Hartmann, and W. Jüptner, Spatially resolved small-angle non-collinear interferometric autocorrelation of ultrashort pulses with micro-axicon arrays, *J. Opt. Soc. Am. A* **18**, 2923–2931 (2001).
38. R. Grunwald, U. Neumann, V. Kebbel, H.-J. Kühn. K. Mann, U. Leinhos, H. Mischke, and D. Wulff-Molder, VUV beam array generation by flat microoptical structures, *Opt. Lett.* **29**, 977–979 (2004).
39. J.-F. Fortin, G. Rousseau, N. McCarthy, and M. Piché, Generation of quasi-Bessel beams and femtosecond optical X-waves with conical mirrors, *Proc. SPIE* **4833**, 877–884 (2002).

40. V. Kebbel, Untersuchungen zur Erzeugung und Propagation ultrakurzer optischer Bessel-Impulse [Investigation of the generation and propagation of ultrashort Bessel pulses], dissertation, University of Bremen, Bremen Germany, 2003; *Strahltechnik*, **25**, BIAS, Bremen, Germany, 2004.

41. H. Sõnajalg and P. Saari, Suppression of temporal spread of ultrashort pulses in dispersive media by Bessel beam generators, *Opt. Lett.* **21**, 1162–1164 (1996).

42. M. Zamboni-Rached, K. Z. Nóbrega, H. E. Hernández-Figueroa, and E. Recami, Localized superluminal solutions to the wave equation in (vacuum or) dispersive media, for arbitrary frequencies and with adjustable bandwidth. *Opt. Commun.* **226**, 15–23 (2003).

43. Y. S. Kivshar and G. P. Agrawal, *Optical Solitons: From Fibers to Photonic Crystals*, Academic Press, San Diego, CA, 2003, pp. 226–228.

44. W. H. F. Talbot, Facts relating to optical science, No. IV, *Philos. Mag.* **9**, 401–407 (1836).

45. J. Fagerholm, A. T. Friberg, J. Huttunen, D. P. Morgan, and M. M. Salomaa, Angular-spectrum representation of nondiffracting X waves, *Phys. Rev. E* **54**, 4347–4352 (1996).

46. R. Grunwald, U. Neumann, S. Langer, G. Stibenz, G. Steinmeyer, V. Kebbel, M. Piché, and J.-L. Néron, Ultrabroadband spectral transfer in extended focal zones: truncated few-cycle Bessel–Gauss beams, CLEO 2005, Baltimore, MD, May 23–27, 2005, CTuAA6.

47. R. Grunwald, U. Griebner, U. Neumann, A. Kummrow, E. T. J. Nibbering, M. Piché, G. Rousseau, M. Fortin, and V. Kebbel, Generation of ultrashort-pulse nondiffracting beams and X-waves with thin-film axicons, in M. Murnane, N. F. Scherer, and A. M. Weiner, Eds., *Ultrafast Phenomena XIII*, Springer-Verlag, New York, Nov. 2002.

48. F. Gori, G. Guattari, and C. Padovani, Bessel–Gauss beams, *Opt. Commun.* **64**, 491 (1987).

49. R. Grunwald, M. Bock, U. Neumann, G. Stibenz, G. Steinmeyer, M. Piché, J.-L. Néron, V. Kebbel, and S. Huferath, Ultrashort-pulsed truncated polychromatic Bessel–Gauss beams (to be published).

CHAPTER TWELVE

Localization and Wannier Wave Packets in Photonic Crystals Without Defects

STEFANO LONGHI AND DAVIDE JANNER
Politecnico di Milano, Milan, Italy

12.1 INTRODUCTION

In the past decade, photonic crystals (PCs) have received tremendous and growing interest, motivated primarily by their unique ability to mold the flow of light on a subwavelength spatial scale, which opens up new perspectives in the realm of photonic technologies and applications. Many pioneering works on PCs highlighted and exploited the deep analogy between electromagnetic wave localization and propagation in periodic media and electron dynamics in periodic potentials (crystals) [1]. For instance, the presence of forbidden propagation bands (bandgaps) in PCs is very similar to the way in which atomic lattices produce energy bands for electrons [1–3]. Such analogies have provided a clear insight into not only photonic bandgap formation, but also into some other important phenomena, such as the possibility of light localization. Many mechanisms of localization long known in solid-state physics, such as localization at interfaces (surface modes), Anderson localization in disordered lattices, and bound states due to lattice defects or impurities (defect modes), have found their optical counterpart in engineered PC devices. One of the most important achievements in this context is perhaps the existence of strongly confined modes with typical exponential decay at frequencies inside the bandgap obtained when the periodicity of the PC is broken by a point or line defect; such modes,

Localized Waves, Edited by Hugo E. Hernández-Figueroa, Michel Zamboni-Rached, and Erasmo Recami

either localized or propagating, allow us to build high-Q resonant cavities and efficient PC-based guiding structures on a subwavelength spatial scale. The prediction of superprism effects [4], anomalous diffraction and refraction [5,6], negative refraction [7], superlensing and superresolution phenomena [8] in PCs has also demonstrated how deeply the canonical laws of optics are affected by the band structure. To date, few efforts have been devoted to a general analysis of spatial and spatiotemporal localization in purely periodic media. Indeed, only quite recently [9–12] has the impact of a photonic band structure on wave localization phenomena in a broad sense been studied, leading to the discovery of nondiffractive localized waves which do not involve evanescent modes inside a bandgap (defect modes) but have a typical subexponential decay. In a rather different context, wave localization in a weaker sense [13,14] has been the subject of an extensive research for many years, since the discovery of nondiffractive and/or nondispersive solutions to the wave equation. Most of the studies related to this subject have been concerned so far with wave localization in vacuum [13,14] or in homogeneous dispersive media [15–19], showing the existence of a wide variety of localized waves traveling undistorted at subluminal, luminal, or superluminal velocities. The most characteristic localized waves are the *X-shaped waves* [14,20,21], which may play a major role in nonlinear wave propagation as well [22,23]. The special X shape and the *spatiotemporal localization* of such waves is essentially related to the *hyperbolic* character of the underlying wave equation. Moreover, considering all different types of traveling X-waves, it can be shown that they *cannot exist at rest* in homogeneous media, since in the monochromatic case the diffractive wave equation—the Helmholtz equation—is always of elliptical type, admitting at most sinc-shaped solutions. Although X-shaped waves can be realized only approximatively in practice, several experiments in acoustic and optics have been reported so far showing nearly undistorted localized wave propagation [24–28]. For the sake of completeness we recall here that in addition to X-shaped waves, other localized waves, such as Bessel beams, focus-wave modes, and pulsed Bessel beams, have been discovered (a review and a unified description of these waves may be found in [17,20,29]). In this chapter, we provide a review of the issue of wave localization in periodic PCs in a weak (nonexponential) sense, with particular emphasis on hyperbolic localization. In particular, we present analytical and numerical results on three-dimensional (3D) wave localization in two-dimensional (2D) PCs without defects considering both stationary (monochromatic) and moving (polychromatic) localized waves. In the latter case, a general procedure to construct a wide class of nonspreading Wannier wave packets propagating with an arbitrary velocity is presented. The chapter is organized as follows. In Section 12.2 we deal with the monochromatic wave localization in PCs and presents the main results on hyperbolic localization in 2D PCs (Section 12.2.2). In Section 12.3 the issue of spatiotemporal localization in periodic structures using a Wannier function technique is addressed, and a wide class of propagating localized waves is introduced. The close connection between group velocity, diffraction tensor, and type of localization is highlighted. Finally, in Section 12.4 the main conclusions are outlined.

12.2 DIFFRACTION AND LOCALIZATION OF MONOCHROMATIC WAVES IN PHOTONIC CRYSTALS

12.2.1 Basic Equations

The starting point of our analysis is provided by a standard model of electromagnetic wave propagation in a nonmagnetic and nondispersive linear dielectric medium whose dielectric constant $\epsilon(x, y, z)$ is assumed to be periodic in one or two spatial dimensions. For monochromatic fields at frequency ω we may write

$$\mathbf{E}(\mathbf{r}, t) = \mathbf{E}(\mathbf{r})e^{-i\omega t}, \qquad \mathbf{H}(\mathbf{r}, t) = \mathbf{H}(\mathbf{r})e^{-i\omega t} \tag{12.1}$$

where the amplitudes of electric $\mathbf{E}$ and magnetic $\mathbf{H}$ fields satisfy vectorial wave equations that are derived from Maxwell's equations. In particular, for the magnetic field $\mathbf{H}$, one has

$$\nabla \times \left(\frac{1}{\epsilon}\nabla \times \mathbf{H}\right) = \left(\frac{\omega}{c}\right)^2 \mathbf{H} \tag{12.2}$$

where c is the speed of light in vacuum. The electric field can then be calculated as $\mathbf{E} = -(i/\omega\epsilon)\nabla \times \mathbf{H}$. Now we consider a PC structure having periodic dielectric function $\epsilon(\mathbf{r} + \mathbf{a}_i) = \epsilon(\mathbf{r})$ defined by the lattice vectors $\mathbf{a}_i$, which can be 1D periodic [typically, $\mathbf{a}_1 = (a, 0, 0)$], 2D periodic [$\mathbf{a}_1 = (a_{11}, a_{12}, 0)$ and $\mathbf{a}_2 = (a_{21}, a_{22}, 0)$], or even 3D periodic [$\mathbf{a}_1 = (a_{11}, a_{12}, a_{13})$, $\mathbf{a}_2 = (a_{21}, a_{22}, a_{23})$, and $\mathbf{a}_3 = (a_{31}, a_{32}, a_{33})$]. Moreover, we assume from now on that the PC structure is *homogeneous and isotropic* along nonperiodic directions. Owing to invariance along these directions, we can expand the solutions to Eq. (12.2) as a superposition of Bloch modes given by

$$\mathbf{H}_{n,\mathbf{k}}(\mathbf{r}) = \mathbf{u}_{n,\mathbf{k}}(\mathbf{r}_{\text{per}}) \exp(i\mathbf{k} \cdot \mathbf{r}) \tag{12.3}$$

with corresponding eigenvalue $\omega = \omega_n(\mathbf{k})$, where $\mathbf{r} = (x, y, z)$, $\mathbf{k} = (k_x, k_y, k_z)$, $\mathbf{r}_{\text{per}}$ is the set of all periodic coordinates selected from $\mathbf{r}$, and $\mathbf{u}_{n,\mathbf{k}}$ is a periodic function with the same periodicity properties of the PC lattice. Furthermore, along periodic directions, we allow the reciprocal vector $\mathbf{k}$ to range in the first Brillouin zone (BZ). To study wave localization—in the broad sense as explained in [13,14] for continuous media—we construct an *exact* wave packet solution to Maxwell's equations by a superposition of Bloch modes with $\mathbf{k}$ in the neighborhood of a given point $\mathbf{k}_0$ and belonging to a given band of index n. Such a general spatiotemporal wave packet can be written as $\mathbf{H}(\mathbf{r}, t) = \mathcal{H}(\mathbf{r}, t) \exp(i\mathbf{k}_0 \cdot \mathbf{r} - i\omega_0 t)$, where

$$\mathcal{H}(\mathbf{r}, t) = \int d\mathbf{k}\, F(\mathbf{k})\mathbf{u}_{n,\mathbf{k}}(\mathbf{r}_{\text{per}}) \exp\{i(\mathbf{k} - \mathbf{k}_0) \cdot \mathbf{r} - i[\omega_n(\mathbf{k}) - \omega_0]t\}, \tag{12.4}$$

$\omega_0 = \omega(\mathbf{k}_0)$ is a reference carrier frequency, and $F(\mathbf{k})$ is a spectral amplitude, narrow around $\mathbf{k} = \mathbf{k}_0$, which is left undetermined at this stage. If we assume that in the neighborhood of $\mathbf{k}_0$ the Bloch mode $\mathbf{u}_{n,\mathbf{k}}$ varies slowly, we can express the wave packet as $\mathcal{H} \simeq \mathbf{u}_{n,\mathbf{k}_0}(\mathbf{r}_{\text{per}})\psi(\mathbf{r}, t)$, where the scalar wave envelope ψ reads explicitly:

$$\psi(\mathbf{r}, t) = \int d\mathbf{k}\, F(\mathbf{k}) \exp\{i(\mathbf{k} - \mathbf{k}_0) \cdot \mathbf{r} - i[\omega_n(\mathbf{k}) - \omega_0]t\}. \tag{12.5}$$

Monochromatic solutions to Maxwell's equations are obtained by restricting the integration domain in integrals (12.4) or (12.5) to values of $\mathbf{k}$ belonging to the isofrequency surface implicitly defined by the equation

$$\omega_n(\mathbf{k}) - \omega_0 = \Omega, \tag{12.6}$$

where Ω is a frequency detuning from the reference carrier frequency ω_0. Note that the properties of the monochromatic solutions, including the degree of localization, will depend strictly on the form of the spectrum $F(\mathbf{k})$. As a limiting case, for the choice $F(\mathbf{k}) = \delta(\mathbf{k} - \mathbf{k}_0)$ we recover the fully delocalized Bloch mode of the PC. Our choice of the spectrum function will be oriented to obtain localized solutions using well-known spectra that generate X-shaped waves or in general, spatially localized solutions [20]. To investigate the possibility for X-type wave localization analytically, it is worth deriving a differential equation for the envelope ψ starting from its integral representation (12.5). To this aim, let us observe that deriving Eq. (12.5) with respect to t, we obtain

$$i\frac{\partial \psi}{\partial t} = \int d\mathbf{k}\, [\omega_n(\mathbf{k}) - \omega_0] F(\mathbf{k}) \exp\{i(\mathbf{k} - \mathbf{k}_0) \cdot \mathbf{r} - i[\omega_n(\mathbf{k}) - \omega_0]t\}. \tag{12.7}$$

On the other hand, if we consider the Taylor expansion of $\omega_n(\mathbf{k})$ at around $\mathbf{k}_0$, we may write

$$\omega_n(\mathbf{k}) = \omega_n(\mathbf{k}_0) + \nabla_{\mathbf{k}}\omega_n(\mathbf{k})|_{\mathbf{k}_0} \cdot (\mathbf{k} - \mathbf{k}_0) + \frac{1}{2}\, \nabla^2_{\mathbf{k}}\omega_n(\mathbf{k})\Big|_{\mathbf{k}_0} (\mathbf{k} - \mathbf{k}_0)^2 + O(\mathbf{k}^3), \tag{12.8}$$

and considering the operator $\omega_n(\mathbf{k}_0 - i\nabla_{\mathbf{r}})$ defined by the straight substitution $\mathbf{k} - \mathbf{k}_0 \rightarrow -i\nabla_{\mathbf{r}}$, that is,

$$\omega_n(\mathbf{k}_0 - i\nabla_{\mathbf{r}}) = \omega_n(\mathbf{k}_0) - i\; \nabla_{\mathbf{k}}\omega_n(\mathbf{k})|_{\mathbf{k}_0} \cdot \nabla_{\mathbf{r}} - \frac{1}{2}\, \nabla^2_{\mathbf{k}}\omega_n(\mathbf{k})\Big|_{\mathbf{k}_0} \nabla^2_{\mathbf{r}} + O(\nabla^3_{\mathbf{r}}), \tag{12.9}$$

we may write

$$\omega_n\left(\mathbf{k}_0 - i\nabla_{\mathbf{r}}\right)\psi = \int d\mathbf{k}\, \omega_n(\mathbf{k}) F(\mathbf{k}) \exp\{i(\mathbf{k} - \mathbf{k}_0) \cdot \mathbf{r} - i[\omega_n(\mathbf{k}) - \omega_0]t\}. \tag{12.10}$$

A comparison of Eqs. (12.7) and (12.10) then yields the following differential equation for the envelope ψ:

$$i\frac{\partial\psi}{\partial t} = [\omega_n(\mathbf{k}_0 - i\nabla_{\mathbf{r}}) - \omega_0]\,\psi. \tag{12.11}$$

Equation (12.11) describes the basic differential equation of an envelope wave packet accounting for the characteristic dispersion curve of the PC in which it propagates. Since we are seeking for monochromatic (stationary) solutions of localized waves, we assume for the envelope ψ the following behavior:

$$\psi(x, y, z; t) = \Phi(x, y, z)e^{-i\Omega t}, \tag{12.12}$$

so that Eq. (12.11) yields the following basic equation for the spatial shape Φ of the localized wave:

$$\{\Omega - [\omega_n(\mathbf{k}_0 - i\nabla_{\mathbf{r}}) - \omega_0]\}\Phi = 0. \tag{12.13}$$

This equation is the starting point of our analysis, which focuss on 2D PC wave localization. It should be noted that the envelope equation (12.13) has been derived under the assumption of a slow variation of Bloch modes around $\mathbf{k}_0$, which implies a slow variation of Φ over the periodic spatial scale of the PC. However, as shown in Section 12.3.1, a different and more general approach to wave localization, based on a Wannier function expansion, leads basically to the same result as that expressed by Eq. (12.13).

12.2.2 Localized Waves

In 2D PCs, many periodic combinations and different types of localized field configurations can be produced by suitable superpositions of Bloch modes. Usually, the PC configuration parameters are (1) the shape of the element to be repeated in the crystal structure (e.g., a cylinder rod), and (2) the lattice vectors that account for the distribution of the original shape (examples are triangular, square, 12-fold). In our analysis [11] we consider a square lattice of cylindrical rods and consider the localization properties of that structure. In that case, the dielectric function can be expressed as $\epsilon(x, y, z) = \epsilon(x, y)$, along with the periodicity conditions $\epsilon(\mathbf{r} + \mathbf{R}) = \epsilon(\mathbf{r})$, where $\mathbf{R} = m\mathbf{a}_1 + n\mathbf{a}_2$ is a vector of the PC lattice, and the vectors $\mathbf{a}_1 = (a_1, 0, 0)$ and $\mathbf{a}_2 = (0, a_2, 0)$ represent the basis lattice vectors. Owing to invariance along the z direction , the most general solution to the vectorial wave equation (12.2) is a superposition of solutions of the form $\mathbf{H}_{n,\mathbf{k}}(\mathbf{r}) = \mathbf{u}_{n,\mathbf{k}}(x, y)\exp(i\mathbf{k}\cdot\mathbf{r})$, as stated in Eq. (12.3). Note that Bloch modes $\mathbf{u}_{n,\mathbf{k}}$ and the dispersion curve ω_n depend parametrically on the out-of-plane wave number k_z. In addition, for symmetry reasons, ω_n is invariant when $k_z \rightarrow -k_z$, so that one has $\partial\omega_n/\partial k_z = 0$ at $k_z = 0$. In particular, for $k_z = 0$, Eq. (12.2) splits into two distinct equations and the eigenmodes can be classified as TE modes, corresponding to $H_x = H_y = E_z = 0$, or TM modes, corresponding to

$E_x = E_y = H_z = 0$. For $k_z \neq 0$, mixed modes, with all field components nonvanishing, are obtained instead. Now if we carry out the same analysis that we made in Section 12.2.1, we recover an envelope equation similar to Eq. (12.11), and finally, if we expand the band curve $\omega_n(\mathbf{k})$ at around $\mathbf{k}_0$ up to second order in k_x, k_y, and k_z (the effective mass approximation), taking into account that $\partial\omega_n/\partial k_z = 0$ at $\mathbf{k} = \mathbf{k}_0$, the envelope equation (12.11) takes the simplified form†

$$\frac{\partial \psi}{\partial t} + \frac{\partial \omega_n}{\partial \mathbf{k}_\perp}\frac{\partial \psi}{\partial \mathbf{r}_{\mathrm{per}}} = \frac{i}{2}\left(\frac{\partial^2 \omega_n}{\partial \mathbf{k}_\perp \partial \mathbf{k}_\perp}\frac{\partial^2 \psi}{\partial \mathbf{r}_{\mathrm{per}} \partial \mathbf{r}_{\mathrm{per}}} + \frac{\partial^2 \omega_n}{\partial k_z^2}\frac{\partial^2 \psi}{\partial z^2}\right), \tag{12.14}$$

where $\mathbf{k}_\perp = (k_x, k_y)$, $\mathbf{r}_{\mathrm{per}} = (x, y)$, and the derivatives $\partial\omega_n/\partial\mathbf{k}_\perp$, $\partial^2\omega_n/\partial\mathbf{k}_\perp\partial\mathbf{k}_\perp$, and $\partial^2\omega_n/\partial k_z^2$ are calculated at $\mathbf{k} = \mathbf{k}_0$. If the right-hand-side terms in Eq. (12.14) were negligible, the wave packet envelope would travel undistorted in the PC plane with a group velocity $\partial\omega_n/\partial\mathbf{k}_\perp$. The inclusion of the right-hand-side term is responsible for wave packet diffraction, which is ruled by the tensorial term $\partial^2\omega_n/\partial\mathbf{k}_\perp\partial\,\mathbf{k}_\perp$ for in-plane diffraction and by the scalar term $\partial\omega_n/\partial k_z^2$ for out-of-plane diffraction. Since we are interested in *hyperbolic* localization of steady waves, we assume that the group velocity is zero (i.e., we choose $\mathbf{k}_0$ to be a stationary point on the band surface) and set $\psi = \Phi(x, y, z)\exp(-i\Omega t)$. In addition, after a suitable rotation of $\mathbf{r}_{\mathrm{per}}$ axes, the symmetric tensor $\partial^2\omega_n/\partial\mathbf{k}_\perp\partial\,\mathbf{k}_\perp$ can always be made diagonal. Under these assumptions and again indicating by (x, y) and (k_x, k_y) the principal directions where the diffraction tensor $\partial^2\omega_n/\partial\mathbf{k}_\perp\,\partial\mathbf{k}_\perp$ is diagonal, the envelope equation (12.14) yields

$$\alpha\frac{\partial^2 \Phi}{\partial x^2} + \beta\frac{\partial^2 \Phi}{\partial y^2} + \gamma\frac{\partial^2 \Phi}{\partial z^2} = -\Omega\Phi, \tag{12.15}$$

where the coefficients α, β, and γ are defined as

$$\alpha \equiv \frac{1}{2}\frac{\partial^2 \omega_n}{\partial k_x^2}\bigg|_{\mathbf{k}_0}, \qquad \beta \equiv \frac{1}{2}\frac{\partial^2 \omega_n}{\partial k_y^2}\bigg|_{\mathbf{k}_0}, \qquad \gamma \equiv \frac{1}{2}\frac{\partial^2 \omega_n}{\partial k_z^2}\bigg|_{\mathbf{k}_0}. \tag{12.16}$$

X-wave localization requires Eq. (12.15) to be hyperbolic (i.e., a sign discordance among the coefficients α, β, and γ should occur). In this case, for $\Omega = 0$ and within the effective mass approximation, the isofrequency $\mathbf{k}$ modes—entering in Eqs. (12.4) or (12.5) and defining the integral representation of X-waves—lie on the pseudocone (hyperbolic) surface of equation:

$$\alpha(k_x - k_{x0})^2 + \beta(k_y - k_{y0})^2 + \gamma k_z^2 = 0. \tag{12.17}$$

Since in general it turns out that the out-of-plane diffraction is always positive (i.e., $\gamma > 0$), X-waves exist provided that both α and β are negative—which is the case

†The envelope equation is analogous to the effective mass theorem for electrons in solids (see, e.g., [3]).

when $\mathbf{k}_0$ is a local top band point—or they have the opposite sign, which is the case when $\mathbf{k}_0$ is a saddle point of the band. In these cases, Eq. (12.15) is a 2D Klein–Gordon equation that admits of a general class of localized solutions (see, e.g., [13,14,30]). For $\alpha < 0$, $\beta < 0$, one of the simplest X-wave solutions to Eq. (12.15) reads explicitly [30]

$$\Phi = \mathrm{Re}\left\{\frac{\exp(-\sqrt{\Omega v})}{\sqrt{v}}\right\} \tag{12.18}$$

with $v \equiv (z_0 - iz)^2/\gamma + x^2/|\alpha| + y^2/|\beta|$, whereas for $\alpha > 0$ and $\beta < 0$, we have

$$\Phi = \mathrm{Re}\left\{\frac{\exp(-\sqrt{-\Omega v})}{\sqrt{v}}\right\} \tag{12.19}$$

with $v \equiv (y_0 - iy)^2/|\beta| + x^2/\alpha + z^2/\gamma$. In the equations above, z_0 and y_0 are arbitrary real-valued parameters that roughly determine, in the two cases, the characteristic spatial extension of the envelope along the z- and y-directions, respectively. The best known X-waves correspond to zero-frequency detuning ($\Omega = 0$), for which the wave packet envelope ψ, modulating the Bloch mode $\mathbf{u}_{n,\mathbf{k}_0}$, describes an X-wave whose pseudoaxis is directed along the z-direction [X-waves of the first kind, Eq. (12.18)], or along the y-direction [X-waves of the second kind, Eq. (12.19)]. X-waves of the first kind need the light to be tuned at a *local* maximum point of the band and are therefore similar to 3D X-waves in 1D lattices [9], apart for a possible anisotropy of the X-wave cone if $\alpha \neq \beta$. However, taking into account the 2D nature of the band structure in our case, the existence of a true bandgap is not required as in the 1D case, and X-waves of the first kind exist even for sufficiently low-contrast indices where a true PC bandgap is not achieved. The second type of X-waves are fully peculiar to the 2D periodic structure since in this case, the in-plane diffraction is made hyperbolic and the frequency of a saddle point is fully internal to the band. These considerations clearly show that the existence of X-type wave localization is not related precisely to the existence of bandgaps, as the 1D case might suggest [9], but on the hyperbolic character of the isofrequency surface emanating from zero-group-velocity band points [i.e., on the hyperbolic nature of the 3D diffraction operator entering in Eq. (12.14)].

We checked the validity of the effective mass approach [Eqs. (12.14) and (12.15)] and constructed localized PC X-waves of the vectorial Maxwell's equations by a direct numerical computation of monochromatic spatial wave packets based on the spectral representation given by Eqs. (12.4) and (12.5). The computation of Bloch modes and corresponding bands of the PC was performed using a standard plane-wave method that accounts for out-of-plane propagation [31]. The PC consists of a square lattice of cylinders with radius $r = 0.2a$ (a is the period of the square lattice) and $\epsilon = 8.9$, embedded in air (see Fig. 12.1). The PC dispersion curves have been computed for the 10 lowest-frequency bands by using 400 plane waves in the Fourier expansion of $\epsilon(x, y)$. A few low-order dispersion bands for $k_z = 0$, corresponding to TE and TM

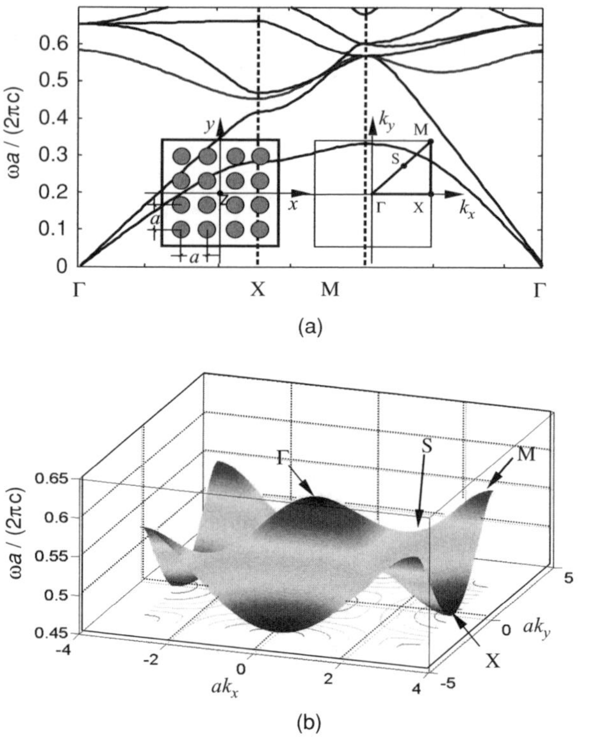

FIGURE 12.1 (a) Dispersion curves for in-plane TE and TM modes of a square lattice PC made of cylinders (radius $r = 0.2a$, $\epsilon = 8.9$) embedded in air. A sketch of the PC structure and first Brillouin zone in the reciprocal plane, with a triangular irreducible zone with high-symmetry points Γ, X, and M, are shown as insets. The blue curve in the figure (the third one from the bottom) corresponds to the second band for TM modes (TM air band). (b) Surface diagram of PC band $\omega(k_x, k_y, 0)$ for in-plane waves ($k_z = 0$) corresponding to the TM air-band of part (a). Note the existence of a relative maximum on the surface at point Γ and of a saddle point at S.

modes, are shown in Fig. 12.1a along the contour ΓXM of the irreducible part of the first Brillouin zone in the reciprocal lattice space (see also Fig. 2 in Chap. 5 of [1]). Figure 12.1b shows, as an example, the entire 2D band surface corresponding to the second TM band of Fig. 12.1a (called the TM air band).

From an inspection of Fig. 12.1b we note that on this surface there is a relative maximum at point Γ (i.e., for $k_x = 0, k_y = 0$) and a saddle point at S along the Γ–M line, corresponding to $k_x = k_y \simeq 1.7957/a$. The principal directions corresponding to the top band point Γ are those shown in the insets of Fig. 12.1a, whereas the principal directions corresponding to the saddle point S turn out to be rotated in both the direct and reciprocal planes by $\pi/4$. To obtain a localized wave, according to the spectral representation (12.4), we construct a superposition of Bloch modes with wave vectors in the neighborhood of either one of the two stationary points Γ

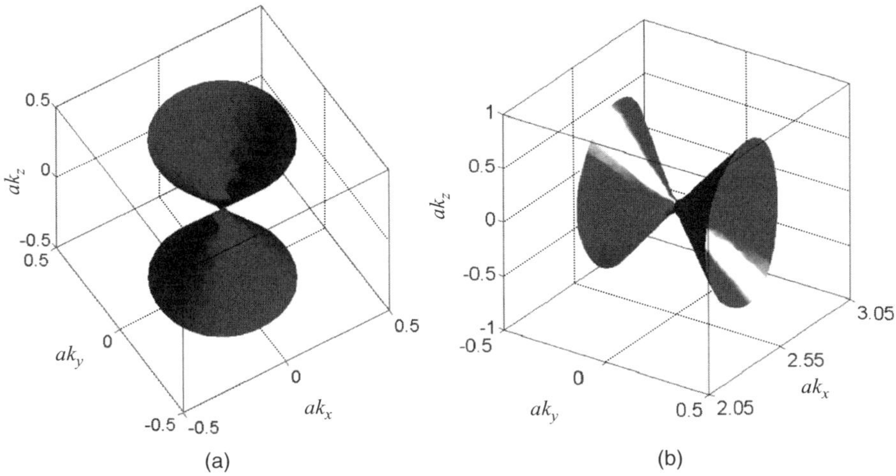

FIGURE 12.2 Numerically computed PC isofrequency surfaces (pseudocones) in the (k_x, k_y, k_z) space at frequencies corresponding to top band point Γ (a), and saddle point S (b) of the TM air band shown in Fig 12.1b.

and S, including off-axis Bloch modes ($k_z \neq 0$). The superposition is performed on the isofrequency surface $\omega_n(k_x, k_y, k_z) = \omega_0$ assuming in all calculations the direct (x, y) and reciprocal (k_x, k_y) axes directed along the principal directions. The isofrequency surfaces corresponding to the top point Γ and saddle point S of the band in Fig. 12.1b, as computed numerically through band calculation using 50 out-of-plane k_z waves, are shown in Fig. 12.2. Note that, according to effective mass approximation, these surfaces have a hyperbolic character and appear as pseudocones with an axis oriented along the k_z (for $\alpha < 0, \beta < 0$) or k_y (for $\alpha > 0, \beta < 0$) directions [see Eq. (12.17)].

In order to construct a localized wave according to Eqs. (12.4) and (12.5), it is worth referring the space (k_x, k_y, k_z) to pseudocylindrical coordinates with a polar axis oriented along the cone axis of the isofrequency surface. Considering, for the sake of definiteness, the case of Fig. 12.2a (but a similar analysis could be done for the saddle point S, Fig. 12.2b), we introduce the pseudopolar coordinates $(k_\perp, \varphi)$ such that $k_x = |\alpha|^{-1/2} k_\perp \cos\varphi$, $k_y = |\beta|^{-1/2} k_\perp \sin\varphi$. Note that owing to the isofrequency constraint $\omega_n(k_\perp, \varphi, k_z) = \omega_0$, only two among the three variables $k_\perp$, k_z, and φ can be assumed as independent in the integrals entering in Eqs. (12.4) and (12.5). Assuming that k_z and φ as independent variables, so that $k_\perp = k_\perp(k_z, \varphi)$, Eq. (12.4) then takes the explicit form suitable for numerical computation:

$$\mathcal{H}(x, y, z) = \int_{-\infty}^{\infty} dk_z \int_0^{2\pi} d\varphi F(k_z, \varphi)\mathbf{u}_{n,\mathbf{k}}(x, y) \times \exp[ix\sqrt{|\alpha|}k_\perp \cos\varphi + iy\sqrt{|\beta|}k_\perp \sin\varphi + ik_z z]. \quad (12.20)$$

A similar integral can be written for the envelope ψ [see Eq. (12.5)]:

$$\psi(x, y, z) = \int_{-\infty}^{\infty} dk_z \int_{0}^{2\pi} d\varphi\, F(k_z, \varphi) \times \exp[ix\sqrt{|\alpha|}k_\perp \cos\varphi + iy\sqrt{|\beta|}k_\perp \sin\varphi + ik_z z]. \quad (12.21)$$

The choice of the spectral amplitude F determines the shape and degree of localization of the X-wave. In order to reproduce the special X-wave solution given by Eq. (12.18) with $\Omega = 0$, in our numerical computations we assumed that $F(k_z, \varphi) = \exp(-|k_z|z_0)$.† Figures 12.3 and 12.4 show examples of numerically computed 3D profiles of the envelope ψ and full vectorial **H**-field, as obtained from Eqs. (12.20) and (12.21), respectively. Note that a characteristic X shape of the envelope ψ is obtained (Fig. 12.3), in agreement with the analytical model based on Eq. (12.15), and that according to the previous analysis, for the top band point Γ, the X-wave profile develops along plane $x =$ const (or $y =$ const), whereas for the saddle point S, it develops along plane $z =$ const (or $x =$ const). Figure 12.4 shows the full exact field amplitude profile $|\mathcal{H}(\mathbf{r})|$ of the localized wave as obtained from numerical computation of Eq. (12.21) (i.e., by accounting for the dispersion of Bloch mode profiles $\mathbf{u}_{n,\mathbf{k}}$ with $\mathbf{k}$). Note that, as expected, the exact field profile shows a rapidly varying and almost periodic pattern that is enveloped by a slowly varying X-shaped wave.

12.3 SPATIOTEMPORAL WAVE LOCALIZATION IN PHOTONIC CRYSTALS

In the following we extend our analysis to polychromatic wave propagation in periodic media, applying the formalism of the Wannier function expansion to predict the existence of undistorted spatiotemporal wave packets in PC structures. In the present analysis, the group velocity at which the wave packet travels may be chosen arbitrarily, and its localization properties may be influenced by this choice, leading to three different localization regimes, which are determined by the parabolic, elliptic, or hyperbolic character of the associated diffraction tensor.

† Within the effective mass approximation, this choice of F allows for an analytical calculation of the integral (12.21) in terms of Bessel functions, yielding precisely the X-wave solution (12.18) with $\Omega = 0$. In fact, in the effective mass limit from Eq. (12.17), one has $k_\perp = \sqrt{\gamma}|k_z|$, and for any spectral amplitude F independent of φ, Eq. (12.21) yields $\psi(x, y, z) = (1/2\pi)\int_{-\infty}^{\infty} dk_z\, F(k_z) J_0((x^2/|\alpha| + y^2/|\beta|)^{1/2}\gamma^{1/2}|k_z|)\exp(-ik_z z)$. If we assume that $F(k_z) \propto \exp(-i|k_z|z_0)$, the integral can be calculated analytically [32], yielding Eq. (12.18) with $\Omega = 0$. Gaussian-like instead of exponential would lead to similar X-shaped waves; however, an analytical form of these waves is not available.

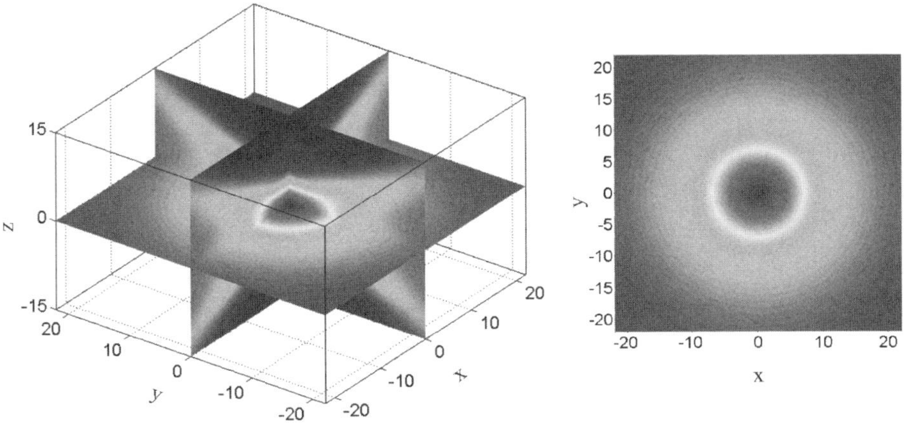

(a)

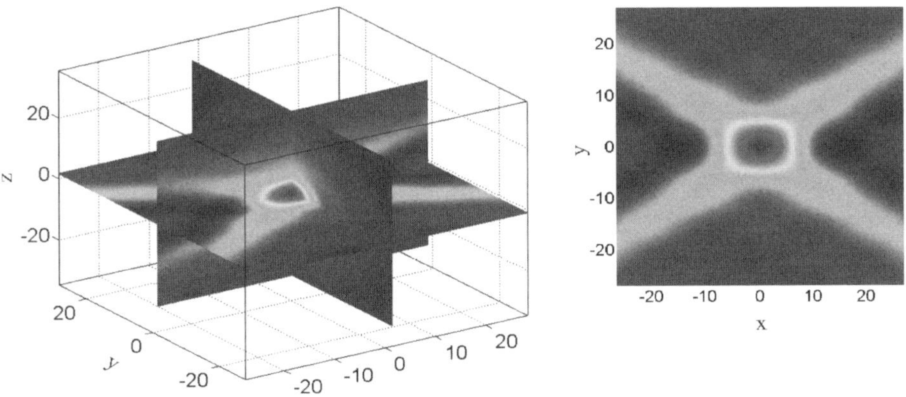

(b)

FIGURE 12.3 Plots of 3D X wave envelope $|\psi(x, y, z)|$ for top band point Γ (a), and saddle point S (b), and for an exponential spectral amplitude F with $z_0 = 5a$ (see the text). The profiles have been calculated by numerical computation of Eq. (12.21). The figures on the right side show the behavior of $|\psi|$ on the $z = 0$ plane.

12.3.1 Wannier Function Technique

In the nonmonochromatic case, the equation describing spatiotemporal propagation is

$$\nabla \times \left(\frac{1}{\epsilon} \nabla \times \mathbf{H} \right) = -\frac{1}{c^2} \frac{\partial^2 \mathbf{H}}{\partial t^2}, \tag{12.22}$$

which replaces the corresponding Eq. (12.2), valid for monochromatic waves. The explicit expression for the electric field, if needed, can be retrieved from the equation

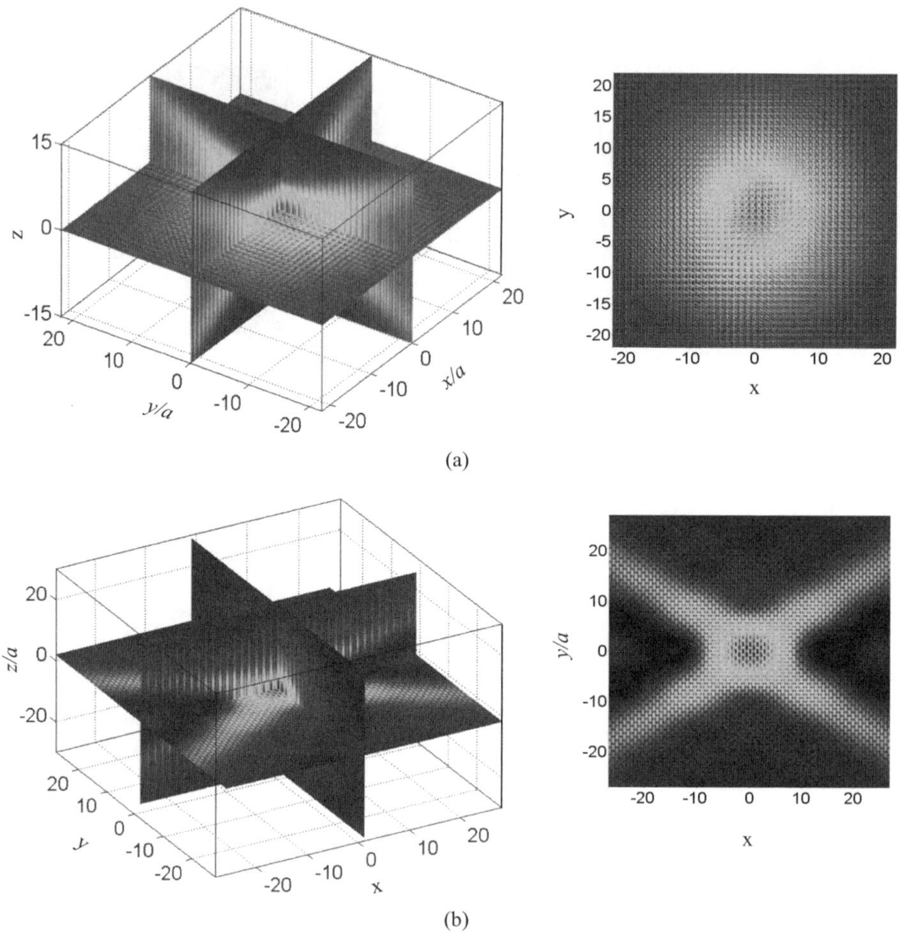

FIGURE 12.4 3D and 2D plots (on the $z = 0$ plane) of the full electric field amplitude $|\mathcal{E}|$ for the two kinds of X-waves shown in Fig. 12.3, computed using Eq. (12.20).

$\partial \mathbf{E}/\partial t = (\epsilon_0 \epsilon)^{-1} \nabla \times \mathbf{H}$. Since we are interested in the space–time behavior of polychromatic waves in periodic media, material dispersion is typically negligible compared to band-induced dispersion, so the former has not been included in Eq. (12.22). In the following we consider the general case of a 3D PC with a structure defined by lattice vectors $\mathbf{a}_1 = (a_{11}, a_{12}, a_{13})$, $\mathbf{a}_2 = (a_{21}, a_{22}, a_{23})$, and $\mathbf{a}_3 = (a_{31}, a_{32}, a_{33})$ (see Section 12.2), along with the periodicity condition $\epsilon(\mathbf{r} + \mathbf{R}) = \epsilon(\mathbf{r})$, where $\mathbf{R} = l\mathbf{a}_1 + m\mathbf{a}_2 + n\mathbf{a}_3$ is a vector of the PC lattice.

In order to study the propagation of a spatiotemporal wave packet, we adopt the method of the Wannier functions, which is commonplace in the study of the quasiclassical electron dynamics in solids [3,33,34] and of superstructured Bragg gratings [35], and has recently been applied to study localized modes and defect structures in PCs [36–38]. In the following analysis, we follow closely the approach presented in [12]. Let us first consider the monochromatic Bloch-type solutions to Eq. (12.22) for a field oscillating at frequency ω, $\mathbf{H}(\mathbf{r},t) = \mathbf{H}_{n,\mathbf{k}}(\mathbf{r})\exp(-i\omega t)$, where $\mathbf{k}$ lies in the first Brillouin zone of the reciprocal $\mathbf{k}$-space, $\omega = \omega_n(\mathbf{k})$ is the dispersion curve for the nth band of the PC, and $\mathbf{H}_{n,\mathbf{k}}(\mathbf{r}) = \mathbf{u}_{n,\mathbf{k}}(\mathbf{r})\exp(i\mathbf{k}\cdot\mathbf{r})$ are the band modes as in Eq. (12.3), satisfying both the periodicity and the normalization conditions:

$$\mathbf{H}_{n,\mathbf{k}}(\mathbf{r}+\mathbf{R}) = \mathbf{H}_{n,\mathbf{k}}(\mathbf{r})\exp(i\mathbf{k}\cdot\mathbf{R}), \tag{12.23}$$

$$\langle \mathbf{H}_{n',\mathbf{k}'}|\mathbf{H}_{n,\mathbf{k}}\rangle = V_{\mathrm{BZ}}\,\delta_{n,n'}\delta(\mathbf{k}'-\mathbf{k}). \tag{12.24}$$

In Eqs. (12.23) and (12.24), $\mathbf{R} = l\mathbf{a}_1 + m\mathbf{a}_2 + n\mathbf{a}_3$ is the general lattice site in real space ($\mathbf{r}$-space), $V_{\mathrm{BZ}} = (2\pi)^3/V$ is the volume of the first Brillouin zone in the reciprocal space, and V is the volume of the $\mathbf{r}$-space unit cell. For each band of the PC, we can construct a Wannier function $\mathbf{W}_n(\mathbf{r})$ as a localized superposition of Bloch functions on the whole band according to the relation

$$\mathbf{W}_n(\mathbf{r}) = \frac{1}{V_{\mathrm{BZ}}}\int_{\mathrm{BZ}} d\mathbf{k}\,\mathbf{H}_{n,\mathbf{k}}(\mathbf{r}). \tag{12.25}$$

In the superposition, the phase of Bloch modes $\mathbf{H}_{n,\mathbf{k}}$ can be chosen such that $\mathbf{W}_n(\mathbf{r})$ is strongly localized around $\mathbf{r} = 0$ with an exponential decay away from that point. Moreover, Wannier functions satisfy orthogonality conditions analogous to Eq. (12.24):

$$\langle \mathbf{W}_{n'}(\mathbf{r}-\mathbf{R}')|\mathbf{W}_n(\mathbf{r}-\mathbf{R})\rangle \equiv \int W^*_{n'}(\mathbf{r}-\mathbf{R}')W_n(\mathbf{r}-\mathbf{R})\,d\mathbf{r} = \delta_{n,n'}\delta_{\mathbf{R},\mathbf{R}'}. \tag{12.26}$$

Exploiting the orthogonality condition and applying the vector wave operator $\nabla\times(\epsilon^{-1}\nabla\times)$ to the Wannier function [Eq. (12.25)], the following relation can be derived:

$$\begin{aligned}&\langle \mathbf{W}_{n'}(\mathbf{r}-\mathbf{R}')\left|\nabla\times(\epsilon^{-1}\nabla\times)\right|\mathbf{W}_n(\mathbf{r}-\mathbf{R})\rangle\\ &\quad= \int W^*_{n'}(\mathbf{r}-\mathbf{R}')\left[\frac{1}{V_{\mathrm{BZ}}}\int d\mathbf{k}\,\frac{\omega_n^2(\mathbf{k})}{c^2}\mathbf{H}_{n,\mathbf{k}}(\mathbf{r}-\mathbf{R})\right]d\mathbf{r}.\end{aligned} \tag{12.27}$$

Let us now introduce the Fourier expansion coefficients $\theta_{n,\mathbf{R}''}$ of the dispersion curve $\omega_n^2(\mathbf{k})$ of the band according to the relations

$$\omega_n^2(\mathbf{k}) = \sum_{\mathbf{R}''}\theta_{n,\mathbf{R}''}e^{i\mathbf{k}\cdot\mathbf{R}''}, \qquad \theta_{n,\mathbf{R}''} \equiv \frac{1}{V_{\mathrm{BZ}}}\int_{\mathrm{BZ}} d\mathbf{k}\,\omega_n^2(\mathbf{k})e^{-i\mathbf{k}\cdot\mathbf{R}''}. \tag{12.28}$$

Substitution of the expansion for $\omega_n^2(\mathbf{k})$ given by Eq. (12.28) into Eq. (12.27) and using the periodicity property of Bloch modes $\mathbf{H}_{n,\mathbf{k}}$, we obtain

$$\frac{1}{V_{\mathrm{BZ}}}\int d\mathbf{k}\,\frac{\omega_n^2(\mathbf{k})}{c^2}\mathbf{H}_{n,\mathbf{k}}(\mathbf{r}-\mathbf{R})=\sum_{\mathbf{R}''}\theta_{n,\mathbf{R}''}W_n(\mathbf{r}+\mathbf{R}''-\mathbf{R}), \tag{12.29}$$

which after some straightforward algebraic calculations leads to the equation

$$\langle \mathbf{W}_{n'}(\mathbf{r}-\mathbf{R}')\left|\nabla\times(\epsilon^{-1}\nabla\times)\right|\mathbf{W}_n(\mathbf{r}-\mathbf{R})\rangle=\delta_{n,n'}\theta_{n,\mathbf{R}'-\mathbf{R}}. \tag{12.30}$$

We now look for a spatiotemporal wave packet, which is a solution to Eq. (12.22), as a superposition of translated Wannier functions localized at different lattice points $\mathbf{R}$ of the periodic structure with amplitudes $f(\mathbf{R},t)$ that depend on the lattice point $\mathbf{R}$ and can vary in time; that is, we set

$$\mathbf{H}(\mathbf{r},t)=\sum_{\mathbf{R}}f(\mathbf{R},t)\mathbf{W}_n(\mathbf{r}-\mathbf{R}). \tag{12.31}$$

Note that since we consider a pure periodic structure without defects and neglect perturbation terms in Eq. (12.22) (e.g., nonlinearities), coupling among different bands (e.g., Landau–Zener tunneling) does not occur and in Eq. (12.31) the sum can be taken over a single band, of index n. Coupled-mode equations for the temporal evolution of the amplitudes $f(\mathbf{R},t)$ of Wannier functions at different lattice points can be obtained after substitution of Eq. (12.31) into Eq. (12.22), taking the scalar product with $\mathbf{W}_n(\mathbf{r}-\mathbf{R})$ and using the orthogonality conditions of Wannier functions [Eq. (12.26)], together with Eq. (12.30). We obtain

$$\frac{\partial^2 f(\mathbf{R},t)}{\partial t^2}+\sum_{\mathbf{R}'}\theta_{n,\mathbf{R}'-\mathbf{R}}f(\mathbf{R}',t)=0. \tag{12.32}$$

The solution to the coupled-mode equations (12.32) can be seen as the restriction to the lattice sites $\mathbf{R}$, explicitly $f(\mathbf{r}=\mathbf{R},t)$, of the *continuous* function $f(\mathbf{r},t)$, depending on the space coordinates $\mathbf{r}$ and time t, which satisfies the partial differential equation

$$\frac{\partial^2 f(\mathbf{r},t)}{\partial t^2}+\omega_n^2(-i\nabla_{\mathbf{r}})f(\mathbf{r},t)=0, \tag{12.33}$$

where $\omega_n^2(-i\nabla_{\mathbf{r}})$ is the operator obtained after the substitution $\mathbf{k}\to -i\nabla_{\mathbf{r}}$ in the Fourier expansion of $\omega_n^2(\mathbf{k})$. It should be noted that the differential equation for the *continuous envelope* $f(\mathbf{r},t)$ of the Wannier function wave packet [Eq. (12.31)], as given by Eq. (12.33), is exact. Thus, for any band of the PC we can write an envelope equation accounting for the specific details of the band solely in the dispersion curve $\omega_n^2(\mathbf{k})$ and in the shape of the corresponding Wannier function $\mathbf{W}_n$ [Eq. (12.25)].

12.3.2 Undistorted Propagating Waves in Two- and Three-Dimensional Photonic Crystals

In the two- and three-dimensional cases, the most general solution to the Wannier function envelope equation (12.33) can be expressed as a superposition of functions $\psi(\mathbf{r}, \pm t)$, where $\psi(\mathbf{r}, t)$ is a solution to the wave equation

$$i\frac{\partial \psi}{\partial t} = \omega_n(-i\nabla_{\mathbf{r}})\psi. \tag{12.34}$$

We are searching for spatiotemporal localized solutions to Eq. (12.34) such that $|\psi|$ corresponds to a wave propagating undistorted with a group velocity v_g. To this aim, let us set

$$\psi(\mathbf{r}, t) = g(\mathbf{r}, t)\exp(i\mathbf{k}_0 \cdot \mathbf{r} - i\Omega t), \tag{12.35}$$

where $\mathbf{k}_0$ is chosen inside the first Brillouin zone in the reciprocal space and the frequency Ω is chosen close to (but not necessarily coincident with) $\omega_0 = \omega_n(\mathbf{k}_0)$. It can easily be shown that the envelope g satisfies the wave equation

$$i\frac{\partial g}{\partial t} = [\omega_n(\mathbf{k}_0 - i\nabla_{\mathbf{r}}) - \Omega]\, g. \tag{12.36}$$

We first note that if g varies slowly with respect to the spatial variables $\mathbf{r}$ over a distance of the PC lattice size, at leading order we can expand $\omega_n(\mathbf{k}_0 - i\nabla_{\mathbf{r}})$ up to first order around $\mathbf{k}_0$; taking $\Omega = \omega_0$, we obtain

$$\partial g/\partial t + \nabla_{\mathbf{k}}\omega_n \cdot \nabla_{\mathbf{r}} g = 0, \tag{12.37}$$

which shows that at leading order, *an arbitrary* 3D spatially localized wave packet travels undistorted with a group velocity given by $\nabla_{\mathbf{k}}\omega_n$. Higher-order terms are generally responsible for wave packet spreading, in both space and time, and should be accounted for in our analysis. To find propagation-invariant envelope waves *even when dispersive terms* are present, let us assume, without loss of generality, that $\partial\omega_n/\partial k_y|_{\mathbf{k}_0} = \partial\omega_n/\partial k_z|_{\mathbf{k}_0} = 0$; that is, let us choose the orientation of the x axis such that the wave packet group velocity $\nabla_{\mathbf{k}}\omega_n$ is directed along this axis, and let us look for a spatiotemporal localized solution to Eq. (12.36) of the form $g = g(x_1, x_2, x_3)$, with $x_1 = x - v_g t$, $x_2 = y$ and $x_3 = z$, traveling along the x axis with a group velocity v_g, which is left undetermined at this stage. The function g then satisfies the equation

$$-iv_g\frac{\partial g}{\partial x_1} = [\omega_n(\mathbf{k}_0 - i\nabla_{\mathbf{x}}) - \Omega]\, g, \tag{12.38}$$

whose solution can be written formally as

$$g(x_1, x_2, x_3) = \int dk_2\, dk_3\; G(k_2, k_3)\exp(i\mathbf{k}\cdot\mathbf{x}). \tag{12.39}$$

In Eq. (12.39), $\mathbf{x} = (x_1, x_2, x_3)$, $\mathbf{k} = (k_1, k_2, k_3)$, G is an arbitrary spectral amplitude, and $k_1 = k_1(k_2, k_3)$ is defined implicitly by the following *dispersion relation*:

$$\omega_n(\mathbf{k}_0 + \mathbf{k}) - \Omega - v_g k_1 = 0. \tag{12.40}$$

To avoid the occurrence of evanescent (exponentially growing) waves, the integral in Eq. (12.39) is extended over the values of (k_2, k_3) such that k_1, obtained after solving Eq. (12.40), turns out to be real valued. We note that for an *arbitrary* spectral amplitude G, Eq. (12.39) always represents an *exact* solution of the Wannier function envelope equation, which propagates *undistorted* with group velocity v_g once the proper band dispersion curve $\omega_n(\mathbf{k})$ of the PC and corresponding dispersion relation (12.40) are computed (e.g., by numerical methods). For *some* specific choices of the spectral amplitude G, in addition to undistorted wave propagation a certain degree of spatiotemporal wave localization can be obtained. It is worth getting some explicit examples, although approximate, of such 3D localized waves, admitting the integral representation given by Eq. (12.39), and relating them to already known localized solutions to canonical wave equations [14]. To this aim we develop an asymptotic analysis of Eq. (12.40) by assuming that the spectral amplitude G is nonvanishing in a narrow interval around $k_2 = k_3 = 0$, so that for Ω close to ω_0, the value of k_1 as obtained from Eq. (12.40) is also close to $k_1 = 0$. In this case, an approximate expression for the dispersion relation $k_1 = k_1(k_2, k_3)$ can be obtained by expanding in Eq. (12.40) the band dispersion curve $\omega_n(\mathbf{k}_0 + \mathbf{k})$ at around $\mathbf{k}_0$. Now, depending on the value of the group velocity v_g, which is basically a free parameter in our analysis, we should distinguish two cases:

1. The group velocity v_g is different from $\partial\omega_n/\partial k_x$ and far enough from this value.
2. The group velocity v_g is chosen to be equal to $\partial\omega_n/\partial k_x$.

In the *first case* (i.e., when $v_g \ll \partial\omega_n/\partial k_x$ or $v_g \gg \partial\omega_n/\partial k_x$), the leading-order terms entering in Eq. (12.40) after a power expansion of $\omega_n(\mathbf{k}_0 + \mathbf{k})$, are quadratic in k_2, k_3 and linear in k_1; precisely, one has

$$\left(\left.\frac{\partial\omega_n}{\partial k_x}\right|_{\mathbf{k}_0} - v_g \right) k_1 + \omega_0 - \Omega + \frac{1}{2}\sum_{i,j=2}^{3} \left.\frac{\partial^2\omega_n}{\partial k_i \partial k_j}\right|_{\mathbf{k}_0} k_i k_j = 0, \tag{12.41}$$

where $k_i = k_{x,y,z}$ for $i = 1, 2, 3$ and the derivatives of the band dispersion curve are calculated at $\mathbf{k}_0$. If, given Eq. (12.41), the approximate expression of k_1 is introduced into Eq. (12.39), it can easily be shown that the envelope $g(x_1, x_2, x_3)$ satisfies the differential equation

$$i\left(\left.\frac{\partial\omega_n}{\partial k_x}\right|_{\mathbf{k}_0} - v_g \right) \frac{\partial g}{\partial x_1} = (\omega_0 - \Omega) g - \frac{1}{2}\sum_{i,j=2}^{3} \left.\frac{\partial^2\omega_n}{\partial k_i\, \partial k_j}\right|_{\mathbf{k}_0} \frac{\partial^2 g}{\partial x_i\, \partial x_j}. \tag{12.42}$$

Since the matrix $\partial^2\omega_n/\partial k_i\,\partial k_j$ is symmetric, after suitable rotation of the (x_2, x_3) axes by the transformation $x'_j = \mathcal{R}_{ji}x_i$ $(i, j = 2, 3)$, where $\mathcal{R}_{ji}$ is the orthogonal matrix that diagonalizes $\partial^2\omega_n/\partial k_i\,\partial k_j$, assuming without loss of generality that $\Omega = \omega_0$, Eq. (12.42) can be written in the canonical Schrödinger-like form

$$i\left(\left.\frac{\partial\omega_n}{\partial k_x}\right|_{\mathbf{k}_0} - v_g\right)\frac{\partial g}{\partial x_1} = -\frac{1}{2}\alpha_2\frac{\partial^2 g}{\partial x_2'^2} - \frac{1}{2}\alpha_3\frac{\partial^2 g}{\partial x_3'^2}, \tag{12.43}$$

where α_2 and α_3 are the eigenvalues of the 2×2 matrix $\partial^2\omega_n/\,\partial k_i\,\partial k_j$ $(i, j = 2, 3)$. Localized 3D wave solutions to Eq. (12.43) are expressed in terms of well-known Gauss–Hermite functions, which are in general anisotropic for $\alpha_2 \neq \alpha_3$. These 3D localized waves, which exist regardless of the sign of α_2 and α_3, represent Gaussian-like beams, with exponential localization in the transverse y–z plane and algebraic localization, determined by the beam Rayleigh range, in the longitudinal x direction (and hence in time). These beams propagate undistorted along the x direction with an *arbitrary* group velocity v_g, either subluminal or superluminal, provided that $v_g \neq \partial\omega_n/\partial k_x$. Such pulsed propagating Gaussian beams represent an extension, in a PC structure, of similar solutions found in vacuum (see [39] and references therein). In particular, the special case $v_g = 0$ leads to stationary (monochromatic) Gaussian-like beams; note that the condition $v_g \neq \partial\omega_n/\partial k_x$ implies that such steady Gaussian beams do not exist in a PC close to a bandgap edge where $\partial\omega_n/\partial k_x$ vanishes. Other solutions to Eq. (12.43), leading to *spatial* 2D localized and monochromatic waves in the transverse (y, z) plane (but delocalized in the longitudinal x direction), can be sought in the form $g(x_1, x_2, x_3) = \varphi(x_2, x_3)\exp(i\lambda x_1)$, where λ is a propagation constant. If α_2 and α_3 have the same sign, the function $\varphi(x_2, x_3)$ satisfies a 2D Helmholtz equation, admitting well-known Bessel-beam solutions in cylindrical coordinates. For $\alpha_2 \neq \alpha_3$, such solutions are anisotropic, and again they represent a generalization to a PC of well-known spatial Bessel beams in vacuum. If α_2 and α_3 have opposite sign, one obtains a hyperbolic 2D equation (or equivalently, a 1D Klein–Gordon equation), which admits of 2D X-type localized solutions involving modified Bessel functions recently studied in [40] (see Eqs. (3a) and (4) of [40]; see also [41]).

In the *second case* (i.e., when $v_g = \partial\omega_n/\partial k_x$), the leading-order approximation to the dispersion relation [Eq. (12.40)] should also include second-order derivatives with respect to x_1 of the band dispersion curve $\omega_n(\mathbf{k}_0 + \mathbf{k})$, yielding

$$\omega_0 - \Omega + \frac{1}{2}\sum_{i,j=1}^{3}\left.\frac{\partial^2\omega_n}{\partial k_i\,\partial k_j}\right|_{\mathbf{k}_0} k_i k_j = 0, \tag{12.44}$$

where the derivatives of the band dispersion curve are calculated at $\mathbf{k}_0$. If the approximate expression of k_1, defined implicitly by the quadratic equation (12.44), is introduced into Eq. (12.39), one can easily show that in this case the envelope

$g(x_1, x_2, x_3)$ satisfies the differential equation

$$(\omega_0 - \Omega)g = \frac{1}{2} \sum_{i,j=1}^{3} \left. \frac{\partial^2 \omega_n}{\partial k_i \, \partial k_j} \right|_{\mathbf{k}_0} \frac{\partial^2 g}{\partial x_i \, \partial x_j}. \tag{12.45}$$

Since the matrix $\partial^2 \omega_n / \partial k_i \, \partial k_j$ is symmetric, after suitable rotation of the (x_1, x_2, x_3) axes by the transformation $x'_j = \mathcal{R}_{ji} x_i$ $(i, j = 1, 2, 3)$, where $\mathcal{R}_{ji}$ is the orthogonal matrix that diagonalizes $\partial^2 \omega_n / \partial k_i \, \partial k_j$, Eq. (12.45) takes the canonical form

$$(\omega_0 - \Omega)g = \frac{1}{2} \left(\alpha_1 \frac{\partial^2 g}{\partial x_1'^2} + \alpha_2 \frac{\partial^2 g}{\partial x_2'^2} + \alpha_3 \frac{\partial^2 g}{\partial x_3'^2} \right), \tag{12.46}$$

where α_i $(i = 1, 2, 3)$ are the eigenvalues of the 3×3 matrix $\partial^2 \omega_n / \partial k_i \, \partial k_j$ $(i, j = 1, 2, 3)$. The sign of the eigenvalues α_i basically determines the elliptic or hyperbolic character of Eq. (12.46), and hence the nature of their solutions (see, e.g., [14]). If the α_i have the same sign (e.g., they are positive), for $\Omega < \omega_0$ Eq. (12.46) reduces, after a scaling of axis length, to a 3D Helmholtz equation which in spherical coordinates admits of localized solutions in the form of sinc-shaped waves (see, e.g., [14,18]). If, conversely, there is sign discordance among the eigenvalues α_i, one obtains a 2D Klein–Gordon equation, which admits of 3D localized X-type waves which have been discussed at length in many works (see, e.g., [14,22,40] and references therein). In some special cases, one of the eigenvalues α_i may vanish, which may yield further nonspreading wave packet solutions. Notably, if $\alpha_1 = 0$, the solution to Eq. (12.46) is given by $g(x_1, x_2, x_3) = h(x_1)\varphi(x_2, x_3)$, where h is an arbitrary function of $x_1 = x - v_g t$ and φ satisfies a 2D Helmholtz equation for $\alpha_2\alpha_3 > 0$, admitting Bessel beam solutions, or a 1D Klein–Gordon equation for $\alpha_2\alpha_3 < 0$, admitting 2D X-type solutions. For these special solutions a cancellation of temporal dispersion is obtained. As the former case $(\alpha_2\alpha_3 > 0)$ extends to a PC structure the pulsed Bessel beams found in homogeneous dispersive media [42], the latter case $(\alpha_2\alpha_3 < 0)$ is rather peculiar for a PC structure, which realizes a bidiffractive propagation regime [40] (i.e., positive and negative diffraction along the two transverse directions y and z). Instead of pulses with a transverse Bessel beam profile, in this case one obtains a transverse X-shaped beam with an arbitrary longitudinal (temporal) profile that propagates without spreading.

We note that although our analysis has focused on a 3D PC, similar results can be obtained *mutatis mutandis* for the lower-dimensional case of a 2D PC, neglecting out-of-plane propagation. In this case the fields depend solely on the two spatial variables x and y defining the PC plane, and the most general solution to the Wannier function envelope equation (12.36) propagating undistorted with a group velocity v_g along the x axis reads

$$g(x_1, x_2) = \int dk_2 \, G(k_2) \exp(ik_1 x_1 + ik_2 x_2), \tag{12.47}$$

TABLE 12.1 Different Regimes of Spatiotemporal Localization in a 2D PC

Group Velocity	Hessian Values	Governing Equation Type in φ	Solution Type (Example)
$v_g \ll \nabla_{k_x}\omega_n$, $v_g \gg \nabla_{k_x}\omega_n$	$H \gtrless 0$	Schrödinger	2D Gaussian
$v_g = \partial\omega_n/\partial k_x$	$H > 0$	Helmholtz	2D Bessel
	$H < 0$	Klein–Gordon	2D X-wave

where $x_1 = x - v_g t$, $x_2 = y$, G is an arbitrary spectral amplitude, and $k_1 = k_1(k_2)$ is again defined implicitly by the dispersion relation given by Eq. (12.40). Depending on the value of the group velocity v_g compared to $\partial\omega_n/\partial k_x$, the scenario discussed for the 3D case still holds. In particular, for v_g different from (and enough far from) $\partial\omega_n/\partial k_x$, the lowest-order differential equation describing undistorted wave propagation is a 1D Schrödinger equation—given by Eq. (12.43), where the last term on the right-hand side is dropped—which admits of 1D Gauss–Hermite solutions (*parabolic* localization regime). Conversely, for $v_g = \partial\omega_n/\partial k_x$, the lowest-order differential equation describing undistorted spatiotemporal wave propagation is given by Eq. (12.46) provided that the term containing the derivative $\partial^2/\partial x_3'^2$ is dropped. In this case, Eq. (12.46) corresponds to either a 2D Helmholtz equation or to a 1D Klein–Gordon equation, depending on whether the Hessian $H = \det(\partial^2\omega_n/\partial k_i \partial k_j)$ is positive or negative.

In the former case (*elliptic* localization regime), typical localized solutions in cylindrical coordinates are expressed by Bessel functions of first kind, whereas in the latter case (*hyperbolic* localization regime) X-type wave localization is achieved (see, e.g., [40,41]). The different types of solutions obtainable in a 2D PC are reviewed in Table 12.1.

As an example of undistorted spatiotemporal wave propagation, we consider the 2D PC structure presented in Section 12.2.2: a square lattice of period a made of alumina rods ($\epsilon = 8.9$) with radius $r = 0.2a$ embedded in air (see Fig. 12.1a). A few low-order dispersion bands for TE and TM modes of the PC, as obtained by a standard plane-wave expansion technique, are shown in Fig. 12.1a. Since the computation and shape of Wannier functions for a 2D square lattice PC have been reported in [37], we focus here on the construction of undistorted spatiotemporal localized waves of the Wannier function envelope by direct numerical computation of the integral representation in Eq. (12.47), which requires only numerical computation of the band dispersion curve $\omega_n(k_x, k_y)$ and—by means of Eq. (12.40)—of the dispersion relation $k_1 = k_1(k_2)$. As an example, let us consider the second TM band in Fig. 12.1b, the TM air band; the entire 2D band surface $\omega_n(k_x, k_y)$ of this band is shown in Fig. 12.5a. Figure 12.5b shows the regions in the reciprocal **k**-space within the first Brillouin zone in which the Hessian $H = \det(\partial^2\omega_n/\partial k_i \partial k_j|_{\mathbf{k}_0})$ is positive (i.e., $H > 0$), indicating elliptic localization, or is negative (i.e., $H < 0$) indicating hyperbolic localization. For a fixed $\mathbf{k}_0$ point, parabolic localization, supporting stationary or propagating 1D Gaussian-like beams, is attained whenever $\mathbf{v}_g$ is far from $\nabla_{\mathbf{k}}\omega_n$, as discussed earlier: whereas for $\mathbf{v}_g = \nabla_{\mathbf{k}}\omega_n$, localization is of either hyperbolic or elliptic types, depending

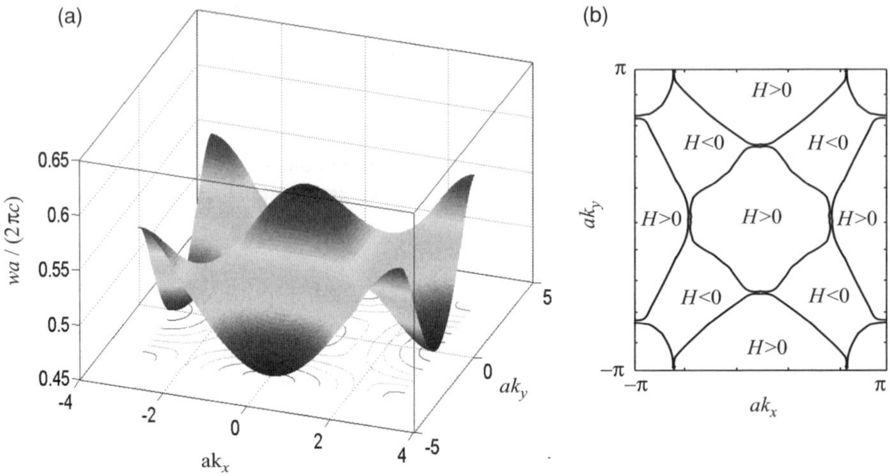

FIGURE 12.5 (a) Surface diagram of PC band $\omega_n(k_x, k_y)$ for the TM air band of Fig. 12.1a. (b) Sign of the Hessian $H = \det(\partial^2\omega_n/\partial k_i \partial k_j)$ for the TM air band (the continuous lines correspond to the boundary $H = 0$).

on whether $H < 0$ or $H > 0$. As an example, Fig. 12.6 shows typical undistorted localized waves as obtained from Eq. (12.47) at point $\mathbf{k}_0 = (1.6/a)(\mathbf{u}_{\mathbf{k}x} + \mathbf{u}_{\mathbf{k}y})$—on the ΓM diagonal, where the group velocity is directed along the bisection line of the x–y plane—for two different values of the group velocity and assuming a Gaussian spectral amplitude profile $G(k_2) = \exp[-(\lambda k_2)^2]$. In the same figure, the dispersion relations $k_1 = k_1(k_2)$, as obtained by numerical solution of Eq. (12.40), are also depicted. Note that at $\mathbf{k} = \mathbf{k}_0$, one has $\nabla_{\mathbf{k}}\omega_n \simeq -0.0532c\mathbf{n}$ and $H < 0$, where $\mathbf{n} = (1/\sqrt{2})(\mathbf{u}_x + \mathbf{u}_y)$ is the unit vector of the bisection line of the x–y plane. Figure 12.6a corresponds to $\mathbf{v}_g = \nabla_{\mathbf{k}}\omega_n$, leading—as expected for a hyperbolic localization regime—to an X-shaped wave. Conversely, in Fig. 12.6b we have chosen the group velocity $\mathbf{v}_g = -c\mathbf{n}$, leading to a Gaussian-like beam propagating at a luminal velocity. To summarize, our asymptotic analysis of the Wannier envelope equation shows that a wide class of localized (either spatial or spatiotemporal) waves exist, including propagating Gaussian beams, 2D and 3D X-waves, sinc-shaped waves, pulsed Bessel beams, and pulsed 2D X-waves.

12.4 CONCLUSIONS

We have reviewed spatial and spatiotemporal wave localization in PCs without defects and demonstrated the existence of spatiotemporal Wannier wave packets that propagate nearly undistorted in such periodic media. In the monochromatic case, we focused our attention on the hyperbolic localization regime in 2D PCs, demonstrating the possibility of obtaining 3D X-wave solutions at rest for particular points in the

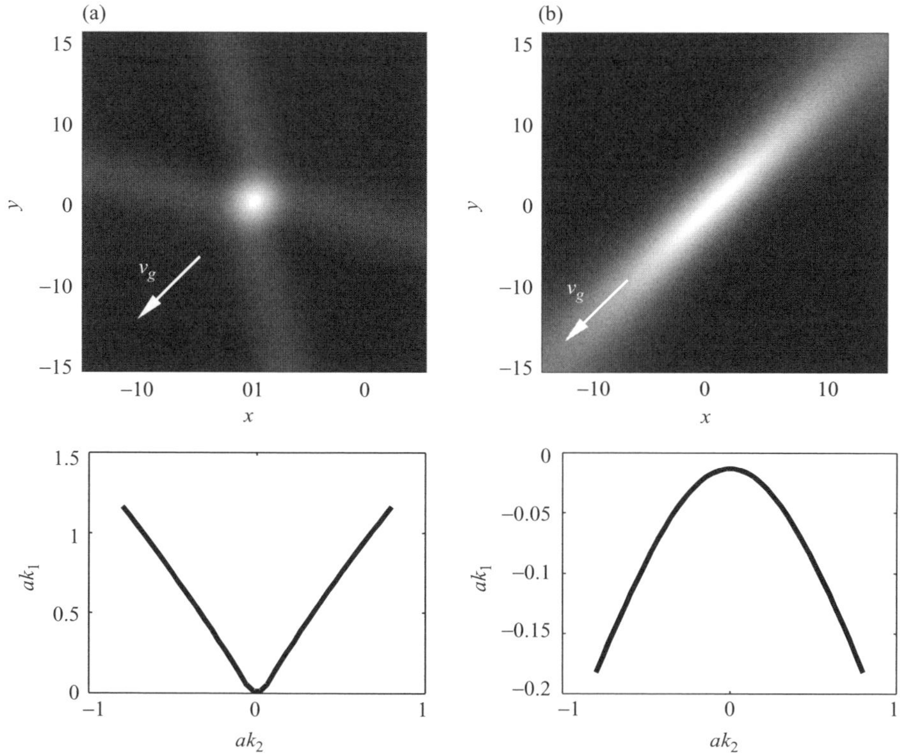

FIGURE 12.6 Gray-scale plots of localized propagating waves as obtained from Eq. (12.47) for a Gaussian spectral amplitude with $\lambda = 2a$ and for the following choices: $\mathbf{k}_0 = (1.6/a)(\mathbf{u}_{\mathbf{k}x} + \mathbf{u}_{\mathbf{k}y})$, $\Omega = \omega_n(\mathbf{k}_0) \simeq 0.5338(2\pi c/a)$ (see the text). In (a), $v_g = |\nabla_{\mathbf{k}}\omega_n| \simeq 0.0532c$ (hyperbolic localization), whereas in (b), $v_g = c$ (parabolic localization). The bottom figures show, for the two cases, the corresponding dispersion relation $k_1 = k_1(k_2)$ as computed numerically by solving Eq. (12.40).

band diagram. The polychromatic case analysis has been carried out using a Wannier function technique which allowed us to demonstrate rigorously the existence of a wide variety of spatiotemporal localized waves.

REFERENCES

1. J. D. Joannopoulos, R. D. Meade, and J. N. Winn, *Photonic Crystals: Molding the Flow of Light*, Princeton University Press, Princeton, NJ, 1995.
2. K. Sakoda, *Optical Properties of Photonic Crystals*, Springer-Verlag, New York, 2002.
3. J. M. Ziman, *Principles of the Theory of Solids*, 2nd ed., Cambridge University Press, New York, 1972.

4. H. Kosaka, T. Kawashima, A. Tomita, M. Notomi, T. Tamamura, T. Sato, and S. Kawakami, Superprism phenomena in photonic prystals, *Phys. Rev. B* **58**, R10096 (1998).
5. H. S. Eisenberg, Y. Silberberg, R. Morandotti, and J. S. Aitchison, Diffraction management, *Phys. Rev. Lett.* **85**, 1863 (2000).
6. T. Pertsch, T. Zentgraf, U. Peschel, A. Bruer, and F. Lederer, Anomalous refraction and diffraction in discrete optical systems, *Phys. Rev. Lett.* **88** 093901 (2002).
7. M. Notomi, Negative refraction in photonic crystals, *Opt. Quantum Electron.* **34**, 133 (2002).
8. E. Cubukcu, K. Aydin, E. Ozbaym, S. Foteinopolou, and C. M. Soukoulis, Subwavelength resolution in a two-dimensional photonic-crystal-based superlens, *Phys. Rev. Lett.* **91**, 207401 (2003).
9. C. Conti and S. Trillo, Nonspreading wave packets in three dimensions formed by an ultracold Bose gas in an optical lattice, *Phys. Rev. Lett.* **92**, 120404 (2004).
10. S. Longhi and D. Janner, Diffraction and localization in low-dimensional photonic bandgaps, *Opt. Lett.* **29**, 2653 (2004).
11. S. Longhi and D. Janner, X-shaped waves in photonic crystals, *Phys. Rev. B* **70**, 235123 (2004).
12. S. Longhi, Localized and nonspreading spatiotemporal wannier wave packets in photonic crystals, *Phys. Rev. E* **71**, 016603 (2005).
13. I. M. Besieris, M. Abdel-Rahman, A. Shaarawi, and A. Chatzipetros, Two fundamental representations of localized pulsed solutions to the scalar wave equation, *Prog. Electromagn. Res.* **19**, 1 (1998).
14. R. Donnelly and R. W. Ziolkowski, Designing localized waves, *Proc. R. Soc. London A* **440**, 541 (1993).
15. H. Sonajalg and P. Saari, Suppression of temporal spread of ultrashort pulses in dispersive media by Bessel beam generators, *Opt. Lett.* **21**, 1162 (1996).
16. M. A. Porras, , S. Trillo, C. Conti, and P. Di Trapani, Paraxial envelope X waves, *Opt. Lett.* **28**, 1090 (2003).
17. M. A. Porras, G. Valiulis, and P. Di Trapani, Unified description of Bessel X waves with cone dispersion and tilted pulses, *Phys. Rev. E* **68**, 016613 (2003).
18. S. Longhi, Spatial-temporal Gauss–Laguerre waves in dispersive media, *Phys. Rev. E* **68**, 066612 (2003).
19. M. Zamboni-Rached, K. Z. Nobrega, H. E. Hernández-Figueroa, and E. Recami, Localized superluminal solutions to the wave equation in (vacuum or) dispersive media, for arbitrary frequencies and with adjustable bandwidth, *Opt. Commun.* **226**, 15–23 (2003).
20. J. Salo, J. Fagerholm, A. T. Friberg, and M. Salomaa, Unified description of nondiffracting X and Y waves, *Phys. Rev. E* **62**, 4261 (2000).
21. E. Recami, M. Zamboni-Rached, K. Z. Nóbrega, C. A. Dartora, and H. E. Hernandez, On the localized superluminal solutions to the maxwell equations, *IEEE J. Sel. Top. Quantum Electron.* **9**, 59 (2003).
22. C. Conti, S. Trillo, P. Di Trapani, G. Valiulis, A. Piskarskas, O. Jedrkiewicz, and J. Trull, Nonlinear electromagnetic X waves, *Phys. Rev. Lett.* **90**, 170406 (2003).
23. P. Di Trapani, G. Valiulis, A. Piskarskas, O. Jedrkiewicz, J. Trull, C. Conti, and S. Trillo, Spontaneously generated X-shaped light bullets, *Phys. Rev. Lett.* **91**, 093904 (2003).

24. J.-Y. Lu and J. F. Greenleaf, Experimental verification of nondiffracting X waves, *IEEE Trans. Ultrason. Ferrelectr. Freq. Control* **39**, 441 (1992).
25. R. W. Ziolkowski, D. K. Lewis, and B. D. Cook, Evidence of localized wave transmission, *Phys. Rev. Lett.* **62**, 147 (1989).
26. P. Saari and K. Reivelt, Evidence of X-shaped propagation-invariant localized light waves, *Phys. Rev. Lett.* **79**, 4135 (1997).
27. H. Sonajalg, M. Ratsep, and P. Saari, Demonstration of the Bessel-X pulse propagating with strong lateral and longitudinal localization in a dispersive medium, *Opt. Lett.* **22**, 310 (1997).
28. O. Jedrkiewicz, J. Trull, G. Valiulis, A. Piskarskas, C. Conti, S. Trillo, and P. Di Trapani, Nonlinear X waves in second-harmonic generation: experimental results, *Phys. Rev. E* **68**, 026610 (2003).
29. P. Saari and K. Reivelt, Generation and classification of localized waves by Lorentz transformations in Fourier space, *Phys. Rev. E* **69**, 036612 (2004).
30. C. Conti, X-wave-mediated instability of plane waves in Kerr media, *Phys. Rev. E* **68**, 016606 (2003).
31. A. A. Maradudin and A. R. McGurn, Out of plane propagation of electromagnetic waves in a two dimensional periodic dielectric medium, *J. Mod. Opt.* **41**, 275 (1994).
32. I. Gradshteyn and I. Ryzhik, *Table of Integrals, Series and Products*, enlarged ed., Academic Press, New York (1980).
33. P. Feuer, Electronic states in crystals under large over-all perturbations, *Phys. Rev.* **88**, 92 (1952).
34. W. Kohn, Analytic properties of Bloch waves and wannier functions, *Phys. Rev.* **115**, **809** (1959).
35. I. Talanina and C. M. de Sterke, Bloch waves and Wannier functions in periodic superstructure Bragg gratings, *Phys. Rev. A* **62**, 043802 (2000).
36. J. P. Albert, C. Jouanin, D. Cassagne, and D. Bertho, Generalized Wannier function method for photonic crystals, *Phys. Rev. B* **61**, 4381 (2000).
37. A. Garcia-Martin, D. Hermann, F. Hagmann, K. Busch, and P. Woelfle, Defect computation in photonic crystals: a solid state theoretical approach, *Nanotechnology* **14**, 177 (2003).
38. D. M. Whittaker and M. P. Croucher, Maximally localized wannier functions for photonic lattices, *Phys. Rev. B* **67**, 085204 (2003).
39. S. Longhi, Gaussian pulsed beams with arbitrary speed, *Opt. Express* **12**, 935 (2004).
40. D. N. Christodoulides, N. K. Efremedis, P. Di Trapani, and B. A. Malomed, Bessel X waves in two and three dimensional bidispersive optical systems, *Opt. Lett.* **29**, 1446 (2004).
41. A. Ciattoni and P. Di Porto, One-dimensional nondiffracting pulses, *Phys. Rev. E* **69**, 056611 (2004).
42. M. A. Porras, Diffraction-free and dispersion-free pulsed beam propagation in dispersive media, *Opt. Lett.* **26**, 1364 (2001).

CHAPTER THIRTEEN

Spatially Localized Vortex Structures

ZDENĚK BOUCHAL AND RADEK ČELECHOVSKÝ
Palacký University, Olomouc, Czech Republic

GROVER A. SWARTZLANDER, JR.
University of Arizona, Tucson, Arizona

13.1 INTRODUCTION

The spatial localization of light is restricted by the fundamental property of optics known as *diffraction*. Owing to dephasing between spatial frequency components as a beam propagates, diffraction prohibits the propagation of a narrow light tube. In the paraxial approximation, the amount of dephasing may be written $\Delta\Phi = zk_t^2/2k$, where z is the propagation distance, k_t the magnitude of the transverse wave vector, and $k = 2\pi/\lambda$, where λ is the wavelength of a monochromatic beam. Spatial frequency components that have greater values of k_t therefore dephase over shorter propagation distances than do those having smaller values of k_t. This may readily be observed in the region behind an aperture. There the intensity profile exhibits short oscillations (corresponding to large values of k_t) before long oscillations appear.

A monochromatic Gaussian beam is another well-known example. The radius of the beam waist w_0 determines the characteristic angular beam divergence $\theta_d = \lambda/\pi w_0$. The value of k_t associated with this angle is equal to $k\theta_d$. The characteristic dephasing may thus be expressed as $\Delta\Phi = 2z/q_0$, where $q_0 = \pi w_0^2/\lambda$ is the Rayleigh range. The area of the beam doubles after propagating a distance q_0, owing

Localized Waves, Edited by Hugo E. Hernández-Figueroa, Michel Zamboni-Rached, and Erasmo Recami

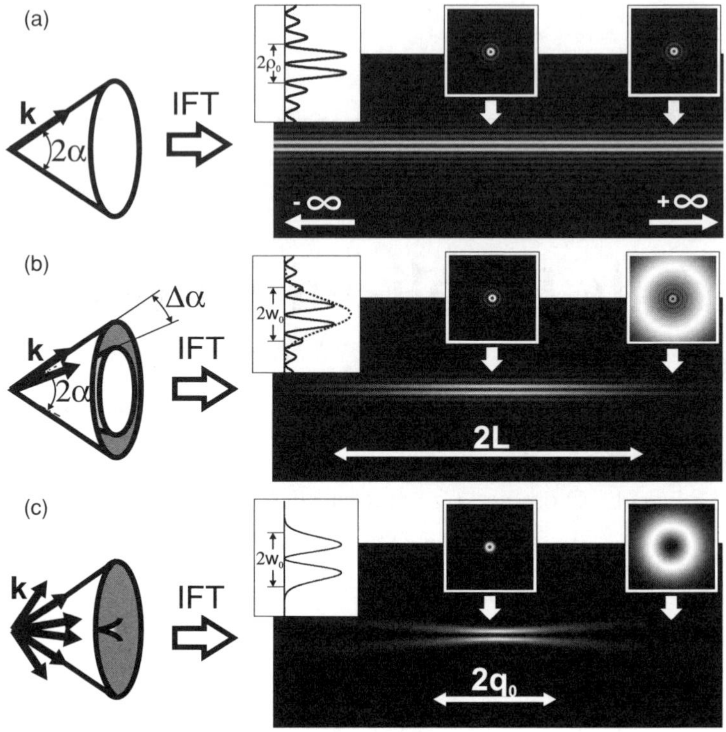

FIGURE 13.1 Free-space evolution of a hollow nondiffracting beam (a), a pseudo-nondiffracting beam (b), and a Laguerre–Gaussian beam (c).

to diffraction. Thus, the beam may be said to be transversely localized over the range $|z| \leq q_0$.

In general, all bounded inhomogeneous waves exhibit diffraction. (Here the term *bounded* implies that the intensity asymptotically vanishes faster than $1/\rho^2$, where ρ is the distance from the center of the beam. The energy of bounded waves is therefore finite.) However, a class of fields called *nondiffracting beams*, which exhibit little or no diffraction, were described by Durnin in 1987 [1] (this type of beam was noted earlier by Sheppard and Wilson [2]). For the ideal unbounded case, a scalar monochromatic nondiffracting beam, $E(\mathbf{r}, t) = U(\mathbf{r}) \exp(i\omega t)$, satisfies the Helmholtz equation $(\nabla^2 + k^2)U(\mathbf{r}) = 0$, and its complex amplitude has a modelike form, $U(\mathbf{r}) = u(x, y, k_t) \exp(-ik_z z)$, where u is a complex amplitude, k_z is a propagation constant, $k_t = (k^2 - k_z^2)^{1/2}$ is the magnitude of a transverse wave vector, and $\mathbf{r} \equiv (x, y, z)$. According to this ansatz, the beam intensity $I = UU^*$ remains invariant under propagation along the z-axis. The invariant propagation range extends from $-\infty$ to $+\infty$, assuming that the initial beam is unbounded.

An example of a nondiffracting beam is shown in Fig. 13.1a). The intensity profile exhibits concentric annuli having peak intensities that diminish with distance from the center. In the ideal unbounded case the peaks do not decrease faster than $1/\rho^2$, and thus the beam has infinite energy. In this example, the beam is an optical vortex, and thus the intensity at the center of the beam ($\rho = 0$) is equal to zero. An optical vortex is characterized by a phase factor, $\exp(im\theta)$, where m is a nonzero integer called the topological charge [3–5], and $\theta = \arctan(y/x)$. Vortex beams belong to the class of fields possessing circularly harmonic wavefronts.

Unbounded nondiffracting beams are not physical because they occupy infinite space and require infinite energy. In practice, pseudo-nondiffracting (P-N) beams may be achieved using bounded beams that have finite energy. For example, a P-N beam in the plane $z = 0$ may be described as the product of an ideal nondiffracting beam and an envelope function such as a Gaussian profile, $G = \exp(-\rho^2/w_0^2)$, where $\rho = (x^2 + y^2)^{1/2}$. Although such beams do not remain localized as they propagate, they may be expected to maintain some nondiffracting attributes over a significant propagation distance. An ansatz for a P-N beam may be written

$$U_{PN}(\mathbf{r}) = u(\mathbf{r}, k_t, w_0)\exp(-ik_z z). \tag{13.1}$$

In this case, the slowly varying amplitude u depends on the propagation coordinate z. The nature of the propagating beam depends on whether the product $w_0 k_t$ is greater or less than unity. If $w_0 k_t >> 1$, the beam possesses P-N propagation properties, whereas for $w_0 k_t \leq 1$ the common diffraction can be observed. For example, Fig. 13.1a may be viewed as an unbounded nondiffracting beam obtained with $w_0 k_t \to \infty$, whereas Fig. 13.1b depicts a P-N beam for the case $w_0 k_t >> 1$. The limited localization range is clearly evident in Fig. 13.1b. Extended localization effects disappear in Fig. 13.1c for the case when $w_0 k_t < 1$.

The full range over which the beam exhibits nondiffracting propagation may be expressed as $2L$, where

$$L \approx \frac{k w_0}{k_t} \approx \frac{2\pi w_0 \rho_0}{\lambda}, \tag{13.2}$$

and $\rho_0 \approx 1/k_t$ characterizes the size of the central intensity lobes in the plane $z = 0$. Thus, for a given size of the beam spot ρ_0, the value of L may be increased by increasing the bounding size, w_0. The latter requires larger optics. Further, more energy is required if constant intensity values are to be maintained in the lobes.

From an experimental point of view, P-N beams may be easily generated for both nonvortex [6] and vortex [7] phase profiles. Demonstrations of the former case have been achieved by use of an annular spatial filter, axicon, hologram, or a Fabry–Perot resonator and abberrating lens [6]. Observations of the latter case require an additional spiral phase mask or a holographic element such as a spatial light modulator (SLM). The resistance of P-N beams against amplitude and phase perturbations has been explored [8,9]. This regenerating property has been used successfully to trap multiple particles along the beam axis [10]. Another attractive attribute of P-N beams is that many intensity distributions are allowed for a given value of k_z. Beam shaping can be achieved by modifying the spatial spectrum [11]. What is more, arrays of nondiffracting beams may be generated [12].

The superposition of nondiffracting vortex beams provides other application opportunities, such as the encoding, transfer, and decoding of information [13,14]. The description of composite vortex fields created by the interference of two or more noncollinear vortices [15] may also be extended to nondiffracting beams. For example, the superposition of nondiffracting beams with different propagation constants k_z may result in the *self-imaging* or *Talbot effect* [16]. This effect manifests itself as a longitudinal periodicity of the field and represents a spatial analog of the temporal effect known as *mode locking*. By means of self-imaging, a controllable 3D spatial localization of light enabling trapping of particles can be achieved [17,18].

13.2 SINGLE AND COMPOSITE OPTICAL VORTICES

Optical vortices are ubiquitous in optical systems. They are most easily recognized in spatially coherent quasi-monochromatic radiation. Under this condition they appear as dark spots in the intensity profile. The intensity vanishes at one or more points in these spots, owing to phase defects. A *phase defect* is a point where the phase of the electromagnetic wave is undefined. For partially coherent light, the dark core appears diffuse and a zero-intensity point does not occur [19,20].

Optical vortices are a class of waves having circularly harmonic phase profiles. In the scalar approximation, a single optical vortex in a coherent quasimonochromatic field may be represented by the complex amplitude in the transverse $x - y$ plane:

$$U(\rho, \theta) = U_0 u(\rho, \theta) \exp(im\theta + i\beta), \tag{13.3}$$

where the amplitude U_0 is a real constant, m is a nonzero integer called the *topological charge*, and β is an arbitrary constant phase. Here the vortex center coincides with the origin, and thus the envelope function $u(\rho, \theta)$ must vanish at $\rho = 0$ (i.e., $\Re\{U\} = \Im\{U\} = 0$).

In this chapter the temporal phase $\exp(i\omega t)$ may be neglected. The fundamental values of topological charge are $m = \pm 1$. All other charges may be constructed by combining the fundamental charges. In the special nonvortex case, $m = 0$, the wavefront is planar and the intensity is not required to vanish, since there is no phase defect. In principle, any field may be decomposed into a series of vortex modes:

$$U(r, \theta) = \sum_{m=-\infty}^{+\infty} g_m(\rho) \exp(im\theta) \tag{13.4}$$

$$g_m(\rho) = \frac{1}{2\pi} \int_{-\pi}^{+\pi} U(r, \theta) \exp(-im\theta)\, d\theta, \tag{13.5}$$

where the g_m are radially dependent Fourier series coefficients corresponding to the vortex modes.

For the ideal case where a single vortex mode exists and the envelope function is radially symmetric, the propagating field may be written

$$U(\rho, \theta, z) = U_0 u(\rho, z) \exp[im\theta + i\beta - ikz + i\Phi(\rho, z)], \tag{13.6}$$

where $\Phi(\rho, z)$ describes the wavefront curvature and $u(\rho, z)$ represents the evolving beam amplitude. The surface of constant phase, $m\theta + \beta - kz + \Phi(\rho, z) = \text{const}$, is helicoidal.

From a ray optics point of view, the wave vector at each point in the beam may be described by the gradient of the phase:

$$\begin{aligned}\mathbf{k}(\mathbf{r}) &= -\nabla[m\theta + \beta - kz + \Phi(\rho, z)] \\ &= -\mathbf{e}_\rho \partial\Phi/\partial\rho - \mathbf{e}_\theta(m/\rho) + \mathbf{e}_z(k - \partial\Phi/\partial z),\end{aligned} \tag{13.7}$$

where $(\mathbf{e}_\rho, \mathbf{e}_\theta, \mathbf{e}_z)$ are unit vectors in cylindrical coordinates. Although this description of the local wave vectors fails near the origin, where the azimuthal component diverges, it highlights the vortex nature of the field: The azimuthal component circulates. Formally, the circulation may be described by a closed line integral,

$$\oint_s \mathbf{k} \cdot d\mathbf{s} = -2\pi m. \tag{13.8}$$

The line integral is equal to zero if the phase defect lies outside the area enclosed by the closed line, **s**.

From an experimental point of view, the detection of optical vortices requires a phase measurement (i.e., interferometry). The detection of zero-intensity points is insufficient. The interference of the vortex field (13.3) with a tilted plane wave having a transverse wave number k_x and field $U' = U'_0 \exp(-ik_x x)$ produces an intensity distribution

$$|U + U'|^2 = {U_0}^2 u^2 + {U'_0}^2 + 2U_0 U'_0 \cos(m\theta + \beta - k_x x). \tag{13.9}$$

Far from the vortex center, the interferogram displays lines of constant phase. At the center of the vortex, the intensity has a value of $U_0'^2$. Above $(x \approx 0,\ y > 0)$ and below $(x \approx 0,\ y < 0)$, the phase singularity, the value of the factor $\cos(m\theta + \beta - k_x x)$, changes sign, creating a forklike pattern. In addition to providing the location of the vortex center, the interferogram also reveals the value of the topological charge [21]. The number of additional fringes corresponds to the magnitude $|m|$, whereas the relative orientation of the fringes with respect to k_x distinguishes the sign, $m/|m|$. If the vortex beam is superimposed with a spherical reference wave, a spiral interference pattern is obtained. In both cases the interferograms may be recorded on film and reexposed to create a holographic image of the vortex [22,23]. In Fig.13.2, a structure composed of three vortices with the topological charges $m = 1,\ -1$, and 2 is visualized by interference with planar and spherical reference waves.

The net field created by the interference of multiple beams may produce *composite vortices*. Applications that utilize the properties of optical vortices require an understanding of composite vortices. Superimposed fields may be the basis for vortex control. On the other hand, undesirable scattered light may perturb the vortex and spoil system performance. In general, a composite vortex represents a complex structure whose resultant zero-intensity point does not coincide with the zeros of the

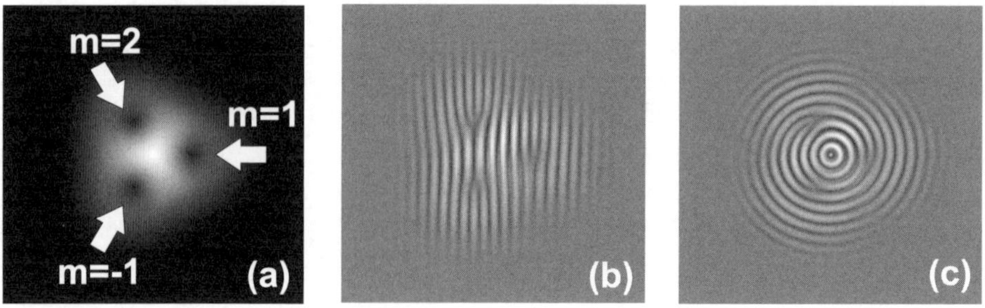

FIGURE 13.2 Intensity spot of a vortex field (a), forklike pattern (b), and spiral pattern (c) obtained by interference of a vortex field by plane and spherical reference waves. The vortex field is composed of three spatially separated vortices with the topological charges $m = 1, -1$, and 2.

composing beams. For example, composing fields that have no vortices may produce composite vortices. Other examples may include cases where the composite field contains no vortices even though one or more of the composing fields contains vortices. In general, the location, number, and charge of composite vortices depend on the relative phase and amplitude of the composing beams [3,15]. As an example, consider the interference of two noncollinear mutually coherent vortex beams. The composite field is then given by

$$U(\rho, \theta) = \sum_{j=1}^{2} U_{0,j} u(\rho_j) \exp(i m_j \theta_j + i\beta_j), \tag{13.10}$$

where (ρ_j, θ_j) are the transverse coordinates measured with respect to the center of the jth beam, and β_j is the phase of the jth beam. For this example we assume that the composing beams have equal charges ($m_1 = m_2 = 1$) that are separated along the x-axis by a distance $\pm\Delta x$. We also assume that each composing beam has amplitude $u(\rho_j)$ with a Laguerre–Gaussian profile. Numerically generated intensity profiles for this case are shown in Fig.13.3. An analytical solution for the location of composite vortices was found by setting the real and imaginary parts of Eq. (13.10) to zero. The coordinates of the vortex centers, (x, y), must satisfy the transcendental equation

$$x + iy = \Delta x \tanh\left(\frac{2\sigma^2 x}{\Delta x} + \frac{1}{2} \ln \frac{U_{0,1}}{U_{0,2}} + \frac{i\Delta\beta}{2}\right), \tag{13.11}$$

where $\sigma = \Delta x / w_0$ and $\Delta\beta = \beta_1 - \beta_2$.

Some properties of composite vortices can be demonstrated for the case where the composing beams have equal amplitudes ($U_{0,1} = U_{0,2}$) and are either in phase ($\Delta\beta = 0$) or out of phase ($\Delta\beta = \pi$). In the in-phase case, the composite vortex structure depends strongly on the relative displacement of the original vortices σ. When it has a certain subcritical value $\sigma < \sigma_{cr} = 2^{-1/2}$, an elongated composite beam having

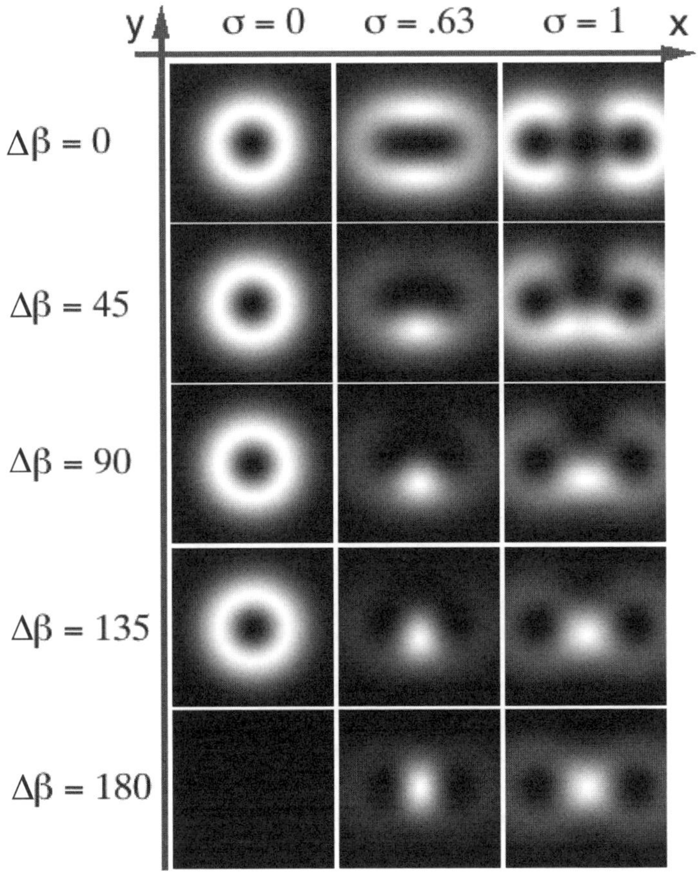

FIGURE 13.3 Intensity patterns of a composite beam created by superimposing two equal-amplitude $m = 1$ vortex beams with different dephasing $\Delta\beta$ and relative spatial separation of vortex centers σ.

a single vortex $m = 1$ appears. For $\sigma > \sigma_{\mathrm{cr}}$, two additional $m = 1$ vortices located on the x-axis emerge. For well-separated original beams $\sigma \gg \sigma_{\mathrm{cr}}$, their position can be determined by

$$\left|\frac{x}{\Delta x}\right| = 1 - 2\exp\left(\frac{-4\sigma^2 x}{\Delta x}\right). \tag{13.12}$$

In the out-of-phase case, two vortices are observable on the x-axis for all values of σ. For $\sigma \gg \sigma_{\mathrm{cr}}$, their position can be estimated by

$$\left|\frac{x}{\Delta x}\right| = 1 + 2\exp\left(\frac{-4\sigma^2 x}{\Delta x}\right). \tag{13.13}$$

These examples demonstrate that composite vortices may be repositioned by changing the relative position, phase, or intensity of the composing beams. From an experimental and application point of view, the variation of the relative phase $\Delta\beta$ is perhaps easiest to control in a linear fashion. For example, the composite field may be constructed from two collinear beams from a Mach–Zehnder interferometer, whereby the phase is varied by introducing an optical delay in one arm of the interferometer.

The propagation of composite vortices may be described in terms of the propagating composing beams. Significant simplification of the composite field occurs when the composing beams exhibit self-similar propagation (e.g., Laguerre–Gaussian beams).

Most experiments involving vortex beams have been conducted with *filtered* rather than *composite* beams. A filtered vortex may be created by transmitting an arbitrary beam through a vortex phase mask. This class of beams includes "doughnut" modes of cavities and waveguides. The field in the plane $z = 0$ for such beams may thus be expressed

$$U(\rho, \theta) = U_0\, u(\rho, \theta) \prod_{j=1}^{N} g_j(\rho_j) \exp(i m_j \theta_j), \tag{13.14}$$

where $g_j(\rho_j)$ is the vortex core function of the jth vortex. The envelope function, $u(\rho, \theta)$, typically has a Gaussian or Bessel profile. The propagation dynamics of the vortices may be described using fluid dynamics analogies. This is possible because the real and imaginary parts of the Helmholtz equation may be decomposed into a Bernoulli and continuity equation. In general, the vortices move according to intensity and phase gradients at the vortex core [24,25].

Increasing attention has been devoted to nondiffracting beams containing vortices. They possess some peculiar properties important for applications. Intriguing opportunities arise when nondiffracting beams are combined to create composite vortices. Indeed, the connection between nondiffracting and singular optics [7] opens exciting new application opportunities.

13.3 BASIC CONCEPT OF NONDIFFRACTING BEAMS

An ideal monochromatic nondiffracting beam represents a modelike field with well-defined propagation direction (z-axis), and the complex amplitude U, is given as

$$U(\mathbf{r}) = u(x, y, k_t) \exp(-ik_z z). \tag{13.15}$$

The transverse amplitude profile u remains unchanged during propagation, and the propagation constant k_z appoints spatial phase oscillations. If Eq. (13.15) is substituted into the Helmholtz equation, it can be rewritten as

$$\nabla_\perp^2 u_m(x, y, k_t) = -k_t^2 u_m(x, y, k_t), \tag{13.16}$$

where $\nabla_\perp^2 \equiv \partial^2/\partial x^2 + \partial^2/\partial y^2$, $k_t^2 = k^2 - k_z^2$, and k denotes the wave number. Admissible profiles of the nondiffracting beam u_m are found as eigenfunctions of the transverse Laplace operator $\nabla_\perp^2$. Its spectrum is degenerated so that a single eigenvalue k_t is related to the spectrum of the eigenfunctions u_m. In this case, k_t serves as a parameter appearing in the argument of the eigenfunctions u_m. It causes a shape-invariant rescaling of the transverse amplitude profile. The transverse amplitude profile of the nondiffracting beam, Eq. (13.15), is not specified unambiguously by the propagation constant k_z but it can be chosen from a spectrum of the eigenfunctions admitted by the Helmholtz equation, $u = u_m$. The homogeneous (source-free) Helmholtz equation (13.16) can be solved by a separation of variables realized in convenient coordinate systems [26]. Since the nondiffracting propagation requires a separation of the complex amplitude U into transverse and longitudinal parts, only the Cartesian, circular cylindrical, elliptical cylindrical, and parabolic cylindrical coordinates are applicable [27]. In these coordinate systems, the transverse amplitude profiles u_m can be expressed by the known functions. In Cartesian coordinates, a homogeneous plane wave can be obtained for $k_t = 0$ and $k_z = k$ as a trivial nondiffracting solution of (13.16). Interference fields created by two or four plane waves with amplitudes given by $u = \cos k_t x$ and $u = \cos(k_t x/\sqrt{2}) \cos(k_t y/\sqrt{2})$, respectively, represent simple examples fulfilling (13.16). Best-known nondiffracting beams can be obtained by using circular cylindrical coordinates (ρ, θ, z). Solutions of (13.16) then arise that adopt the restricting assumption that u_m can be expressed as a product of functions R_m and Φ_m, depending on the coordinates $\rho \in (0, \infty)$ and $\theta \in \langle 0, 2\pi\rangle$, respectively. The function Φ_m must be periodic, and usually it is assumed to be of the form $\Phi_m = \exp(im\theta)$, $m = 0, 1, 2, \ldots$. The mode numer m therefore represents the topological charge of the vortex. The Helmholtz equation (13.16) then reduces to the Bessel equation whose solutions are the mth-order Bessel functions of the first kind, $R_m = J_m$ (the mth-order Neumann functions are usually disregarded). The nondiffracting beams are called *Bessel beams* and their complex amplitude can be written as

$$U_m(\mathbf{r}) = J_m(k_t\rho) \exp(im\theta - ik_z z), \qquad m = 0, 1, 2, \ldots. \tag{13.17}$$

Although these solutions satisfy the Helmholtz equation, they do not necessary satisfy Maxwell's equations. An exact solution requires an examination of the vector fields. For paraxial beams, however, Eq. (13.17) is a valid approximation.

For $m = 0$, a beam with a bright central spot and a planar wavefront is obtained. The nondiffracting vortex beams related to $m \neq 0$ are dark at the axis and their wavefront is helical. The indices m represent the topological charges of the vortex beams. A free-space propagation of the first-order Bessel beam J_1 possessing the topological charge $m = 1$ is illustrated in Fig. 13.1a). In elliptic cylindrical coordinates, Mathieu nondiffracting beams can be derived. Their transverse amplitude profiles u_m are described by even and odd Mathieu functions of the mth order. Applying parabolic cylindrical coordinates, parabolic nondiffracting beams can be earned. Their transverse amplitude profile is given by even and odd solutions to the parabolic cylindrical differential equation [27].

The nondiffracting propagation of light beams can be explained clearly by means of their Fourier representation. The Fourier spectrum of a beam U can be written as

$$\overline{U}(\mathbf{f}) = \iiint_{-\infty}^{+\infty} U(\mathbf{r}) \exp(i2\pi\mathbf{f}\cdot\mathbf{r})\, d\mathbf{r}, \tag{13.18}$$

where $\mathbf{f} \equiv (f_x, f_y, f_z)$ denotes the vector of spatial frequencies. For the complex amplitude of the nondiffracting beam (13.15), the spatial spectrum can be expressed by the Dirac delta function as

$$\overline{U}(\mathbf{f}) = \overline{u}(\mathbf{f}_\perp)\delta\left(f_z - \frac{k_z}{2\pi}\right), \tag{13.19}$$

where $\mathbf{r}_\perp \equiv (x, y)$, $\mathbf{f}_\perp \equiv (f_x, f_y)$, and

$$\overline{u}(\mathbf{f}_\perp) = \iint_{-\infty}^{+\infty} u(\mathbf{r}_\perp) \exp(i2\pi\mathbf{f}_\perp\cdot\mathbf{r}_\perp)\, d\mathbf{r}_\perp. \tag{13.20}$$

Since the wave number and the spatial frequencies are related as $\mathbf{f}\cdot\mathbf{f} = k^2/(2\pi)^2$, the spatial frequency f_z can be excluded and the spatial spectrum of the nondiffracting beam (13.19) becomes

$$\overline{U}(\mathbf{f}_\perp) = \overline{u}(\mathbf{f}_\perp)\delta\left(|\mathbf{f}_\perp| - \frac{k_t}{2\pi}\right). \tag{13.21}$$

At the plane of spatial frequencies (f_x, f_y), the spectrum of the nondiffracting beam is represented by a circle with radius $k_t/2\pi$. This means that the Fourier spectrum of the transverse amplitude profile $\overline{u}(\mathbf{f}_\perp) \equiv \overline{u}(|\mathbf{f}_\perp|, \psi)$ is restricted to the single radial spatial frequency $|\mathbf{f}_\perp| = k_t/2\pi$, but its dependence on the azimuthal angle $\psi = \arctan(f_y/f_x)$ is arbitrary. The structure of the spatial spectrum (13.21) represents a sufficient condition of the nondiffracting propagation. It can be demonstrated simply if a free-space propagation is described by the optical transfer function. We assume that the 2D spatial spectrum representing a nondiffracting beam at an initial plane $z = 0$ is known and can be written in the form (13.21), $\overline{U}_0(\mathbf{f}_\perp) \equiv \overline{U}(\mathbf{f}_\perp)$. A spatial spectrum representing the beam at a distance z from the initial plane is obtained as a product of the initial spectrum and the optical transfer function of a free space,

$$\overline{U}(\mathbf{f}_\perp, z) = \overline{U}_0(\mathbf{f}_\perp)H(\mathbf{f}_\perp, z), \tag{13.22}$$

where

$$H(\mathbf{f}_\perp, z) = \exp\left[-i2\pi\left(\frac{1}{\lambda^2} - \mathbf{f}_\perp \cdot \mathbf{f}_\perp\right)^{1/2} z\right]. \tag{13.23}$$

The complex amplitude of the beam at the distance z is earned as an inverse 2D Fourier transform of $\overline{U}(\mathbf{f}_\perp, z)$,

$$U(\rho, \theta, z) = \frac{k_t}{2\pi}\exp(-ik_z z)\int_0^{2\pi} \overline{u}(k_t, \psi)\exp[-ik_t\rho\cos(\theta - \psi)]\, d\psi. \tag{13.24}$$

As is obvious, the spatial spectrum (13.21) results in a beam whose complex amplitude has an oscillatory phase dependence on the propagation coordinate z, so that it represents a nondiffracting beam specified by $\overline{u}$. With a choice $\overline{u} = (2\pi/k_t)A(\psi)$, we obtain an integral representation of nondiffracting beams,

$$U(\rho, \theta, z) = \exp(-ik_z z)\int_0^{2\pi} A(\psi)\exp[-ik_t\rho\cos(\theta - \psi)]\, d\psi. \tag{13.25}$$

The complex amplitude of the beam is given by an arbitrary periodical function A, so there exist an infinite number of nondiffracting beams with different transverse amplitude profiles. In the integral representation (13.25), the nondiffracting beam is given as an interference field of plane waves whose wave vectors $\mathbf{k}$ create a conical surface with the vertex angle 2α. In this case, all wave vectors have the same projection to the direction of propagation, $k_z = k\cos\alpha$. The projections of wave vectors to the transverse plane have the same magnitude $k_t = k\sin\alpha$, but their azimuthal orientations ψ are different. A diffraction spread of standard beams is caused by propagation changes of relative phases of plane-wave components of the spatial spectrum. A composition of the spatial spectrum of the nondiffracting beams ensures that the relative phases of the plane-wave components remain constant during propagation and the diffraction is eliminated. The function $A(\psi)$ represents a complex amplitude of interfering plane waves distinguished by the azimuthal angle ψ. By its choice, a connection with nondiffracting solutions of the Helmholtz equation (13.16) can be established. If all interfering plane-waves have the same amplitudes and phases, the nondiffracting interference pattern has a shape described by the zero-order Bessel function J_0. The beamlike field has a significant intensity maximum at its axis because the plane wave components interfere constructively at the axial points. If the plane waves of the spectrum have a constant amplitude but their phases change in a dependence on the azimuthal angle as $A(\psi) = \exp(im\psi)$, the interference field created represents a higher-order Bessel beam. It belongs to a class of nondiffracting vortex beams and propagates with a helical wavefront specified by the topological charge m. A superposition of the plane waves (13.25) can be continuous or discrete. The nondiffracting field given by a superposition of M plane waves can be obtained with

$$A(\psi) = \sum_{m=1}^{M} A_m(\psi)\delta(\psi - \psi_m) \tag{13.26}$$

substituted into (13.25). An admissible variability of $A(\psi)$ enables generation of the nondiffracting beam, whose amplitude profile approximates a predetermined form and can also be used for the creation of arrays of nondiffracting beams.

13.4 ENERGETICS OF NONDIFFRACTING VORTEX BEAMS

For applications, the energetics of nondiffracting vortex beams is important. In the Maxwell theory framework, the density of the flow of electromagnetic energy, known as the Poynting vector $\mathbf{S}$, and the volume density of electromagnetic energy w can be adopted as basic energetical quantities. They are mutually coupled in the energy conservation law. In nonconducting media, the law can be expressed as

$$\nabla \cdot \mathbf{S} + \frac{\partial w}{\partial t} = 0. \tag{13.27}$$

In real situations, the time-averaged quantities $\langle \mathbf{S} \rangle$ and $\langle w \rangle$ are of particular importance. For a monochromatic nondiffracting field propagating along the direction given by the unit vector $\mathbf{e}$, it is useful to decompose the Poynting vector into transverse and longitudinal parts $\langle \mathbf{S} \rangle = \langle \mathbf{S}_\perp \rangle + \langle \mathbf{S}_\parallel \rangle$, fulfilling relations $\langle \mathbf{S}_\perp \rangle \cdot \mathbf{e} = 0$ and $\langle \mathbf{S}_\parallel \rangle \times \mathbf{e} = 0$. The energy conservation law formulated for the nondiffracting beams then becomes

$$\nabla \cdot \langle \mathbf{S}_\perp \rangle = 0. \tag{13.28}$$

This requires zero divergence of the transverse component of the Poynting vector, but the transverse energy flow itself can be nonzero. The fact that the monochromatic nondiffracting beam admits a nonzero energy flow orthogonal to the propagation direction originates from a wavefront helicity. Although a detailed study of the energy flow requires a vectorial electromagnetic description of the nondiffracting beams, an approximate analysis dealing with a scalar complex amplitude (13.15) is also possible. By simple manipulations of the Helmholtz equation, the energy conservation law (13.27) can be obtained with quantities given by the scalar complex amplitude,

$$\mathbf{S} = i\omega(\nabla U U^* - \nabla U^* U) \tag{13.29}$$

$$w = \nabla U \cdot \nabla U^* + k^2 |U|^2. \tag{13.30}$$

If the transverse component of the Poynting vector is decomposed into the radial and azimuthal components, $\langle \mathbf{S}_\perp \rangle = \langle \mathbf{S}_\rho \rangle + \langle \mathbf{S}_\theta \rangle$, and the position vector is expressed by the circular cylindrical coordinates $\mathbf{r} \equiv (\rho, \theta, z)$, we obtain

$$\langle S_\rho \rangle = i\omega \left(u^* \frac{\partial u}{\partial \rho} - u \frac{\partial u^*}{\partial \rho} \right) \tag{13.31}$$

$$\langle S_\theta \rangle = i \frac{\omega}{\rho} \left(u^* \frac{\partial u}{\partial \theta} - u \frac{\partial u^*}{\partial \theta} \right) \tag{13.32}$$

$$\langle S_\parallel \rangle = 2\omega k_z |u|^2. \tag{13.33}$$

The quantity approximating the volume energy density can be written as

$$\langle w \rangle = \left| \frac{\partial u}{\partial \rho} \right|^2 + \frac{1}{\rho^2} \left| \frac{\partial u}{\partial \theta} \right|^2 + \left(k^2 + k_z^2 \right) |u|^2. \tag{13.34}$$

From the energy conservation law (13.28) it can be concluded that the radial component of the Poynting vector $\langle S_\rho \rangle$ must be equal to zero, and the conditions $\Im\{u^* \partial^2 u / \partial j^2\} = 0,\ j \equiv \rho, \theta$, must be fulfilled. For nondiffracting Bessel beams (13.17), the complex amplitude becomes $u \equiv u_m = J_m(k_t \rho) \exp(im\theta)$. In this case, $\langle S_\rho \rangle = 0$ and $\langle S_\theta \rangle$ is given by

$$\langle S_\theta \rangle = -\frac{2\omega m J_m^2(k_t \rho)}{\rho}. \tag{13.35}$$

As is obvious, the azimuthal energy flow is nonzero only for the higher-order Bessel beams $m \neq 0$ possessing a helical wavefront typical for vortex beams. Superposition of the azimuthal and longitudinal components of the Poynting vector results in the helical total energy flow associated with the angular momentum, which can be transferred to atoms and microscopic particles [28]. The mechanical consequence of that interaction is rotation of the particles. The angular momentum has two components: the orbital angular momentum and the spin. The spin depends on the polarization state of the beam and is equal to zero for the linear polarization. If the beam interacts with particles, the spin causes rotation of the particles around their own axis. The orbital angular momentum is a consequence of the spiral flow of the electromagnetic energy and causes rotation of the particles around the vortex center [29]. The angular momentum $\langle \mathbf{J} \rangle$ can be written as a vectorial product of the position vector $\mathbf{r}$ and the linear momentum $\langle \mathbf{p} \rangle$,

$$\langle \mathbf{J} \rangle = \mathbf{r} \times \langle \mathbf{p} \rangle. \tag{13.36}$$

If the beam propagates in vacuum with the phase velocity c, $\langle \mathbf{p} \rangle$ can be expressed by means of the Poynting vector as

$$\langle \mathbf{p} \rangle = \frac{\langle \mathbf{S} \rangle}{c^2}. \tag{13.37}$$

Assuming beam propagation along the z-axis, its orbital angular momentum is given by the z-component of $\langle \mathbf{J} \rangle$. In circular cylindrical coordinates, we obtain

$$\langle J_z \rangle = \frac{\rho \langle S_\theta \rangle}{c^2}. \tag{13.38}$$

Performing normalization of the orbital angular momentum by the volume energy density,

$$\langle J_z \rangle = \frac{\langle J_z \rangle}{\langle w \rangle}, \tag{13.39}$$

the quantity $\langle J_z \rangle$ can be interpreted as the magnitude of the orbital angular momentum carried by the photon of the vortex beam [4]. For paraxial Bessel beams, we can adopt $k_z \approx k >> k_t$, and $\langle w \rangle$ given by (13.34) can be approximated by

$$\langle w \rangle = 2k^2 J_m^2(k_t \rho). \tag{13.40}$$

By means of (13.35) and (13.40), it can be shown that the photon of the Bessel beam with the topological charge m carries the orbital angular momentum given as

$$\langle J_z \rangle = -\frac{m}{\omega}. \tag{13.41}$$

13.5 VORTEX ARRAYS AND MIXED VORTEX FIELDS

Due to the linearity of the Helmholtz equation (13.16), its new nondiffracting solution U can be obtained as a superposition of M nondiffracting beams possessing different transverse amplitude profiles u_m and the same propagation constant k_z. The weights of separate beams are given by the coefficients c_m. Furthermore, the centra of the nondiffracting beams can be shifted to the off-axis positions specified by coordinates $(\Delta x_m, \Delta y_m)$ so that an array of nondiffracting beams is created:

$$U(\mathbf{r}) = \exp(-ik_z z) \sum_{m=1}^{M} c_m u_m(x - \Delta x_m, y - \Delta y_m). \tag{13.42}$$

If the vortex beams with different topological charges m are superimposed, the nondiffracting composite vortex field is obtained. If a coaxial superposition of beams is used ($\Delta x_m = \Delta y_m = 0$), a mixed vortex field applicable to information encoding is created [13]. The nondiffracting arrays (13.42) can be realized by direct modulation of the complex amplitude of the light beam or by manipulation of its spatial spectrum.The required form of the spatial spectrum can be obtained by the Fourier transform of (13.42):

$$\overline{U}(\mathbf{f}_\perp) = \delta\left(|\mathbf{f}_\perp| - \frac{k_t}{2\pi}\right)\sum_{m=1}^{M} c_m \overline{u}_m(\mathbf{f}_\perp)\exp(i2\pi\mathbf{f}_\perp \cdot \Delta\mathbf{r}_{\perp m}), \tag{13.43}$$

where $\Delta\mathbf{r}_{\perp m} \equiv (\Delta x_m, \Delta y_m)$ and $\overline{u}_m$ is given by

$$\overline{u}_m(\mathbf{f}_\perp) = \iint_{-\infty}^{+\infty} u_m(\mathbf{r}_\perp)\exp(i2\pi\mathbf{f}_\perp \cdot \mathbf{r}_\perp)\, d\mathbf{r}_\perp. \tag{13.44}$$

The components of the spatial spectrum of the nondiffracting array are localized on a circle with radius $k_t/2\pi$ at the plane of spatial frequencies (f_x, f_y). At the points of the circle, the Fourier components are modified by the terms of the sum appearing in (13.43). The spatial shape of the mth nondiffracting beam is formed by $\overline{u}_m$, and its transverse shift is caused by the term $\exp(i2\pi\mathbf{f}_\perp \cdot \Delta\mathbf{r}_{\perp m})$. Applying the free-space transfer function, the complex amplitude of the single nondiffracting beam (13.25) can be generalized to describe the array of nondiffracting beams,

$$U(x, y, z) = \exp(-ik_z z)\sum_{m=1}^{M} c_m T(x - \Delta x_m, y - \Delta y_m), \tag{13.45}$$

where

$$T = \int_0^{2\pi} A_m(\psi)\exp\{-ik_t[(x - \Delta x_m)\cos\psi + (y - \Delta y_m)\sin\psi]\}\, d\psi. \tag{13.46}$$

The transverse profile of the separate nondiffracting beams creating the array is given by T. Independent shaping of the beam spots is enabled by the periodical function A_m. For example, with the choice $A_m(\psi) = \exp(il_m\psi)$, an array composed of Bessel beams representing optical vortices with the different topological charges l_m is created. The centra of the beams are localized at the positions given by $(\Delta x_m, \Delta y_m)$. The spot size of the beams can be scaled by the parameter k_t. For $\Delta x_m = \Delta y_m = 0$, the coaxial superposition of the higher-order Bessel beams is obtained.

Admissible transverse profiles u_m of the nondiffracting beam with the propagation constant k_z can be obtained as solutions of (13.16). They depend on the parameter k_t, whose change causes scaling of the beam profile. For applications, a superposition of N nondiffracting beams with different propagation constants is also important. If we confine ourselves to one of the possible solutions of (13.16), $u = u_m$, the superposition can be written as

$$U(x, y, z) = \sum_{n=1}^{N} u(x, y, k_{tn})\exp(-ik_{zn}z), \tag{13.47}$$

where $k_{tn}^2 + k_{zn}^2 = k^2$. The transverse profile of the nondiffracting beams has the same shape but its size is different for separate beams, due to a dependence on k_{tn}. The total field is not nondiffracting and its amplitude U exhibits a periodical repetition if the propagation constants of the nondiffracting beams k_{zn} fulfill a coupling condition of the form

$$k_{zn} = \frac{2n\pi}{L}, \qquad n = 0, 1, 2, \ldots, \; n < \frac{L}{\lambda}. \tag{13.48}$$

During propagation, the transverse amplitude profile reappears with a longitudinal period L, $U(x, y, z) = U(x, y, z + nL)$. This effect known as *self-imaging* represents a spatial analogy of the mode locking observable in the temporal domain. If the transverse profiles of the nondiffracting beams used in (13.47) have the same axial amplitudes, $u(0, 0, k_{tn}) \equiv u_0$, a longitudinal evolution of the total field can be demonstrated in a dependence on the number of beams N. The axial intensity $I_0(z) \equiv |U(0, 0, z)|^2$ can be written as

$$I_0(z) = |u_0|^2 \frac{\sin^2(\pi Nz/L)}{\sin^2(\pi z/L)}. \tag{13.49}$$

Due to a constructive interference of the nondiffracting beams, the axial intensity of the total field exhibits apparent peaks appearing periodically with the distance L. Their maximal intensity is given by $I_{\max} = |u_0|^2 N^2$. The width of the peaks Δz is inversely proportional to N and can be expressed as $\Delta z = L/N$. By (13.45), an array composed of M transversaly shifted nondiffracting beams is defined. As all beams possess the same propagation constant k_z, the array created is also nondiffracting. The self-imaging effect is induced by a superposition of N nondiffracting arrays with the propagation constants k_{zn} given by (13.48). In this way, the self-imaging can be used for 3D light shaping [17,30] potentially applicable to the design of the laser tweezer. Recently, the principle of self-imaging was used for a controllable formation of the axial intensity of light beams [31].

13.6 PSEUDO-NONDIFFRACTING VORTEX FIELDS

The ideal nondiffracting field whose spatial spectrum contains a single radial frequency is not exactly realizable. Its propagation range is unlimited, and the energy covered by the cross section is infinite. An experimentally realizable approximation of the nondiffracting beam is P-N beam. It is obtained if the ideal nondiffracting field is transversally bounded by an integrable envelope. In this case, both the propagation range and the energy carried become finite. If the bounding envelope has a Gaussian form, the P-N beams can be earned as solutions of the paraxial Helmholtz equation. In circular cylindrical coordinates, P-N Bessel–Gauss beams are obtained as an approximation of the nondiffracting Bessel beams. P-N beams provide a variability similar to the transverse intensity profile as the nondiffracting beams. It can be demonstrated by their integral representation. A P-N beam can be comprehended as an interference field of inclined monochromatic Gaussian beams G whose axes create a conical surface with the vertex angle 2α [32]. The axes of the separate Gaussian beams are specified by the angles α and ψ, and their amplitudes are given by $A(\psi)$. The complex amplitude of the P-N field can then be expressed as

$$U_{\mathrm{PN}}(\rho, \theta, z) = \int_0^{2\pi} A(\psi) G(\rho, \alpha, \psi)\, d\psi. \tag{13.50}$$

The Gaussian beams under the integral can be specified by the complex parameter q defined by the confocal parameter q_0 as $q = z + iq_0$. The complex amplitude U_{PN} then becomes

$$U_{\mathrm{PN}}(\rho, \theta, z) = \exp\left(-i\frac{k_t^2 z^2}{2kq}\right) U_G(\rho, z) U(Q\rho, \theta, z), \tag{13.51}$$

where U_G denotes the Gaussian envelope and U is given as

$$U(Q\rho, \theta, z) = \exp(-ik_{zp}z) \int_0^{2\pi} A(\psi) \exp[-ik_t Q\rho \cos(\psi - \theta)]\, d\psi. \tag{13.52}$$

As is obvious, U has a form of the nondiffracting beam (13.25), but the propagation constant k_z is replaced by its paraxial approximation,

$$k_{zp} = k\left(1 - \frac{k_t{}^2}{2k^2}\right), \tag{13.53}$$

and the variable ρ is replaced by a scaled variable $Q\rho$. The field intensity is not propagation invariant because except for the oscillating term, the complex scaling factor Q depends on the propagation coordinate,

$$Q = 1 + \frac{z}{q}. \tag{13.54}$$

The Gaussian envelope can be expressed by means of the standard parameters w_0, R, and Φ, representing the beam waist radius, the wavefront radius, and the phase shift, respectively,

$$U_G(\rho, z) = \frac{w_0}{w} \exp\left(-\frac{\rho^2}{w^2} - i\frac{k\rho^2}{2R} + i\Phi\right), \tag{13.55}$$

where

$$w^2 = w_0^2 \left(1 + \frac{z^2}{q_0^2 \cos^2\alpha}\right)^{1/2} \tag{13.56}$$

$$R = \frac{z}{\cos\alpha} + \frac{q_0^2 \cos\alpha}{z} \tag{13.57}$$

$$\Phi = \arctan\left(\frac{z}{q_0 \cos\alpha}\right). \tag{13.58}$$

An arbitrary periodical function $A(\psi)$ appearing in (13.50) can be used for a controllable formation of the transverse amplitude profile of P-N beams similar to the case of ideal nondiffracting beams. If a phase of the inclined Gaussian beams creating the P-N field is changed in dependence on the azimuthal angle ψ as $A(\psi) = \exp(il_m\psi)$, an mth-order Bessel–Gauss beam is obtained. It represents a P-N vortex beam with the topological charge l_m. The P-N array of M vortices is created by azimuthal modulation, given as

$$A(\psi) = \sum_{m=1}^{M} \exp[il_m\psi + ik_t(\Delta x_m \cos\psi + \Delta y_m \sin\psi)]. \tag{13.59}$$

The total field represents a coherent superposition of P-N vortices with various topological charges l_m localized at the positions $(\Delta x_m, \Delta y_m)$ of the transverse plane. Its complex amplitude U_{PN} is still given by (13.51) where U is obtained substituting (13.59) into (13.52),

$$U(Q\rho, \theta, z) = 2\pi \exp(-ik_{zp}z) \sum_{m=1}^{M} i^{-m} J_{l_m}(k_t Q\rho_m) \exp(il_m\theta_m). \tag{13.60}$$

The used complex variables ρ_m and θ_m can be comprehended as generalized circular cylindrical coordinates depending on the off-axis shifts,

$$\rho_m = \left(\overline{x}_m^2 + \overline{y}_m^2\right)^{1/2} \tag{13.61}$$

$$\theta_m = \arctan\left(\frac{\overline{y}_m}{\overline{x}_m}\right), \tag{13.62}$$

where

$$\overline{x}_m = x - \frac{q}{z+q}\Delta x_m \tag{13.63}$$

$$\overline{y}_m = y - \frac{q}{z+q}\Delta y_m. \tag{13.64}$$

The vortex array carried by the P-N beam is not invariant under free propagation. The separate vortices change their transverse positions so that their centra create curved trajectories [33,34]. Additional vortex pairs can also appear during propagation. A stable propagation of the vortex array is realized under a convenient combination of the parameters k_t, w_0 and the off-axis shifts. The azimuthal modulation (13.59) used with $\Delta x_m = \Delta y_m = 0$ provides a coaxial superposition of vortex beams representing a multihelix mixed vortex field applicable to information encoding.

13.7 EXPERIMENTS

Experiments providing P-N fields are based on a formation of the spatial spectrum by the Fourier methods or on a modulation of the light complex amplitude by the holographical methods. In advanced experiments, the holograms are realized by the spatial light modulator (SLM) enabling dynamical formation of the beams.

13.7.1 Fourier Methods

An ideal nondiffracting beam can be thought of as an interference field of plane waves whose amplitudes and relative phases are arbitrary but whose wave vector directions are sharply restricted by condition (13.21), representing a conical surface in the k-space shown in Fig. 13.1a. At the Fourier plane, the plane-wave spectrum is focused on a circle with a radius depending on the spot size of the nondiffracting beam related to the angle α. To simulate a realizable P-N beam, the ideal nondiffracting beam must be bounded by a Gaussian or hard-edged aperture. While the interference field creating the ideal nondiffracting beam ranges from $-\infty$ to $+\infty$, the bounded beams behind the aperture create an interference region whose length can be approximated by $L \approx D/2\alpha$, where D is the aperture diameter. The field behind the aperture can still be represented by the plane-wave spectrum, but due to diffraction effects, the angles of significant plane-wave components exhibit an angular spread $\Delta\alpha$ and their wave vectors are confined between two conical surfaces with vertex angles $2(\alpha - \Delta\alpha)$ and $2(\alpha + \Delta\alpha)$. The angular spectrum of the P-N beam is illustrated in Fig. 13.1b. Due to the angular spread of the plane-wave components, condition (13.21) is violated and the ideal beam becomes P-N. At the Fourier plane, the spatial spectrum is represented by an annular ring whose mean radius and width are inversely proportional to α and directly proportional to $\Delta\alpha$, respectively.

In simulation models, an aperture with a Gaussian shape is commonly used. In this case the beam spot size is defined by the parameter k_t similar to the case of the nondiffracting beam, while the Gaussian envelope is specified by the waist radius w_0. The spatial spectrum is then represented by an annular ring whose mean radius and width depend on k_t and w_0, respectively. The propagation range where the P-N beam approximates the ideal nondiffracting beam depends on the width of the annular ring. If the width of the ring is narrowed and its mean radius preserved, the propagation range of the beam is extended while its spot size remains unchanged. The composition required for the spatial spectrum of the P-N beam can be simply realized by means of the Fourier filter of an annular shape placed at the front focal plane of a lens. If the thin annular ring is illuminated by the laser beam, its points generate spherical waves which are transformed by the lens to the inclined plane waves. For an annular ring with a negligible width and a lens with an infinite aperture, the wave vectors of the plane waves create a cone with the vertex angle $2\alpha \approx 2r_0/f$, where r_0 is the ring radius and f denotes the lens focal length. In a real situation, the width of the annular ring and the lens finite aperture cause an angular spread, so that the wave vectors of significant plane-wave components are confined between two conical surfaces. The angular spread of the wave vectors related to a difference in the vertex

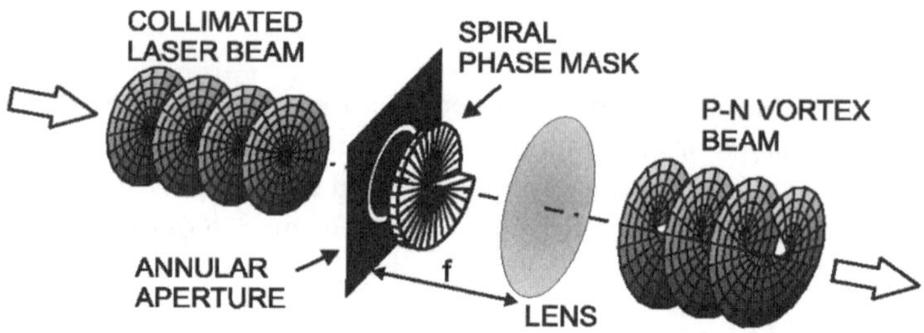

FIGURE 13.4 Experimental setup enabling conversion of a collimated Gaussian beam to a pseudo-nondiffracting vortex beam.

angles of the cones influences the length of the propagation range of the P-N beam. Amplitude and phase changes at separate points of the annular ring enable shaping of the transverse intensity profile of the P-N beam generated. If a spiral phase plate is adjacent to the annular ring, the P-N vortex beam appears behind the lens [35]. The experimental setup enabling conversion of the collimated laser beam with a planar wavefront to the P-N beam with a helical wavefront is illustrated in Fig. 13.4. The spiral phase mask used introduces an azimuthal phase shift $\exp(im\theta)$ and can be prepared photolithographically. In this case, a good approximation of the mth-order Bessel–Gauss beam representing the P-N vortex beam with the topological charge m is obtained. Experiments using the annular ring clearly demonstrate generation of the P-N beam, but this is not convenient for applications because the efficiency is low. It can be increased significantly if the Gaussian beam is converted to a P-N beam by means of an axicon [6] or Fresnel axicon [36].

13.7.2 Spatial Light Modulation

The shape variability of nondiffracting fields is enabled by degrees of freedom provided by their spatial spectrum. Nondiffracting beams and arrays of the form (13.45) can be obtained by proper azimuthal modulation of the spatial spectrum realized by Fourier methods. Another possibility is to encode the required changes in the spatial spectrum directly into the beam complex amplitude using holographical methods. In advanced experiments they are realized by an SLM driven by a personal computer (Fig. 13.5). Depending on its design, the SLM transmits or reflects the laser beam and modulates its amplitude or phase. After Fourier filtering, the required field is obtained. In Fig. 13.5, a coaxial superposition of vortex beams with the topological charges 1, 2, 3, and 4, respectively, is shown.

The amplitude modulation follows directly from principles of holography. The mathematical description of a required pattern enables creation of a computer-generated hologram representing a superposition of the simulated field with a reference beam. The hologram is sent to the amplitude SLM illuminated by a collimated laser beam. As the hologram represents a cosinusoidal grating, the spatial spectrum includes the zeroth and the positive and negative first diffraction orders. If we select

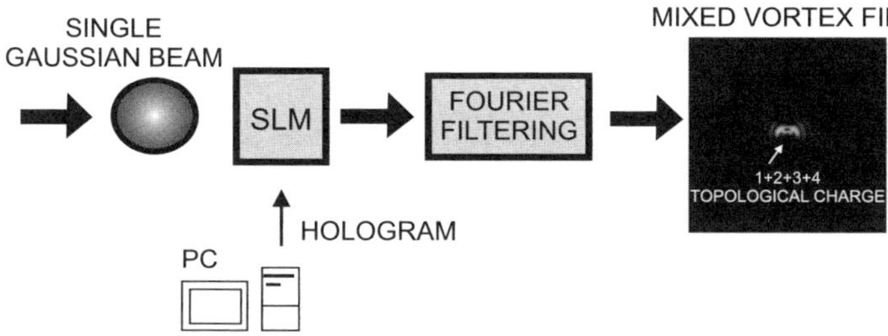

FIGURE 13.5 Principle of the spatial modulation of laser beams. A collimated beam illuminates the computer-generated hologram sent to the SLM and splits into several diffraction orders. After Fourier filtering, the beam required is earned from the positive first diffraction order. As an example, a mixed vortex field composed of four coaxial vortices with different topological charges is shown.

the positive first diffraction order, a good approximation of the required field encoded into the hologram is obtained.

Because the hologram is written on an SLM composed of separate pixels of liquid crystals, additional diffraction takes place. A periodical structure of the SLM represents a binary amplitude grating with spatial period Λ and the width of the transparent and nontransparent regions given by $p\Lambda$ and $(1 - p)\Lambda$, respectively. Its diffraction efficiency for the nth diffraction order is given as $\eta_n = p^2 \text{sinc}^2 \pi np$. The diffraction orders caused by the cosinusoidal grating then appear in each diffraction order of the binary amplitude grating. If we operate with the field encoded into the positive first diffraction order of the cosinusoidal grating of the contrast V created by the zeroth diffraction order of the binary grating, the total diffraction efficiency can be written as $\eta = (pV/4)^2$. Applying the amplitude modulator CRL OPTO (1024×768 pixels, $p = 0.65$, and $V = 1$), the Gaussian laser beam is converted to the required field with the very low efficiency $\eta \approx 2.6\%$.

The conversion efficiency can be improved significantly if a phase SLM is used. Its structure is also pixeled so that the diffraction on the amplitude binary grating takes place but the beam created is encoded into a phase grating. For a cosinusoidal grating with the phase stroke $\Delta\Phi$, the diffraction efficiency of the nth diffraction order is given as $\eta_n = J_n^2(\Delta\Phi/2)$. For the positive first diffraction order used in the experiment, the diffraction efficiency is maximal for $\Delta\Phi = \pi/2$. If the zeroth diffraction order of the amplitude grating and the positive first diffraction order of the phase cosinusoidal grating are used, the total conversion efficiency is given as $\eta \approx p^2 J_1^2(\pi/4)$. For a phase SLM with $p = 0.78$ (BOULDER, 512×512 pixels), we obtain $\eta \approx 20.5\%$. The efficiency can be increased if the phase map sent to the SLM has the form of a blazed phase grating. In a one-dimensional simplification, it can be written as $T(x) = \exp[ikd \bmod(x, \Lambda)]$, where d is a parameter that defines the maximal phase stroke. In this case, the diffraction efficiency for the nth diffraction order can be written as $\eta_n = \text{sinc}^2[k(n\lambda - d\Lambda)/2]$. For the positive first diffraction order used, the maximal diffraction efficiency is obtained for $d = \lambda/\Lambda$, providing the phase stroke

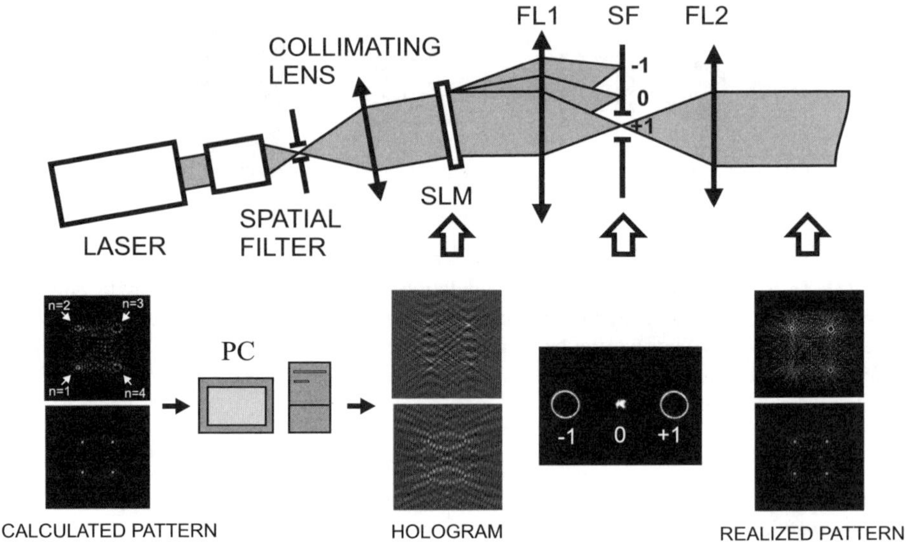

FIGURE 13.6 Setup enabling creation of the variable nondiffracting fields by means of the phase SLM. As examples, transformation of a single Gaussian beam to the composite vortex field containing four vortices with different topological charges and to the array of spatially separated P-N beams is demonstrated.

2π. In this case the efficiency of the blazed grating is 100% and the beam conversion efficiency is given by the diffraction on the amplitude binary grating representing the structure of the SLM. For the phase SLM, specified, its value is $\eta \approx 60\%$. The single P-N beam (13.51) can be obtained by a phase-only modulation of an input laser beam. To generate a multihelix vortex structure (13.60), both amplitude and phase modulation must be used. Although this complex modulation can be realized by the SLM [37], its efficiency is relatively low. Recently, a new technique enabling realization of mixed vortex beams by phase-only modulation has been proposed and verified experimentally [38]. By this method the fields required have been realized in a very good approximation and with a high efficiency.

The optical scheme using a phase SLM is illustrated in Fig. 13.6. The phase map representing the optical field required is prepared on a PC and sent to the phase SLM BOULDER, illuminated by the spatially filtered and collimated beam of an He–Ne laser. The laser beam modulated by the SLM splits into diffraction orders separated spatially at the back focal plane of Fourier lens FL1. After selection of the positive first diffraction order and realization of the inverse Fourier transform by lens FL2, a good approximation of the field desired is obtained. In Fig. 13.6, two examples of realized fields are shown. In a former case, generation of the nondiffracting array of four vortex beams with different topological charges is illustrated. In the latter example, an array composed of four spatially separated Bessel–Gauss beams of zero order is demonstrated.

13.8 APPLICATIONS AND PERSPECTIVES

A concept of wave dislocations including vortex beams was introduced three decades ago and has been explored extensively both theoretically and experimentally. In recent years, attention has been focused on propagation effects resulting in looped, knotted, linked, or braided vortices [39], on colored vortices, on dynamical interaction between vortices in optically nonlinear materials, on quantum effects associated with optical singularities, and in particular on dynamic aspects of vortices associated with angular momentum [4]. The peculiar properties of optical vortices seem to be prospective for applications in metrology [40,41], data storage [42], nonlinear optics [43], and photolithography and have been utilized for guiding, manipulation, and rotation of particles and atoms [44] and for realization of the optical vortex coronograph [45]. The spatially localized vortex structures examined in this chapter include P-N vortex beams and their superpositions and provide additional effects interesting for theoretical research, experiments, and applications. Both bright and dark P-N beams were used successfully in the design of multiple optical traps [10,46] enabled by the self-healing property of P-N beams [8,9]. A collinear superposition of two vortex modes of reverse topological charges was used to construct the trapped structures, which can be rotated and translated in three dimensions [47], and to realize size-selective trapping [48]. Multimode superpositions of P-N beams resulting in self-imaging provide an interesting alternative to methods of three-dimensional light shaping. In particular, the self-imaging realized with vortex fields provides real research potential for particle manipulation.

Coaxial superposition of P-N vortex beams with different topological charges has been utilized for realization of a new principle of information encoding that is potentially applicable to free-space communications [13]. In conventional information channels based on a temporal modulation, a sequence of N optical pulses is used to transfer N information bits. Adopting an optical mixed vortex field, additional degrees of freedom provided by its peculiar spatial structure are available to increase the density of the information carried. The information bits are represented by the weight coefficients of the superimposed P-N vortex beams. The different topological charges of the vortices are used as markers, enabling information decoding. Separate steps of the process are illustrated in Fig. 13.7. In the first step, an encoding base must be chosen, $U_B = \sum_{m=1}^{M} u_m \exp(il_m\theta)$. It represents the superposition of M vortices with the topological charges l_m. Applying an SLM, an input laser beam is converted into the mixed vortex field $U_{c_1,c_2,\ldots,c_M} = \sum_{m=1}^{M} c_m u_m \exp(il_m\theta)$. The weight coefficients c with values 0 or 1 create an information code $c_1, c_2, \ldots, c_M$ representing M bits of information. The P-N mixed vortex field propagates toward a receiver without significant changes in the structure. For free-space communications, its self-healing ability is a very useful property. In decoding, the mixed vortex field is transmitted through the complex mask $t = \sum_{m=1}^{M} u_m \exp(-il_m\theta + \Omega_m)$. After the Fourier transform, the value 0 or 1 of the weight coefficients is detected at positions defined by Ω_m. For the value 0, these positions remain dark, but for the coefficient 1 they become bright. In this way, the information is decoded by intensity detection. The intensity spots illustrated in the upper and lower rows of Fig. 13.7 demonstrate encoding,

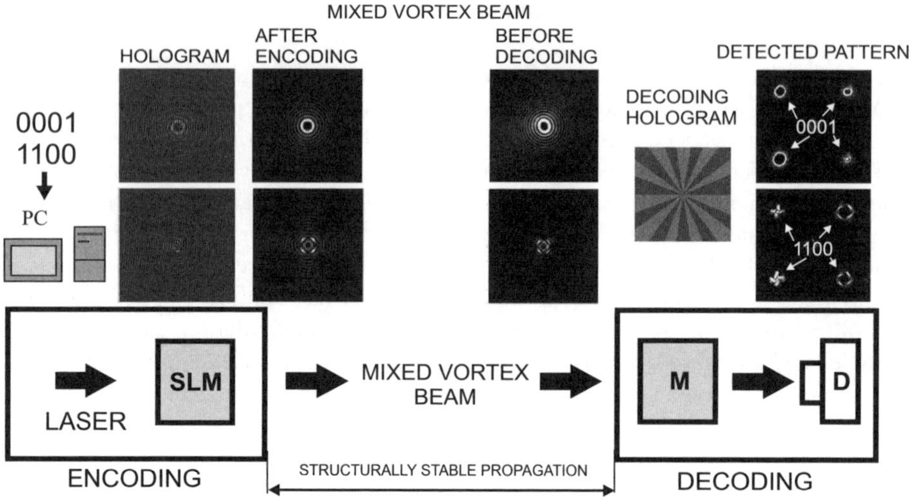

FIGURE 13.7 Channel enabling information transmission by means of P-N mixed vortex beams.

transfer, and decoding of the information codes 0001 and 1100, respectively. Direct applications of the method are restricted by the low repetition rate of SLMs. Current research is focused on the possibility of using a static hologram and replacing the dynamics of an SLM by a standard temporal modulation of an array of light sources [49].

Recently, an SLM was used for spatial modulation of laser beams focused on fibers [50]. By this method, selective excitation of vortex fiber modes has been verified in 100-m-long few-mode fiber. The method has been demonstrated on a dynamical switching, phase coupling, and weighted mixing of LP_{01}, LP_{11}, and LP_{21} fiber modes.

Except for optical applications, the multidisciplinary character of spatially localized vortex structures is also important. The methods and approaches presented, based on physical optics, can be used for investigation of phase singularities in waves of a different nature. For example, very perspective areas are quantum physics, where the vortices can be adopted as higher-dimensional states of the orbital angular momentum [51], and in matter physics, where the vortices are realized in Bose–Einstein condensates [52].

Acknowledgments

This work was supported by the Research Projects Measurement and Information in Optics MSM 6198959213, Center of Modern Optics LC06007, and project FT-TA2/059 of the Czech Ministry of Industry and Trade.

REFERENCES

1. J. Durnin, Exact solutions for nondiffracting beams: I. The scalar theory, *J. Opt. Soc. Am. A* **4**, 651 (1987).
2. C. J. R. Sheppard and T. Wilson, Gaussian-beam theory of lenses with annular aperture, *Microwaves Opti. acoust.* **2**, 105 (1978).
3. M. S. Soskin and M. V. Vasnetsov, *Singular Optics*, Progress in Optics, Vol. 42, Elsevier, Amsterdam, The Netherlands, 2001.
4. L. Allen, S. Barnett, and M. J. Padgett, *Optical Angular Momentum*, Institute of Physics Publishing, Bristol, England, 2003.
5. J. Yin, W. Gao, and Y. Zhu, *Generation of Dark Hollow Beams and Their Applications*, Progress in Optics, Vol. 45, Elsevier, Amsterdam,The Netherlands 2003.
6. M. R. Lapointe, Review of non-diffracting Bessel beam experiments, *Opt. Laser Technol.* **24**, 315 (1992).
7. Z. Bouchal and J. Courtial, The connection of singular and nondiffracting optics, *J. Opt. A Pure Appl. Opt.* **6**, S184 (2004).
8. Z. Bouchal, J. Wagner, and M. Chlup, Self-reconstruction of a distorted nondiffracting beam, *Opt. Commun.* **151**, 207 (1988).
9. Z. Bouchal, Resistance of nondiffracting vortex beam against amplitude and phase perturbations, *Opt. Commun.* **210**, 155 (2002).
10. V. Garcés-Chávez, D. McGloin, M. D. Summers, A. Fernandez-Nieves, G. C. Spalding, G. Cristobal, and K. Dholakia, The reconstruction of optical angular momentum after distortion in amplitude, phase and polarization, *J. Opt. A Pure Appl. Opt.* **6**, S235 (2004).
11. Z. Bouchal, Nondiffracting optical beams: physical properties, experiments, and applications, *Czech. J. Phys.* **53**, 537 (2003).
12. Z. Bouchal, Controlled spatial shaping of nondiffracting patterns and arrays, *Opt. Lett.* **27**, 1376 (2002).
13. Z. Bouchal and R. Čelechovský, Mixed vortex states of light as information carriers, *New J. Phys.* **6**, 131 (2004).
14. G. Gibson, J. Courtial, M. J. Padgett, M. Vasnetsov, V. Pas'ko, S. M. Barnett, and S. Franke-Arnold, Free-space information transfer using light beams carrying orbital angular momentum, *Opt. Express.* **12**, 5448 (2004).
15. I. D. Maleev and G. A. Swartzlander, Jr., Composite optical vortices, *J. Opt. Soc. Am. B* **20**, 1169 (2003).
16. K. Patorski, *progress in Optics*, Vol. 27, North-Holland, Amsterdam, The Netherlands, 1989.
17. Z. Bouchal and J. Kyvalský, Controllable 3D spatial localization of light fields synthesized by nondiffracting modes, *J. Mod. Opt.* **51**, 157 (2004).
18. T. Čižmár, V. Kollárová, Z. Bouchal, and P. Zemánek, Sub-micron particle organization by self-imaging of nondiffracting beams, *New J. Phys.* **8**, 43 (2006).
19. I. D. Maleev, D. M. Palacios, A. S. Marathay, and G. A. Swartzlander, Jr., Spatial correlation vortices in partially coherent light: theory, *J. Opt. Soc. Am. B* **21**, 1895 (2004).
20. D. M. Palacios, I. D. Maleev, A. S. Marathay, and G. A. Swartzlander, Jr., Spatial correlation singularity of a vortex field, *Phys. Rev. Lett.* **92**, 143905, 1–4 (2004).

21. I. V. Basistiy, M. S. Soskin, and M. V. Vasnetsov, Optical wavefront dislocations and their properties, *Opt. Commun.* **119**, 604 (1995).
22. N. R. Heckenberg, R. McDuff, C. P. Smith, and A. G. White, Generation of optical phase singularities by computer-generated holograms, *Opt. Lett.* **17**, 221 (1992).
23. V. Y. Bazhenov, M. S. Soskin, and M. V. Vasnetsov, Screw dislocations in light wavefronts, *J. Mod. Opt.* **39**, 985 (1992).
24. D. Rozas, C. T. Law, and G. A. Swartzlander, Jr. Propagation dynamics of optical vortices, *J. Opt. Soc. Am. B* **14**, 3054 (1997).
25. D. Rozas, Z. S. Sacks, and G. A. Swartzlander, Jr., Experimental observation of fluidlike motion of optical vortices, *Phys. Rev. Lett.* **79**, 3399 (1997).
26. P. M. Morse and H. Feshbach, *Methods of Theoretical Physics*, McGraw-Hill, New York, 1953.
27. J. C. Gutiérrez-Vega and M. A. Bandres, Helmholtz–Gauss waves, *J. Opt. Soc. Am. A* **22**, 289 (2005).
28. L. Allen, M. V. Beijersbergen, R. J. C. Spreeuv, and J. P. Woerdman, Orbital angular momentum of light and the transformation of Laguerre–Gaussian laser modes, *Phys. Rev. A* **45**, 8185 (1992).
29. H. He, M. E. J. Friese, N. R. Heckenberg, and H. Rubinsztein-Dunlop, Direct observation of transfer of angular momentum to absorptive particles from a laser beam with a phase singularity, *Phys. Rev. Lett.* **75**, 826 (1995).
30. C. A. Dartora, K. Z. Nobrega, A. Dartora, and H. E. Hernández-Figueroa, Superposition of monochromatic Bessel beams in (k_ρ, k_z)-plane to obtain wave focusing: spatial localized waves, *Opt. Commun.* **249**, 407 (2005).
31. M. Zamboni-Rached, Stationary optical wave fields with arbitrary longitudinal shape by superposing equal frequency Bessel beams: frozen waves, *Opt. Express.* **12**, 4001 (2004).
32. V. Bagini, F. Frezza, M. Santarsiero, G. Schettini, and G. Schirripa Spagnolo, Generalized Bessel–Gauss beams, *J. Mod. Opt.* **43**, 1155 (1996).
33. U. T. Schwarz, S. Sogomonian, and M. Maier, Propagation dynamics of phase dislocations embedded in a Bessel light beam, *Opt. Commun.* **208**, 255 (2002).
34. Z. Bouchal, Vortex array carried by a pseudo-nondiffracting beam, *J. Opt. Soc. Am. A* **21**, 1694 (2004).
35. V. V. Kotlyar, A. A. Almazov, S. N. Khonina, V. A. Soifer, H. Elfstrom, and J. Turunen, Generation of phase singularity through diffracting a plane or Gaussian beam by a spiral phase plate, *J. Opt. Soc. Am. A* **22**, 849 (2005).
36. I. Golub, Fresnel axicon, *Opt. Lett.* **31**, 1890 (2006).
37. V. Arrizón, Optimum on-axis computer-generated hologram encoded into low-resolution phase-modulation devices, *Opt. Lett.* **28**, 2521 (2003).
38. J. Lin, X. C. Yuan, S. H. Tao, and R. E. Burge, Collinear superposition of multiple helical beams generated by a single azimuthally modulated phase-only element, *Opt. Lett.* **30**, 3266 (2005).
39. J. Leach, M. Dennis, J. Courtial, and M. J. Padgett, Vortex knots in light, *New J. Phys.* **7**, 55 (2005).
40. O. Angelsky, D. Burkovets, P. Maksimyak, and G. Demyanovskii, Singular-optics approach for classification of the rough surfaces, *Elektronika* **45**, 259 (2004).

41. J. Masajada, Small-angle rotations measurement using optical vortex interferometer, *Opt. Commun.* **239**, 373 (2004).
42. R. J. Voogd, R. J. Singh, and J. J. M. Braat, The use of orbital angular momentum of light beams for optical data storage, *Proc. SPIE* **5380**, 387 (2004).
43. G. A. Swartzlander, Jr., and C. T. Law, Optical vortex solitons observed in Kerr nonlinear media, *Phys. Rev. Lett.* **69**, 2503 (1992).
44. K. Ladavac and D. G. Grier, Microoptomechanical pumps assembled and driven by holographic optical vortex arrays, *Opt. Express.* **12** (2004).
45. G. Foo, D. M. Palacios, and G. A. Swartzlander, Jr., Optical vortex coronograph, *Opt. Lett.* **30**, 3308 (2005).
46. V. Garcés-Chávez, D. McGloin, H. Melville, W. Sibbett, and K. Dholakia, Simultaneous micromanipulation in multiple planes using a self-reconstructing light beam, *Nature.* **419**, 145 (2002).
47. M. P. MacDonald, L. Paterson, K. Volke-Sepulveda, J. Arlt, W. Sibbett, and K. Dholakia, Creation and manipulation of three-dimensional optically trapped structures, *Science.* **296**, 1101 (2002).
48. A. Jesacher, S. Furhapter, S. Bernet, and M. Ritsch-Marte, Size selective trapping with optical "cogwheel" tweezers, *Opt. Express.* **12**, 4129 (2004).
49. R. Čelechovský and Z. Bouchal, Generation of variable mixed vortex fields by a single static hologram, *J. Mod. Opt.* **53**, 473 (2006).
50. Z. Bouchal, O. Haderka, and R. Čelechovský, Selective excitation of vortex fibre modes using a spatial light modulator, *New J. Phys.* **7**, 125 (2005).
51. A. Mair, A. Vaziri, G. Weish, and A. Zeilinger, Entanglement of orbital angular momentum states of photons, *Nature.* **412**, 3123 (2001).
52. P. G. Kevrekidis, R. Carretero-Gonzales, D. J. Frantzeskakis, and I.G. Kevrekidis, Vortices in Bose–Einstein condensates: some recent developments, *Mod. Phys. Lett. B* **18**, 1481 (2004).

Index

WILEY SERIES IN MICROWAVE AND OPTICAL ENGINEERING

KAI CHANG, Editor
Texas A&M University

FIBER-OPTIC COMMUNICATION SYSTEMS, Third Edition • *Govind P. Agrawal*

ASYMMETRIC PASSIVE COMPONENTS IN MICROWAVE INTEGRATED CIRCUITS • *Hee-Ran Ahn*

COHERENT OPTICAL COMMUNICATIONS SYSTEMS • *Silvello Betti, Giancarlo De Marchis, and Eugenio Iannone*

PHASED ARRAY ANTENNAS: FLOQUET ANALYSIS, SYNTHESIS, BFNs, AND ACTIVE ARRAY SYSTEMS • *Arun K. Bhattacharyya*

HIGH-FREQUENCY ELECTROMAGNETIC TECHNIQUES: RECENT ADVANCES AND APPLICATIONS • *Asoke K. Bhattacharyya*

RADIO PROPAGATION AND ADAPTIVE ANTENNAS FOR WIRELESS COMMUNICATION LINKS: TERRESTRIAL, ATMOSPHERIC, AND IONOSPHERIC • *Nathan Blaunstein and Christos G. Christodoulou*

COMPUTATIONAL METHODS FOR ELECTROMAGNETICS AND MICROWAVES • *Richard C. Booton, Jr.*

MICROWAVE RING CIRCUITS AND ANTENNAS • *Kai Chang*

MICROWAVE SOLID-STATE CIRCUITS AND APPLICATIONS • *Kai Chang*

RF AND MICROWAVE WIRELESS SYSTEMS • *Kai Chang*

RF AND MICROWAVE CIRCUIT AND COMPONENT DESIGN FOR WIRELESS SYSTEMS • *Kai Chang, Inder Bahl, and Vijay Nair*

MICROWAVE RING CIRCUITS AND RELATED STRUCTURES, Second Edition • *Kai Chang and Lung-Hwa Hsieh*

MULTIRESOLUTION TIME DOMAIN SCHEME FOR ELECTROMAGNETIC ENGINEERING • *Yinchao Chen, Qunsheng Cao, and Raj Mittra*

DIODE LASERS AND PHOTONIC INTEGRATED CIRCUITS • *Larry Coldren and Scott Corzine*

RADIO FREQUENCY CIRCUIT DESIGN • *W. Alan Davis and Krishna Agarwal*

MULTICONDUCTOR TRANSMISSION-LINE STRUCTURES: MODAL ANALYSIS TECHNIQUES • *J. A. Brandão Faria*

PHASED ARRAY-BASED SYSTEMS AND APPLICATIONS • *Nick Fourikis*

FUNDAMENTALS OF MICROWAVE TRANSMISSION LINES • *Jon C. Freeman*

OPTICAL SEMICONDUCTOR DEVICES • *Mitsuo Fukuda*

MICROSTRIP CIRCUITS • *Fred Gardiol*

HIGH-SPEED VLSI INTERCONNECTIONS, Second Edition • *Ashok K. Goel*

FUNDAMENTALS OF WAVELETS: THEORY, ALGORITHMS, AND APPLICATIONS • *Jaideva C. Goswami and Andrew K. Chan*

HIGH-FREQUENCY ANALOG INTEGRATED CIRCUIT DESIGN • *Ravender Goyal (ed.)*

ANALYSIS AND DESIGN OF INTEGRATED CIRCUIT ANTENNA MODULES • *K. C. Gupta and Peter S. Hall*

PHASED ARRAY ANTENNAS • *R. C. Hansen*

LOCALIZED WAVES • *Hugo E. Hernández-Figueroa, Michel Zamboni-Rached, and Erasmo Recami (eds.)*

MICROSTRIP FILTERS FOR RF/MICROWAVE APPLICATIONS • *Jia-Sheng Hong and M. J. Lancaster*

MICROWAVE APPROACH TO HIGHLY IRREGULAR FIBER OPTICS • *Huang Hung-Chia*

NONLINEAR OPTICAL COMMUNICATION NETWORKS • *Eugenio Iannone, Francesco Matera, Antonio Mecozzi, and Marina Settembre*

FINITE ELEMENT SOFTWARE FOR MICROWAVE ENGINEERING • *Tatsuo Itoh, Giuseppe Pelosi, and Peter P. Silvester (eds.)*

INFRARED TECHNOLOGY: APPLICATIONS TO ELECTROOPTICS, PHOTONIC DEVICES, AND SENSORS • *A. R. Jha*

SUPERCONDUCTOR TECHNOLOGY: APPLICATIONS TO MICROWAVE, ELECTRO-OPTICS, ELECTRICAL MACHINES, AND PROPULSION SYSTEMS • *A. R. Jha*

OPTICAL COMPUTING: AN INTRODUCTION • *M. A. Karim and A. S. S. Awwal*

INTRODUCTION TO ELECTROMAGNETIC AND MICROWAVE ENGINEERING • *Paul R. Karmel, Gabriel D. Colef, and Raymond L. Camisa*

MILLIMETER WAVE OPTICAL DIELECTRIC INTEGRATED GUIDES AND CIRCUITS • *Shiban K. Koul*

MICROWAVE DEVICES, CIRCUITS AND THEIR INTERACTION • *Charles A. Lee and G. Conrad Dalman*

ADVANCES IN MICROSTRIP AND PRINTED ANTENNAS • *Kai-Fong Lee and Wei Chen (eds.)*

SPHEROIDAL WAVE FUNCTIONS IN ELECTROMAGNETIC THEORY • *Le-Wei Li, Xiao-Kang Kang, and Mook-Seng Leong*

ARITHMETIC AND LOGIC IN COMPUTER SYSTEMS • *Mi Lu*

OPTICAL FILTER DESIGN AND ANALYSIS: A SIGNAL PROCESSING APPROACH • *Christi K. Madsen and Jian H. Zhao*

THEORY AND PRACTICE OF INFRARED TECHNOLOGY FOR NONDESTRUCTIVE TESTING • *Xavier P. V. Maldague*

OPTOELECTRONIC PACKAGING • *A. R. Mickelson, N. R. Basavanhally, and Y. C. Lee (eds.)*

OPTICAL CHARACTER RECOGNITION • *Shunji Mori, Hirobumi Nishida, and Hiromitsu Yamada*

ANTENNAS FOR RADAR AND COMMUNICATIONS: A POLARIMETRIC APPROACH • *Harold Mott*

INTEGRATED ACTIVE ANTENNAS AND SPATIAL POWER COMBINING • *Julio A. Navarro and Kai Chang*

ANALYSIS METHODS FOR RF, MICROWAVE, AND MILLIMETER-WAVE PLANAR TRANSMISSION LINE STRUCTURES • *Cam Nguyen*

FREQUENCY CONTROL OF SEMICONDUCTOR LASERS • *Motoichi Ohtsu (ed.)*

WAVELETS IN ELECTROMAGNETICS AND DEVICE MODELING • *George W. Pan*

SOLAR CELLS AND THEIR APPLICATIONS • *Larry D. Partain (ed.)*

ANALYSIS OF MULTICONDUCTOR TRANSMISSION LINES • *Clayton R. Paul*

INTRODUCTION TO ELECTROMAGNETIC COMPATIBILITY, Second Edition • *Clayton R. Paul*

ADAPTIVE OPTICS FOR VISION SCIENCE: PRINCIPLES, PRACTICES, DESIGN AND APPLICATIONS • *Jason Porter, Hope Queener, Julianna Lin, Karen Thorn, and Abdul Awwal (eds.)*

ELECTROMAGNETIC OPTIMIZATION BY GENETIC ALGORITHMS • *Yahya Rahmat-Samii and Eric Michielssen (eds.)*

INTRODUCTION TO HIGH-SPEED ELECTRONICS AND OPTOELECTRONICS • *Leonard M. Riaziat*

NEW FRONTIERS IN MEDICAL DEVICE TECHNOLOGY • *Arye Rosen and Harel Rosen (eds.)*

ELECTROMAGNETIC PROPAGATION IN MULTI-MODE RANDOM MEDIA • *Harrison E. Rowe*

ELECTROMAGNETIC PROPAGATION IN ONE-DIMENSIONAL RANDOM MEDIA • *Harrison E. Rowe*

HISTORY OF WIRELESS • *Tapan K. Sarkar, Robert J. Mailloux, Arthur A. Oliner, Magdalena Salazar-Palma, and Dipak L. Sengupta*

SMART ANTENNAS • *Tapan K. Sarkar, Michael C. Wicks, Magdalena Salazar-Palma, and Robert J. Bonneau*

NONLINEAR OPTICS • *E. G. Sauter*

APPLIED ELECTROMAGNETICS AND ELECTROMAGNETIC COMPATIBILITY • *Dipak L. Sengupta and Valdis V. Liepa*

COPLANAR WAVEGUIDE CIRCUITS, COMPONENTS, AND SYSTEMS • *Rainee N. Simons*

ELECTROMAGNETIC FIELDS IN UNCONVENTIONAL MATERIALS AND STRUCTURES • *Onkar N. Singh and Akhlesh Lakhtakia (eds.)*

ELECTRON BEAMS AND MICROWAVE VACUUM ELECTRONICS • *Shulim E. Tsimring*

FUNDAMENTALS OF GLOBAL POSITIONING SYSTEM RECEIVERS: A SOFTWARE APPROACH, Second Edition • *James Bao-yen Tsui*

RF/MICROWAVE INTERACTION WITH BIOLOGICAL TISSUES • *André Vander Vorst, Arye Rosen, and Youji Kotsuka*

InP-BASED MATERIALS AND DEVICES: PHYSICS AND TECHNOLOGY • *Osamu Wada and Hideki Hasegawa (eds.)*

COMPACT AND BROADBAND MICROSTRIP ANTENNAS • *Kin-Lu Wong*

DESIGN OF NONPLANAR MICROSTRIP ANTENNAS AND TRANSMISSION LINES • *Kin-Lu Wong*

PLANAR ANTENNAS FOR WIRELESS COMMUNICATIONS • *Kin-Lu Wong*

FREQUENCY SELECTIVE SURFACE AND GRID ARRAY • *T. K. Wu (ed.)*

ACTIVE AND QUASI-OPTICAL ARRAYS FOR SOLID-STATE POWER COMBINING • *Robert A. York and Zoya B. Popović (eds.)*

OPTICAL SIGNAL PROCESSING, COMPUTING AND NEURAL NETWORKS • *Francis T. S. Yu and Suganda Jutamulia*

SiGe, GaAs, AND InP HETEROJUNCTION BIPOLAR TRANSISTORS • *Jiann Yuan*

ELECTRODYNAMICS OF SOLIDS AND MICROWAVE SUPERCONDUCTIVITY • *Shu-Ang Zhou*